Probability Theory and Stochastic Modelling

Volume 86

The **Probability Theory and Stochastic Modelling** series is a merger and continuation of Springer's two well established series Stochastic Modelling and Applied Probability and Probability and Its Applications series. It publishes research monographs that make a significant contribution to probability theory or an applications domain in which advanced probability methods are fundamental. Books in this series are expected to follow rigorous mathematical standards, while also displaying the expository quality necessary to make them useful and accessible to advanced students as well as researchers. The series covers all aspects of modern probability theory including

- Gaussian processes
- Markov processes
- Random fields, point processes and random sets
- Random matrices
- Statistical mechanics and random media
- Stochastic analysis

as well as applications that include (but are not restricted to):

- Branching processes and other models of population growth
- Communications and processing networks
- Computational methods in probability and stochastic processes, including simulation
- Genetics and other stochastic models in biology and the life sciences
- Information theory, signal processing, and image synthesis
- Mathematical economics and finance
- Statistical methods (e.g. empirical processes, MCMC)
- Statistics for stochastic processes
- Stochastic control
- Stochastic models in operations research and stochastic optimization
- Stochastic models in the physical sciences

More information about this series at http://www.springer.com/series/13205

Jianfeng Zhang

Backward Stochastic Differential Equations

From Linear to Fully Nonlinear Theory

 Springer

Jianfeng Zhang
University of Southern California
Department of Mathematics
Los Angeles, CA, USA

ISSN 2199-3130   ISSN 2199-3149   (electronic)
Probability Theory and Stochastic Modelling
ISBN 978-1-4939-8432-9   ISBN 978-1-4939-7256-2   (eBook)
DOI 10.1007/978-1-4939-7256-2

Mathematics Subject Classification: 60H10, 60H30

Printed on acid-free paper

This Springer imprint is published by Springer Nature
The registered company is Springer Science+Business Media LLC
The registered company address is: 233 Spring Street, New York, NY 10013, U.S.A.

To my family

张发生　　赵玉铭
何荼娥　　牛福华
Ying　　　*Albert*

Preface

Why this book was written: Initiated by Pardoux & Peng [167], the theory of Backward Stochastic Differential Equations has been extensively studied in the past decades, and their applications have been found in many areas. While there are a few excellent monographs and book chapters on the subject, see, e.g., El Karoui & Mazliak [80], Peng [175], Ma & Yong [148] (on forward-backward SDEs), and Pardoux & Rascanu [170], there is an increasing demand for a textbook which is accessible to graduate students and junior researchers interested in this important and fascinating area. In the meantime, there is a strong need for a book that includes the more recent developments, e.g., the fully nonlinear theory and path-dependent PDEs. The aim in this book is to introduce up-to-date developments in the field in a systematic and "elementary" way.

There is often a trade-off between generality and clarity. While it is convenient to have the most general results for the purpose of direct applications, in many situations, one may need the ideas rather than the results. The high technicality involved due to the generality may unfortunately obscure the key ideas, even for experts. In this book, the focus is on ideas, so that readers may have a comprehensive taste of the main results, the required conditions, and the techniques involved. Almost all results in the book have been proven from scratch, and the arguments have been made to look as natural as possible. As such, the generality has been sacrificed in many places.

Who is it for: Ph.D. students and junior researchers majoring in Stochastic Analysis are the main target audience. However, it is the author's hope that it (at least Part I) proves useful for graduate students majoring in Engineering and Quantitative Finance. The material from Part I of the book was used for regular Ph.D. courses at USC, with students majoring in various fields.

The last part of the book on fully nonlinear theory is more advanced. The material has been used for a special topics course at USC as well as for some short courses in other places. It is hoped that junior researchers interested in this area will find it helpful.

Prerequisites: A solid knowledge of graduate-level Stochastic Calculus and Real Analysis is required, and basic knowledge on second-order PDEs and financial derivatives will also be helpful. However, the book has been written to be as self-contained as possible (except some limited material in Part III), with the more advanced prerequisite material presented in the Chapter 1. Therefore, readers with less knowledge but with a good sense of general mathematics theory may also find the book accessible, by skipping some technical proofs when necessary.

Structure of the book: The book is divided into three main parts: basic theory of SDEs and BSDEs, further theory of BSDEs, and more recent developments in fully nonlinear theory. References for related topics are given at the end of each chapter.

Part I is basic and is more or less mature. It starts with the basics of stochastic calculus, such as stochastic integration and martingale representation theorem, which can be viewed as linear SDEs and linear Backward SDEs, respectively. In contrast with the fully nonlinear theory in Part III, these materials can be viewed as the linear theory. Then, the general (nonlinear) SDEs and Backward SDEs are dealt with in the same spirit. In particular, BSDE theory is described as a semilinear theory because it is associated with semilinear PDE in the Markovian case and semilinear path-dependent PDE in the non-Markovian case. Such a connection in the Markovian case is established rigorously in this part, and the non-Markovian case is studied in Part III.

Part II covers three important extensions of the theory: reflected BSDEs, BSDEs with quadratic growth in the Z component, and coupled forward-backward SDEs. This part can be expanded drastically, for example, Peng's g-expectation, BSDEs with non-smooth coefficients, reflected BSDEs with two barriers and Dynkin games, BSDEs with general constraints, infinite-dimensional BSDEs, BSDEs with random or infinite horizon and their connection with elliptic PDEs, weak solutions of BSDEs, and backward stochastic PDEs, and BSDEs with jumps, to mention a few. These topics are not covered here, following the aim to keep the book within a reasonable size. However, they are briefly discussed for the benefit of the readers.

Part III has been developing very dynamically in recent years. It covers three topics: nonlinear expectation, path dependent PDEs, and second-order BSDEs, together with a preparation chapter on weak formulation upon which all the three subjects are built. The theory is far from mature. Nevertheless, it is intrinsically a continuation of Parts I and II, and we have received strong feedback to provide an introduction on its recent developments.

The book is aimed at theory rather than application. While there are numerous publications on applications of BSDEs, including a few excellent books (see, e.g., Yong & Zhou [242], Pham [190], Cvitanic & Zhang [52], Touzi [227], Crepey [43], Delong [60], and Carmona [29]), the author's opinion is that most fall into three categories: pricing and hedging financial derivatives, stochastic optimization and games, and connections with nonlinear PDEs. These applications are scattered in this book, mainly to motivate the theories.

Another major topic in the field is probabilistic numerical methods for nonlinear PDEs. The book provides only a very brief introduction to it. It is believed that this deserves a separate research monograph.

Exercises are an important part of the book. There are mainly four types of problems: (i) technical preparations for some main results, (ii) alternative proofs of the main results, (iii) extensions of some results proved in the book, and (iv) practice of some important techniques used in the book. Hints and/or solutions to some selective problems are available in the author's website: http://www-bcf.usc.edu/~jianfenz/.

A few additional notes: As already mentioned, clarity has been given precedence over generalization. There are occasions where we impose conditions stronger than necessary so as to focus on the main ideas and make the proofs more accessible to the readers. So generalized results which require rather sophisticated techniques are not included, but some of them are listed in the Bibliographical Notes section for interested readers.

To make the book more readable, for most proofs, only one-dimensional notation has been used. For results where the dimension is crucial, it is mentioned explicitly.

The field has grown rapidly. It is impossible to exhaust the references. The author admits that they have inevitably missed many very important and highly relevant works.

Acknowledgments A very large portion of my research was related to the material presented in this book. I take this opportunity to thank all my teachers, collaborators, students, and friends who have helped me along my career. In particular, I am very grateful to the following people who have had the greatest impact on my academic career: Jaksa Cvitanic, Jin Ma, Shige Peng, and Nizar Touzi.

While some presentations in the book might be new, all credits of the results should go to the original papers, and thus I am indebted to all the authors whose results and/or ideas I have borrowed from. I am grateful to several reviewers who provided many constructive suggestions and corrected numerous typos in the earlier versions of the book. Of course, I am solely responsible for the remaining errors in the book, and I would truly appreciate any comments or suggestions the readers may have.

I am grateful for the support from the Springer staff, especially Donna Chernyk for her endless patience throughout the writing of this book. The book was partially supported by the National Science Foundation grants DMS 1008873 and 1413717. I am also grateful for the support of my home institution: the University of Southern California.

Finally, my foremost gratitude goes to my family, for their understanding, support, and love.

Los Angeles, CA, USA   Jianfeng Zhang
April 2017

Contents

Chapter 1
Preliminaries

In this chapter we introduce the very basic materials in probability theory, which will be used throughout the book without mentioning. We shall also introduce the notation we are going to use in the rest of the book.

1.1 Probability Spaces and Random Variables

1.1.1 Probability Spaces

In this subsection we introduce the probability space, denoted as $(\Omega, \mathscr{F}, \mathbb{P})$. Here the *sample space* Ω denotes the set of all possible outcomes of a random event, and its elements are denoted by ω. One typical example we will consider in this book is $\Omega = \{\omega \in C([0, T]) : \omega_0 = 0\}$, the set of all continuous functions on $[0, T]$ with initial value 0 and some finite time horizon $T > 0$.

$\mathscr{F}$ is a *σ-field* on Ω, also called *σ-algebra*, which consists of subsets of Ω and satisfies:

- The empty set ϕ is an element of $\mathscr{F}$;
- For any $A \in \mathscr{F}$, its complement $A^c := \Omega \backslash A$ is also in $\mathscr{F}$;
- For a countable (or finite) sequence $\{A_n, n \geq 1\} \subset \mathscr{F}$, its union $\bigcup_{n \geq 1} A_n \in \mathscr{F}$.

From above it is clear that $\Omega \in \mathscr{F}$ and $\bigcap_{n \geq 1} A_n \in \mathscr{F}$ for $\{A_n, n \geq 1\} \subset \mathscr{F}$. The smallest σ-field is $\{\phi, \Omega\}$, and the largest σ-field is the collection of all subsets of Ω, often denoted as 2^Ω. We call a set in $\mathscr{F}$ an $\mathscr{F}$-measurable *event*, and call $(\Omega, \mathscr{F})$ a *measurable space*.

We remark that, if $\{A_i, i \in I\} \subset \mathscr{F}$ and I is *uncountable*, in general it is possible that $\bigcup_{i \in I} A_i \notin \mathscr{F}$ or $\bigcap_{i \in I} A_i \notin \mathscr{F}$. See Problem 1.4.1.

© Springer Science+Business Media LLC 2017

J. Zhang, *Backward Stochastic Differential Equations*, Probability Theory and Stochastic Modelling 86, DOI 10.1007/978-1-4939-7256-2_1

Let $\mathscr{A} \subset 2^{\Omega}$ be a family of subsets of Ω, not necessarily a σ-field. Denote

$$\sigma(\mathscr{A}) := \bigcap \Big[\mathscr{G} : \mathscr{G} \text{ is a } \sigma\text{-field on } \Omega \text{ and } \mathscr{A} \subset \mathscr{G} \Big]. \qquad (1.1.1)$$

Then $\sigma(\mathscr{A})$ is the smallest σ-field containing $\mathscr{A}$, and is called the σ-field *generated* by $\mathscr{A}$. In particular, if we set $\Omega = \mathbb{R}^d$, $\mathscr{A} = $ the set of all open sets in $\mathbb{R}^d$, then $\sigma(\mathscr{A})$ are the Borel sets in $\mathbb{R}^d$, and denoted as $\mathscr{B}(\mathbb{R}^d)$ in this book.

A *probability* $\mathbb{P}$ on $(\Omega, \mathscr{F})$, also called *probability measure*, is a mapping $\mathscr{F} \to [0, 1]$ satisfying:

- $\mathbb{P}(\phi) = 0$, $\mathbb{P}(\Omega) = 1$;
- If a countable sequence $\{A_n, n \geq 1\} \subset \mathscr{F}$ are *disjoint*, that is, $A_i \cap A_j = \phi$ for all $i \neq j$, then

$$\mathbb{P}\Big(\bigcup_{n \geq 1} A_n \Big) = \sum_{n \geq 1} \mathbb{P}(A_n).$$

We call the triplet $(\Omega, \mathscr{F}, \mathbb{P})$ a probability space. We remark that, given Ω, there can be many different σ-fields $\mathscr{F}$ on Ω; and given $(\Omega, \mathscr{F})$, one can consider many different probability measures $\mathbb{P}$ on $(\Omega, \mathscr{F})$.

We say an event $A \in \mathscr{F}$ occurs $\mathbb{P}$-*almost surely* if $\mathbb{P}(A) = 1$, abbreviated as $\mathbb{P}$-a.s., and an event $A \in \mathscr{F}$ is a $\mathbb{P}$-*null event* if $\mathbb{P}(A) = 0$. We may omit the $\mathbb{P}$ and simply call them a.s. and null event, respectively, when there is no confusion on the probability measure $\mathbb{P}$. Given $(\Omega, \mathscr{F}, \mathbb{P})$, denote

$$\mathscr{N}(\mathscr{F}, \mathbb{P}) := \{A \subset \Omega : \text{there exists a } \mathbb{P}\text{-null set } \tilde{A} \in \mathscr{F} \text{ such that } A \subset \tilde{A}\}. \quad (1.1.2)$$

We say $(\Omega, \mathscr{F}, \mathbb{P})$ is a *complete* probability space if $\mathscr{N}(\mathscr{F}, \mathbb{P}) \subset \mathscr{F}$. We remark that, for any probability space $(\Omega, \mathscr{F}, \mathbb{P})$, we can complete it following a standard procedure, see Problem 1.4.3:

$$\overline{\mathscr{F}}^{\mathbb{P}} := \mathscr{F} \vee \mathscr{N}(\mathscr{F}, \mathbb{P}) := \sigma\big(\mathscr{F} \cup \mathscr{N}(\mathscr{F}, \mathbb{P})\big) \qquad (1.1.3)$$

1.1.2 Random Variables

Given a measurable space $(\Omega, \mathscr{F})$. We say a mapping $X : \Omega \to \mathbb{R}$ is an $\mathscr{F}$-measurable *random variable*, denoted as $X \in \mathbb{L}^0(\mathscr{F})$, if

$$X^{-1}(A) := \{X \in A\} := \{\omega \in \Omega : X(\omega) \in A\} \in \mathscr{F} \quad \text{for all } A \in \mathscr{B}(\mathbb{R}).$$

Given a random variable $X \in \mathbb{L}^0(\mathscr{F})$, denote

$$\mathscr{F}^X := \sigma(X) := \Big\{X^{-1}(A) : A \in \mathscr{B}(\mathbb{R})\Big\}. \qquad (1.1.4)$$

Then $\mathscr{F}^X$ is a sub-σ-field of $\mathscr{F}$. We call it the σ-field generated by X. Notice that X is always $\mathscr{F}^X$-measurable. The following Doob-Dynkin's lemma is important.

Lemma 1.1.1 *Let $X, Y \in \mathbb{L}^0(\mathscr{F})$. Then $Y \in \mathbb{L}^0(\mathscr{F}^X)$ if and only if there exists a Borel-measurable function $g : \mathbb{R} \to \mathbb{R}$ such that $Y = g(X)$.*

We remark that the concepts above do not involve any probability measure $\mathbb{P}$. Now fix a $\mathbb{P}$. Denote

$$\mu_X(A) := (\mathbb{P} \circ X^{-1})(A) := \mathbb{P}(X \in A), \quad A \in \mathscr{B}(\mathbb{R}).$$

Then $(\mathbb{R}, \mathscr{B}(\mathbb{R}), \mu_X)$ is a probability space, and we call μ_X the distribution of X. In particular, $F_X(x) := \mathbb{P}(X \leq x)$ is called the cumulative distribution function (abbreviated as cdf) of X. When F_X is absolutely continuous with respect to the Borel measure dx, we call $f_X(x) := F'_X(x)$ the probability density function (abbreviated as pdf) of X.

When $\int_{\mathbb{R}} |x| \mu_X(dx) < \infty$, we say X is integrable and define its *expectation* by

$$\mathbb{E}[X] := \int_{\mathbb{R}} x \mu_X(dx) = \int_{\mathbb{R}} x dF_X(x) \text{ and } \mathbb{E}[X] = \int_{\mathbb{R}} x f_X(x) dx \text{ when the pdf exists.}$$

In particular, if g is a Borel-measurable function and $g(X)$ is integrable, we have

$$\mathbb{E}[g(X)] = \int_{\mathbb{R}} g(x) \mu_X(dx).$$

For any $p \geq 1$, let $\mathbb{L}^p(\mathscr{F})$ denote the set of $X \in \mathbb{L}^0(\mathscr{F})$ such that $\mathbb{E}[|X|^p] < \infty$, and $\mathbb{L}^\infty(\mathscr{F})$ the set of bounded $X \in \mathbb{L}^0(\mathscr{F})$, namely there exists a constant $C > 0$ such that $|X| \leq C$, a.s. Moreover, we define the variance and covariance of random variables in $\mathbb{L}^2(\mathscr{F})$ by

$$Var(X) := \mathbb{E}\left[(X - \mathbb{E}[X])^2\right] = \mathbb{E}[X^2] - (\mathbb{E}[X])^2,$$

$$Cov(X, Y) := \mathbb{E}\left[(X - \mathbb{E}[X])(Y - \mathbb{E}[Y])\right] = \mathbb{E}[XY] - \mathbb{E}[X]\mathbb{E}[Y].$$

Let $X \in \mathbb{L}^1(\mathscr{F})$ and $\mathscr{G} \subset \mathscr{F}$ be a sub-σ-field. A random variable in $\mathbb{L}^1(\mathscr{G})$ is called the *conditional expectation* of X given $\mathscr{G}$, denoted as $\mathbb{E}[X|\mathscr{G}]$, if it satisfies

$$\mathbb{E}\left[\mathbb{E}[X|\mathscr{G}]Y\right] = \mathbb{E}[XY] \text{ for all } Y \in \mathbb{L}^\infty(\mathscr{G}).$$

For any $X \in \mathbb{L}^1(\mathscr{F})$ and any sub-σ-field $\mathscr{G}$, there exists a unique $\mathbb{E}[X|\mathscr{G}]$. Here uniqueness is in the a.s. sense, that is, if another random variable $Z \in \mathbb{L}^1(\mathscr{G})$ also satisfies the above property, then $\mathbb{E}[X|\mathscr{G}] = Z$, a.s. We notice that, if $Y \in \mathbb{L}^0(\mathscr{F})$ and $\mathscr{G} = \mathscr{F}^Y$, by definition $\mathbb{E}[X|Y] := \mathbb{E}[X|\mathscr{F}^Y]$ is $\mathscr{F}^Y$-measurable, and thus by Doob-Dynkin's Lemma 1.1.1 we have $\mathbb{E}[X|Y] = g(Y)$ for some Borel measurable function g.

At below we collect some important properties of conditional expectations:

- $\mathbb{E}\big[\mathbb{E}[X|\mathcal{G}]\big] = \mathbb{E}[X]$ for any $X \in \mathbb{L}^1(\mathcal{F})$;
- $\mathbb{E}[XY|\mathcal{G}] = Y\mathbb{E}[X|\mathcal{G}]$ for any $X \in \mathbb{L}^1(\mathcal{F})$ and $Y \in \mathbb{L}^0(\mathcal{G})$ such that $XY \in \mathbb{L}^1(\mathcal{F})$. In particular, $\mathbb{E}[X|\mathcal{G}] = X$ for any $X \in \mathbb{L}^1(\mathcal{G})$.
- If $\mathcal{G}_1 \subset \mathcal{G}_2 \subset \mathcal{F}$ are two sub-σ-fields and $X \in \mathbb{L}^1(\mathcal{F})$, then

$$\mathbb{E}\big[\mathbb{E}[X|\mathcal{G}_1]\big|\mathcal{G}_2\big] = \mathbb{E}\big[\mathbb{E}[X|\mathcal{G}_2]\big|\mathcal{G}_1\big] = \mathbb{E}[X|\mathcal{G}_1].$$

We note that the integrability condition in conditional expectation can be weakened if we use the conditional probability distribution, which will be introduced in Section 9.3 later. Moreover, we say $X \in \mathbb{L}^0(\mathcal{F})$ is *independent* of $\mathcal{G}$ if $\mathbb{E}[g(X)|\mathcal{G}] = \mathbb{E}[g(X)]$ for all bounded Borel measurable function g. In particular, if $\mathcal{G} = \{\phi, \Omega\}$, then all $X \in \mathbb{L}^0(\mathcal{F})$ are independent of $\mathcal{G}$.

We remark that all the concepts and operators in above three paragraphs depend on the probability measure $\mathbb{P}$. When there are different probability measures involved, we use $\mathbb{E}^{\mathbb{P}}$, $\mathbb{L}^p(\mathcal{F}, \mathbb{P})$, etc. to indicate their dependence on $\mathbb{P}$. Moreover, all the above equalities involving conditional probabilities should be understood in the $\mathbb{P}$-a.s. sense.

1.1.3 Random Vectors

In this book, all vectors are considered to be column vectors, namely we take the convention that $\mathbb{R}^d = \mathbb{R}^{d\times 1}$ for some dimension d. A d-dimensional *random vector* is a mapping $X = (X_1, \cdots, X_d)^\top : \Omega \to \mathbb{R}^d$ such that $X_i \in \mathbb{L}^0(\mathcal{F})$, $i = 1, \cdots, d$, where $^\top$ stands for transpose. When there is no confusion on dimension, we may still call it a random variable. On the other hand, when there is a need to emphasize the dimension, we denote $X \in \mathbb{L}^0(\mathcal{F}, \mathbb{R}^d)$.

Given $(\Omega, \mathcal{F}, \mathbb{P})$ and $X \in \mathbb{L}^0(\mathcal{F}, \mathbb{R}^d)$, its joint distribution and joint cdf are defined by

$$\mu_X(A) := \mathbb{P}(X \in A), \ \ A \in \mathcal{B}(\mathbb{R}^d), \text{ and}$$

$$F_X(x) := \mathbb{P}(X \le x) := \mathbb{P}(X_i \le x_i, i = 1, \cdots, d), \ \ x \in \mathbb{R}^d.$$

We say random variables $X_1, \cdots, X_n$ (each with arbitrary dimension) are *independent* if

$$F_{X_1, \cdots, X_n}(x_1, \cdots, x_n) = F_{X_1}(x_1) \cdots F_{X_n}(x_n),$$

for all $x_1, \cdots, x_n$ with appropriate dimensions. The following statements are equivalent:

- $X_1, \cdots, X_n$ are independent;
- Each X_i is independent of the σ-algebra $\sigma(X_j, j \neq i)$;
- $\mathbb{E}\left[\Pi_{i=1}^n g_i(X_i)\right] = \Pi_{i=1}^n \mathbb{E}[g_i(X_i)]$ for any bounded scalar Borel measurable functions $g_1, \cdots, g_n$.

1.1.4 Normal Distribution

Normal distribution is probably the most important distribution, both in theory and in applications. We say a random variable X has *normal distribution* with mean μ and variance σ^2, denoted as $X \sim N(\mu, \sigma^2)$, if it has pdf:

$$f(x) = \frac{1}{\sqrt{2\pi}\sigma} e^{-\frac{(x-\mu)^2}{2\sigma^2}}, \quad x \in \mathbb{R}.$$

One can check straightforwardly that

$$\mathbb{E}[X] = \mu, \quad Var(X) = \sigma^2, \quad \text{and} \quad Z := \frac{X - \mu}{\sigma} \sim N(0, 1).$$

We say random variables $X_1, \cdots, X_n$ have *Gaussian distribution*, also called *multivariate normal distribution*, if any linear combination of $X_1, \cdots, X_n$ has normal distribution. In particular, if $X_1, \cdots, X_n$ are independent and each has normal distribution, then they have Gaussian distribution. If $X_1, \cdots, X_n$ have Gaussian distribution, then they are independent if and only if $Cov(X_i, X_j) = 0$ for all $i \neq j$. Moreover, if X_n (with arbitrary dimension) has Gaussian distribution for each $n \geq 1$ and $\lim_{n\to\infty} \mathbb{E}[|X_n - X|] = 0$, then X also has Gaussian distribution.

The following *central limit theorem* explains why normal distribution is so important.

Theorem 1.1.2 *Assume X_n, $n \geq 1$, are independent and identically distributed with* $\mathbb{E}[X_n] = \mu$ *and* $Var(X_n) = \sigma^2 > 0$. *Denote* $\bar{X}_n := \frac{1}{n}\sum_{i=1}^n X_i$ *and* $Z_n := \frac{\sqrt{n}(\bar{X}_n - \mu)}{\sigma}$. *The Z_n converges to $N(0, 1)$ in distribution, that is, for any $x \in \mathbb{R}$,*

$$\lim_{n\to\infty} \mathbb{P}(Z_n \leq x) = \int_{-\infty}^x \frac{1}{\sqrt{2\pi}} e^{-\frac{y^2}{2}} \, dy.$$

1.1.5 Product Spaces

Let $(\Omega_i, \mathscr{F}_i, \mathbb{P}_i)$, $1 \leq i \leq n$, be probability spaces. We define the product probability space as follows:

- $\Omega_1 \times \cdots \times \Omega_n := \{(\omega_1, \cdots, \omega_n) : \omega_i \in \Omega_i, 1 \leq i \leq n\};$

- $\mathscr{F}_1 \times \cdots \times \mathscr{F}_n := \sigma\Big(\{A_1 \times \cdots \times A_n : A_i \in \mathscr{F}_i\}\Big);$
- $\mathbb{P}_1 \times \cdots \times \mathbb{P}_n$ is the unique probability measure on $\mathscr{F}_1 \times \cdots \times \mathscr{F}_n$ satisfying

$$(\mathbb{P}_1 \times \cdots \times \mathbb{P}_n)(A_1 \times \cdots \times A_n) = \Pi_{i=1}^{n}\mathbb{P}_i(A_i), \quad \text{for all } A_i \in \mathscr{F}_i, i = 1, \cdots, n.$$

The following is a version of the Fubini's theorem. Let $n = 2$, $X_i \in \mathbb{L}^0(\mathscr{F}_i)$, and $\varphi : \mathbb{R}^2 \to \mathbb{R}$ be Borel measurable such that $\varphi(X_1, X_2) \in \mathbb{L}^1(\mathscr{F}_1 \times \mathscr{F}_2, \mathbb{P}_1 \times \mathbb{P}_2)$. Then

$$\mathbb{E}^{\mathbb{P}_1 \times \mathbb{P}_2}\big[\varphi(X_1, X_2)\big] = \mathbb{E}^{\mathbb{P}_1}[\varphi_1(X_1)] = \mathbb{E}^{\mathbb{P}_2}[\varphi_2(X_2)]$$

$$\text{where} \quad \varphi_1(x) := \mathbb{E}^{\mathbb{P}_2}[\varphi(x, X_2)], \quad \varphi_2(x) := \mathbb{E}^{\mathbb{P}_1}[\varphi(X_1, x)].$$

1.1.6   The Essential Supremum

The concept of essential supremum is slightly more advanced, and will play an important role in Chapters 6 and 12. Readers who are not interested in those chapters can skip this part.

Definition 1.1.3 *Let $(\Omega, \mathscr{F}, \mathbb{P})$ be a probability space, $\mathscr{X} := \{X_i, i \in I\} \subset \mathbb{L}^0(\mathscr{F}, \mathbb{R})$, where I is a possibly uncountable index set.*

 (i) *We say $X \in \mathbb{L}^0(\mathscr{F})$ is a $\mathbb{P}$-essential upper bound of $\mathscr{X}$ if $X \geq X_i$, $\mathbb{P}$-a.s. for all $i \in I$;*
 (ii) *We say $X \in \mathbb{L}^0(\mathscr{F})$ is a $\mathbb{P}$-essential supremum of $\mathscr{X}$, denoted as $X = \text{ess sup}_{i\in I}^{\mathbb{P}}X_i$, if it is a $\mathbb{P}$-essential upper bound of $\mathscr{X}$ and $X \leq \tilde{X}$, $\mathbb{P}$-a.s. for all $\mathbb{P}$-essential upper bound $\tilde{X}$.*
 (iii) *We say $X \in \mathbb{L}^0(\mathscr{F})$ is a $\mathbb{P}$-essential infimum of $\mathscr{X}$, denoted as $X = \text{ess inf}_{i\in I}^{\mathbb{P}} X_i$, if $-X = \text{ess sup}_{i\in I}^{\mathbb{P}}(-X_i)$.*

It is clear that ess sup and ess inf are unique in $\mathbb{P}$-a.s. sense. Moreover, we have the following existence result, whose proof is sketched in Problem 1.4.12.

Theorem 1.1.4 *Let $\mathscr{X} = \{X_i, i \in I\} \subset \mathbb{L}^0(\mathscr{F}, \mathbb{R})$ with a $\mathbb{P}$-essential upper bound Y.*

 (i) *The essential supremum $X := \text{ess sup}_{i\in I}^{\mathbb{P}}X_i$ exists and there exists a countable sequence $\{i_n, n \geq 1\} \subset I$ such that $X = \sup_{n\geq 1} X_{i_n}$, $\mathbb{P}$-a.s.*
 (ii) *If the class $\mathscr{X}$ satisfies the so-called filtrating family condition:*

$$\text{for any } i, j \in I, \text{ there exists } k \in I \text{ such that } X_i \leq X_k, X_j \leq X_k, \mathbb{P}\text{-a.s.} \quad (1.1.5)$$

then there exist $\{i_n, n \geq 1\} \subset I$ such that $X_{i_n} \uparrow X$, $\mathbb{P}$-a.s., as $n \to \infty$.

1.2 Stochastic Processes

In this book we shall fix a time duration $[0, T]$. While many concepts and results in Chapters 1–3 remain valid in the case $T = \infty$ (after some appropriate modifications when necessary), most results for later chapters require T to be finite. For simplicity we assume $T < \infty$ throughout the book.

1.2.1 Filtrations

Let $(\Omega, \mathscr{F}, \mathbb{P})$ be a probability space. We say $\mathbb{F} := \{\mathscr{F}_t, 0 \le t \le T\}$ is a *filtration* if

- for any $t \in [0, T]$, $\mathscr{F}_t \subset \mathscr{F}$ is a sub-σ-field;
- for any $0 \le t_1 < t_2 \le T$, $\mathscr{F}_{t_1} \subset \mathscr{F}_{t_2}$.

We remark that the above definition does not involve the probability measure $\mathbb{P}$. The quadruple $(\Omega, \mathscr{F}, \mathbb{F}, \mathbb{P})$ is called a filtered probability space. For notational simplicity, we shall always denote $\mathscr{F}_t := \mathscr{F}_T$ for $t > T$ and $\mathscr{F}_t := \{\emptyset, \Omega\}$ for $t < 0$.
   Denote

$$\mathscr{F}_t^+ := \mathscr{F}_{t+} := \bigcap_{s > t} \mathscr{F}_s, \quad \mathbb{F}^+ := \{\mathscr{F}_t^+ : 0 \le t \le T\}. \tag{1.2.1}$$

We say $\mathbb{F}$ is *right continuous* if $\mathbb{F}^+ = \mathbb{F}$. Notice that $\mathbb{F}^+$ is always right continuous. Given the probability measure $\mathbb{P}$, we define its *completion* $\mathbb{F}^{\mathbb{P}} := \{\mathscr{F}_t^{\mathbb{P}}, 0 \le t \le T\}$ and *augmentation* $\overline{\mathbb{F}}^{\mathbb{P}} := \{\overline{\mathscr{F}}_t^{\mathbb{P}}, 0 \le t \le T\}$ as follows:

$$\mathscr{F}_t^{\mathbb{P}} := \mathscr{F}_t \vee \mathscr{N}(\mathscr{F}_t, \mathbb{P}), \quad \overline{\mathscr{F}}_t^{\mathbb{P}} := \mathscr{F}_t \vee \mathscr{N}(\mathscr{F}, \mathbb{P}). \tag{1.2.2}$$

In the literature, typically one assumes that the filtered probability space $(\Omega, \mathscr{F}, \mathbb{F}, \mathbb{P})$ satisfies the following *usual hypotheses*:

- $\mathbb{F}$ is right continuous, that is, $\mathbb{F}^+ = \mathbb{F}$.
- $\mathbb{F}$ is augmented, that is, $\overline{\mathbb{F}}^{\mathbb{P}} = \mathbb{F}$ or say $\mathscr{N}(\mathscr{F}, \mathbb{P}) \subset \mathscr{F}_0$.

Given arbitrary $(\Omega, \mathscr{F}, \mathbb{F}, \mathbb{P})$, note that $\overline{\mathbb{F}^+}^{\mathbb{P}}$ always satisfies the usual hypotheses. However, for the purpose of the fully nonlinear theory in Part III, we emphasize that in this book we shall *not* assume the usual hypotheses, and we will mention these two hypotheses explicitly when we need them. This relaxation is mild in most cases, see Proposition 1.2.1 below. However, the issue can be very subtle in some cases, see, e.g., Proposition 1.2.5 (ii) and Section 12.3 below.

1.2.2 Stochastic Processes

A stochastic process X is a mapping from $[0, T] \times \Omega \to \mathbb{R}$. One may view X in three different ways:

- as a function of two variables (t, ω);
- as a family of random variables $\{X_t, 0 \leq t \leq T\}$;
- as a family of paths $\{X.(\omega), \omega \in \Omega\}$.

We say X is $\mathbb{F}$-*progressively measurable*, or simply called $\mathbb{F}$-*measurable* and denoted as $X \in \mathbb{L}^0(\mathbb{F})$, if the restriction of X on $[0, t]$ is $\mathscr{B}([0, t]) \times \mathscr{F}_t$-measurable, for any $0 \leq t \leq T$. We note that we are abusing the notation by using X for both random variables and stochastic processes. By definition, $X \in \mathbb{L}^0(\mathscr{F})$ implies X is a random variable while $X \in \mathbb{L}^0(\mathbb{F})$ implies it is a process.

Given $X \in \mathbb{L}^0(\mathbb{F})$, recall that $\mathscr{F}^{X_t}$ is the σ-field generated by the random variable X_t. Let $\mathscr{F}_t^X$ denote the σ-field generated by $\bigcup_{0 \leq s \leq t} \mathscr{F}^{X_s}$, and $\mathbb{F}^X := \{\mathscr{F}_t^X, 0 \leq t \leq T\}$. Then $\mathbb{F}^X$ is the smallest filtration to which X is progressively measurable. We call $\mathbb{F}^X$ the filtration generated by the process X. Roughly speaking, if a random variable $Y \in \mathbb{L}^0(\mathscr{F}_t^X)$, then the value of Y is determined by the path of X on $[0, t]$, that is, if $X_s(\omega_1) = X_s(\omega_2), 0 \leq s \leq t$, then we should have $Y(\omega_1) = Y(\omega_2)$.

Given two processes $X, Y \in \mathbb{L}^0(\mathbb{F})$, we say they are $\mathbb{P}$-*modifications*, or say Y is a $\mathbb{P}$-*modified version* of X, if $\mathbb{P}(X_t = Y_t) = 1$ for all $t \in [0, T]$. We say X and Y are $\mathbb{P}$-*indistinguishable*, or say Y is a $\mathbb{P}$-*indistinguishable version* of X, if $\mathbb{P}(X_t = Y_t, 0 \leq t \leq T) = 1$. Clearly an indistinguishable version must be a modified version, but the converse is not true in general. See Problem 1.4.5.

Since we do not assume the usual hypotheses in this book, the following results will be crucial.

Proposition 1.2.1 *Let $(\Omega, \mathscr{F}, \mathbb{F}, \mathbb{P})$ be a filtered probability space.*

(i) *For any $X \in \mathbb{L}^0(\overline{\mathscr{F}}^{\mathbb{P}})$, there exists $(\mathbb{P}$-a.s.$)$ unique $\tilde{X} \in \mathbb{L}^0(\mathscr{F})$ such that $\tilde{X} = X$, $\mathbb{P}$-a.s.*

(ii) *For any $X \in \mathbb{L}^0(\overline{\mathbb{F}}^{\mathbb{P}})$, there exists $(dt \times d\mathbb{P}$-a.s.$)$ unique $\tilde{X} \in \mathbb{L}^0(\mathbb{F})$ such that $\tilde{X} = X$, $dt \times d\mathbb{P}$-a.s. Moreover, if X is càdlàg $($that is, right continuous with left limits$)$, $\mathbb{P}$-a.s., then so is $\tilde{X}$ and $\tilde{X}_t = X_t$, $0 \leq t \leq T$, $\mathbb{P}$-a.s.*

(iii) *If $X \in \mathbb{L}^0(\mathbb{F}^+)$ is $($left$)$ continuous, $\mathbb{P}$-a.s, then there exists $(\mathbb{P}$-a.s.$)$ unique $\tilde{X} \in \mathbb{L}^0(\mathbb{F})$ such that $\tilde{X}_t = X_t$, $0 \leq t \leq T$, $\mathbb{P}$-a.s.*

For an arbitrary filtration $\mathbb{F}$, recall that $\overline{\mathbb{F}^+}^{\mathbb{P}}$ always satisfies the usual hypotheses. Given the above proposition, we may derive many results in this book by first finding $X \in \mathbb{L}^0(\overline{\mathbb{F}^+}^{\mathbb{P}})$, as done in standard literature, and then choose the version $\tilde{X} \in \mathbb{L}^0(\mathbb{F})$. We shall emphasize though, the version $\tilde{X}$ typically depends on $\mathbb{P}$.

We next introduce the distribution of a process X. We first note that the marginal distributions $\{\mu_{X_t}, 0 \leq t \leq T\}$ do not provide enough information for the process X. On the other hand, since there are uncountably many t, it is difficult to study the joint distribution of $\{X_t, 0 \leq t \leq T\}$. We thus consider the finite distribution of X as follows. For any n and any time partition $0 \le t_1 < \cdots < t_n \le T$, let $\mu^X_{t_1, \cdots, t_n}$ denote the joint distribution of $(X_{t_1}, \cdots, X_{t_n})$. We call the family of all these distributions $\mu^X_{t_1, \cdots, t_n}$ the finite distribution of X. In particular, we say X and Y have the same distribution if they have the same finite distribution. One can easily check that the finite distribution satisfies the following property: for any n, any $0 \le t_1 < \cdots < t_n \le T$, any i, and any $A_j \in \mathscr{B}(\mathbb{R}), j \ne i$,

$$
\begin{aligned}
&\mu^X_{t_1, \cdots, t_n}\big(A_1 \times \cdots \times A_{i-1} \times \mathbb{R} \times A_{i+1} \times \cdots \times A_n\big) \\
&= \mu^X_{t_1, \cdots, t_{i-1}, t_{i+1}, \cdots, t_n}\big(A_1 \times \cdots \times A_{i-1} \times A_{i+1} \times \cdots \times A_n\big).
\end{aligned}
\tag{1.2.3}
$$

Theorem 1.2.2 (Kolmogorov's Extension Theorem) *Let $\mu_{t_1, \cdots, t_n}$ be a family of distributions on $\mathscr{B}(\mathbb{R}^n)$ satisfying (1.2.3). Then there exist $(\Omega, \mathscr{F}, \mathbb{P})$ and $X \in \mathbb{L}^0(\mathscr{F})$ such that $\mu^X_{t_1, \cdots, t_n} = \mu_{t_1, \cdots, t_n}$ for all n and $(t_1, \cdots, t_n)$.*

We now introduce a few convenient notations for X which will be used throughout the book:

$$
X^*_t := \sup_{0 \le s \le t} |X_s|, \quad X_{s,t} := X_t - X_s \;\forall 0 \le s < t \le T; \quad OSC_\delta(X) := \sup_{0 \le s < t \le T, t-s \le \delta} |X_{s,t}|. \tag{1.2.4}
$$

One important issue for stochastic processes is to derive pathwise properties from its distribution. At below is a result in this direction.

Theorem 1.2.3 (Kolmogorov's Continuity Theorem) *Let $X \in \mathbb{L}^0(\mathbb{F})$. Assume there exist constants $\alpha, \beta, C > 0$ such that*

$$
\mathbb{E}\Big[|X_{s,t}|^\alpha\Big] \le C|t - s|^{1+\beta}, \quad \text{for all } s, t \in [0, T]. \tag{1.2.5}
$$

Then for any $\gamma \in (0, \frac{\beta}{\alpha})$, X is Hölder-γ continuous, a.s.
We shall emphasize that, the condition (1.2.5) involves only finite distributions of X, and thus is invariant under modifications. So the pathwise property has to be understood in terms of modifications. To be precise, the conclusion of the theorem should be understood as:

$$
\begin{aligned}
&\text{there exists a } \mathbb{P}\text{-modification } \tilde{X} \in \mathbb{L}^0(\mathbb{F}) \text{ of } X \text{ such that} \\
&\tilde{X}.(\omega) \text{ is Hölder} - \gamma \text{ continuous for } \mathbb{P} - a.e. \; \omega.
\end{aligned}
\tag{1.2.6}
$$

All the pathwise properties in the book, unless the processes are constructed explicitly, should be understood in this way.

We say $X = (X^1, \cdots, X^d)^\top$ is a d-dimensional process if $X^1, \cdots, X^d$ are processes, and we denote $X \in \mathbb{L}^0(\mathbb{F}, \mathbb{R}^d)$ if each $X^i \in \mathbb{L}^0(\mathbb{F})$. We shall still call it a process when there is no need to emphasize the dimension. We remark that all the notations in (1.2.4) can be extended to multidimensional setting.

Finally, quite often we will need random field $f : [0, T] \times \Omega \times \mathbb{R}^{d_1} \to \mathbb{R}^{d_2}$ for appropriate dimensions d_1, d_2. That is, for any fixed $x \in \mathbb{R}^{d_1}$, $f(x)$ is a d_2-dimensional process. By abusing the terminology slightly we say the random field f is $\mathbb{F}$-measurable in its all variables if, for any $t \in [0, T]$, the restriction of f on $[0, t]$ is $\mathscr{B}([0, t]) \times \mathscr{F}_t \times \mathscr{B}(\mathbb{R}^{d_1})$-measurable.

1.2.3 Stopping Times

Given a filtered probability space $(\Omega, \mathscr{F}, \mathbb{F}, P)$, we say a mapping $\tau : \Omega \to [0, T]$ is an $\mathbb{F}$-*stopping time*, also called *Markov time*, if $\{\tau \leq t\} \in \mathscr{F}_t$ for all $t \in [0, T]$. Let $\mathscr{T}(\mathbb{F})$ denote the set of all $\mathbb{F}$-stopping times, and we shall omit $\mathbb{F}$ when the filtration is clear.

Intuitively, when $\mathbb{F} = \mathbb{F}^X$ for some process X, $\tau \in \mathscr{T}(\mathbb{F})$ means that, for any t, once we observe the paths of X up to time t, then we know whether or not $\tau \leq t$. In other words, for any $\omega_1, \omega_2 \in \Omega$, if $X_s(\omega_1) = X_s(\omega_2)$, $0 \leq s \leq t$, then either $\tau(\omega_1) = \tau(\omega_2) \leq t$ or both $\tau(\omega_1)$ and $\tau(\omega_2)$ are greater than t. In the latter case, $\tau(\omega_1) = \tau(\omega_2)$ is not necessarily true. At below is a typical example of stopping times.

Example 1.2.4 *Let* $D \subset \mathbb{R}^d$, $X \in \mathbb{L}^0(\mathbb{F}, \mathbb{R}^d)$ *be continuous, and* $\tau := \inf\{t \geq 0 : X_t \in D\}$. *Then* $\tau \in \mathscr{T}(\mathbb{F})$ *when* D *is closed, and* $\tau \in \mathscr{T}(\mathbb{F}^+)$ *when* D *is open.* Note that $\mathbb{F}^+$-stopping times are also called $\mathbb{F}$-optional time, and

$$\tau \in \mathscr{T}(\mathbb{F}^+) \text{ if and only if } \{\tau < t\} \in \mathscr{F}_t \text{ for all } t.$$

We now collect some important properties of stopping times.

- The deterministic time $\tau = t$ is an $\mathbb{F}$-stopping time.
- Let $\tau_n \in \mathscr{T}(\mathbb{F})$, $n \geq 1$. Then $\sup_{n \geq 1} \tau_n \in \mathscr{T}(\mathbb{F})$ and $\inf_{n \geq 1} \tau_n \in \mathscr{T}(\mathbb{F}^+)$.
- Let $\tau_1, \tau_2 \in \mathscr{T}(\mathbb{F})$, then $(\tau_1 + \tau_2) \wedge T \in \mathscr{T}(\mathbb{F})$. However, even if $\tau_1 \leq \tau_2$, in general $\tau_2 - \tau_1$ is not a stopping time.

We next introduce the σ-field $\mathscr{F}_\tau$:

$$\mathscr{F}_\tau := \Big\{ A \subset \Omega : A \cap \{\tau \leq t\} \in \mathscr{F}_t \text{ for all } 0 \leq t \leq T \Big\}. \tag{1.2.7}$$

Intuitively, a random variable $Y \in \mathbb{L}^0(\mathscr{F}_\tau)$ means that, for any $\omega \in \Omega$, the value $Y(\omega)$ is determined by $\tau(\omega)$ and $\{X_t(\omega), 0 \leq t \leq \tau(\omega)\}$. At below are some basic properties of this σ-field.

- If $X \in \mathbb{L}^0(\mathbb{F})$, then $X_\tau \in \mathbb{L}^0(\mathscr{F}_\tau)$. In particular, by setting $X_t = t$, we have $\tau \in \mathbb{L}^0(\mathscr{F}_\tau)$.
- If $\tau_1 \leq \tau_2$, then $\mathscr{F}_{\tau_1} \subset \mathscr{F}_{\tau_2}$.
- For any τ_1, τ_2, $\{\tau_1 \leq \tau_2\} \in \mathscr{F}_{\tau_1} \wedge \mathscr{F}_{\tau_2}$ and $\tau_1 \mathbf{1}_{\{\tau_1 \leq \tau_2\}} \in \mathbb{L}^0(\mathscr{F}_{\tau_1} \wedge \mathscr{F}_{\tau_2})$.

Note that all the results above in this subsection do not involve $\mathbb{P}$. The next result involves $\mathbb{P}$ and is analogous to Proposition 1.2.1.

Proposition 1.2.5

(i) *Assume* $\tau \in \mathscr{T}(\overline{\mathbb{F}^+}^{\mathbb{P}})$. *Then there exists* $\tilde{\tau} \in \mathscr{T}(\mathbb{F}^+)$ *such that* $\tilde{\tau} = \tau$, $\mathbb{P}$-a.s.

(ii) *Assume* $\tau \in \mathscr{T}(\mathbb{F}^+)$ *is previsible, namely there exist* $\tau_n \in \mathscr{T}(\mathbb{F}^+)$, $n \geq 1$, *such that* $\tau_n < \tau$ *and* $\tau_n \uparrow \tau$, $\mathbb{P}$-*a.s. Then there exists* $\tilde{\tau} \in \mathscr{T}(\mathbb{F})$ *such that* $\tilde{\tau} = \tau$, $\mathbb{P}$-*a.s.*

We emphasize that the measurability of stopping times is very subtle, and the previsibility is crucial in (ii) above. Finally, we say a process $X \in \mathbb{L}^0(\mathbb{F})$ is *uniformly integrable*, abbreviated as u.i., if $\{X_\tau, \tau \in \mathscr{T}(\mathbb{F})\}$ are uniformly integrable, namely

$$\lim_{R \to \infty} \sup_{\tau \in \mathscr{T}(\mathbb{F})} \mathbb{E}\left[|X_\tau| \mathbf{1}_{\{|X_\tau| \geq R\}}\right] = 0. \tag{1.2.8}$$

1.2.4   Martingales

We say $M \in \mathbb{L}^0(\mathbb{F})$ is a $(\mathbb{P}, \mathbb{F})$-*martingale*, or simply a martingale when there is no confusion, if

- $\mathbb{E}[|M_t|] < \infty$ for all $t \in [0, T]$.
- $\mathbb{E}[M_t | \mathscr{F}_s] = M_s$, $\mathbb{P}$-a.s. for all $0 \leq s < t \leq T$.

For a martingale M, we have $\mathbb{E}[M_t] = \mathbb{E}[M_0]$ for all $t \in [0, T]$. On the converse side, assume $M \in \mathbb{L}^0(\mathbb{F})$ is u.i., then M is a martingale (and hence a u.i. martingale) if and only if $\mathbb{E}[M_\tau] = \mathbb{E}[M_0]$ for all $\tau \in \mathscr{T}(\mathbb{F})$. Moreover, if M is u.i. and right continuous, then $\mathbb{E}[M_{\tau_2} | \mathscr{F}_{\tau_1}] = M_{\tau_1}$ for any stopping times $\tau_1 \leq \tau_2$.

Let M be a u.i. martingale and $\tau \in \mathscr{T}(\mathbb{F})$, then $M_{\tau \wedge \cdot}$ is also a martingale. For a process $M \in \mathbb{L}^0(\mathbb{F})$, we say M is a *local martingale* if there exist $\{\tau_n, n \geq 1\} \subset \mathscr{T}(\mathbb{F})$ such that

- τ_n is increasing and $\tau_n = T$ when n is large enough, a.s.
- $M_{\tau_n \wedge \cdot}$ is a martingale for all $n \geq 1$.

Clearly a martingale is always a local martingale. On the other hand, if $M \in \mathbb{L}^0(\mathbb{F})$ is u.i., then M is a martingale (and hence a u.i. martingale) if and only if it is a local martingale.

We remark that, when M is a continuous martingale, which is typically the case in this book, for the results we are interested in we do not need the right continuity of $\mathbb{F}$. However, the right continuity of $\mathbb{F}$ could be crucial to construct a right continuous version of an arbitrarily given martingale.

Finally, we say a process $X \in \mathbb{L}^0(\mathbb{F})$ is a $(\mathbb{P}, \mathbb{F})$-submartingale (resp. supermartingale) if

- $\mathbb{E}[|X_t|] < \infty$ for all $t \in [0, T]$.
- $\mathbb{E}[X_t | \mathscr{F}_s] \geq$ (resp. $\leq$) X_s, $\mathbb{P}$-a.s. for all $0 \leq s < t \leq T$.

1.2.5 Markov Processes

Let $X \in \mathbb{L}^0(\mathbb{F})$. We say X is *Markov* (under $\mathbb{P}$) if, for any $0 \leq s < t \leq T$ and any bounded Borel measurable function φ,

- $\mathbb{E}[\varphi(X_t)|\mathscr{F}_s^X] = \mathbb{E}[\varphi(X_t)|\mathscr{F}^{X_s}]$, a.s.

Roughly speaking, this means that, given X_s, $\{X_t, t \leq s\}$ and $\{X_t, t \geq s\}$ are conditionally independent. By Lemma 1.1.1, we have $\mathbb{E}[\varphi(X_t)|\mathscr{F}_s^X] = \psi(X_s)$ for some Borel measurable function ψ.

Moreover, we say X is *strong Markov* if the above Markov property holds for any stopping time. That is, for any $\tau, \tau' \in \mathscr{T}(\mathbb{F}^X)$ with $\tau \leq \tau'$ and any bounded Borel measurable function φ,

- $\mathbb{E}[\varphi(X_{\tau'})|\mathscr{F}_\tau^X] = \mathbb{E}[\varphi(X_{\tau'})|\mathscr{F}^{\tau, X_\tau}]$, a.s.

Note that in this case $\mathbb{E}[\varphi(X_{\tau'})|\mathscr{F}_\tau^X] = \psi(\tau, X_\tau)$ for some Borel measurable function ψ. In particular, ψ may depend on τ as well. We remark that the strong Markov property typically requires some regularity, such as X and/or $\mathbb{F}$ is right continuous.

1.3 Some Inequalities and Convergence Theorems

One major theme of this book is the a priori estimates for various equations we will study. In this section we introduce some inequalities and convergence theorems which will be used frequently in the book. Fix a filtered probability space $(\Omega, \mathscr{F}, \mathbb{F}, \mathbb{P})$.

1.3.1 Some Norms and Spaces

For any $p, q \geq 1$, we introduce the following norms and spaces:

- $\mathbb{L}^p(\mathscr{F}) := \left\{ X \in \mathbb{L}^0(\mathscr{F}) : \|X\|_p^p := \mathbb{E}[|X|^p] < \infty \right\}$.
- $\mathbb{L}^{p,q}(\mathbb{F}) := \left\{ X \in \mathbb{L}^0(\mathbb{F}) : \|X\|_{p,q}^q := \mathbb{E}\left[\left(\int_0^T |X_t|^p dt\right)^{\frac{q}{p}}\right] < \infty \right\}$. In particular, when $q = p$ we abbreviate it as $\mathbb{L}^p(\mathbb{F}) := \mathbb{L}^{p,p}(\mathbb{F})$ and $\|X\|_p := \|X\|_{p,p}$ for $X \in \mathbb{L}^p(\mathbb{F})$.
- $\mathbb{S}^p(\mathbb{F}) := \left\{ X \in \mathbb{L}^0(\mathbb{F}) : X \text{ is continuous, a.s. and } \|X\|_{\infty,p}^p := \mathbb{E}\left[\sup_{0 \leq t \leq T} |X_t|^p\right] < \infty \right\}$.
- $\mathbb{I}^p(\mathbb{F}) := \left\{ X \in \mathbb{S}^p(\mathbb{F}, \mathbb{R}) : X \text{ is increasing, a.s. and } X_0 = 0 \right\}$.

- $\mathbb{L}^\infty(\mathscr{F})$ and $\mathbb{L}^\infty(\mathbb{F})$ are the spaces of bounded random variables $X \in \mathbb{L}^0(\mathscr{F})$ and bounded processes $X \in \mathbb{L}^0(\mathbb{F})$, respectively, with their $\mathbb{L}^\infty$-norm denoted by $\|X\|_\infty$.
- $\mathbb{L}^p_{loc}(\mathbb{F}) := \{X \in \mathbb{L}^0(\mathbb{F}) : \int_0^T |X_t|^p dt < \infty, \text{a.s.}\}$.
- We use $\mathbb{L}^p(\mathscr{F}, \mathbb{P})$ and $\|X\|_{\mathbb{P},p}$, etc. to denote the dependence on the probability $\mathbb{P}$, when there is a need to emphasize it.
- If the random variables or processes take vector values, say in $\mathbb{R}^n$, we use the notations $\mathbb{L}^p(\mathscr{F}, \mathbb{P}, \mathbb{R}^n)$, etc. If the dimension is not specified, by default they are 1-dimensional. In particular, $\mathbb{P}(\mathbb{F})$ are always scalar processes.

We note that here we abuse the notation $\mathbb{S}$ a little bit since $\mathbb{S}^d$ is used to denote the set of symmetric matrices.

1.3.2 Some Inequalities

We first present the Jensen's inequality.

Proposition 1.3.1 *(Jensen's Inequality) Let $X \in \mathbb{L}^1(\mathscr{F}, \mathbb{R}^d)$ and $\varphi : \mathbb{R}^d \to \mathbb{R}$ convex. Then*

$$\varphi(\mathbb{E}[X]) \leq \mathbb{E}[\varphi(X)].$$

Recall that $p, q \in [1, \infty]$ are called *conjugates* if $\frac{1}{p} + \frac{1}{q} = 1$.

Proposition 1.3.2 *Let p, q be conjugates.*

(i) *(Young's Inequality) Assume $p, q > 1$. For any $x, y \in \mathbb{R}$, it holds:*

$$|xy| \leq \frac{|x|^p}{p} + \frac{|y|^q}{q}.$$

In particular, for $p = q = 2$ and for any $\varepsilon > 0$, we have

$$2xy \leq \varepsilon x^2 + \varepsilon^{-1} y^2 \quad \text{and} \quad (x + y)^2 \leq (1 + \varepsilon)x^2 + (1 + \varepsilon^{-1})y^2.$$

(ii) *(Hölder's Inequality) Let $X \in \mathbb{L}^p(\mathscr{F})$, $Y \in \mathbb{L}^q(\mathscr{F})$, or $X \in \mathbb{L}^p(\mathbb{F})$, $Y \in \mathbb{L}^q(\mathbb{F})$. Then*

$$\|XY\|_1 \leq \|X\|_p \|Y\|_q.$$

Proposition 1.3.3 (Gronwall Inequalities)

(i) *(Continuous Version) Assume $a : [0, T] \to [0, \infty)$ satisfies*

$$a_t \leq C_0 + C_1 \int_0^t a_s ds,$$

for some $C_0, C_1 \geq 0$. Then $a_t \leq C_0 e^{C_1 t}$, $0 \leq t \leq T$.

(ii) (Discrete Version) Let $n \geq 1$, $h := \frac{T}{n}$, and $a_i, b_i \geq 0$, $i = 1, \cdots, n$. Assume

$$a_0 = 0 \quad and \quad a_i \leq (1 + C_0 h)a_{i-1} + C_1 b_i h, \quad i = 1, \cdots, n.$$

Then $\max_{0 \leq i \leq n} a_i \leq C_1 e^{C_0 T} \sum_{i=1}^{n} b_i h.$

1.3.3   Some Convergence Theorems

We first introduce three types of convergence. Let $X, X_n \in \mathbb{L}^0(\mathscr{F})$, $n \geq 1$.

- We say $X_n \to X$, $\mathbb{P}$-a.s., if $\mathbb{P}\big(\lim_{n \to \infty} X_n = X\big) = 1$;
- We say $X_n \to X$ in probability $\mathbb{P}$ if $\lim_{n \to \infty} \mathbb{P}\big(|X_n - X| \geq \varepsilon\big) = 0$ for any $\varepsilon > 0$;
- We say $X_n \to X$ in $\mathbb{L}^p(\mathscr{F}, \mathbb{P})$ if $\lim_{n \to \infty} \mathbb{E}^{\mathbb{P}}\big[|X_n - X|^p\big] = 0$.

Similarly we define the convergence for processes. The following convergence theorems are standard in real analysis.

Theorem 1.3.4 (Dominated Convergence Theorem) *Let $1 \leq p < \infty$.*

(i) Let $X, X_n \in \mathbb{L}^0(\mathscr{F})$, $n \geq 1$, and assume $X_n \to X$ in $\mathbb{P}$. If $\{|X_n|^p\}_{n \geq 1}$ are uniformly integrable, in particular if $|X_n| \leq Y$, $n \geq 1$, for some $Y \in \mathbb{L}^p(\mathscr{F})$, then $X_n \to X$ in $\mathbb{L}^p(\mathscr{F})$.

(ii) Let $X, X^n \in \mathbb{L}^0(\mathbb{F})$, $n \geq 1$ and assume $X^n \to X$ in $dt \times d\mathbb{P}$. If $\{|X_n|^p\}_{n \geq 1}$ are uniformly integrable, in particular if $|X_n| \leq Y$, $n \geq 1$, for some $Y \in \mathbb{L}^p(\mathbb{F})$, then $X^n \to X$ in $\mathbb{L}^p(\mathbb{F})$.

Theorem 1.3.5 (Monotone Convergence Theorem)

(i) Let $X_n \in \mathbb{L}^1(\mathscr{F})$, $n \geq 1$, and $X_n \uparrow X$, $\mathbb{P}$-a.s. Then $\mathbb{E}[X] = \lim_{n \to \infty} \mathbb{E}[X_n]$.

(ii) Let $X^n \in \mathbb{L}^1(\mathbb{F})$, $n \geq 1$, and $X^n \uparrow X$, $dt \times d\mathbb{P}$-a.s. Then $\mathbb{E}[\int_0^T X_t dt] = \lim_{n \to \infty} \mathbb{E}[\int_0^T X_t^n dt]$.

Theorem 1.3.6 (Fatou Lemma)

(i) Let $0 \leq X_n \in \mathbb{L}^1(\mathscr{F})$, $n \geq 1$. Then $\mathbb{E}\Big[\liminf_{n \to \infty} X_n\Big] \leq \liminf_{n \to \infty} \mathbb{E}[X_n]$.

(ii) Let $0 \leq X^n \in \mathbb{L}^1(\mathbb{F})$, $n \geq 1$. Then $\mathbb{E}\Big[\int_0^T \liminf_{n \to \infty} X_t^n dt\Big] \leq \liminf_{n \to \infty} \mathbb{E}\Big[\int_0^T X_t^n dt\Big]$.

1.3.4   Weak Convergence

Let $X, X^n \in \mathbb{L}^2(\mathbb{F})$, $n \geq 1$. We say X^n converges to X *weakly* in $\mathbb{L}^2(\mathbb{F})$ if

$$\lim_{n \to \infty} \mathbb{E}\Big[\int_0^T X_t^n \eta_t dt\Big] = \mathbb{E}\Big[\int_0^T X_t \eta_t dt\Big], \quad \forall \eta \in \mathbb{L}^2(\mathbb{F}). \tag{1.3.1}$$

Note that $\mathbb{L}^2(\mathbb{F})$ is a Hilbert space. The first result concerns the compactness under weak convergence.

Theorem 1.3.7 *Any bounded subset of $\mathbb{L}^2(\mathbb{F})$ is compact under weak convergence. That is, assume $X^n \in \mathbb{L}^2(\mathbb{F})$, $n \geq 1$, such that $\sup_{n \geq 1} \mathbb{E}\left[\int_0^T |X_t^n|^2 dt\right] < \infty$. Then there exists a subsequence n_k such that X^{n_k} converges weakly to some $X \in \mathbb{L}^2(\mathbb{F})$.*

The next theorem connects weak convergence and strong convergence.

Theorem 1.3.8 (Mazur's Lemma) *Assume $X^n \to X$ weakly in $\mathbb{L}^2(\mathbb{F})$. Then there exist $\hat{X}^n$ converging to X (strongly) in $\mathbb{L}^2$, where for each n, $\hat{X}^n$ is a finite convex combination of $(X^m)_{m \geq n}$.*

1.3.5 Monotone Class Theorem

The first theorem is in the form of events.

Theorem 1.3.9 *Let $\mathscr{A} \subset \overline{\mathscr{A}} \subset 2^{\Omega}$. Assume*

(i) $\mathscr{A}$ is a π-system, namely $A_1 \cap A_2 \in \mathscr{A}$ for any $A_1, A_2 \in \mathscr{A}$;
(ii) $\overline{\mathscr{A}}$ is a λ-system, namely

- $\Omega \in \overline{\mathscr{A}}$;
- *If $A_1, A_2 \in \overline{\mathscr{A}}$ and $A_1 \subset A_2$, then $A_2 - A_1 \in \overline{\mathscr{A}}$;*
- *If $A_n \in \overline{\mathscr{A}}$, $n \geq 1$, and $A_n \uparrow A$, then $A \in \overline{\mathscr{A}}$.*

Then $\sigma(\mathscr{A}) \subset \overline{\mathscr{A}}$.

The next theorem is in the form of random variables.

Theorem 1.3.10 *Let $\mathscr{H} \subset \overline{\mathscr{H}} \subset \mathbb{L}^{\infty}(\mathscr{F}, \mathbb{R})$. Assume*

(i) $\mathscr{H}$ is a multiplicative class, namely $X_1 X_2 \in \mathscr{H}$ for any $X_1, X_2 \in \mathscr{H}$;
(ii) $\overline{\mathscr{H}}$ is a monotone vector space, namely

- *constant random variable 1 is in $\overline{\mathscr{H}}$;*
- *$\overline{\mathscr{H}}$ is a vector space, that is, $\alpha_1 X_1 + \alpha_2 X_2 \in \overline{\mathscr{H}}$ for any $X_1, X_2 \in \overline{\mathscr{H}}$ and $\alpha_1, \alpha_2 \in \mathbb{R}$;*
- *If $X_n \in \overline{\mathscr{H}}$, $n \geq 1$, and $0 \leq X_n \uparrow X \in \mathbb{L}^{\infty}(\mathscr{F})$, then $X \in \overline{\mathscr{H}}$.*

Then $\mathbb{L}^{\infty}(\sigma(\mathscr{H})) \subset \overline{\mathscr{H}}$.

1.4 Exercises

Problem 1.4.1 Construct a measurable space $(\Omega, \mathscr{F})$ and an uncountable sequence $\{A_i, i \in I\} \subset \mathscr{F}$ such that $\bigcup_{i \in I} A_i \notin \mathscr{F}$. ∎

Problem 1.4.2

(i) Let $\{\mathscr{F}_i : i \in I\}$ be σ-fields, where I can be uncountable. Show that $\mathscr{G} :=$ $\cap_{i \in I}\mathscr{F}_i$ is still a σ-field. Consequently, the $\sigma(\mathscr{U})$ introduced in (1.1.1) is a σ-field.

(ii) Construct a (countable) sequence of σ-fields $\mathscr{F}_n$, $n = 1, 2, \cdots$, such that $\mathscr{F}_n$ is increasing (namely $\mathscr{F}_n \subset \mathscr{F}_{n+1}$ for each n) but $\cup_n \mathscr{F}_n$ is *not* a σ-field.

(iii) Let $\mathscr{F}_n \subset \mathscr{F}$ be an increasing sequence of σ-fields, denote by $\mathscr{F}_\infty := \bigvee_n \mathscr{F}_n$ the σ-field generated by $\cup_n \mathscr{F}_n$. For any $\xi \in \mathbb{L}^1(\mathscr{F})$, show that $\mathbb{E}[\xi|\mathscr{F}_n] \to$ $\mathbb{E}[\xi|\mathscr{F}_\infty]$ in $\mathbb{L}^1(\mathscr{F})$. ∎

Problem 1.4.3 Let $(\Omega, \mathscr{F}, \mathbb{P})$ be a probability space and recall the $\overline{\mathscr{F}}^{\mathbb{P}}$ in (1.1.3). Show that

(i) For any $A \in \overline{\mathscr{F}}^{\mathbb{P}}$, there exists $\tilde{A} \in \mathscr{F}$ such that $A \subset \tilde{A}$ and $\tilde{A}\backslash A \in \mathscr{N}(\mathscr{F}, \mathbb{P})$.

(ii) Define $\tilde{\mathbb{P}}(A) := \mathbb{P}(\tilde{A})$ for all $A \in \overline{\mathscr{F}}^{\mathbb{P}}$, where $\tilde{A}$ is introduced in (i). Show that $\tilde{\mathbb{P}}$ is a probability on $(\Omega, \overline{\mathscr{F}}^{\mathbb{P}})$. (Note: you should first show that $\tilde{P}$ is independent of the choice of $\tilde{A}$.)

(iii) Show that $(\Omega, \overline{\mathscr{F}}^{\mathbb{P}}, \tilde{\mathbb{P}})$ is a complete probability space and $\tilde{P} = \mathbb{P}$ on $\mathscr{F}$. ∎

Problem 1.4.4 Let $A, B \in \mathscr{F}$ be two events with $\mathbb{P}(B) > 0$, and $\mathscr{G} := \{\emptyset, B, B^c, \Omega\}$ be the σ-field generated by 1_B. Recall the conditional probability $\mathbb{P}(A|B) := \frac{\mathbb{P}(AB)}{\mathbb{P}(B)}$. Show that $\mathbb{E}[1_A|\mathscr{G}] = \mathbb{P}(A|B)1_B + \mathbb{P}(A|B^c)1_{B^c}$. ∎

Problem 1.4.5

(i) Let $X, Y \in \mathbb{L}^0(\mathbb{F})$ be càdlàg , $\mathbb{P}$-a.s., namely $\mathbb{P}(\omega : X.(\omega)$ is càdlàg$) = 1$, and similarly for Y. Assume X and Y are modifications. Show that they are indistinguishable.

(iii) Find a counterexample such that X and Y are modifications, but not indistinguishable. ∎

Problem 1.4.6

(i) Let $X \in \mathbb{L}^0(\mathbb{F}, \mathbb{R})$ be càdlàg , $\mathbb{P}$-a.s. Assume $X_t \le 0$, $\mathbb{P}$-a.s. for all t. Show that $X_t \le 0, 0 \le t \le T$, $\mathbb{P}$-a.s.

(ii) Let $b : [0, T] \times \Omega \times \mathbb{R}^d \to \mathbb{R}$ be progressively measurable (namely jointly measurable in all variables (t, ω, x)) and uniformly continuous in $x \in \mathbb{R}^d$. Assume $b(x) \le 0$, $dt \times d\mathbb{P}$-a.s. for all $x \in \mathbb{R}^d$. Show that $b(x) \le 0$, for all $x \in \mathbb{R}^d$, $dt \times d\mathbb{P}$-a.s.

(iii) Let $b_n : [0, T] \times \Omega \times \mathbb{R}^d \to \mathbb{R}$ be progressively measurable and uniformly continuous in $x \in \mathbb{R}^d$, uniformly in n, and $X \in \mathbb{L}^0(\mathbb{F}, \mathbb{R}^d)$. Assume $b_n(x) \to 0$ in measure $dt \times d\mathbb{P}$, as $n \to \infty$, for all $x \in \mathbb{R}^d$. Show that $b_n(X) \to 0$ in measure $dt \times d\mathbb{P}$, as $n \to \infty$. ∎

Problem 1.4.7 Prove Propositions 1.2.1 and 1.2.5. ∎

Problem 1.4.8

(i) Let $\tau \in \mathscr{T}(\mathbb{F})$. For each $n \geq 1$, define $\tau_n := \sum_{i=1}^{2^n} t_i \mathbf{1}_{(t_{i-1}, t_i]}(\tau)$, where $t_i := i2^{-n}T$. Show that $\tau_n \in \mathscr{T}(\mathbb{F})$, $\tau_n \downarrow \tau$ and $\tau_n - \tau \leq 2^{-n}T$.

(ii) Let $X \in \mathbb{L}^0(\mathbb{F})$ be continuous and define $\tau_n := \inf\{t : |X_t| \geq n\} \wedge T$, $n \geq 1$. Show that $\tau_n \in \mathscr{T}(\mathbb{F})$ satisfy the first property required for local martingales: τ_n is increasing and $\tau_n = T$ when n is large enough, a.s. ∎

Problem 1.4.9

(i) Prove Example 1.2.4.

(ii) Let $\tau_n \in \mathscr{T}(\mathbb{F})$, $n \geq 1$. Then $\sup_n \tau_n \in \mathscr{T}(\mathbb{F})$ and $\inf_n \tau_n \in \mathscr{T}(\mathbb{F}^+)$.

(iii) Assume $\tau_1, \tau_2 \in \mathscr{T}(\mathbb{F})$. Show that $(\tau_1 + \tau_2) \wedge T \in \mathscr{T}(\mathbb{F})$. ∎

Problem 1.4.10

(i) Let $M \in \mathbb{L}^0(\mathbb{F})$ be uniformly integrable. Then M is a martingale if and only if it is a local martingale.

(ii) Let $X \in \mathbb{L}^0(\mathbb{F})$ be such that $\mathbb{E}[X_T^*] < \infty$, where $X_t^* := \sup_{0 \leq s \leq t} |X_s|$. Then X is uniformly integrable.

(iii) Let $M \geq 0$ be a local martingale. Then M is a supermartingale. ∎

Problem 1.4.11 Let $X^n \in \mathbb{L}^\infty(\mathbb{F})$ with uniform bound, and $Y^n \in \mathbb{L}^2(\mathbb{F})$, $n \geq 1$. Assume $X^n \to X$, $dt \times d\mathbb{P}$-a.s. and $Y^n \to Y$ weakly in $\mathbb{L}^2(\mathbb{F})$. Show that

(i) There exists a subsequence n_k such that $X^{n_k} Y^{n_k} \to XY$ weakly in $\mathbb{L}^2(\mathbb{F})$.

(ii) $\mathbb{E}\left[\int_0^T |X_t Y_t|^2 dt \right] \leq \liminf_{n \to \infty} \mathbb{E}\left[\int_0^T |X_t^n Y_t^n|^2 dt \right]$. ∎

Problem 1.4.12 This problem proves Theorem 1.1.4 in following steps.

(i) First assume $|X_i| \leq C$ for all $i \in I$ and $\mathscr{X}$ is closed under maximization. Show that $\sup_n X_{i_n}$ is the essential supremum where i_n is a supremum sequence of $\{\mathbb{E}^\mathbb{P}[X_i], i \in I\}$, namely $\mathbb{E}^\mathbb{P}[X_{i_n}] \uparrow \sup_{i \in I} \mathbb{E}^\mathbb{P}[X_i]$ as $n \to \infty$.

(ii) Next assume $|X_i| \leq C$ for all $i \in I$, but without assuming closeness of $\mathscr{X}$ under maximization. Denote by $\overline{\mathscr{X}}$ the set of random variables taking the form $\max_{n \geq 1} X_{i_n}$ for arbitrary $\{i_n, n \geq 1\} \subset I$. Then clearly $\overline{\mathscr{X}}$ is closed under maximization. Show that ess $\sup_{X \in \overline{\mathscr{X}}}^\mathbb{P} X$ is the desired essential supremum ess $\sup_{i \in I}^\mathbb{P} X_i$.

(iii) Complete the proof of the general case by certain approximations. ∎

Problem 1.4.13

(i) Let $\mathscr{X} = \{X_i, i \in I\} \subset \mathbb{L}^0(\mathscr{F})$. If I is countable, then ess $\sup_{i \in I}^\mathbb{P} X_i = \sup_{i \in I} X_i$, $\mathbb{P}$-a.s.

(ii) Construct a class $\mathscr{X} = \{X_i, i \in I\} \subset \mathbb{L}^0(\mathscr{F})$ such that $\sup_{i \in I} X_i$ is not $\mathscr{F}$-measurable.

(iii) Construct a class $\mathscr{X} = \{X_i, i \in I\} \subset \mathbb{L}^0(\mathscr{F})$ such that $\sup_{i \in I} X_i$ is $\mathscr{F}$-measurable but $\sup_{i \in I} X_i > $ ess $\sup_{i \in I}^\mathbb{P} X_i$, $\mathbb{P}$-a.s.

(iv) Let $X \in \mathbb{L}^0(\mathbb{F}, \mathbb{R})$ be càdlàg, $\mathbb{P}$-a.s. Show that ess $\sup_{0 \leq t \leq T}^\mathbb{P} X_t = \sup_{0 \leq t \leq T} X_t$, $\mathbb{P}$-a.s. ∎

Problem 1.4.14 Let $\eta \in C^\infty(\mathbb{R}^d)$ such that: $\eta \geq 0$, $\eta(x) = 0$ for $|x| \geq 1$, and $\int_{\mathbb{R}^d} \eta(x)dx = 1$. For any Borel measurable function $\varphi : \mathbb{R}^d \to \mathbb{R}$ with $\int_{\mathbb{R}^d} |\varphi(y)|dy < \infty$, define its smooth mollifier:

$$\varphi_\varepsilon(x) := \int_{\mathbb{R}^d} \varphi(x - \varepsilon y)\eta(y)dy = \varepsilon^{-d} \int_{\mathbb{R}^d} \varphi(y)\eta(\frac{x-y}{\varepsilon})dy, \qquad (1.4.1)$$

for all $\varepsilon > 0$. Then

  (i) For each $\varepsilon > 0$, $\varphi_\varepsilon \in C^\infty(\mathbb{R}^d)$ with bounded derivatives (where the bounds may depend on ε);
 (ii) $\lim_{\varepsilon \to 0} \varphi_\varepsilon(x) = \varphi(x)$, for Lebesgue-a.e. $x \in \mathbb{R}^d$; and the convergence holds for all $x \in \mathbb{R}^d$ if φ is continuous;
(iii) If $|\varphi| \leq C$, then $|\varphi_\varepsilon| \leq C$ for any $\varepsilon > 0$;
 (iv) If φ is uniformly continuous with modulus of continuity function ρ, then for all $\varepsilon > 0$, φ_ε is also uniformly continuous with the same modulus of continuity function ρ; Moreover, φ_ε converges to φ uniformly;
  (v) If φ is uniformly Lipschitz continuous with Lipschitz constant L, then for all $\varepsilon > 0$, φ_ε is also uniformly Lipschitz continuous with the same Lipschitz constant L;
 (vi) If φ is differentiable, then $\partial_x\varphi_\varepsilon$ is the modifier of $\partial_x\varphi$, and thus the results (i)–(v) apply to $\partial_x\varphi_\varepsilon$ as well; In particular, if $\varphi \in C^n(\mathbb{R}^d)$ with bounded derivatives, then the derivatives of φ_ε up to order n are uniformly bounded, uniformly in ε, and they converge to the corresponding derivatives of φ.   ∎

Part I
The Basic Theory of SDEs and BSDEs

Chapter 2
Basics of Stochastic Calculus

Let $(\Omega, \mathscr{F}, \mathbb{F}, \mathbb{P})$ be a filtered probability space. We remark again that, unlike in standard literature, we do not assume $\mathbb{F} = \{\mathscr{F}_t\}_{0 \le t \le T}$ satisfy the usual hypothesis. This will be crucial for the fully nonlinear theory in Part III, and for fixed $\mathbb{P}$ this is a very mild relaxation due to Proposition 1.2.1.

2.1 Brownian Motion

2.1.1 Definition

Definition 2.1.1 *We say a process* $B : [0, T] \times \Omega \to \mathbb{R}$ *is a (standard) Brownian motion if*

- $B_0 = 0$, *a.s.*
- *For any* $0 = t_0 < \cdots < t_n \le T$, $B_{t_1}, B_{t_1, t_2}, \cdots, B_{t_{n-1}, t_n}$ *are independent.*
- *For any* $0 \le s < t \le T$, $B_{s,t} \sim N(0, t - s)$.

Moreover, we call B *an* $\mathbb{F}$-*Brownian motion if* $B \in \mathbb{L}^0(\mathbb{F})$ *and*

- *For any* $0 \le s < t \le T$, $B_{s,t}$ *and* $\mathscr{F}_s$ *are independent.*

We note that as in the previous chapter we restrict B to a finite horizon $[0, T]$. But the definition can be easily extended to $[0, \infty)$, by first extending the filtration $\mathbb{F}$ to $[0, \infty)$. When necessary, we may interpret B as a Brownian motion on $[0, \infty)$ without mentioning it explicitly. Moreover, when there is a need to emphasize the dependence on the probability measure $\mathbb{P}$ and/or the filtration $\mathbb{F}$, we call B a $\mathbb{P}$-Brownian motion or $(\mathbb{P}, \mathbb{F})$-Brownian motion. Since B has independent increments,

© Springer Science+Business Media LLC 2017
J. Zhang, *Backward Stochastic Differential Equations*, Probability Theory and Stochastic Modelling 86, DOI 10.1007/978-1-4939-7256-2_2

clearly $(B_{t_1}, \cdots, B_{t_n})$ have Gaussian distribution, or say B is a Gaussian process. Moreover, from the definition we can easily compute the finite distribution of B. Then by the Kolmogorov's Extension Theorem we know that Brownian motion does exist. The following properties are immediate and left to the readers.

Proposition 2.1.2 *Let B be a standard Brownian motion. For any t_0 and any constant $c > 0$, the processes $B_t^{t_0} := B_{t_0, t+t_0}$ and $\hat{B}_t^c := \frac{1}{\sqrt{c}} B_{ct}$ are also standard Brownian motions.*

Proposition 2.1.3 *A Brownian motion is Markov, and an $\mathbb{F}$-Brownian motion is an $\mathbb{F}$-martingale.*

In the multidimensional case, we call $B = (B^1, \cdots, B^d)^{\top}$ a d-dimensional Brownian motion if $B^1, \cdots, B^d$ are independent Brownian motions. In most cases we do not emphasize the dimension and thus still call it a Brownian motion.

From now on, throughout this chapter, B is a d-dimensional $\mathbb{F}$-Brownian motion. All our results hold true in multidimensional setting. However, while we shall state the results in multidimensional case, for notional simplicity quite often we will carry out the proofs only in the case $d = 1$. The readers may extend the arguments to multidimensional cases straightforwardly.

2.1.2   Pathwise Properties

We start with its pathwise continuity. Notice that Brownian motion is defined via its distribution. As mentioned in the paragraph after Theorem 1.2.3, the pathwise properties should be understood for a version of B.

Theorem 2.1.4 *For any $\varepsilon \in (0, \frac{1}{2})$, B is Hölder-$(\frac{1}{2} - \varepsilon)$ continuous, a.s. In particular, B is continuous, a.s.*

Proof For notational simplicity, assume $d = 1$. For any $s < t$, since $B_{s,t} \sim N(0, t - s)$, we have

$$\mathbb{E}\left[|B_{s,t}|^p\right] = C_p |t - s|^{\frac{p}{2}}, \text{ for all } p \geq 1.$$

Apply the Kolmogorov's Continuity Theorem 1.2.3, by considering a modification if necessary, B is Hö-γ continuous for $\gamma := \frac{\frac{p}{2} - 1}{p - 1}$. Since p is arbitrary, one can always find p large enough so that $\gamma > \frac{1}{2} - \varepsilon$.   ∎

From now on, we shall always consider a continuous version of B. We next study the quadratic variation of B. For a time partition $\pi : 0 = t_0 < \cdots < t_n = T$, denote $|\pi| := \max_{1 \leq i \leq n}(t_i - t_{i-1})$. We recall that the *total variation* of a process $X \in \mathbb{L}^0(\mathbb{F}, \mathbb{R}^d)$ is defined pathwise by: for $0 \leq a < b \leq T$,

$$\bigvee_a^b (X) := \sup_{\pi} \sum_{i=1}^n |X_{a \vee t_{i-1} \wedge b, \, a \vee t_i \wedge b}|, \text{ in particular }, \bigvee_0^T (X) := \sup_{\pi} \sum_{i=1}^n |X_{t_{i-1}, t_i}|. \quad (2.1.1)$$

Definition 2.1.5 *Let* $X \in \mathbb{L}^0(\mathbb{F}, \mathbb{R}^d)$. *We say* X *has quadratic variation if the following limit exists :*

$$\langle X \rangle_t := \lim_{|\pi| \to 0} \sum_{i=1}^{n} X_{t_{i-1} \wedge t, \, t_i \wedge t} X_{t_{i-1} \wedge t, \, t_i \wedge t}^{\mathsf{T}}, \quad \text{in the sense of convergence in probability.}$$

$$(2.1.2)$$

In this case we call $\langle X \rangle$ *the quadratic variation process of* X.

Note that $\langle X \rangle$ takes values in $\mathbb{S}^d$, the set of $d \times d$-symmetric matrices. Its (i, j)-th component is:

$$\langle X \rangle_t^{i,j} := \lim_{|\pi| \to 0} \sum_{k=1}^{n} X_{t_{k-1} \wedge t, \, t_k \wedge t}^{i} X_{t_{k-1} \wedge t, \, t_k \wedge t}^{j}.$$

We also remark that, unlike total variation, the quadratic variation is not defined in a pathwise manner. It is interesting to understand the pathwise definition of quadratic variation, which we will study in Part III. See also Remark 2.2.6.

Theorem 2.1.6 *It holds that*

$$\lim_{|\pi| \to 0} \mathbb{E}\Big[\Big(\sum_{i=1}^{n} B_{t_{i-1} \wedge t, \, t_i \wedge t} B_{t_{i-1} \wedge t, \, t_i \wedge t}^{\mathsf{T}} - t I_d\Big)^2\Big] = 0, \quad \text{and consequently,} \quad \langle B \rangle_t = t I_d.$$

Proof For notational simplicity we assume $d = 1$, and without loss of generality we prove the theorem only at T. Fix a partition $\pi : 0 = t_0 < \cdots < t_n = T$, and denote

$$\Delta t_i := t_i - t_{i-1}, \quad \eta_i := |B_{t_{i-1}, t_i}|^2 - \Delta t_i, \quad i = 1, \cdots, n.$$

Then η_i, $i = 1, \cdots, n$, are independent. Since $B_{t_{i-1}, t_i} \sim N(0, \Delta t_i)$, we have $\mathbb{E}[\eta_i] = 0$ and

$$Var(\eta_i) = Var(|B_{t_{i-1}, t_i}|^2) = \mathbb{E}[|B_{t_{i-1}, t_i}|^4] - \Big(\mathbb{E}[|B_{t_{i-1}, t_i}|^2]\Big)^2 = 3(\Delta t_i)^2 - (\Delta t_i)^2 = 2(\Delta t_i)^2.$$

Notice also that $\sum_{i=1}^{n} \Delta t_i = T$. Then

$$\mathbb{E}\Big[\Big(\sum_{i=1}^{n} |B_{t_{i-1}, t_i}|^2 - T\Big)^2\Big] = \mathbb{E}\Big[\Big(\sum_{i=1}^{n} \eta_i\Big)^2\Big] = Var\Big(\sum_{i=1}^{n} \eta_i\Big)$$

$$= \sum_{i=1}^{n} Var(\eta_i) = 2\sum_{i=1}^{n} (\Delta t_i)^2 \leq 2|\pi| \sum_{i=1}^{n} \Delta t_i = 2T|\pi| \to 0, \quad \text{as } |\pi| \to 0.$$

Since $\mathbb{L}^2$ convergence implies convergence in probability, we conclude that $\langle B \rangle_T = T$. ∎

As a corollary of Theorems 2.1.4 and 2.1.6, we have

Corollary 2.1.7 $\bigvee_a^b(B) = \infty$ *for any* $0 \le a < b \le T$, *a.s.*

Proof We proceed in two steps, again assuming $d = 1$.

Step 1. Fix $0 \le a < b \le T$. For any partition π, denote $\tilde{t}_i := a \vee t_i \wedge b$ and notice that

$$\sum_{i=1}^n |B_{\tilde{t}_{i-1}, \tilde{t}_i}|^2 \le \left(\sup_{1 \le i \le n} |B_{\tilde{t}_{i-1}, \tilde{t}_i}| \right) \sum_{i=1}^n |B_{\tilde{t}_{i-1}, \tilde{t}_i}| \le \bigvee_a^b(B) \times \sup_{1 \le i \le n} |B_{\tilde{t}_{i-1}, \tilde{t}_i}|$$

Send $|\pi| \to 0$, by Theorems 2.1.4 and 2.1.6 we have

$$\sup_{1 \le i \le n} |B_{\tilde{t}_{i-1}, \tilde{t}_i}| \to 0, \quad \text{a.s.} \quad \text{and} \quad \sum_{i=1}^n |B_{\tilde{t}_{i-1}, \tilde{t}_i}|^2 \to \langle B \rangle_b - \langle B \rangle_a = b - a > 0, \text{ in } \mathbb{P}.$$

This clearly implies that

$$\bigvee_a^b(B) = \infty, \quad \text{a.s.}$$

Step 2. For any $0 \le a < b \le T$, by Step 1 we have

$$\mathbb{P}(\mathcal{N}(a,b)) = 0, \quad \text{where} \quad \mathcal{N}(a,b) := \left\{ \omega : \bigvee_a^b(B(\omega)) < \infty \right\}.$$

Denote

$$\mathcal{N} := \bigcup \left[\mathcal{N}(r_1, r_2) : 0 \le r_1 < r_2 \le T, \ r_1, r_2 \in \mathbb{Q} \right]$$

Then $\mathbb{P}(\mathcal{N}) = 0$. Now for any $\omega \notin \mathcal{N}$, and for any $0 \le a < b \le T$, there exist $r_1, r_2 \in \mathbb{Q}$ such that $a \le r_1 < r_2 \le b$. Then

$$\bigvee_a^b(B(\omega)) \ge \bigvee_{r_1}^{r_2}(B(\omega)) = \infty.$$

The proof is complete now. ∎

Remark 2.1.8

(i) Corollary 2.1.7 implies that B is nowhere absolutely continuous with respect to dt. Let $d = 1$. We actually have the following so-called *Law of Iterated Logarithm*: for any $t \in [0, T)$,

$$\limsup_{\delta \downarrow 0} \frac{B_{t,t+\delta}}{\sqrt{2\delta \ln \ln \frac{1}{\delta}}} = 1, \quad \liminf_{\delta \downarrow 0} \frac{B_{t,t+\delta}}{\sqrt{2\delta \ln \ln \frac{1}{\delta}}} = -1, \quad \text{a.s.} \quad (2.1.3)$$

This implies that B is nowhere Hölder-$\frac{1}{2}$ continuous. In particular, B is nowhere differentiable.

(ii) The regularity (2.1.3) is right regularity. The left regularity $B_{t-\delta,t}$ is less clear. Moreover, the null set in (2.1.3) depends on t. Indeed, the uniform regularities $\sup_t B_{t,t+\delta}$ and $\sup_{0 \leq s < t \leq T, t-s \leq \delta} |B_{s,t}|$ are more involved. ∎

2.1.3 The Augmented Filtration

Let $\mathbb{F}^B$ denote the filtration generated by B. We notice that neither $\mathbb{F}^B$ nor its completed filtration is right continuous. For example,

$$\left\{ \omega \in \Omega : \limsup_{\delta \downarrow 0} \frac{B_\delta(\omega)}{\sqrt{2\delta \ln \ln \frac{1}{\delta}}} = 1 \right\} \in \mathscr{F}^B_{0+} \backslash \mathscr{F}^B_0.$$

However, the augmented filtration, denoted as $\overline{\mathbb{F}}^B$, is right continuous. We first have the Blumenthal 0–1 law.

Theorem 2.1.9 *For any random variable* $X \in \mathbb{L}^0(\mathscr{F}^B_{0+})$, *we have* $X = \mathbb{E}[X]$, *a.s. Consequently, For any event* $A \in \mathscr{F}^B_{0+}$, *we have* $\mathbb{P}(A) = 0$ *or* 1.

Proof Let $X \in \mathbb{L}^0(\mathscr{F}^B_{0+})$. For any $n \geq 1$, let $\mathscr{G}_n := \sigma(B_{n^{-1},s}, n^{-1} \leq s \leq T)$, the σ-field generated by $\{B_{n^{-1},s}, n^{-1} \leq s \leq T\}$. Since B has independent increments, $\mathscr{G}_n$ and $\mathscr{F}^B_{0+} \subset \mathscr{F}^B_{n^{-1}}$ are independent. Thus $\mathbb{E}[X|\mathscr{G}_n] = \mathbb{E}[X]$, a.s. for all $n \geq 1$. On the other hand, denote $\mathscr{G} := (\vee_n \mathscr{G}_n) \vee \mathcal{N}(\mathscr{F})$. For any $t > 0$, $B_t = B_{0,t} = \lim_{n \to \infty} B_{n^{-1},t}$, a.s. Since $B_{n^{-1},s} \in \mathbb{L}^0(\mathscr{G}_{n^{-1}}) \subset \mathbb{L}^0(\mathscr{G})$, we see that $B_t \in \mathbb{L}^0(\mathscr{G})$ for any $t > 0$. Thus $\mathscr{F}^B_{0+} \subset \mathscr{G}$. Note that $\mathscr{G}_n$ is increasing in n, then by Problem 1.4.2 (iii) we obtain

$$X = \mathbb{E}[X|\mathscr{G}] = \lim_{n \to \infty} \mathbb{E}[X|\mathscr{G}_n] = \mathbb{E}[X], \quad a.s.$$

Finally, for any $A \in \mathscr{F}^B_{0+}$, set $X := \mathbf{1}_A$, we see that $\mathbb{P}(A) = \mathbf{1}_A$, a.s. and thus $\mathbb{P}(A) = 0$ or 1. ∎

Corollary 2.1.10 *The augmented filtration* $\overline{\mathbb{F}}^B$ *satisfies the usual hypotheses.*

Proof It suffices to show that $\overline{\mathbb{F}}^B$ is right continuous. Theorem 2.1.9 implies that $\mathscr{F}^B_{0+} \subset \mathcal{N}(\mathscr{F}) \subset \overline{\mathscr{F}}^B_0$. Then $\overline{\mathscr{F}}^B_{0+} = \overline{\mathscr{F}}^B_0$. Similarly, for any t, we have $\overline{\mathscr{F}}^B_{t+} = \overline{\mathscr{F}}^B_t$. ∎

Remark 2.1.11

(i) In this book, we shall use $\mathbb{F}^B$. When $\mathbb{P}$ is given, in most cases this is equivalent to using the augmented filtration $\overline{\mathbb{F}}^B$ as in standard literature, in the spirit of Proposition 1.2.1.

(ii) As we will see, all $\mathbb{F}^B$-local martingales are continuous. If we consider more general càdlàg martingales, it is more convenient to use right continuous filtration. ∎

2.2 Stochastic Integration

2.2.1 Some Heuristic Arguments

In this subsection we assume $d = 1$. We first recall the Rieman-Stieltjes integral. Let $A : [0, T] \to \mathbb{R}$ be a function with bounded variation, and $b : [0, T] \to \mathbb{R}$ be continuous. For a partition $\pi : 0 = t_0 < \cdots < t_n = T$, define the Rieman-Stieltjes partial sum:

$$\sum_{i=0}^{n-1} b(\hat{t}_i) A_{t_i, t_{i+1}} \quad \text{where } \hat{t}_i \in [t_i, t_{i+1}] \text{ is arbitrary.}$$

It is well known that, as $|\pi| \to 0$, the above partial sum converges and the limit is independent of the choices of π and $\hat{t}_i$, and thus is defined as the integral of b with respect to A:

$$\int_0^T b_t dA_t := \lim_{|\pi| \to 0} \sum_{i=0}^{n-1} b(\hat{t}_i) A_{t_i, t_{i+1}}. \tag{2.2.1}$$

Now assume $A, b \in \mathbb{L}^0(\mathbb{F})$ such that A has bounded variation and b is continuous, a.s. Then clearly we can define the integral pathwise:

$$\left(\int_0^T b_t dA_t \right)(\omega) := \int_0^T b_t(\omega) d(A_t(\omega)).$$

We next discuss stochastic integrals with respect to B. Let $\sigma \in \mathbb{L}^0(\mathbb{F})$ be continuous, a.s. We first notice that in this case the limits of the Rieman-Stieltjes partial sum may depend on the choices of $\hat{t}_i$. Indeed, let $\sigma = B$ and set $\hat{t}_i$ as the left end point and right end point respectively, we have

$$S_L(\pi) := \sum_{i=0}^{n-1} B_{t_i} B_{t_i, t_{i+1}}, \quad S_R(\pi) := \sum_{i=0}^{n-1} B_{t_{i+1}} B_{t_i, t_{i+1}}.$$

Then, by Theorem 2.1.6,

$$S_R(\pi) - S_L(\pi) = \sum_{i=0}^{n-1} |B_{t_i, t_{i+1}}|^2 \to T, \quad \text{in } \mathbb{P} \text{ as } |\pi| \to 0.$$

So $S_R(\pi)$ and $S_L(\pi)$ cannot converge to the same limit, and therefore, it is important to choose appropriate points $\hat{t}_i$. As in standard literature, we shall study the Itô integral, which uses the left end points. The main reason is that, among others, in this case we use σ_{t_i} to approximate σ on the interval $[t_i, t_{i+1})$ and thus the approximating process σ^{π} defined below is still $\mathbb{F}$-measurable:

$$\sigma^{\pi} := \sum_{i=0}^{n-1} \sigma_{t_i} \mathbf{1}_{[t_i, t_{i+1})}. \qquad (2.2.2)$$

Remark 2.2.1 If we use $\hat{t}_i := \frac{t_i + t_{i+1}}{2}$, the corresponding limit is called the Stratonovic Integral. We shall study Itô integral in this book, which has the following advantages:

- The $\mathbb{F}$-measurability of the σ^{π} in (2.2.2) is natural in many applications, see, e.g., Section 2.8;
- As we will see soon, the Itô integral has martingale property and thus allows us to use the martingale theory;
- Unlike Stratonovic Integral, the Itô integral does not require any regularity on the integrand σ.

However, Stratonovic Integral is more convenient for pathwise analysis. In particular, under Stratonovic Integral, the chain rule same as the deterministic case remains true. See Problem 2.10.13. ∎

2.2.2 Itô Integral for Elementary Processes

Definition 2.2.2 We say $\sigma \in \mathbb{L}^2(\mathbb{F})$ is an elementary process, denote as $\sigma \in \mathbb{L}_0^2(\mathbb{F})$, if there exist a partition $0 = t_0 < \cdots < t_n = T$ such that $\sigma_t = \sigma_{t_i}$ for all $t \in [t_i, t_{i+1})$, $i = 0, \cdots, n - 1$.

Clearly, for $\sigma \in \mathbb{L}_0^2(\mathbb{F}, \mathbb{R}^d)$, we may define the stochastic integral in a pathwise manner:

$$\int_0^t \sigma_s \cdot dB_s := \sum_{i=0}^{n-1} \sigma_{t_i} \cdot B_{t_i \wedge t, \, t_{i+1} \wedge t}, \quad 0 \le t \le T. \qquad (2.2.3)$$

Lemma 2.2.3 *Let $\sigma \in \mathbb{L}^2_0(\mathbb{F}, \mathbb{R}^d)$ and denote $M_t := \int_0^t \sigma_s \cdot dB_s$.*

(i) *M is an $\mathbb{F}$-martingale. In particular, $\mathbb{E}[M_t] = 0$.*

(ii) *$M \in \mathbb{L}^2(\mathbb{F})$ and $N_t := M_t^2 - \int_0^t |\sigma_s|^2 ds$ is a martingale. In particular,*

$$\mathbb{E}\left[|M_t|^2\right] = \mathbb{E}\left[\int_0^t |\sigma_s|^2 ds\right]. \qquad (2.2.4)$$

(iii) *For any $\sigma^i \in \mathbb{L}^2_0(\mathbb{F}, \mathbb{R}^d)$, $\lambda_i \in \mathbb{L}^\infty(\mathscr{F}_0, \mathbb{R})$, $i = 1, 2$, we have $\lambda_1 \sigma^1 + \lambda_2 \sigma^2 \in \mathbb{L}^2_0(\mathbb{F}, \mathbb{R}^d)$ and*

$$\int_0^t [\lambda_1 \sigma_s^1 + \lambda_2 \sigma_s^2] \cdot dB_s = \lambda_1 \int_0^t \sigma_s^1 \cdot dB_s + \lambda_2 \int_0^t \sigma_s^2 \cdot dB_s.$$

(iv) *M is continuous, a.s.*

Proof

(i) It suffices to show that, for any i,

$$M_t = \mathbb{E}[M_{t_{i+1}}|\mathscr{F}_t], \quad t_i \le t \le t_{i+1}.$$

Indeed, note that $\sigma_{t_i} \in \mathscr{F}_{t_i} \subset \mathscr{F}_t$ and B has independent increments, then

$$\mathbb{E}[M_{t,t_{i+1}}|\mathscr{F}_t] = \mathbb{E}\left[\sigma_{t_i} \cdot B_{t,t_{j+1}}\Big|\mathscr{F}_t\right] = \sigma_{t_i} \cdot \mathbb{E}\left[B_{t,t_{j+1}}\Big|\mathscr{F}_t\right] = \sigma_{t_i} \cdot \mathbb{E}[B_{t,t_{j+1}}] = 0.$$

(ii) The square integrability of M follows directly from (2.2.4). Then it suffices to show that

$$N_t = \mathbb{E}[N_{t_{i+1}}|\mathscr{F}_t], \quad t_i \le t \le t_{i+1}.$$

To illustrate the arguments, in this proof we use multidimensional notations. Note that

$$N_{t,t_{i+1}} = M_{t_{i+1}}^2 - M_t^2 - |\sigma_{t_i}|^2(t_{i+1} - t) = |M_{t,t_{i+1}}|^2 + 2M_t M_{t,t_{i+1}} - |\sigma_{t_i}|^2(t_{i+1} - t)$$

$$= [\sigma_{t_i}\sigma_{t_i}^\top] : [B_{t,t_{i+1}} B_{t,t_{i+1}}^\top - (t_{i+1} - t)I_d] + 2M_t\sigma_{t_i} \cdot B_{t,t_{i+1}}.$$

Then, similar to (i) we have

$$\mathbb{E}[N_{t,t_{i+1}}|\mathscr{F}_t] = [\sigma_{t_i}\sigma_{t_i}^\top] : \mathbb{E}\left[B_{t,t_{i+1}} B_{t,t_{i+1}}^\top - (t_{i+1} - t)I_d\right] + 2M_t\sigma_{t_i} \cdot \mathbb{E}[B_{t,t_{i+1}}] = 0.$$

(iii) and (iv) are obvious. ∎

The following estimates are important, and we leave a more general result in Problem 2.10.3 below. Recall the notation X^* in (1.2.4).

Lemma 2.2.4 (Doob's Maximum Inequality) *Let $\sigma \in \mathbb{L}_0^2(\mathbb{F}, \mathbb{R}^d)$, $M_t := \int_0^t \sigma_s \cdot dB_s$. Then*

$$\mathbb{E}[|M_T|^2] \leq \mathbb{E}[|M_T^*|^2] \leq 4\mathbb{E}[|M_T|^2]. \tag{2.2.5}$$

Proof The left inequality is obvious. We prove the right inequality in two steps.

Step 1. We first prove it under an additional assumption:

$$\mathbb{E}[|M_T^*|^2] < \infty. \tag{2.2.6}$$

Given $\lambda > 0$, denote

$$\tau_\lambda := \inf\{t \geq 0 : |M_t| \geq \lambda\} \wedge T. \tag{2.2.7}$$

Since M is continuous, we see that

$$\tau_\lambda \in \mathscr{T}(\mathbb{F}), \quad |M_{\tau_\lambda}| \leq \lambda, \quad \text{and} \quad \{M_T^* \geq \lambda\} = \{|M_{\tau_\lambda}| = \lambda\}. \tag{2.2.8}$$

Moreover, by (2.2.6) M is a u.i. martingale, then

$$|M_{\tau_\lambda}| = \left|\mathbb{E}[M_T|\mathscr{F}_{\tau_\lambda}]\right| \leq \mathbb{E}\left[|M_T|\Big|\mathscr{F}_{\tau_\lambda}\right].$$

This implies

$$\mathbb{P}(M_T^* \geq \lambda) = \mathbb{E}\left[\mathbf{1}_{\{|M_{\tau_\lambda}|=\lambda\}}\right] = \mathbb{E}\left[\frac{|M_{\tau_\lambda}|}{\lambda}\mathbf{1}_{\{|M_{\tau_\lambda}|=\lambda\}}\right] \leq \frac{1}{\lambda}\mathbb{E}\left[\mathbb{E}(|M_T||\mathscr{F}_{\tau_\lambda})\mathbf{1}_{\{|M_{\tau_\lambda}|=\lambda\}}\right]$$

$$= \frac{1}{\lambda}\mathbb{E}\left[\mathbb{E}(|M_T|\mathbf{1}_{\{|M_{\tau_\lambda}|=\lambda\}}|\mathscr{F}_{\tau_\lambda})\right] = \frac{1}{\lambda}\mathbb{E}\left[|M_T|\mathbf{1}_{\{|M_{\tau_\lambda}|=\lambda\}}\right] = \frac{1}{\lambda}\mathbb{E}\left[|M_T|\mathbf{1}_{\{M_T^*\geq\lambda\}}\right]. \tag{2.2.9}$$

Thus

$$\mathbb{E}[|M_T^*|^2] = 2\int_0^\infty \lambda\mathbb{P}(M_T^* \geq \lambda)d\lambda \leq 2\int_0^\infty \mathbb{E}\left[|M_T|\mathbf{1}_{\{M_T^*\geq\lambda\}}\right]d\lambda$$

$$= 2\mathbb{E}\left[\int_0^\infty |M_T|\mathbf{1}_{\{M_T^*\geq\lambda\}}d\lambda\right] = 2\mathbb{E}\left[|M_T|M_T^*\right] \leq 2\left(\mathbb{E}[|M_T|^2]\right)^{\frac{1}{2}}\left(\mathbb{E}[|M_T^*|^2]\right)^{\frac{1}{2}},$$

where the last inequality thanks to the Hölder's inequality. This implies (2.2.5) immediately.

Step 2. In the general case, for each $n \geq 1$, let τ_n be defined by (2.2.7) and denote

$$\sigma^n := \sigma\mathbf{1}_{[0,\tau_n]}, \quad M_t^n := \int_0^t \sigma_s^n \cdot dB_s, \quad M_t^{n,*} := \sup_{0 \leq s \leq t} |M_s^n|.$$

Since M is continuous, by Problem 1.4.8 (ii) we see that τ_n is increasing, and $\tau_n = T$ when n is large enough. Then

$$M_t^n = M_{\tau_n \wedge t}, \quad M_T^{n,*} \le n, \quad \text{and} \quad M_T^{n,*} \uparrow M_T^*.$$

By Step 1 and (2.2.4) we have

$$\mathbb{E}\Big[|M_T^{n,*}|^2\Big] \le 4\mathbb{E}\Big[|M_T^n|^2\Big] = 4\mathbb{E}\Big[\int_0^{\tau_n} |\sigma_s|^2 ds\Big] \le 4\mathbb{E}\Big[\int_0^T |\sigma_s|^2 ds\Big] = 4\mathbb{E}\Big[|M_T|^2\Big].$$

Now applying the Monotone Convergence Theorem we obtain (2.2.5).   ∎

2.2.3 Itô Integral in $\mathbb{L}^2(\mathbb{F})$ and $\mathbb{L}^2_{loc}(\mathbb{F})$

We now extend the Itô stochastic integration to all processes in $\mathbb{L}^2(\mathbb{F})$. We first need a lemma.

Lemma 2.2.5 *For any* $\sigma \in \mathbb{L}^2(\mathbb{F}, \mathbb{R}^d)$, *there exist* $\sigma^n \in \mathbb{L}^2_0(\mathbb{F}, \mathbb{R}^d)$ *such that* $\lim_{n\to\infty} \|\sigma^n - \sigma\|_2 = 0$.

Proof We proceed in three steps.

Step 1. We first assume σ is continuous and bounded. For each n, define

$$\sigma_t^n := \sum_{i=0}^{n-1} \sigma_{t_i} \mathbf{1}_{[t_i, t_{i+1})} \text{ where } t_i := \frac{i}{n} T, \ i = 0, \cdots, n.$$

Then by the Dominated Convergence Theorem we obtain the result immediately.

Step 2. We now assume only that $|\sigma| \le C$. For each $\delta > 0$, define $\sigma_t^\delta := \frac{1}{\delta} \int_{(t-\delta)\vee 0}^t \sigma_s ds$. Clearly $|\sigma^\delta| \le C$, σ^δ is continuous, and by real analysis, in the spirit of Problem 1.4.14, we have $\lim_{\delta\to 0} \int_0^T |\sigma_t^\delta - \sigma_t|^2 dt = 0$, a.s. By the Dominated Convergence Theorem again, we have $\lim_{\delta\to 0} \|\sigma^\delta - \sigma_t\|_2 = 0$. Now for each n, there exists δ_n such that $\|\sigma^\delta - \sigma_t\|_2 \le \frac{1}{2n}$. Moreover, by Step 1, there exists $\sigma^n \in \mathbb{L}^2_0(\mathbb{F}, \mathbb{R}^d)$ such that $\|\sigma^n - \sigma_t^\delta\|_2 \le \frac{1}{2n}$. This implies $\|\sigma^n - \sigma_t\|_2 \le \frac{1}{n} \to 0$, as $n \to \infty$.

Step 3. For the general case, for each n, denote $\tilde{\sigma}^n := (-n) \vee \sigma \wedge n$, where the truncation is component wise. Then $\tilde{\sigma}^n \to \sigma$ and $|\sigma^n| \le |\sigma|$. Applying the Dominated Convergence Theorem we get $\lim_{n\to\infty} \|\tilde{\sigma}^n - \sigma\|_2 = 0$. Moreover, since $|\sigma^n| \le n\sqrt{d}$, by Step 2 there exists $\sigma^n \in \mathbb{L}^2_0(\mathbb{F}, \mathbb{R}^d)$ such that $\|\tilde{\sigma}^n - \sigma^n\|_2 \le \frac{1}{n}$. Thus

$$\|\sigma^n - \sigma\|_2 \le \|\tilde{\sigma}^n - \sigma\|_2 + \|\tilde{\sigma}^n - \sigma^n\|_2 \le \|\tilde{\sigma}^n - \sigma\|_2 + \frac{1}{n} \to 0.$$

The proof is complete now.   ∎

For the above $\sigma^n \in \mathbb{L}_0^2(\mathbb{F}, \mathbb{R}^d)$, we have defined $M_t^n := \int_0^t \sigma_s^n \cdot dB_s$ by (2.2.3). Applying Lemma 2.2.4 we get

$$\mathbb{E}\left[|(M^n - M^m)_T^*|^2\right] \le 4\|\sigma^n - \sigma^m\|_2^2 \to 0, \quad \text{as } m, n \to \infty.$$

Thus there exists a ($\mathbb{P}$-a.s.) unique continuous process $M \in \mathbb{L}^0(\mathbb{F}, \mathbb{R})$ such that

$$\lim_{n \to \infty} \mathbb{E}\left[|(M^n - M)_T^*|^2\right] = 0. \tag{2.2.10}$$

Moreover, if there exist another sequence $\tilde{\sigma}^n \in \mathbb{L}_0^2(\mathbb{F}, \mathbb{R}^d)$ such that $\lim_{n \to \infty} \|\tilde{\sigma}^n - \sigma\|_2 = 0$, then $\lim_{n \to \infty} \|\tilde{\sigma}^n - \sigma^n\|_2 = 0$. This implies that, for $\tilde{M}_t^n := \int_0^t \tilde{\sigma}_s^n \cdot dB_s$,

$$0 \le \mathbb{E}\left[|(M^n - \tilde{M}^n)_T^*|^2\right] \le 4\|\sigma^n - \tilde{\sigma}^n\|_2^2 \to 0, \quad \text{as } n \to \infty.$$

Thus $\tilde{M}^n$ also converges to M. That is, the process M does not depend on the choices of σ^n. Therefore, we may define M as the stochastic integral of σ: for each $t \in [0, T]$,

$$\int_0^t \sigma_s \cdot dB_s := \lim_{n \to \infty} \int_0^t \sigma_s^n \cdot dB_s, \quad \text{where the convergence is in the sense of (2.2.10).}$$

$$\tag{2.2.11}$$

Remark 2.2.6 We emphasize that the convergence in (2.2.11) is in $\mathbb{L}^2$-sense, and thus the above definition of stochastic integral is not in a pathwise manner. That is, given $\sigma(\omega)$ and $B(\omega)$, in general we cannot determine $\left(\int_0^t \sigma_s \cdot dB_s\right)(\omega)$. The theory on pathwise stochastic integration is important and challenging, see some discussion along this line in Sections 2.8.3 and 12.1.1, and Problem 2.10.14. ∎

By the uniform convergence in (2.2.10), it follows immediately that

Theorem 2.2.7 Let $\sigma \in \mathbb{L}^2(\mathbb{F}, \mathbb{R}^d)$ and $M_t := \int_0^t \sigma_s \cdot dB_s$. All the results in Lemmas 2.2.3 and 2.2.4 still hold true.

We finally extend the stochastic integration to all processes $\sigma \in \mathbb{L}_{loc}^2(\mathbb{F}, \mathbb{R}^d)$. For $n \ge 1$, define

$$\tau_n := \inf\{t \ge 0 : \int_0^t |\sigma_s|^2 ds \ge n\} \wedge T, \quad \sigma_t^n := \sigma_t \mathbf{1}_{[0,\tau_n)}(t).$$

Then $\sigma^n \in \mathbb{L}^2(\mathbb{F}, \mathbb{R}^d)$, τ_n is increasing and $\tau_n = T$ for n large enough, a.s. Denote $M_t^n := \int_0^t \sigma_s^n \cdot dB_s$. One can easily check that, for $n < m$,

$$M_t^n = M_t^m \quad \text{for } t \le \tau_n.$$

Thus we may define

$$\int_0^t \sigma_s \cdot dB_s := M_t^n \text{ for } t \le \tau_n. \tag{2.2.12}$$

So $M_t := \int_0^t \sigma_s \cdot dB_s$ is well defined for all $t \in [0, T]$. By Theorem 2.2.7 it is obvious that

Theorem 2.2.8 *For any* $\sigma \in \mathbb{L}_{loc}^2(\mathbb{F}, \mathbb{R}^d)$, $M_t := \int_0^t \sigma_s \cdot dB_s$ *is a continuous local martingale.*

2.3   The Itô Formula

The Itô formula is the extension of the chain rule in calculus to stochastic calculus, and plays a key role in stochastic calculus. In particular, it will be crucial to build the connection between the martingale theory and partial differential equations, see, e.g., Section 5.1 below.

2.3.1   Some Heuristic Arguments

Assume $A \in \mathbb{L}^0(\mathbb{F}, \mathbb{R})$ has bounded variation, a.s. and $f \in C^1(\mathbb{R})$ is a deterministic function. The standard chain rule tells that

$$df(A_t) = f'(A_t)dA_t. \tag{2.3.1}$$

The following simple example shows that the above formula fails if we replace A with the Brownian motion B and thus dA becomes stochastic integration dB.

Example 2.3.1 *Let* $d = 1$ *and set* $f(x) := x^2$. *Then*

$$|B_T|^2 = 2 \int_0^T B_t dB_t + T.$$

Proof For any partition $\pi : 0 = t_0 < \cdots < t_n = T$, we have

$$|B_T|^2 = \sum_{i=0}^{n-1} \left[|B_{t_{i+1}}|^2 - |B_{t_i}|^2 \right] = \sum_{i=0}^{n-1} \left[|B_{t_i,t_{i+1}}|^2 + 2B_{t_i} B_{t_i,t_{i+1}} \right].$$

Send $|\pi| \to 0$, we have

$$\sum_{i=0}^{n-1} |B_{t_i,t_{i+1}}|^2 \to T \text{ in } \mathbb{L}^2(\mathscr{F}_T). \tag{2.3.2}$$

Moreover, denote $B_t^\pi := \sum_{i=0}^{n-1} B_{t_i} \mathbf{1}_{[t_i, t_{i+1})}$. Then $B^\pi \in \mathbb{L}_0^2(\mathbb{F})$, and one can easily check that

$$\lim_{|\pi| \to 0} \|B^\pi - B\|_2 = 0.$$

This implies that

$$\sum_{i=0}^{n-1} B_{t_i} B_{t_i, t_{i+1}} \to \int_0^T B_t dB_t, \quad \text{in } \mathbb{L}^2(\mathscr{F}_T),$$

which, together with (2.3.2), proves the result.   ∎
    Note that

$$f'(B_t) = 2B_t, \quad f''(B_t) = 2, \quad \langle B \rangle_t = t.$$

Then Example 2.3.1 implies

$$f(B_T) - f(B_0) = \int_0^T f'(B_t) dB_t + \frac{1}{2} \int_0^T f''(B_t) d\langle B \rangle_t. \tag{2.3.3}$$

This is a special case of the Itô formula. We see that there is a correction term $\frac{1}{2} \int_0^T f''(B_t) d\langle B \rangle_t$ for stochastic integrations. We prove the general case in the next subsection.

2.3.2 The Itô Formula

In this subsection we focus on one-dimensional case. The multidimensional case will be introduced in detail in the next subsection. Let $b \in \mathbb{L}_{loc}^1(\mathbb{F}), \sigma \in \mathbb{L}_{loc}^2(\mathbb{F})$, and denote

$$X_t = X_0 + \int_0^t b_s ds + \int_0^t \sigma_s dB_s \quad \text{and} \quad \langle X \rangle_t := \int_0^t |\sigma_s|^2 ds. \tag{2.3.4}$$

Theorem 2.3.2 (Ito Formula) *Let $f \in C^{1,2}([0,T] \times \mathbb{R}, \mathbb{R})$. Then*

$$df(t, X_t) = \partial_t f(t, X_t) dt + \partial_x f(t, X_t) dX_t + \frac{1}{2} \partial_{xx} f(t, X_t) d\langle X \rangle_t \tag{2.3.5}$$

$$= \left[\partial_t f + \partial_x f b_t + \frac{1}{2} \partial_{xx} f |\sigma_t|^2 \right] (t, X_t) dt + \partial_x f(t, X_t) \sigma_t dB_t.$$

Or equivalently,

$$f(t, X_t) = f(0, X_0) + \int_0^t \left[\partial_t f + \partial_x f b_s + \frac{1}{2} \partial_{xx} f |\sigma_s|^2 \right] (s, X_s) ds + \int_0^t \partial_x f(s, X_s) \sigma_s dB_s.$$

$$\tag{2.3.6}$$

Proof We first note that, since X is continuous and $f \in C^{1,2}$, for $\varphi = \partial_t f, \partial_x f, \partial_{xx} f$, we know that $\varphi(t, X_t)$ is continuous and thus $\sup_{0 \le t \le T} |\varphi(t, X_t)| < \infty$, a.s. This implies that

$$\left[\partial_t f + \partial_x f b + \frac{1}{2}\partial_{xx} f |\sigma|^2\right](\cdot, X) \in \mathbb{L}^1_{loc}(\mathbb{F}), \quad \partial_x f(\cdot, X)\sigma \in \mathbb{L}^2_{loc}(\mathbb{F}),$$

and thus the right side of (2.3.6) is well defined.

Without loss of generality, we prove (2.3.6) only for $t = T$. We proceed in several steps.

Step 1. We first assume that $b_t = b_0$, $\sigma_t = \sigma_0$ are $\mathscr{F}_0$-measurable and bounded, and f is smooth enough with all related derivatives bounded.

For an arbitrary partition $\pi : 0 = t_0 < \cdots < t_n = T$, we have

$$f(T, X_T) - f(0, X_0) = \sum_{i=0}^{n-1} \left[f(t_{i+1}, X_{t_{i+1}}) - f(t_i, X_{t_i})\right]. \tag{2.3.7}$$

Denote $\Delta t_{i+1} := t_{i+1} - t_i$ and note that $X_{t_i, t_{i+1}} = b_0 \Delta t_{i+1} + \sigma_0 B_{t_i, t_{i+1}}$. Then, by Taylor expansion,

$$f(t_{i+1}, X_{t_{i+1}}) - f(t_i, X_{t_i}) = f(t_i + \Delta t_{i+1}, X_{t_i} + X_{t_i, t_{i+1}}) - f(t_i, X_{t_i})$$

$$= \partial_t f(t_i, X_{t_i})\Delta t_{i+1} + \partial_x f(t_i, X_{t_i})X_{t_i, t_{i+1}}$$

$$+ \frac{1}{2}\partial_{tt} f(t_i, X_{t_i})|\Delta t_{i+1}|^2 + \partial_{tx} f(t_i, X_{t_i})\Delta t_{i+1}X_{t_i, t_{i+1}} + \frac{1}{2}\partial_{xx} f(t_i, X_{t_i})|X_{t_i, t_{i+1}}|^2 + R^{\pi}_{i+1}$$

$$= \left[\partial_t f + b_0 \partial_x f + \frac{1}{2}\partial_{xx} f |\sigma_0|^2\right](t_i, X_{t_i})\Delta t_{i+1} + \sigma_0 \partial_x f(t_i, X_{t_i})B_{t_i, t_{i+1}} \tag{2.3.8}$$

$$+ \frac{1}{2}\partial_{xx} f(t_i, X_{t_i})|\sigma_0|^2[|B_{t_i, t_{i+1}}|^2 - \Delta t_{i+1}] + I^{\pi}_{i+1}$$

where

$$I^{\pi}_{i+1} := \frac{1}{2}\left[\partial_{tt} f + \partial_{xx} f |b_0|^2\right](t_i, X_{t_i})|\Delta t_{i+1}|^2 + \left[\partial_{tx} f + b_0 \sigma_0 \partial_{xx} f\right](t_i, X_{t_i})\Delta t_{i+1}B_{t_i, t_{i+1}} + R^{\pi}_{i+1},$$

$$\text{and} \quad |R^{\pi}_{i+1}| \le C[|\Delta t_{i+1}|^3 + |X_{t_i, t_{i+1}}|^3] \le C[|\Delta t_{i+1}|^3 + |B_{t_i, t_{i+1}}|^3].$$

Send $|\pi| \to 0$. First, applying the Dominated Convergence Theorem we have:

$$\sum_{i=0}^{n-1} \left[\partial_t f + b_0 \partial_x f + \frac{1}{2}\partial_{xx} f |\sigma_0|^2\right](t_i, X_{t_i})\Delta t_{i+1}$$

$$\to \int_0^T \left[\partial_t f + b_0 \partial_x f + \frac{1}{2}|\sigma_0|^2 \partial_{xx} f\right](t, X_t)dt, \quad \text{in } \mathbb{L}^2(\mathscr{F}_T). \tag{2.3.9}$$

Next, applying the Dominated Convergence Theorem again we have

$$\mathbb{E}\Big[\sum_{i=0}^{n-1}\int_{t_i}^{t_{i+1}}|\partial_x f(t,X_t)-\partial_x f(t_i,X_{t_i})|^2 dt\Big]\to 0,$$

and thus

$$\sum_{i=0}^{n-1}\sigma_0\partial_x f(t_i,X_{t_i})B_{t_i,t_{i+1}}\to\int_0^T\sigma_0\partial_x f(t,X_t)dB_t,\quad\text{in }\mathbb{L}^2(\mathscr{F}_T).\quad(2.3.10)$$

Moreover, note that, for any $p\geq 1$ and some constant $c_p>0$,

$$\mathbb{E}[|B_{t_i,t_{i+1}}|^p]=c_p|\Delta t_{i+1}|^{\frac{p}{2}}.$$

Then

$$\mathbb{E}\Big[\Big(\sum_{i=0}^{n-1}I_{i+1}^{\pi}\Big)^2\Big]\leq C\mathbb{E}\Big[\sum_{i=0}^{n-1}\big(|\Delta t_{i+1}|^2+\Delta t_{i+1}|B_{t_i,t_{i+1}}|+|B_{t_i,t_{i+1}}|^3\big)\Big]$$

$$\leq C\sum_{i=0}^{n-1}|\Delta t_{i+1}|^{\frac{3}{2}}\leq C|\pi|^{\frac{1}{2}}\to 0.\quad(2.3.11)$$

Finally, by Example 2.3.1 we see that

$$|B_{t_i,t_{i+1}}|^2-\Delta t_{i+1}=2\int_{t_i}^{t_{i+1}}B_{t_i,t}dB_t.$$

Clearly

$$\mathbb{E}\Big[\int_0^T\Big|\sum_{i=0}^{n-1}\partial_{xx}f(t_i,X_{t_i})B_{t_i,t}\mathbf{1}_{[t_i,t_{i+1})}\Big|^2 dt\Big]=\mathbb{E}\Big[\sum_{i=0}^{n-1}\int_{t_i}^{t_{i+1}}\Big|\partial_{xx}f(t_i,X_{t_i})B_{t_i,t}\Big|^2 dt\Big]$$

$$\leq C\sum_{i=0}^{n-1}\int_{t_i}^{t_{i+1}}(t-t_i)dt=C\sum_{i=0}^{n-1}|\Delta t_{i+1}|^2\leq C|\pi|\to 0.$$

Then

$$\sum_{i=0}^{n-1}\partial_{xx}f(t_i,X_{t_i})[|B_{t_i,t_{i+1}}|^2-\Delta t_{i+1}]=\int_0^T\sum_{i=0}^{n-1}\partial_{xx}f(t_i,X_{t_i})B_{t_i,t}\mathbf{1}_{[t_i,t_{i+1})}dB_t\to 0,\text{ in }\mathbb{L}^2(\mathscr{F}_T).$$

$$(2.3.12)$$

Plug (2.3.9)–(2.3.12) into (2.3.7) and (2.3.8), we prove (2.3.6).

Step 2. Assume that $b_t = b_0$, $\sigma_t = \sigma_0$ are $\mathscr{F}_0$-measurable and bounded, and $f \in C^{1,2}$ with all related derivatives bounded. Let f^n be a smooth mollifier of f, see Problem 1.4.14. Then f^n is smooth with all the related derivatives bounded with a constant C_n which may depend on n, and for $\varphi = \partial_t f, \partial_x f, \partial_{xx} f$,

$$\varphi^n \to \varphi \quad \text{and} \quad |\varphi^n| \le C \text{ where } C \text{ is independent of } n.$$

By Step 1, we have

$$f^n(T, X_T) = f^n(0, X_0) + \int_0^T \left[\partial_t f^n + \partial_x f^n b_0 + \frac{1}{2} \partial_{xx} f^n |\sigma_0|^2 \right](t, X_t) dt + \int_0^T \partial_x f^n(t, X_t) \sigma_0 dB_t.$$

Send $n \to \infty$, we prove (2.3.6) for f immediately.

Step 3. Assume $b = \sum_{i=0}^{n-1} b_{t_i} \mathbf{1}_{[t_i, t_{i+1})} \in \mathbb{L}_0^2(\mathbb{F})$, $\sigma = \sum_{i=0}^{n-1} \sigma_{t_i} \mathbf{1}_{[t_i, t_{i+1})} \in \mathbb{L}_0^2(\mathbb{F})$ are bounded, and $f \in C^{1,2}$ with all related derivatives bounded. Applying Step 2 on $[t_i, t_{i+1}]$ one can easily see that

$$f(t_{i+1}, X_{t_{i+1}}) = f(t_i, X_{t_i}) + \int_{t_i}^{t_{i+1}} \left[\partial_t f + \partial_x f b_{t_i} + \frac{1}{2} \partial_{xx} f |\sigma_{t_i}|^2 \right](t, X_t) dt + \int_{t_i}^{t_{i+1}} \partial_x f(t, X_t) \sigma_{t_i} dB_t.$$

Sum over all i we obtain the result.

Step 4. Assume $b \in \mathbb{L}^1(\mathbb{F})$, $\sigma \in \mathbb{L}^2(\mathbb{F})$, and $f \in C^{1,2}$ with all related derivatives bounded. Analogous to Lemma 2.2.5, one can easily show that there exist bounded $b^n, \sigma^n \in \mathbb{L}_0^2(\mathbb{F})$ such that

$$\lim_{n \to \infty} \|b^n - b\|_1 = 0, \quad \lim_{n \to \infty} \|\sigma^n - \sigma\|_2 = 0.$$

Denote

$$X_t^n := X_0 + \int_0^t b_s^n ds + \int_0^t \sigma_s^n dB_s,$$

and note that

$$(X^n - X)_T^* \le \int_0^T |b_t^n - b_t| dt + \sup_{0 \le t \le T} \left| \int_0^t [\sigma_s^n - \sigma_s] dB_s \right|.$$

Then by (2.2.11) we have

$$\lim_{n \to \infty} \mathbb{E}\left[(X^n - X)_T^* \right] = 0, \quad \text{and thus} \quad (X^n - X)_T^* \to 0 \text{ in probability.}$$

By Step 3, we have

$$f(T, X_T^n) = f(0, X_0) + \int_0^T \left[\partial_t f + \partial_x f b_t^n + \frac{1}{2} \partial_{xx} f |\sigma_t^n|^2 \right](t, X_t^n) dt + \int_0^T \partial_x f(t, X_t^n) \sigma_t^n dB_t.$$

Send $n \to \infty$. Note that

$$\mathbb{E}\left[\int_0^T \left|\partial_x f(t, X_t^n)\sigma_t^n - \partial_x f(t, X_t)\sigma_t\right|^2 dt\right]$$

$$\leq C\mathbb{E}\left[\int_0^T \left[|\sigma_t^n - \sigma_t|^2 + |\partial_x f(t, X_t^n) - \partial_x f(t, X_t)|^2|\sigma_t|^2\right]dt\right] \to 0,$$

thanks to the Dominated Convergence Theorem. Then

$$\int_0^T \partial_x f(t, X_t^n)\sigma_t^n dB_t \to \int_0^T \partial_x f(t, X_t)\sigma_t dB_t \quad \text{in } \mathbb{L}^2(\mathscr{F}_T).$$

Similarly,

$$\int_0^T \left[\partial_t f + \partial_x f b_t^n + \frac{1}{2}\partial_{xx} f|\sigma_t^n|^2\right](t, X_t^n)dt \to \int_0^T \left[\partial_t f + \partial_x f b_t + \frac{1}{2}\partial_{xx} f|\sigma_t|^2\right](t, X_t)dt, \text{ in } \mathbb{L}^1(\mathscr{F}_T).$$

We thus obtain the result.

Step 5. We now show the general case, namely $b \in \mathbb{L}_{loc}^1(\mathbb{F}), \sigma \in \mathbb{L}_{loc}^2(\mathbb{F})$ and $f \in C^{1,2}$. For each $n \geq 1$, define

$$\tau_n := \inf\left\{t \geq 0 : \int_0^t |b_s|ds + \int_0^t |\sigma_s|^2 ds + |X_t| + \int_0^t |\partial_x f(s, X_s)\sigma_s|^2 ds \geq n\right\} \wedge T,$$

$$(2.3.13)$$

and denote

$$b^n := b\mathbf{1}_{[0,\tau_n]}, \quad \sigma^n := \sigma\mathbf{1}_{[0,\tau_n]}, \quad X^n := X_{\tau_n \wedge \cdot},$$

and $f^n \in C^{1,2}$ with bounded derivatives such that

$$f^n(t, x) = f(t, x), \quad \text{for all } 0 \leq t \leq T, |x| \leq n.$$

Then

$$X_t^n = X_0 + \int_0^t b_s^n ds + \int_0^t \sigma_s^n dB_s \quad \text{and} \quad |X^n| \leq n.$$

By Step 4, we have

$$f^n(T, X_T^n) = f^n(0, X_0) + \int_0^T \left[\partial_t f^n + \partial_x f^n b_t^n + \frac{1}{2}\partial_{xx} f^n|\sigma_t^n|^2\right](t, X_t^n)dt$$

$$+ \int_0^T \partial_x f^n(t, X_t^n)\sigma_t^n dB_t.$$

This is equivalent to

$$f(T, X_{\tau_n}) = f(0, X_0) + \int_0^T \left[\partial_t f + \partial_x f b_t^n + \frac{1}{2} \partial_{xx} f |\sigma_t^n|^2 \right](t, X_t^n) dt + \int_0^T \partial_x f(t, X_t^n) \sigma_t^n dB_t.$$

$$(2.3.14)$$

Recall (2.2.12) for stochastic integration in $\mathbb{L}^2_{loc}(\mathbb{F})$ and notice that (2.3.13) include the term $\int_0^t |\partial_x f(s, X_s) \sigma_s|^2 ds$, then

$$\int_0^T \partial_x f(t, X_t^n) \sigma_t^n dB_t = \int_0^{\tau_n} \partial_x f(t, X_t^n) \sigma_t^n dB_t = \int_0^{\tau_n} \partial_x f(t, X_t) \sigma_t dB_t.$$

Plug this into (2.3.14) and send $n \to \infty$. Note that, for n large enough, $\tau_n = T$, $b^n = b, \sigma^n = \sigma, X^n = X$, a.s. This implies that (2.3.6) holds a.s.   ∎

2.3.3   Itô Formula in Multidimensional Case

Let $B = (B^1, \cdots, B^d)^\top$ be a d-dimensional $\mathbb{F}$-Brownian Motion, $b^i \in \mathbb{L}^1_{loc}(\mathbb{F})$, $\sigma^{i,j} \in \mathbb{L}^2_{loc}(\mathbb{F})$, $1 \le i \le d_1$, $1 \le j \le d$. Set $b := (b^1, \cdots, b^{d_1})^\top$ and $\sigma := (\sigma^{i,j})_{1 \le i \le d_1, 1 \le j \le d}$ which take values in $\mathbb{R}^{d_1}$ and $\mathbb{R}^{d_1 \times d}$, respectively. Let $X = (X^1, \cdots, X^{d_1})^\top$ satisfy

$$dX_t^i := b_t^i dt + \sum_{j=1}^d \sigma_t^{i,j} dB_t^j, \ i = 1, \cdots, d_1; \quad \text{or equivalently,} \quad dX_t = b_t dt + \sigma_t dB_t.$$

$$(2.3.15)$$

Denote

$$\langle X \rangle_t := \int_0^t \sigma_s \sigma_s^\top ds \text{ taking values in } \mathbb{S}^{d_1}.$$ $$(2.3.16)$$

We have the following multidimensional Itô formula whose proof is analogous to that of Theorem 2.3.2 and is omitted.

Theorem 2.3.3 *Assume* $f : [0, T] \times \mathbb{R}^{d_1} \to \mathbb{R}$ *is in* $C^{1,2}$. *Then*

$$df(t, X_t) = \partial_t f(t, X_t) dt + \partial_x f(t, X_t) dX_t + \frac{1}{2} \partial_{xx} f(t, X_t) : d\langle X \rangle_t$$

$$= \left[\partial_t f + \partial_x f b_t + \frac{1}{2} \partial_{xx} f : (\sigma_t \sigma_t^\top) \right](t, X_t) dt + \partial_x f(t, X_t) \sigma_t dB_t \quad (2.3.17)$$

$$= \left[\partial_t f + \sum_{i=1}^{d_1} \partial_{x_i} f b_t^i + \frac{1}{2} \sum_{i,j=1}^{d_1} \sum_{k=1}^d \partial_{x_i x_j} f \sigma_t^{i,k} \sigma_t^{j,k} \right](t, X_t) dt + \sum_{i=1}^{d_1} \sum_{j=1}^d \partial_{x_i} f(t, X_t) \sigma_t^{i,j} dB_t^j.$$

Throughout the book, we take the convention that $\partial_x f = (\partial_{x_1} f, \cdots, \partial_{x_{d_1}} f)$ is a row vector, and we note that $\partial_{xx} f$ takes values in $\mathbb{S}^{d_1}$.

2.3.4   An Extended Itô Formula

For future purpose, we need to extend the Itô formula to the case where the drift term $b_t dt$ is replaced with a bounded variational process A. For simplicity, we state the result only for the case $d_1 = 1$, but one may easily generalize it to multidimensional cases. Let B be a d-dimensional $\mathbb{F}$-Brownian Motion, $\sigma \in \mathbb{L}^2_{loc}(\mathbb{F}, \mathbb{R}^{1\times d})$, $A \in$ · $\mathbb{L}^0(\mathbb{F}, \mathbb{R})$ is continuous in t and $\bigvee_0^T A < \infty$, a.s. Denote

$$dX_t := \sigma_t dB_t + dA_t, \quad \langle X \rangle_t := \int_0^t \sigma_s \sigma_s^{\top} ds. \tag{2.3.18}$$

We have the following extended Itô formula whose proof is left to the readers in Problem 2.10.4.

Theorem 2.3.4 *Assume $f : [0, T] \times \mathbb{R} \to \mathbb{R}$ is in $C^{1,2}$. Then*

$$df(t, X_t) = \partial_t f(t, X_t) dt + \partial_x f(t, X_t) dX_t + \frac{1}{2} \partial_{xx} f(t, X_t) : d\langle X \rangle_t$$

$$= \Big[\partial_t f + \frac{1}{2} \partial_{xx} f(\sigma_t \sigma_t^{\top}) \Big](t, X_t) dt + \partial_x f(t, X_t) \sigma_t dB_t + \partial_x f(t, X_t) dA_t, \tag{2.3.19}$$

where the last term in understood in the sense of (2.2.1).

2.4   The Burkholder-Davis-Gundy Inequality

As an application of the Itô formula, we prove the following important inequality due to Burkholder-Davis-Gundy. For any $p > 0$ and $\sigma \in \mathbb{L}^{2,p}(\mathbb{F}, \mathbb{R}^d) \subset \mathbb{L}^2_{loc}(\mathbb{F}, \mathbb{R}^d)$, define $M_t := \int_0^t \sigma_s \cdot dB_s$ and M^* by (1.2.4).

Theorem 2.4.1 (Burkholder-Davis-Gundy Inequality) *For any $p > 0$, there exist universal constants $0 < c_p < C_p$, depending only on p and d, such that*

$$c_p \mathbb{E}\Big[\Big(\int_0^T |\sigma_t|^2 dt \Big)^{\frac{p}{2}} \Big] \le \mathbb{E}[|M_T^*|^p] \le C_p \mathbb{E}\Big[\Big(\int_0^T |\sigma_t|^2 dt \Big)^{\frac{p}{2}} \Big]. \tag{2.4.1}$$

Proof We again assume $d = 1$. The case $p = 2$ is exactly the Doob's maximum inequality in Theorem 2.2.7 and Lemma 2.2.4. Note that $\langle M \rangle_t = \int_0^t \sigma_s^2 ds$. Following the truncation arguments in Step 2 of Lemma 2.2.4, we may assume without loss of generality that

$$M_T^* \text{ and } \langle M \rangle_T \text{ are bounded.} \tag{2.4.2}$$

However, we shall emphasize that the constants C_p, c_p in the proof below will not depend on this bound. We proceed in several steps.

Step 1. We first prove the left inequality by using the right inequality. Apply Itô formula, we have

$$d|M_t|^2 = |\sigma_t|^2 dt + 2M_t\sigma_t dB_t. \tag{2.4.3}$$

Then

$$\langle M \rangle_T = \int_0^T |\sigma_t|^2 dt = M_T^2 - M_0^2 - 2\int_0^T M_t\sigma_t dB_t.$$

Thus, by the right inequality and noting that $ab \le \frac{1}{2}[a^2 + b^2]$, we have

$$\mathbb{E}[\langle M \rangle_T^{\frac{p}{2}}] \le C_p\mathbb{E}[|M_T^*|^p] + C_p\mathbb{E}\Big[\big|\int_0^T M_t\sigma_t dB_t\big|^{\frac{p}{2}}\Big] \le C_p\mathbb{E}[|M_T^*|^p] + C_p\mathbb{E}\Big[\big(\int_0^T |M_t\sigma_t|^2 dt\big)^{\frac{p}{4}}\Big]$$

$$\le C_p\mathbb{E}[|M_T^*|^p] + C_p\mathbb{E}\Big[|M_T^*|^{\frac{p}{2}}\langle M \rangle_T^{\frac{p}{4}}\Big] \le C_p\mathbb{E}[|M_T^*|^p] + \frac{1}{2}\mathbb{E}\Big[\langle M \rangle_T^{\frac{p}{2}}\Big].$$

This, together with (2.4.2), implies the left inequality.

Step 2. We next prove the right inequality for $p \ge 2$. By the same arguments as in (2.2.9), we have

$$\mathbb{E}[|M_T^*|^p] = p\int_0^\infty \lambda^{p-1}\mathbb{P}(M_T^* \ge \lambda)d\lambda \le p\int_0^\infty \lambda^{p-2}\mathbb{E}\Big[|M_T|\mathbf{1}_{\{M_T^*\ge\lambda\}}\Big]d\lambda$$

$$= \mathbb{E}\Big[p\int_0^\infty \lambda^{p-2}|M_T|\mathbf{1}_{\{M_T^*\ge\lambda\}}d\lambda\Big] = \mathbb{E}\Big[p|M_T|\int_0^{M_T^*}\lambda^{p-2}d\lambda\Big] = \frac{p}{p-1}\mathbb{E}\Big[|M_T||M_T^*|^{p-1}\Big].$$

Note that p and $\frac{p}{p-1}$ are conjugates. Then by Hölder inequality we have

$$\mathbb{E}[|M_T^*|^p] \le \frac{p}{p-1}\Big(\mathbb{E}[|M_T|^p]\Big)^{\frac{1}{p}}\Big(\mathbb{E}[|M_T^*|^p]\Big)^{\frac{p-1}{p}}.$$

This, together with (2.4.2), implies

$$\mathbb{E}[|M_T^*|^p] \le \Big(\frac{p}{p-1}\Big)^p\mathbb{E}[|M_T|^p]. \tag{2.4.4}$$

On the other hand, by (2.4.3) and applying the Itô formula, we have

$$d(|M_t|^p) = d\big([|M_t|^2]^{\frac{p}{2}}\big) = \frac{1}{2}p(p-1)|M_t|^{p-2}|\sigma_t|^2 dt + |M_t|^{p-2}M_t\sigma_t dB_t.$$

By (2.4.2), clearly $|M|^{p-2}M\sigma \in \mathbb{L}^2(\mathbb{F})$. Then

$$\mathbb{E}\Big[\int_0^T |M_t|^{p-2}M_t\sigma_t dB_t\Big] = 0.$$

Thus

$$\mathbb{E}[|M_T|^p] = \frac{1}{2}p(p-1)\mathbb{E}\Big[\int_0^T |M_t|^{p-2}|\sigma_t|^2 dt\Big] \le C_p \mathbb{E}\Big[|M_T^*|^{p-2}\langle M\rangle_T\Big].$$

Note that $\frac{p}{p-2}$ and $\frac{p}{2}$ are conjugates. Applying Hölder inequality again we obtain

$$\mathbb{E}[|M_T|^p] \le C_p \mathbb{E}\Big[|M_T^*|^{p-2}\langle M\rangle_T\Big] \le C_p\Big(\mathbb{E}[|M_T^*|^p]\Big)^{\frac{p-2}{p}}\Big(\mathbb{E}\big[\langle M\rangle_T^{\frac{p}{2}}\big]\Big)^{\frac{2}{p}}.$$

This, together with (2.4.4) and (2.4.2), implies the right inequality in (2.4.1) immediately.

Step 3. We finally prove the right inequality for $0 < p < 2$. Note that

$$\mathbb{E}\Big[\int_0^T |\langle M\rangle_t^{\frac{p-2}{4}}\sigma_t|^2 dt\Big] = \mathbb{E}\Big[\int_0^T \langle M\rangle_t^{\frac{p-2}{2}} d\langle M\rangle_t\Big] = \frac{2}{p}\mathbb{E}[\langle M\rangle_T^{\frac{p}{2}}] < \infty.$$

Then $N_t := \int_0^t \langle M\rangle_t^{\frac{p-2}{4}}\sigma_t dB_t$ is a square integrable martingale and $\mathbb{E}[N_T^2] = \frac{2}{p}\mathbb{E}[\langle M\rangle_T^{\frac{p}{2}}]$. Apply Itô formula, we have

$$M_t = \int_0^t \langle M\rangle_s^{\frac{2-p}{4}} dN_s = \langle M\rangle_t^{\frac{2-p}{4}} N_t - \int_0^t N_s d\langle M\rangle_s^{\frac{2-p}{4}}.$$

Note that $\langle M\rangle$ is increasing in t. Then

$$M_T^* \le \langle M\rangle_T^{\frac{2-p}{4}} N_T^* + \int_0^T |N_s| d\langle M\rangle_s^{\frac{2-p}{4}} \le C N_T^* \langle M\rangle_T^{\frac{2-p}{4}}.$$

Note that $\frac{2}{p}$ and $\frac{2}{2-p}$ are conjugates. Applying the Hölder inequality and then the Doob's maximum inequality Lemma 2.2.4, we have

$$\mathbb{E}[|M_T^*|^p] \le C_p\mathbb{E}\Big[|N_T^*|^p \langle M\rangle_T^{\frac{p(2-p)}{4}}\Big] \le C_p\Big(\mathbb{E}[|N_T^*|^2]\Big)^{\frac{p}{2}}\Big(\mathbb{E}[\langle M\rangle_T^{\frac{p}{2}}]\Big)^{\frac{2-p}{2}}$$

$$\le C_p\Big(\mathbb{E}[|N_T|^2]\Big)^{\frac{p}{2}}\Big(\mathbb{E}[\langle M\rangle_T^{\frac{p}{2}}]\Big)^{\frac{2-p}{2}} = C_p\Big(\mathbb{E}[\langle M\rangle_T^{\frac{p}{2}}]\Big)^{\frac{p}{2}}\Big(\mathbb{E}[\langle M\rangle_T^{\frac{p}{2}}]\Big)^{\frac{2-p}{2}} = C_p\mathbb{E}[\langle M\rangle_T^{\frac{p}{2}}].$$

This completes the proof. ∎

Corollary 2.4.2 *Let* $\sigma \in \mathbb{L}^{2,1}(\mathbb{F},\mathbb{R}^d) \subset \mathbb{L}^2_{loc}(\mathbb{F},\mathbb{R}^d)$. *Then* $M_t := \int_0^t \sigma_s \cdot dB_s$ *is a u.i. martingale.*

Proof Apply the Burkholder-Davis-Gundy Inequality Theorem 2.4.1 with $p = 1$, we have

$$\mathbb{E}[M_T^*] \le C\mathbb{E}\Big[\big(\int_0^T |\sigma_t|^2 dt\big)^{\frac{1}{2}}\Big] < \infty.$$

Then the local martingale M is a u.i. martingale. ∎

2.5   The Martingale Representation Theorem

Given $\sigma \in \mathbb{L}^2(\mathbb{F}, \mathbb{R}^d)$, it is known that $M_t := \int_0^t \sigma_s \cdot dB_s$ is a square integrable $\mathbb{F}$-martingale. The Martingale Representation Theorem deals with the opposite issue: given a square integrable $\mathbb{F}$-martingale M, does there exist $\sigma \in \mathbb{L}^2(\mathbb{F}, \mathbb{R}^d)$ such that $M_t = M_0 + \int_0^t \sigma_s \cdot dB_s$?

The answer to the above question is in general negative.

Example 2.5.1 *Let $d = 1$ and $B, \tilde{B}$ be independent $\mathbb{F}$-Brownian Motion. Then $\tilde{B}$ is a square integrable $\mathbb{F}$-martingale, but there is no $\sigma \in \mathbb{L}^2(\mathbb{F})$ such that $\tilde{B}_t = \int_0^t \sigma_s dB_s$.*

Proof We prove by contradiction. Assume $\tilde{B}_t = \int_0^t \sigma_s dB_s$ for some $\sigma \in \mathbb{L}^2(\mathbb{F})$. On one hand, for $X_t^1 := \int_0^t \sigma_s dB_s$ and $X_t^2 := \int_0^t 1 d\tilde{B}_s$, applying Itô formula (2.3.17) we have

$$d|\tilde{B}_t|^2 = d(X_t^1 X_t^2) = X_t^1 d\tilde{B}_t + X_t^2 \sigma_t dB_t$$

and thus $|\tilde{B}_t|^2$ is a local martingale. On the other hand, applying Itô formula (2.3.5) directly on $|\tilde{B}|^2$ we obtain

$$d|\tilde{B}_t|^2 = 2\tilde{B}_t d\tilde{B}_t + dt$$

and thus it is not a local martingale. Contradiction. ∎

The key issue here is that $\tilde{B}$ is independent of B and thus is not $\mathbb{F}^B$-measurable. We have the following important result by using the filtration $\mathbb{F}^B$.

Theorem 2.5.2 *For any $\xi \in \mathbb{L}^2(\mathscr{F}_T^B)$, there exists unique $\sigma \in \mathbb{L}^2(\mathbb{F}^B, \mathbb{R}^d)$ such that*

$$\xi = \mathbb{E}[\xi] + \int_0^T \sigma_t \cdot dB_t. \tag{2.5.1}$$

Consequently, for any $\mathbb{F}^B$-martingale M such that $\mathbb{E}[|M_T|^2] < \infty$, there exists unique $\sigma \in \mathbb{L}^2(\mathbb{F}^B, \mathbb{R}^d)$ such that

$$M_t = M_0 + \int_0^t \sigma_s \cdot dB_s. \tag{2.5.2}$$

Proof Again we assume $d = 1$ for simplicity. First note that (2.5.2) is a direct consequence of (2.5.1). Indeed, for any $\mathbb{F}^B$-martingale M such that $\mathbb{E}[|M_T|^2] < \infty$, by (2.5.1) there exists unique $\sigma \in \mathbb{L}^2(\mathbb{F}^B)$ such that

$$M_T = \mathbb{E}[M_T] + \int_0^T \sigma_t dB_t.$$

Denote

$$\tilde{M}_t := \mathbb{E}[M_T] + \int_0^t \sigma_s dB_s.$$

Then $\tilde{M}$ is an $\mathbb{F}^B$-martingale and $\tilde{M}_T = M_T$. Thus

$$M_t = \mathbb{E}[M_T|\mathscr{F}_t^B] = \mathbb{E}[\tilde{M}_T|\mathscr{F}_t^B] = \tilde{M}_t.$$

In particular,

$$M_0 = \tilde{M}_0 = \mathbb{E}[M_T].$$

This implies (2.5.2) immediately.

We next prove the uniqueness of σ in (2.5.1). If there is another $\tilde{\sigma} \in \mathbb{L}^2(\mathbb{F}^B)$ satisfying (2.5.1). Then

$$\int_0^T (\sigma_t - \tilde{\sigma}_t)dB_t = 0.$$

Square both sides and take expectations, we get

$$\mathbb{E}\left[\int_0^T |\sigma_t - \tilde{\sigma}_t|^2 dt\right] = 0.$$

That is,

$$\tilde{\sigma} = \sigma, \quad dt \times dP - \text{a.s.}$$

It remains to prove the existence in (2.5.1). We proceed in several steps.
Step 1. Assume $\xi = g(B_T)$, where $g \in C_b^2(\mathbb{R})$. Define

$$u(t,x) := \mathbb{E}\Big[g(x + B_{T-t})\Big] = \int_{\mathbb{R}} g(y)p(T-t, y-x)dy, \quad \text{where } p(t,x) := \frac{1}{\sqrt{2\pi t}}e^{-\frac{x^2}{2t}}.$$

$$(2.5.3)$$

Note that

$$\partial_t p(t,x) = \frac{1}{\sqrt{2\pi}}e^{-\frac{x^2}{2t}}\left[-\frac{1}{2}t^{-\frac{3}{2}} + \frac{x^2}{2}t^{-\frac{5}{2}}\right]$$

$$\partial_x p(t,x) = \frac{1}{\sqrt{2\pi t}}e^{-\frac{x^2}{2t}}(-\frac{x}{t}), \quad \partial_{xx}p(t,x) = \frac{1}{\sqrt{2\pi t}}e^{-\frac{x^2}{2t}}\left[\frac{x^2}{t^2} - \frac{1}{t}\right].$$

Then

$$\partial_t p(t,x) - \frac{1}{2}\partial_{xx}p(t,x) = 0.$$

One can easily check that $u \in C_b^{1,2}([0,T] \times \mathbb{R})$ and

$$\partial_t u(t,x) + \frac{1}{2}\partial_{xx}u(t,x) = 0, \quad u(T,x) = g(x). \tag{2.5.4}$$

Now define

$$M_t := u(t, B_t), \quad \sigma_t := \partial_x u(t, B_t). \tag{2.5.5}$$

Apply Itô formula we have

$$du(t, B_t) = u_x(t, B_t)dB_t + [\partial_t u + \frac{1}{2}\partial_{xx} u](t, B_t)dt = \sigma_t dB_t.$$

Thus

$$g(B_T) = u(T, B_T) = u(0, 0) + \int_0^T \sigma_t dB_t = \mathbb{E}[g(B_T)] + \int_0^T \sigma_t dB_t.$$

Since $\partial_x u$ is bounded, we see that $\sigma \in \mathbb{L}^2(\mathbb{F}^B)$, and therefore, (2.5.1) holds.

Step 2. Assume $\xi = g(B_T)$ where $g : \mathbb{R} \to \mathbb{R}$ is Borel measurable and bounded. Let g_n be a smooth mollifier of g as in Problem 1.4.14. Then $g_n \in C_b^2(\mathbb{R})$ for each n, $|g_n| \le C$ for all n, and $g_n(x) \to g(x)$ for dx-a.e. x. Since B_T has density, the probability that B_T lies in a Lebesgue null set is 0. Then $g_n(B_T) \to g(B_T)$ a.s. Applying the Dominated Convergence Theorem we get $\lim_{n\to\infty} \mathbb{E}[|g_n(B_T) - g(B_T)|^2] = 0$. Now for each n, by Step 1 there exists $\sigma^n \in \mathbb{L}^2(\mathbb{F})$ such that $g_n(B_T) = \mathbb{E}[g_n(B_T)] + \int_0^T \sigma_t^n dB_t$. Then (2.5.1) follows from Problem 2.10.5.

Step 3. Assume $\xi = g(B_{t_1}, \cdots, B_{t_n})$, where $0 < t_1 < \cdots < t_n \le T$ and $g : \mathbb{R}^n \to \mathbb{R}$ is Borel measurable and bounded. Denote $g_n(x_1, \cdots, x_n) := g(x_1, \cdots, x_n)$. Apply Step 2 on $[t_{n-1}, t_n]$, there exists $\sigma^n \in \mathbb{L}^2(\mathbb{F}^B)$ such that

$$g_n(B_{t_1}, \cdots, B_{t_n}) = \mathbb{E}\Big[g_n(B_{t_1}, \cdots, B_{t_n})|\mathscr{F}_{t_{n-1}}^B\Big] + \int_{t_{n-1}}^{t_n} \sigma_t^n dB_t$$

$$= g_{n-1}(B_{t_1}, \cdots, B_{t_{n-1}}) + \int_{t_{n-1}}^{t_n} \sigma_t^n dB_t,$$

where, since B has independent increments,

$$g_{n-1}(x_1, \cdots, x_{n-1}) := \mathbb{E}\Big[g_n(x_1, \cdots, x_{n-1}, x_{n-1} + B_{t_{n-1}, t_n})\Big]$$

is also Borel measurable and bounded. Repeating the arguments backwardly in time, we obtain

$$g_{i+1}(B_{t_1}, \cdots, B_{t_{i+1}}) = g_i(B_{t_1}, \cdots, B_{t_i}) + \int_{t_i}^{t_{i+1}} \sigma_t^{i+1} dB_t,$$

where

$$g_i(x_1, \cdots, x_i) := \mathbb{E}\Big[g_{i+1}(x_1, \cdots, x_i, x_i + B_{t_i, t_{i+1}})\Big].$$

Define

$$\sigma := \sum_{i=1}^{n} \sigma^i \mathbf{1}_{[t_{i-1},t_i)}.$$

Then one can easily see that $\sigma \in \mathbb{L}^2(\mathbb{F}^B)$ and satisfies the requirement.

Step 4. Assume $\xi \in \mathbb{L}^\infty(\mathscr{F}_T^B)$. For each n, denote $t_i^n := \frac{iT}{2^n}$, $i = 0, \cdots, 2^n$. Let $\mathscr{F}_T^n$ be the σ-field generated by $\{B_{t_i^n}, 0 \le i \le 2^n\}$ and define $\xi_n := \mathbb{E}[\xi|\mathscr{F}_T^n]$. By the Doob-Dynkin lemma we have

$$\xi_n = g_n(B_{t_1^n}, \cdots, B_{t_{2^n}^n}) \quad \text{for some Borel measurable function } g_n.$$

Since ξ is bounded, then so is ξ_n and thus g_n is bounded. By Step 3 we get

$$\xi_n = E[\xi_n] + \int_0^T \sigma_t^n dB_t \quad \text{for some } \sigma^n \in \mathbb{L}^2(\mathbb{F}^B).$$

Since B is continuous, it is clear that $\mathscr{F}_T^B := \vee_n \mathscr{F}_T^n$. Note that $\mathbb{E}[\xi|\mathscr{F}_T^B] = \xi$. Then by Problem 1.4.2 (iii) and the Dominated Convergence Theorem we have

$$\lim_{n\to\infty} \mathbb{E}\left[|\xi_n - \xi|^2\right] = 0.$$

Now (2.5.1) again follows from Problem 2.10.5.

Step 5. In the general case, for each n, let $\xi_n := (-n) \vee \xi \wedge n$. Then $|\xi_n| \le n$ and thus by Step 4, there exists $\sigma^n \in \mathbb{L}^2(\mathbb{F}^B)$ such that

$$\xi_n = \mathbb{E}[\xi_n] + \int_0^T \sigma_t^n dB_t.$$

Clearly $\xi_n \to \xi$ for all ω. Moreover, $|\xi_n| \le |\xi|$. Then by the Dominated Convergence Theorem we have

$$\lim_{n\to\infty} \mathbb{E}\left[|\xi_n - \xi|^2\right] = 0,$$

and thus (2.5.1) follows from Problem 2.10.5 again.   ∎

Remark 2.5.3 In the financial application in Section 2.8, the stochastic integrand σ is related to the hedging portfolio. In particular, from (2.5.5) we see that σ is the derivative of M with respect to B, and thus is closely related to the so-called delta hedging. In fact, this connection is true even in non-Markov case, by introducing the path derivatives in Section 9.4.   ∎

Remark 2.5.4 The condition that ξ is $\mathscr{F}_T^B$-measurable is clearly crucial in Theorem 2.5.2. When $\xi \in \mathbb{L}^2(\mathscr{F}_T)$ and $\mathbb{F}$ is larger than $\mathbb{F}^B$, we may have the following extended martingale representation theorem: there exists unique $\sigma \in \mathbb{L}^2(\mathbb{F}, \mathbb{R}^d)$ such that

$$\xi = \mathbb{E}[\xi] + \int_0^T \sigma_t dB_t + N_T, \tag{2.5.6}$$

where $N \in \mathbb{L}^2(\mathbb{F})$ is a martingale orthogonal to B, in the sense that the quadric covariation $\langle N, B \rangle = 0$, or equivalently that NB is also a martingale. See, e.g., Protter [196]. ∎

2.6 The Girsanov Theorem

In this section we shall derive another probability measure from $\mathbb{P}$. To distinguish the two probability measures, we shall write $\mathbb{P}$ explicitly. Recall that B is a d-dimensional $(\mathbb{P}, \mathbb{F})$-Brownian motion. Let $\theta \in \mathbb{L}^2_{loc}(\mathbb{F}, \mathbb{P}, \mathbb{R}^d)$, and define

$$M_t^\theta := \exp\left(\int_0^t \theta_s \cdot dB_s - \frac{1}{2}\int_0^t |\theta_s|^2 ds\right), \text{ which implies } M_t^\theta = 1 + \int_0^t M_s^\theta \theta_s \cdot dB_s. \tag{2.6.1}$$

Then M^θ is a $\mathbb{P}$-local martingale. Moreover, we have

Lemma 2.6.1 *Assume* $\theta \in \mathbb{L}^\infty(\mathbb{F}, \mathbb{P}; \mathbb{R}^d)$. *Then* $M^\theta \in \bigcap_{1 \le p < \infty} \mathbb{L}^{\infty, p}(\mathbb{F}, \mathbb{P})$. *In particular,* M^θ *is a u.i.* $(\mathbb{P}, \mathbb{F})$-*martingale.*

Proof For simplicity again we assume $d = 1$. Denote $X_t := \int_0^t \theta_s dB_s$. Since $|\theta| \le C_0$ for some constant $C_0 > 0$, by the Burkholder-Davis-Gundy Inequality we see that $X \in \bigcap_{n \ge 1} \mathbb{L}^{\infty, n}(\mathbb{F}, \mathbb{P})$. For $n \ge 1$, applying Itô formula we have

$$X_t^{2n} = 2n \int_0^t X_s^{2n-1}\theta_s dB_s + n(2n-1)\int_0^t X_s^{2n-2}|\theta_s|^2 ds.$$

Then

$$\mathbb{E}^{\mathbb{P}}[|X_t|^{2n}] = n(2n-1)\mathbb{E}^{\mathbb{P}}\left[\int_0^t X_s^{2n-2}|\theta_s|^2 ds\right] \le \frac{1}{2}C_0^2(2n)(2n-1)\int_0^t \mathbb{E}^{\mathbb{P}}[|X_s|^{2n-2}]ds.$$

By induction one can easily check that

$$\mathbb{E}^{\mathbb{P}}[|X_t|^{2n}] \le \frac{C_0^{2n}}{2^n}t^{2n} \le \left(\frac{C_0 T}{\sqrt{2}}\right)^{2n}. \tag{2.6.2}$$

Then clearly $\mathbb{E}[|X_t|^n] \le C_1^n$, $n \ge 1$, for some constant $C_1 > 0$. Note that

$$|M_t^\theta|^p = \exp\left(pX_t - \frac{p}{2}\int_0^t |\theta_s|^2 ds\right) \le \exp(pX_t) = \sum_{n=0}^\infty \frac{p^n X_t^n}{n!}.$$

Then

$$\mathbb{E}^{\mathbb{P}}[|M_t^\theta|^p] \le \sum_{n=0}^{\infty} \frac{p^n}{n!} \mathbb{E}^{\mathbb{P}}[|X_t|^n] \le \sum_{n=0}^{\infty} \frac{p^n C_1^n}{n!} = e^{pC_1} < \infty.$$

Now it follows from the Bukholder-Davis-Gunday inequality that $M^\theta \in \bigcap_{1 \le p < \infty} \mathbb{L}^{\infty, p}(\mathbb{F}, \mathbb{P})$. ∎

Clearly $M^\theta > 0$, and the above lemma implies $\mathbb{E}^{\mathbb{P}}[M_T^\theta] = M_0^\theta = 1$. Then one can easily check that the following $\mathbb{P}^\theta$ is a probability measure equivalent to $\mathbb{P}$:

$$\mathbb{P}^\theta(A) := \mathbb{E}^{\mathbb{P}}[M_T^\theta \mathbf{1}_A], \quad \forall A \in \mathscr{F}_T, \quad \text{or equivalently,} \quad d\mathbb{P}^\theta := M_T^\theta d\mathbb{P}. \quad (2.6.3)$$

We have the following lemma whose proof is left to the exercise.

Lemma 2.6.2 *Let* $\xi \in \mathbb{L}^0(\mathscr{F}_T)$. *Then* $\mathbb{E}^{\mathbb{P}^\theta}[|\xi|] < \infty$ *if and only if* $\mathbb{E}^{\mathbb{P}}[M_T^\theta |\xi|] < \infty$. *Moreover,*

$$\mathbb{E}^{\mathbb{P}^\theta}[\xi] = \mathbb{E}^{\mathbb{P}}[M_T^\theta \xi].$$

The next result is crucial.

Lemma 2.6.3 *Let* $X \in \mathbb{L}^0(\mathbb{F})$ *such that* $\mathbb{E}^{\mathbb{P}^\theta}[|X_t|] < \infty$ *for each t. Then X is a* $\mathbb{P}^\theta$*-martingale if and only if* $M^\theta X$ *is a* $\mathbb{P}$*-martingale. In particular,* $(M^\theta)^{-1}$ *is a* $\mathbb{P}^\theta$*-martingale.*

Proof First, by Lemma 2.6.2 we see that $\mathbb{E}^{\mathbb{P}^\theta}[|X_t|] < \infty$ implies

$$\mathbb{E}^{\mathbb{P}}[M_t^\theta |X_t|] = \mathbb{E}^{\mathbb{P}}\Big[\mathbb{E}^{\mathbb{P}}[M_T^\theta |\mathscr{F}_t]|X_t|\Big] = \mathbb{E}^{\mathbb{P}}[M_T^\theta |X_t|] = \mathbb{E}^{\mathbb{P}^\theta}[|X_t|] < \infty.$$

We claim that, for any $\xi \in \mathbb{L}^1(\mathscr{F}_T, \mathbb{P}^\theta)$,

$$\mathbb{E}^{\mathbb{P}^\theta}[\xi|\mathscr{F}_t] = (M_t^\theta)^{-1} \mathbb{E}^{\mathbb{P}}[M_T^\theta \xi|\mathscr{F}_t], \quad (2.6.4)$$

Notice that X is a $\mathbb{P}^\theta$-martingale if and only if $X_t = \mathbb{E}^{\mathbb{P}^\theta}[X_T|\mathscr{F}_t]$. By (2.6.4), this is equivalent to $M_t^\theta X_t = \mathbb{E}^{\mathbb{P}}[M_T^\theta \xi|\mathscr{F}_t]$, which amounts to saying that $M^\theta X$ is a $\mathbb{P}$-martingale.

We now prove (2.6.4). For any $\eta \in \mathbb{L}^\infty(\mathscr{F}_t, \mathbb{P}) = \mathbb{L}^\infty(\mathscr{F}_t, \mathbb{P}^\theta)$, applying Lemma 2.6.2 twice and noting that M^θ is a $\mathbb{P}$-martingale we have

$$\mathbb{E}^{\mathbb{P}^\theta}\Big[(M_t^\theta)^{-1} \mathbb{E}^{\mathbb{P}}[M_T^\theta \xi|\mathscr{F}_t]\eta\Big] = \mathbb{E}^{\mathbb{P}}\Big[M_T^\theta (M_t^\theta)^{-1} \mathbb{E}^{\mathbb{P}}[M_T^\theta \xi|\mathscr{F}_t]\eta\Big]$$

$$= \mathbb{E}^{\mathbb{P}}\Big[\mathbb{E}^{\mathbb{P}}[M_T^\theta \xi|\mathscr{F}_t]\eta\Big] = \mathbb{E}^{\mathbb{P}}\Big[M_T^\theta \xi\eta\Big] = \mathbb{E}^{\mathbb{P}^\theta}\Big[\xi\eta\Big],$$

which implies (2.6.4) immediately. ∎

We now prove the main result of this section.

Theorem 2.6.4 *Let* $\theta \in \mathbb{L}^\infty(\mathbb{F}, \mathbb{P}, \mathbb{R}^d)$. *The following* B^θ *is a* $(\mathbb{P}^\theta, \mathbb{F})$-*Brownian motion:*

$$B_t^\theta := B_t - \int_0^t \theta_s ds. \tag{2.6.5}$$

Proof For simplicity we assume $d = 1$. Apply Itô formula, we have

$$d(M_t^\theta B_t^\theta) = M_t^\theta[\theta_t B_t^\theta + 1]dB_t; \quad d\Big(M_t^\theta[|B_t^\theta|^2 - t]\Big) = M_t^\theta\Big[2B_t^\theta + (|B_t^\theta|^2 - t)\theta_t\Big]dB_t.$$

By Lemmas 2.6.1 and 2.6.3 we see that B^θ and $|B_t^\theta|^2 - t$ are $\mathbb{P}^\theta$-martingales.

To show that B^θ is a $(\mathbb{P}^\theta, \mathbb{F})$-Brownian motion, we follow the arguments of the so-called Levy's characterization theorem. Fix $0 \le s < T$. By the martingale properties we have

$$\mathbb{E}^{\mathbb{P}^\theta}[B_{s,t}^\theta | \mathscr{F}_s] = 0, \quad \mathbb{E}^{\mathbb{P}^\theta}[(B_{s,t}^\theta)^2 | \mathscr{F}_s] = t - s, \quad s \le t \le T. \tag{2.6.6}$$

Denote $N_t^\theta := (M_s^\theta)^{-1} M_t^\theta$. For each $n \ge 2$, applying Itô formula we have

$$d\Big[N_t^\theta(B_{s,t}^\theta)^n\Big] = [\cdots]dB_t + \frac{n(n-1)}{2}N_t^\theta(B_{s,t}^\theta)^{n-2}dt.$$

Then

$$\mathbb{E}^{\mathbb{P}^\theta}\Big[(B_{s,t}^\theta)^n\Big|\mathscr{F}_s\Big] = \mathbb{E}^{\mathbb{P}}\Big[N_t^\theta(B_{s,t}^\theta)^n\Big|\mathscr{F}_s\Big] = \frac{n(n-1)}{2}\int_s^t \mathbb{E}^{\mathbb{P}}\Big[N_r^\theta(B_{s,r}^\theta)^{n-2}\Big|\mathscr{F}_s\Big]dr$$

$$= \frac{n(n-1)}{2}\int_s^t \mathbb{E}^{\mathbb{P}^\theta}\Big[(B_{s,r}^\theta)^{n-2}\Big|\mathscr{F}_s\Big]dr.$$

By induction one can easily derive from (2.6.6) that

$$\mathbb{E}^{\mathbb{P}^\theta}[(B_{s,t}^\theta)^{2n+1}|\mathscr{F}_s] = 0, \quad \mathbb{E}^{\mathbb{P}^\theta}[(B_{s,t}^\theta)^{2n}|\mathscr{F}_s] = \frac{(2n)!}{2^n n!}(t-s)^n. \tag{2.6.7}$$

Then, for any $\alpha \in \mathbb{R}$,

$$\mathbb{E}^{\mathbb{P}^\theta}[e^{\alpha B_{s,t}^\theta}|\mathscr{F}_s] = \sum_{n=0}^\infty \frac{\alpha^n}{n!}\mathbb{E}^{\mathbb{P}^\theta}[(B_{s,t}^\theta)^n|\mathscr{F}_s] = \sum_{n=0}^\infty \frac{\alpha^n(t-s)^n}{2^n n!} = e^{\frac{\alpha(t-s)}{2}}. \tag{2.6.8}$$

This implies that, under $\mathbb{P}^\theta$, $B_{s,t}^\theta$ is independent of $\mathscr{F}_s$ and has distribution $N(0, t-s)$. That is, B^θ is a $(\mathbb{P}^\theta, \mathbb{F})$-Brownian motion. ∎

Remark 2.6.5 The above theorem is a special case of the Levy's martingale characterization of Brownian motion (see, e.g., Karatzas & Shreve [117]):

Let M be a continuous process with $M_0 = 0$ and denote $N_t := M_t^2 - t$.
Then M is a Brownian motion if and only if both M and N are martingales.

$$(2.6.9)$$

The result follows similar arguments, but involves the general martingale theory, and we omit it. ∎

We conclude the section with the martingale representation theorem for $(\mathbb{P}^\theta, B^\theta)$. For $\xi \in \mathbb{L}^2(\mathscr{F}_T^{B^\theta}, \mathbb{P}^\theta)$, the result follows from the standard martingale representation Theorem 2.5.2. For $\xi \in \mathbb{L}^2(\mathscr{F}_T, \mathbb{P}^\theta)$, as seen in Example 2.5.1, the result is in general not true. The nontrivial interesting case is $\xi \in \mathbb{L}^2(\mathscr{F}_T^B, \mathbb{P}^\theta)$. We note that

for $\theta \in \mathbb{L}^\infty(\mathbb{F}^B, \mathbb{P})$, we have $\mathbb{F}^{B^\theta} \subset \mathbb{F}^B$, but in general $\mathbb{F}^{B^\theta} \neq \mathbb{F}^B$. (2.6.10)

A counterexample for $\mathbb{F}^{B^\theta} \neq \mathbb{F}^B$ is provided by Tsirelson [229]. Nevertheless, we still have

Theorem 2.6.6 *Assume* $\theta \in \mathbb{L}^\infty(\mathbb{F}^B, \mathbb{P}, \mathbb{R}^d)$. *Then for any* $\xi \in \mathbb{L}^2(\mathscr{F}_T^B, \mathbb{P}^\theta)$, *there exists* ($\mathbb{P}^\theta$-*a.s.*) *unique* $\sigma \in \mathbb{L}^2(\mathbb{F}_T^B, \mathbb{P}^\theta, \mathbb{R}^d)$ *such that*

$$\xi = \mathbb{E}^{\mathbb{P}^\theta}[\xi] + \int_0^T \sigma_t \cdot dB_t^\theta.$$

We remark that in general we cannot expect σ to be $\mathbb{F}^{B^\theta}$-measurable.

Proof Assume for simplicity that $d = 1$. By the truncation arguments in Step 5 of Theorem 2.5.2, we may assume without loss of generality that ξ is bounded. Denote $X_t := \mathbb{E}^{\mathbb{P}^\theta}[\xi | \mathscr{F}_t^B]$. Then X is a bounded $(\mathbb{P}^\theta, \mathbb{F}^B)$-martingale. By Lemmas 2.6.1 and 2.6.3, $M^\theta X$ is a $(\mathbb{P}, \mathbb{F}^B)$-square integrable martingale. By Theorem 2.5.2, there exists $\tilde\sigma \in \mathbb{L}^2(\mathbb{F}^B, \mathbb{P})$ such that

$$d(M_t^\theta X_t) = \tilde\sigma_t dB_t.$$

Apply Itô formula, we have

$$d(M_t^\theta)^{-1} = -(M_t^\theta)^{-2}M_t^\theta \theta_t dB_t + (M_t^\theta)^{-3}|M_t^\theta \theta_t|^2 dt = (M_t^\theta)^{-1}\Big[-\theta_t dB_t + |\theta_t|^2 dt\Big];$$

$$dX_t = d\Big[(M_t^\theta)^{-1}(M_t^\theta X_t)\Big]$$

$$= (M_t^\theta)^{-1}\tilde\sigma_t dB_t + M_t^\theta X_t (M_t^\theta)^{-1}\Big[-\theta_t dB_t + |\theta_t|^2 dt\Big] - \tilde\sigma_t (M_t^\theta)^{-1}\theta_t dt$$

$$= \Big[(M_t^\theta)^{-1}\tilde\sigma_t - X_t\theta_t\Big]dB_t^\theta.$$

This proves the result with $\sigma_t := (M_t^\theta)^{-1}\tilde\sigma_t - X_t\theta_t$. ∎

Remark 2.6.7

(i) In the option pricing theory in Section 2.8, Girsanov theorem is a convenient tool to find the so-called risk neutral probability measure.

(ii) In stochastic control theory, see Section 4.5.2, Girsanov theorem is a powerful tool to stochastic optimization problem with drift control in weak formulation.

(iii) Note that $\mathbb{P}^\theta$ is equivalent to $\mathbb{P}$. For stochastic optimization problem with diffusion control in weak formulation, the involved probability measures are typically mutually singular. Then Girsanov theorem is not enough. We shall introduce new tools in Part III to address these problems. ∎

Remark 2.6.8 The Girsanov theorem holds true under weaker assumptions on θ, see Theorem 7.2.3 and Problem 7.5.2 below. ∎

2.7 The Doob-Meyer Decomposition

The result in this section actually holds for general setting and under much weaker conditions, see, e.g., Karatzas & Shreve [117]. However, for simplicity we shall only present a special case.

Theorem 2.7.1 *Assume* $\mathbb{F} = \mathbb{F}^B$ *and let* $X \in \mathbb{S}^2(\mathbb{F})$ *be a continuous submartingale. Then there exists unique decomposition* $X_t = X_0 + \int_0^t Z_s \cdot dB_s + K_t$, *where* $Z \in \mathbb{L}^2(\mathbb{F}, \mathbb{R}^d)$, $K \in \mathbb{I}^2(\mathbb{F})$ *with* $K_0 = 0$. *Moreover, there exists a constant* $C > 0$, *depending only on d, such that*

$$\mathbb{E}\left[\int_0^T |Z_t|^2 dt + |K_T|^2\right] \le C\mathbb{E}[|X_T^*|^2]. \tag{2.7.1}$$

Proof For simplicity we assume $d = 1$. We first prove the uniqueness. Assume $Z' \in \mathbb{L}^2(\mathbb{F})$ and $K' \in \mathbb{I}^2(\mathbb{F})$ with $K_0' = 0$ provide another decomposition. Then, denoting $\Delta Z := Z - Z'$, $\Delta K := K - K'$,

$$\int_0^t \Delta Z_s dB_s = -\Delta K_t, \quad 0 \le t \le T.$$

For each $n \ge 1$, denote $t_i := t_i^n := \frac{i}{n}T$, $i = 0, \cdots, n$. Then, noting that K, K' are increasing,

$$\mathbb{E}\left[\int_0^T |\Delta Z_t|^2 dt\right] = \sum_{i=0}^{n-1} \mathbb{E}\left[\left(\int_{t_i}^{t_{i+1}} \Delta Z_t dB_t\right)^2\right] = \sum_{i=0}^{n-1} \mathbb{E}\left[|\Delta K_{t_{i+1}} - \Delta K_{t_i}|^2\right]$$

$$= \sum_{i=0}^{n-1} \mathbb{E}\left[|K_{t_i,t_{i+1}} - K_{t_i,t_{i+1}}'|^2\right] \le \sum_{i=0}^{n-1} \mathbb{E}\left[|K_{t_i,t_{i+1}} + K_{t_i,t_{i+1}}'|^2\right]$$

$$\le \mathbb{E}\left[\sup_{0 \le i \le n-1} [K_{t_i,t_{i+1}} + K_{t_i,t_{i+1}}'][K_T + K_T']\right].$$

Since K, K' are continuous, send $n \to \infty$ and apply the Dominated Convergence Theorem, we obtain $\mathbb{E}\left[\int_0^T |\Delta Z_t|^2 dt\right] = 0$. Then $Z = Z'$, which implies further that $K = K'$.

We now prove the existence. Let t_i be as above, $M_{t_0}^n := K_{t_0}^n := 0$, and for $i = 0, \cdots, n-1$,

$$M_{t_{i+1}}^n := M_{t_i}^n + X_{t_{i+1}} - \mathbb{E}_{t_i}[X_{t_{i+1}}], \quad K_{t_{i+1}}^n := K_{t_i}^n + \mathbb{E}_{t_i}[X_{t_{i+1}}] - X_{t_i}. \quad (2.7.2)$$

Then clearly $M_{t_i}^n$ is an $\{\mathscr{F}_{t_i}\}_{0 \le i \le n}$-martingale and, since X is a submartingale, $K_{t_i}^n \in \mathbb{L}^0(\mathscr{F}_{t_{i-1}})$ is increasing in i. Note that

$$\mathbb{E}\left[|X_{t_{i+1}}|^2 - |X_{t_i}|^2\right] = \mathbb{E}\left[|M_{t_i,t_{i+1}}^n + K_{t_i,t_{i+1}}^n + X_{t_i}|^2 - |X_{t_i}|^2\right]$$

$$= \mathbb{E}\left[|M_{t_i,t_{i+1}}^n|^2 + |K_{t_i,t_{i+1}}^n|^2 + 2X_{t_i}K_{t_i,t_{i+1}}^n\right]$$

$$\ge \mathbb{E}\left[|M_{t_i,t_{i+1}}^n|^2 - 2X_T^* K_{t_i,t_{i+1}}^n\right].$$

This implies, noting that M^n is a martingale,

$$\mathbb{E}[|M_T^n|^2] = \sum_{i=0}^{n-1} \mathbb{E}\left[|M_{t_i,t_{i+1}}^n|^2\right] \le \mathbb{E}\left[|X_T|^2 - |X_0|^2 + 2X_T^* K_T^n\right]$$

$$= \mathbb{E}\left[|X_T|^2 - |X_0|^2 + 2X_T^*[X_T - X_0 - M_T^n]\right] \le \mathbb{E}\left[C|X_T^*|^2 + \frac{1}{2}|M_T^n|^2\right].$$

Then

$$\mathbb{E}[|M_T^n|^2] \le C\mathbb{E}[|X_T^*|^2], \text{ which implies further that } \mathbb{E}[|K_T^n|^2] \le C\mathbb{E}[|X_T^*|^2]. \quad (2.7.3)$$

Now by the martingale representation Theorem 2.5.2, for each n there exists $Z^n \in \mathbb{L}^2(\mathbb{F})$ such that $M_T^n = \int_0^T Z_t^n dB_t$. Denote $K_t^n := \sum_{i \ge 0} K_{t_i}^n \mathbf{1}_{[t_i,t_{i+1})}$. By (2.7.3) and applying Theorem 1.3.7, we may assume without loss of generality that (Z^n, K^n) converges weakly to certain $(Z, K) \in \mathbb{L}^2(\mathbb{F})$. Applying Problem 2.10.11 (ii) and (iii) we see that M^n converges weakly to $M. := \int_0^{\cdot} Z_s dB_s$ and $\mathbb{E}[\int_0^T |Z_t|^2 dt] \le C\mathbb{E}[|X_T^*|^2]$. Moreover, since $X_{t_i} = X_0 + M_{t_i}^n + K_{t_i}^n$ and X is continuous. By Problem 2.10.11 (i) it is clear that $X_t = X_0 + M_t + K_t$. In particular, this implies that K is continuous and $\mathbb{E}[|K_T|^2] \le C\mathbb{E}[|X_T^*|^2]$.

It remains to show that K is increasing. Note that each K^n is increasing. Let $\hat{K}^n$ be the convex combination of K^n as in Theorem 1.3.8, then $\hat{K}^n$ is also increasing and $\lim_{n\to\infty} \mathbb{E}\left[\int_0^T |\hat{K}_t^n - K_t|^2 dt\right] = 0$. By otherwise choosing a further subsequence, we have $\int_0^T |\hat{K}_t^n - K_t|^2 dt \to 0$, a.s. This clearly implies that K is increasing, a.s. ∎

2.8 A Financial Application

Consider the Black-Scholes model on a filtered probability space $(\Omega, \mathscr{F}, \mathbb{F}, \mathbb{P})$ with a one-dimensional $(\mathbb{P}, \mathbb{F})$-Brownian motion B. The financial market consists of two assets: a bank account (or bond) with constant interest rate r (continuously compounded), and a stock with price S_t:

$$dS_t = S_t\big[\mu dt + \sigma dB_t\big], \text{ or equivalently } S_t = S_0 \exp\Big(\sigma B_t + (\mu - \frac{1}{2}\sigma^2)t\Big), \quad (2.8.1)$$

where the constants μ and $\sigma > 0$ stand for the appreciation and volatility of the stock, respectively. Let $\xi \in \mathbb{L}^2(\mathscr{F}_T)$ be a European option with maturity time T, namely at time T the option is worth ξ. Now our goal is to find the *fair* price Y_0 of ξ at time 0, or more generally the fair price Y_t at time $t \in [0, T]$. Clearly $Y_T = \xi$.

2.8.1 Pricing via Risk Neutral Measure

We first note that, due to the presence of the interest, we should consider the discounted prices:

$$\bar{S}_t = e^{-rt}S_t, \quad \bar{Y}_t = e^{-rt}Y_t. \tag{2.8.2}$$

One natural guess for the option price is that

$$Y_0 = \bar{Y}_0 = \mathbb{E}^{\mathbb{P}}[\bar{Y}_T] = \mathbb{E}^{\mathbb{P}}[e^{-rT}\xi]. \tag{2.8.3}$$

However, the above guess cannot be true in general. Indeed, if we set $\xi = S_T$, then following (2.8.3) we should have $\bar{Y}_0 = \mathbb{E}^{\mathbb{P}}[\bar{S}_T]$, or more generally $\bar{Y}_t = \mathbb{E}^{\mathbb{P}}[\bar{S}_T|\mathscr{F}_t]$. That is, $\bar{Y}$ should be a $\mathbb{P}$-martingale. However, obviously in this case we should have $Y_t = S_t$ and thus $\bar{Y}_t = \bar{S}_t$. Applying Itô formula we have

$$d\bar{S}_t = \bar{S}_t\big[(\mu - r)dt + \sigma dB_t\big]. \tag{2.8.4}$$

Then $\bar{S}$ is not a $\mathbb{P}$-martingale unless $\mu = r$.

If we want to use price formula in the form of (2.8.3), from the above discussion it seems necessary that $\bar{S}$ needs to be martingale. We thus introduce the following concept.

Definition 2.8.1 *A probability measure $\bar{\mathbb{P}}$ on Ω is called a risk neutral measure, also called martingale measure, if*

 (i) *$\bar{\mathbb{P}}$ is equivalent to $\mathbb{P}$;*
 (ii) *$\bar{S}$ is a $\bar{\mathbb{P}}$-martingale.*

In contrast to $\bar{\mathbb{P}}$, we call the original $\mathbb{P}$ the market measure.

To construct $\overline{\mathbb{P}}$, our main tool is the Girsanov theorem. By (2.8.4), it is clear that

$$d\overline{S}_t = \overline{S}_t \sigma dB_t^{-\theta}, \tag{2.8.5}$$

where $\theta := \frac{\mu-r}{\sigma}$ is the *Sharpe ratio* of the stock, and $dB_t^{-\theta} := dB_t + \theta dt$. Consider the $\mathbb{P}^{-\theta}$ in Section 2.6. Then $\mathbb{P}^{-\theta} \sim \mathbb{P}$ and $B^{-\theta}$ is a $\mathbb{P}^{-\theta}$-Brownian motion. Now it follows from (2.8.5) that $\overline{S}$ is a $\mathbb{P}^{-\theta}$-martingale, and thus $\overline{\mathbb{P}} = \mathbb{P}^{-\theta}$ is a risk neutral measure.

We will justify in the next subsection that $\overline{Y}$ should also be a $\overline{\mathbb{P}}$-martingale. Then we obtain the following pricing formula, in the spirit of (2.8.3) but under the risk neutral measure $\overline{\mathbb{P}}$ instead of the market measure $\mathbb{P}$:

$$\overline{Y}_t = \mathbb{E}^{\overline{\mathbb{P}}}[e^{-rT}\xi|\mathscr{F}_t], \quad \text{or equivalently,} \quad Y_t = \mathbb{E}^{\overline{\mathbb{P}}}[e^{-r(T-t)}\xi|\mathscr{F}_t]. \tag{2.8.6}$$

2.8.2   Hedging the Option

Assume an investor invests in the market with portfolio $(\lambda_t, h_t)_{0 \le t \le T}$. The corresponding portfolio value is:

$$V_t := \lambda_t e^{rt} + h_t S_t. \tag{2.8.7}$$

Note that $\mathbb{F}$ stands for the information flow, thus it is natural to require (λ, h) to be $\mathbb{F}$-measurable. Moreover, we shall assume the investor invests only in this market, which induces the following concept:

Definition 2.8.2 *An $\mathbb{F}$-measurable portfolio (λ, h) is called self-financing if, in addition to certain integrability conditions which we do not discuss in detail,*

$$dV_t = \lambda_t de^{rt} + h_t dS_t. \tag{2.8.8}$$

The fairness of the price is based on the following arbitrage free principle.

Definition 2.8.3

(i) We say a self-financing portfolio (λ, h) has arbitrage opportunity if

$$V_0 = 0, \quad V_T \ge 0, \quad \mathbb{P}\text{-a.s.,} \quad \text{and} \quad \mathbb{P}(V_T > 0) > 0. \tag{2.8.9}$$

(ii) We say the market consisting of the bond and stock is arbitrage free if there is no self-financing portfolio (λ, h) admitting arbitrage opportunity.

The following theorem is called the first fundamental theorem of mathematical finance, which holds true in much more general models.

Theorem 2.8.4 *The market is arbitrage free if and only if there exists a risk neutral measure* $\overline{\mathbb{P}}$.

By the previous subsection, the Black-Scholes market is arbitrage free. We remark that, since $\overline{\mathbb{P}}$ is equivalent to $\mathbb{P}$, so (2.8.9) holds under $\overline{\mathbb{P}}$ as well.

Now given an option ξ, let Y_t denote its market price. We may consider an extended market (e^{rt}, S_t, Y_t), and we can easily extend the concept of arbitrage free to this market.

Definition 2.8.5 *We say Y is a fair price, also called arbitrage free price, if the market (e^{rt}, S_t, Y_t) is arbitrage free.*

Definition 2.8.6 *Given ξ, we say a self-financing portfolio (λ, h) is a hedging portfolio of ξ if $V_T = \xi$, $\mathbb{P}$-a.s.*

Proposition 2.8.7 *If (λ, h) is a hedging portfolio of ξ, then $Y_t := V_t$ is the unique fair price.*

Proof The fairness of V involves the martingale properties and we leave the proof to interested readers. To illustrate the main idea, we prove only that, if $Y_0 > V_0$, then there will be arbitrage opportunity in the extended market (e^{rt}, S_t, Y_t). Indeed, in this case, consider the portfolio: $(\lambda_t + Y_0 - V_0, h_t, -1)$, with value

$$\tilde{V}_t := [\lambda_t + Y_0 - V_0]e^{rt} + h_t S_t - Y_t = V_t - Y_t + [Y_0 - V_0]e^{rt}.$$

Note that

$$d\tilde{V}_t = dV_t - dY_t + [Y_0 - V_0]de^{rt} = \lambda_t de^{rt} + h_t S_t - dY_t + [Y_0 - V_0]de^{rt}$$

$$= [\lambda_t + Y_0 - V_0]de^{rt} + h_t dS_t + (-1)dY_t.$$

That is, the portfolio is self-financing. Note that

$$\tilde{V}_0 = V_0 - Y_0 + [Y_0 - V_0]e^{r0} = 0;$$

$$\tilde{V}_T = V_T - Y_T + [Y_0 - V_0]e^{rT} = \xi - \xi + [Y_0 - V_0]e^{rT} = [Y_0 - V_0]e^{rT} > 0, \quad \mathbb{P}\text{-a.s.}$$

Then the portfolio $(\lambda_t + Y_0 - V_0, h_t, -1)$ has arbitrage opportunity.   ∎

We next find the hedging portfolio in the Black-Scholes model. Our main tool is the martingale representation theorem. Consider the discounted portfolio value $\overline{V}_t := e^{-rt}V_t$. By (2.8.8) and (2.8.5) we have

$$d\overline{V}_t = h_t d\overline{S}_t = h_t \overline{S}_t \sigma dB_t^{-\theta}. \tag{2.8.10}$$

That is, $\overline{V}$ is a $\overline{\mathbb{P}}$-martingale, where, again, $\overline{\mathbb{P}} := \mathbb{P}^{-\theta}$. Note that $\overline{V}_T = e^{-rT}\xi$. Assume

$$\xi \in \mathbb{L}^2(\mathscr{F}_T^B, \overline{\mathbb{P}}). \tag{2.8.11}$$

Then by the generalized martingale representation Theorem 2.6.6, there exists $Z \in$ $\mathbb{L}^2(\mathbb{F}^B, \overline{\mathbb{P}})$ such that

$$e^{-rT}\xi = \mathbb{E}^{\overline{\mathbb{P}}}[e^{-rT}\xi] + \int_0^T Z_t dB_t^{-\theta}. \tag{2.8.12}$$

This induces the hedging portfolio (and the price) immediately:

$$\overline{V}_t := \mathbb{E}^{\overline{\mathbb{P}}}[e^{-rT}\xi|\mathscr{F}_t^B], \quad h_t := \frac{Z_t}{\overline{S}_t\sigma}, \quad \lambda_t := \overline{V}_t - h_t\overline{S}_t = \overline{V}_t - \frac{Z_t}{\sigma}. \tag{2.8.13}$$

The hedging portfolio is closely related to the important notion of completeness of the market.

Definition 2.8.8 *The market is called complete if all option* $\xi \in \mathbb{L}^0(\mathscr{F}_T)$ *satisfying appropriate integrability condition can be hedged.*
From the above analysis we see that the Black-Scholes market is complete if

$$\mathbb{F} = \mathbb{F}^B. \tag{2.8.14}$$

We conclude this subsection with the second fundamental theorem of mathematical finance, which also holds true in much more general models.

Theorem 2.8.9 *Assume the market is arbitrage free. Then the market is complete if and only if the risk neutral measure* $\overline{\mathbb{P}}$ *is unique.*

2.8.3   Some Further Discussion

We first note that one rationale of using Brownian motion to model the stock price lies in the central limit Theorem 1.1.2. As a basic principle in finance, the supply and demand have great impact on the price. That is, the buy orders will push the stock price up, while the sell orders will push the stock price down. Assume there are many small investors in the market and they place their order independently. Then by the central limit Theorem 1.1.2, the accumulative price impact of their trading induces the normal distribution. In the rest of this subsection we discuss two subtle issues.

First, as we see in (2.8.14), even for Black-Scholes model, the completeness relies on the information setting. In a more general model, $\mathbb{F}$, $\mathbb{F}^B$, and $\mathbb{F}^S$ can be all different. The investor's portfolio (λ, h) has to be measurable with respect to the filtration the investor actually observes. While in different situation the real information can be different, typically the investor indeed observes S and thus $\mathbb{F}^S$ is accessible to the investor. As discussed in the previous paragraph, observing B essentially means the investor observes numerous other (small) investors (and possibly other random factors). This is not that natural in practice. Moreover, note that in Theorem 1.1.2, the convergence is in distribution sense, not in pointwise sense. Then even one observes a path of the portfolios of all small investors, one typically does not know a corresponding path of B. So in this sense, at least in some applications, it makes more sense to use $\mathbb{F}^S$ than to use $\mathbb{F}^B$. This implies that in these applications one should use weak formulation, as we will do in Part III. In Parts I and II, however, we will nevertheless use strong formulation, namely use $\mathbb{F}^B$. This could be reasonable in some other applications, and still makes perfect sense in this particular application when $\mathbb{F}^S = \mathbb{F}^B$, which is true in, e.g., Black-Scholes model.

The next is the pathwise stochastic integration. Recall that for an elementary process $\sigma \in \mathbb{L}_0^2(\mathbb{F})$, the Itô integral $(\int_0^T \sigma_t dB_t)(\omega) = \int_0^T \sigma_t(\omega) dB_t(\omega)$ is defined in pathwise manner. For general $\sigma \in \mathbb{L}^2(\mathbb{F})$, however, $\int_0^T \sigma_t dB_t$ is defined as the $\mathbb{L}^2$-limit of $\int_0^T \sigma_t^n dB_t$, where $\sigma^n \in \mathbb{L}_0^2(\mathbb{F})$ is an approximation of σ. As a consequence, $\int_0^T \sigma_t dB_t$ is defined only in a.s. sense, with the null set arbitrary and up to the particular version we want to choose. In particular, for any given ω, since $\mathbb{P}(\{\omega\}) = 0$, the value $(\int_0^T \sigma_t dB_t)(\omega)$ is arbitrary. In other words, in our application, assume we have observed a path $S_t(\omega)$ and decided a path $h_t(\omega)$, the value of $\int_0^T h_t dS_t$ at this particular observed ω is actually arbitrary. This is of course not desirable. We shall mention that in real practice, the portfolio h should be discrete, and thus the issue does not exist. But nevertheless, theoretically this is a subtle issue we face in such applications.

One way to get around of this difficulty is to use pathwise integration. Assume, under certain conditions, $\lim_{n \to \infty} \int_0^T \sigma_t^n dB_t = \int_0^T \sigma_t dB_t$ in a.s. sense, with a common exceptional null set E_0 independent of our choice of the approximation σ^n. Then we may fix a version: $\left(\int_0^T \sigma_t dB_t\right)(\omega) := \lim_{n \to \infty} \left(\int_0^T \sigma_t^n dB_t\right)(\omega) \mathbf{1}_{E_0^c}(\omega)$. If we are lucky that the observed path ω is not in E_0, then we may use the limit of $\left(\int_0^T \sigma_t^n dB_t\right)(\omega)$ as the value of $\left(\int_0^T \sigma_t dB_t\right)(\omega)$. Another powerful tool to study pathwise analysis is the rough path theory, which approximates $B(\omega)$ by smooth paths. We have some discussion along this line in Problem 2.10.14.

2.9 Bibliographical Notes

The materials in this section are very standard in the literature. We refer to the classical reference Karatzas & Shreve [117] for a comprehensive presentation of properties of Brownian motions, some of which are more general or deeper than the results here. We also refer to Revuz & Yor [206] for a more general continuous martingale theory, and Protter [196] for a general semimartingale theory, including semimartingales with jumps.

For the financial application in Section 2.8, Shreve [209, 210] provides an excellent exposition. For the pathwise stochastic integration, we refer to Wong & Zakai [236, 237], Bichteler [17], Follmer [91], Willinger & Taqqu [235], Karandikar [119], and Nutz [160]. The rough path theory was initiated by Lyons [140]. We refer interested readers to the book Friz & Hairer [94]. We also note that the pathwise stochastic integration is closely related to the quasi-sure stochastic integration in Section 12.1.1.

2.10 Exercises

Problem 2.10.1 Prove Propositions 2.1.2 and 2.1.3.   ∎

Problem 2.10.2

(i) Let $X_t := \int_0^t b_s ds$ for some $b \in \mathbb{L}_{loc}^1(\mathbb{F})$ and $0 \le s < t \le T$. Show that $\bigvee_s^t(X) = \int_s^t |b_r| dr$, a.s.

(ii) Let X be as in Definition 2.1.5. Show that $\langle X \rangle$ is increasing in t, a.s. That is, $\langle X \rangle_t - \langle X \rangle_s \in \mathbb{S}^d$ is nonnegatively definite for all $0 \le s < t \le T$.

(iii) Let $x_i \in \mathbb{R}$, $b^i \in \mathbb{L}^{1,2}(\mathbb{F}, \mathbb{R})$, $\sigma^i \in \mathbb{L}^2(\mathbb{F}, \mathbb{R}^d)$, and $X_t^i := x_i + \int_0^t b_s^i ds + \int_0^t \sigma_s^i \cdot dB_s$, $i = 1, 2$. For any $\pi : 0 = t_0 < \cdots < t_n = T$, denote

$$\langle X^1, X^2 \rangle_T^\pi := \sum_{i=0}^{n-1} X_{t_i, t_{i+1}}^1 X_{t_i, t_{i+1}}^2.$$

Show that $\langle X^1, X^2 \rangle_T^\pi \to \int_0^T \sigma_t^1 \cdot \sigma_t^2 dt$ in $\mathbb{L}^1(\mathscr{F}_T)$, as $|\pi| \to 0$.   ∎

Problem 2.10.3 This problem concerns the general Doob's maximum inequality, extending Lemma 2.2.4. Let $X \in \mathbb{L}^1(\mathbb{F})$ be a right continuous nonnegative submartingale. Then

$$\mathbb{P}(X_T^* \ge \lambda) \le \frac{1}{\lambda^p} \mathbb{E}\Big[|X_T|^p \mathbf{1}_{\{X_T^* \ge \lambda\}}\Big], \quad \text{for all } \lambda > 0, p \ge 1;$$

$$\mathbb{E}[|X_T^*|^p] \le (\tfrac{p}{p-1})^p \mathbb{E}[|X_T|^p], \text{ for all } p > 1; \quad \text{and} \quad \mathbb{E}[X_T^*] \le \tfrac{e}{e-1} \mathbb{E}\Big[1 + X_T\big(\ln(X_T)\big)^+\Big].$$

We remark that the $|M|$ in Lemma 2.2.4 is a nonnegative submartingale, thanks to Jensen's inequality. Thus Lemma 2.2.4 is indeed a special case here.   ∎

Problem 2.10.4 Prove the extended Itô formula Theorem 2.3.4.   ∎

Problem 2.10.5 Let $\eta_n \in \mathbb{L}^2(\mathscr{F}_0)$, $\sigma^n \in \mathbb{L}^2(\mathbb{F}, \mathbb{R}^d)$, and denote $\xi_n := \eta_n + \int_0^T \sigma_t^n \cdot dB_t$, $n \ge 1$. Assume $\lim_{n\to\infty} \mathbb{E}[|\xi_n - \xi|^2] = 0$ for some $\xi \in \mathbb{L}^2(\mathscr{F}_T)$. Then there exists unique $\sigma \in \mathbb{L}^2(\mathbb{F}, \mathbb{R}^d)$ such that $\xi = \mathbb{E}[\xi|\mathscr{F}_0] + \int_0^T \sigma_t \cdot dB_t$, and $\lim_{n\to\infty} \mathbb{E}\Big[|\eta_n - E[\xi|\mathscr{F}_0]|^2 + \int_0^T |\sigma_t^n - \sigma_t|^2 dt\Big] = 0$.   ∎

Problem 2.10.6 Let $\sigma \in \mathbb{L}^2(\mathbb{F}^B, \mathbb{S}^d)$ such that $\sigma > 0$, and $X_t := \int_0^t \sigma_s dB_s$. Show that the augmented filtrations of X and B are equal: $\overline{\mathbb{F}^X}^\mathbb{P} = \overline{\mathbb{F}^B}^\mathbb{P}$.   ∎

Problem 2.10.7 Let $p, q \in [1, \infty]$ be conjugates.

(i) Assume $X \in \mathbb{L}^{\infty,p}(\mathbb{F})$, $Y \in \mathbb{L}^{2,q}(\mathbb{F})$ with appropriate dimensions so that XY takes values in $\mathbb{R}^d$. Show that $M_t := \int_0^t (X_s Y_s) \cdot dB_s$ is a u.i. $\mathbb{F}$-martingale.
(ii) Find a counterexample such that $X \in \mathbb{L}^{2,p}(\mathbb{F})$, $Y \in \mathbb{L}^{2,q}(\mathbb{F})$, but $M_t := \int_0^t (X_s Y_s) \cdot dB_s$ is not uniformly integrable.
(iii) Find a counterexample such that M is a local martingale, but not a martingale.
(iv) Find a counterexample such that M is a martingale, but not uniformly integrable.

Note that the M in (ii) is a local martingale, so it serves as a counterexample either for (iii) or for (iv).　∎

Problem 2.10.8 Let $d = 1$ (for simplicity). Prove the following stochastic Fubini theorem:

$$\int_0^T \left[u_t \int_0^t v_s ds \right] dB_t = \int_0^T \left[v_s \int_s^T u_t dB_t \right] ds, \quad \forall u, v \in \mathbb{L}^\infty(\mathbb{F}, \mathbb{R}).$$

We remark that, unless u is deterministic, the following result is not true:

$$\int_0^T \left[u_t \int_0^t v_s dB_s \right] dt = \int_0^T \left[v_s \int_s^T u_t dt \right] dB_s.$$

In fact, the stochastic integrand in the right side above is in general not $\mathbb{F}$-adapted.　∎

Problem 2.10.9 This problem concerns general martingale theory. Let $d = 1$, M a continuous $\mathbb{F}$-martingale with $\mathbb{E}[|M_T|^2] < \infty$, $K \in \mathbb{I}^1(\mathbb{F})$ with $K_0 = 0$, and $M^2 - K$ is also a martingale.

(i) For any bounded $\sigma = \sum_{i=0}^{n-1} \sigma_{t_i} 1_{[t_i, t_{i+1})} \in \mathbb{L}^0(\mathbb{F})$, denote $\int_0^T \sigma_s dM_s := \sum_{i=0}^{n-1} \sigma_{t_i} M_{t_i, t_{i+1}}$. Show that

$$\mathbb{E}\left[\left| \int_0^T \sigma_s dM_s \right|^2 \right] = \mathbb{E}\left[\int_0^T |\sigma_s|^2 dK_s \right].$$

(ii) For any $\sigma \in \mathbb{L}^0(\mathbb{F})$ such that $\mathbb{E}\left[\int_0^T |\sigma_s|^2 dK_s \right] < \infty$, show that there exist bounded elementary processes $\sigma^n \in \mathbb{L}^0(\mathbb{F})$ such that $\lim_{n\to\infty} \mathbb{E}\left[\int_0^T |\sigma_s^n - \sigma_s|^2 dK_s \right] = 0$.
(iii) For σ and σ^n as in (ii), show that $\int_0^T \sigma^n dM_s$ converges in $\mathbb{L}^2$, and the limit is independent of the choices of σ^n. Thus we may define $\int_0^T \sigma_s dM_s := \lim_{n\to\infty} \int_0^T \sigma_s^n dM_s$.
(iv) For σ as in (iii), define $Y_t := \int_0^t \sigma_s dM_s$ similarly. Show that Y is still an $\mathbb{F}$-martingale.

We remark that the above process K is called the quadratic variation of M, and is also denoted as $\langle M \rangle$. Its existence can actually be proved. ∎

Problem 2.10.10 Prove Lemma 2.6.2. ∎

Problem 2.10.11 Assume $X_n \to X$, $Y_n \to Y$ weakly in $\mathbb{L}^2(\mathbb{F})$ and have appropriate dimensions.

(i) $X_n + Y_n \to X + Y$ weakly in $\mathbb{L}^2(\mathbb{F})$.
(ii) $\int_0^{\cdot} X_s^n \cdot dB_s \to \int_0^{\cdot} X_s \cdot dB_s$ weakly in $\mathbb{L}^2(\mathbb{F})$.
(iii) $\mathbb{E}\big[\int_0^T |X_t|^2 dt \big] \le \liminf_{n \to \infty} \mathbb{E}\big[\int_0^T |X_t^n|^2 dt \big]$. ∎

Problem 2.10.12 We note that Theorem 2.7.1 does not hold true for semimartingales in the following sense. Let $d = 1$. For any n, find a counterexample $X_t = \int_0^t b_s ds + \int_0^t \sigma_s dB_s$, where $b \in \mathbb{L}^{1,2}(\mathbb{F})$ and $\sigma \in \mathbb{L}^2(\mathbb{F})$ such that

$$\mathbb{E}\Big[\Big(\int_0^T |b_t| dt \Big)^2 + \int_0^T |\sigma_t|^2 dt \Big] > n \mathbb{E}[|X_T^*|^2].$$

∎

Problem 2.10.13 This problem concerns the Stratonovich integral $\int_0^T X_t \circ dB_t$, for which the integrand X requires some regularity. To be specific, let $X_t := x + \int_0^t b_s ds + \int_0^t \sigma_s dB_s$, where $x \in \mathbb{R}^d$, $b \in \mathbb{L}^{1,2}(\mathbb{F}, \mathbb{R}^d)$, and $\sigma \in \mathbb{L}^2(\mathbb{F}, \mathbb{R}^{d \times d})$.

(i) For any $\pi : 0 = t_0 < \cdots < t_n = T$, denote

$$S(\pi) := \sum_{i=0}^{n-1} X_{\frac{t_i + t_{i+1}}{2}} \cdot B_{t_i, t_{i+1}}.$$

Show that $S_M(\pi) \to \int_0^T X_t \cdot dB_t + \frac{1}{2} \int_0^T \mathrm{tr}(\sigma_t) dt$ in $\mathbb{L}^2(\mathscr{F}_T)$, as $|\pi| \to 0$. We thus define the Stratonovich integral as

$$\int_0^T X_t \circ dB_t := \lim_{|\pi| \to 0} S_M(\pi) = \int_0^T X_t \cdot dB_t + \frac{1}{2} \int_0^T \mathrm{tr}(\sigma_t) dt. \quad (2.10.1)$$

(ii) The Stratonovich integral can be approximated in a different way. For each π, let B^{π} denote the linear interpolation of $(t_i, B_{t_i})_{0 \le i \le n}$, namely

$$B_t^{\pi} := \sum_{i=0}^{n-1} \Big[B_{t_i} \frac{t_{i+1} - t}{t_{i+1} - t_i} + B_{t_{i+1}} \frac{t - t_i}{t_{i+1} - t_i} \Big] \mathbf{1}_{(t_i, t_{i+1}]}.$$

Then B^{π} is absolutely continuous in t and thus the following integration is well defined:

$$\tilde{S}(\pi) := \int_0^T X_t \cdot dB_t^{\pi}.$$

Show that $\lim_{|\pi| \to 0} \tilde{S}(\pi) = \int_0^T X_t \circ dB_t$ in $\mathbb{L}^2$-sense.

(iii) Similarly we may define $Y_t := \int_0^t X_s \circ dB_s = \int_0^t X_s \cdot dB_s + \frac{1}{2} \int_0^t \text{tr}(\sigma_s)ds$. We shall note that Y is in general not a martingale. Prove the following chain rule for Stratonovich integral:

$$df(t, Y_t) = \partial_t f(t, Y_t)dt + [\partial_y f(t, Y_t)X_t] \circ dB_t,$$

for any $f : [0, T] \times \mathbb{R} \to \mathbb{R}$ smooth enough.   ∎

Problem 2.10.14 This problem concerns a.s. convergence of stochastic integration. Given $X \in \mathbb{L}^2(\mathbb{F}, \mathbb{R}^d)$, denote $Y_T := \int_0^T X_t \cdot dB_t$, and, for a partition $\pi : 0 = t_0 < \cdots < t_n = T$,

$$Y_T^\pi := \sum_{i=0}^{n-1} X_{t_i} \cdot B_{t_i, t_{i+1}}. \tag{2.10.2}$$

Let $\alpha \in (0, 1]$ and $\beta > 0$ be two constants, and $\{\pi_m\}_{m \geq 1}$ a sequence of partitions such that $|\pi_m| \leq m^{-\beta}$. At below, all limits are in the sense of a.s. convergence.

(i) Assume X is uniformly Hölder-α continuous and $\beta > \frac{1}{2\alpha}$. Show that $\lim_{m \to \infty} Y_T^{\pi_m} = Y_T$, a.s. (Hint: show that $\mathbb{E}\left[\sum_{m=1}^\infty |Y_T^{\pi_m} - Y_T|^2\right] < \infty$.)

(ii) Assume $dX_t = \sigma_t dB_t$, $\sigma \in \mathbb{L}^2(\mathbb{F}, \mathbb{R}^{d \times d})$, and $\beta > 1$. Show that $\lim_{m \to \infty} Y_T^{\pi_m} = Y_T$, a.s.

(iii) Assume $dX_t = \sigma_t dB_t$, $\sigma \in \mathbb{L}^\infty(\mathbb{F}, \mathbb{R}^{d \times d})$, and $\beta > \frac{1}{2}$. Show that $\lim_{m \to \infty} Y_T^{\pi_m} = Y_T$, a.s.

(iv) Assume $d = 1$, X is as in (ii), σ is uniformly Hölder-α continuous, and $\beta > \frac{1}{1+2\alpha}$. Denote

$$Y_T^{2,\pi} := \sum_{i=0}^{n-1} \left[X_{t_i} B_{t_i, t_{i+1}} + \sigma_{t_i} \frac{|B_{t_i, t_{i+1}}|^2 - (t_{i+1} - t_i)}{2}\right], \tag{2.10.3}$$

which we call the second order approximation. Show that $\lim_{m \to \infty} Y_T^{2,\pi_m} = Y_T$, a.s.

(v) Consider the same setting as in (iv). Assume further that $d\sigma_t = \theta_t dB_t$, θ is uniformly Hölder-α continuous, and $\beta > \frac{1}{1+3\alpha}$. Denote

$$Y_t^{3,\pi} := \sum_{i=0}^{n-1} \left[X_{t_i} B_{t_i, t_{i+1}} + \sigma_{t_i} \frac{|B_{t_i, t_{i+1}}|^2 - (t_{i+1} - t_i)}{2} + \theta_{t_i} \frac{(B_{t_i, t_{i+1}})^3 - 3B_{t_i, t_{i+1}} (t_{i+1} - t_i)}{6}\right], \tag{2.10.4}$$

which we call the third order approximation. Show that $\lim_{m \to \infty} Y_T^{3,\pi_m} = Y_T$, a.s.

We remark that in all the above cases, $|\pi_m|$ converges to 0 with a rate β, and the exceptional null set of the a.s. convergence depends on $\{\pi_m\}_{m\geq 1}$. In the setting of (iv), by rough path theory one can show that there is a common null set E_0 such that $\lim_{|\pi|\to 0} Y_T^{2,\pi}(\omega)$ exists for all $\omega \notin E_0$. ∎

Chapter 3
Stochastic Differential Equations

In this chapter we fix a filtered probability space $(\Omega, \mathscr{F}, \mathbb{F}, \mathbb{P})$, on which is defined a d-dimensional $(\mathbb{P}, \mathbb{F})$-Brownian motion B. Our objective is the following *Stochastic Differential Equation* (SDE, for short):

$$X_t = \eta + \int_0^t b_s(\omega, X_s)ds + \int_0^t \sigma_s(\omega, X_s)dB_s, \quad 0 \le t \le T, \quad \text{a.s.} \quad (3.0.1)$$

where X is d_1-dimensional, $\eta \in \mathbb{L}^0(\mathscr{F}_0, \mathbb{R}^{d_1})$, and $b : [0, T] \times \Omega \times \mathbb{R}^{d_1} \to \mathbb{R}^{d_1}$, $\sigma : [0, T] \times \Omega \times \mathbb{R}^{d_1} \to \mathbb{R}^{d_1 \times d}$ are $\mathbb{F}$-measurable with respect to (t, ω, x). When there is no confusion, we omit ω in the coefficients. We say $X \in \mathbb{L}^0(\mathbb{F}, \mathbb{R}^{d_1})$ is a solution of SDE (3.0.1) if $b(X) \in \mathbb{L}^1_{loc}(\mathbb{F}, \mathbb{R}^{d_1}), \sigma(X) \in \mathbb{L}^2_{loc}(\mathbb{F}, \mathbb{R}^{d_1 \times d})$ and the SDE (3.0.1) holds a.s.

Most results in this chapter hold true in multidimensional case. Again, we shall state the results in multidimensional notations, but for notational simplicity quite often we will carry out the proofs only for the case $d_1 = d = 1$. For the results which hold true only in 1-dimensional case, we will emphasize the dimension in the statements of the results.

3.1 Linear Stochastic Differential Equations

In this section we study the following linear SDE:

$$X_t = \eta + \int_0^t [b^1_s X_s + b^0_s]ds + \int_0^t [\sigma^1_s X_s + \sigma^0_s] \cdot dB_s, \quad 0 \le t \le T, \quad \text{a.s.} \quad (3.1.1)$$

where B can be multidimensional, but X is scalar, namely

$$d_1 = 1. \quad (3.1.2)$$

© Springer Science+Business Media LLC 2017
J. Zhang, *Backward Stochastic Differential Equations*, Probability Theory and
Stochastic Modelling 86, DOI 10.1007/978-1-4939-7256-2_3

We shall assume

$$\eta \in L^0(\mathscr{F}_0, \mathbb{R}), \ b^1 \in L^\infty(\mathbb{F}, \mathbb{R}), \ \sigma^1 \in L^\infty(\mathbb{F}, \mathbb{R}^d), \ b^0 \in \mathbb{L}^1_{loc}(\mathbb{F}, \mathbb{R}), \ \sigma^0 \in \mathbb{L}^2_{loc}(\mathbb{F}, \mathbb{R}^d). \quad (3.1.3)$$

In this case we can solve the equation explicitly.

Case 1. $b^1 = 0$ and $\sigma^1 = 0$.

In this case clearly the SDE (3.1.1) has a unique solution:

$$X_t = \eta + \int_0^t b^0_s ds + \int_0^t \sigma^0_s \cdot dB_s, \ \ 0 \le t \le T. \quad (3.1.4)$$

Case 2. $b^0 = 0$ and $\sigma^0 = 0$.

In this case, it follows directly from Itô formula that the following X is a solution:

$$X_t := \eta \exp\left(\int_0^t \sigma^1_s \cdot dB_s + \int_0^t [b^1_s - \frac{1}{2}|\sigma^1_s|^2] ds \right). \quad (3.1.5)$$

The uniqueness will be proved as a special case of the Case 3 below.

Case 3. In general case, for some $\alpha \in L^\infty(\mathbb{F}, \mathbb{R})$ and $\beta \in L^\infty(\mathbb{F}, \mathbb{R}^d)$ which will be specified later, introduce the adjoint process

$$\Gamma_t := \exp\left(\int_0^t \beta_s \cdot dB_s + \int_0^t [\alpha_s - \frac{1}{2}|\beta_s|^2] ds \right). \quad (3.1.6)$$

By (3.1.5), or applying Itô formula directly, Γ satisfies the following adjoint equation:

$$d\Gamma_t = \Gamma_t[\alpha_t dt + \beta_t \cdot dB_t]. \quad (3.1.7)$$

Apply Itô formula, we get

$$d\left(\Gamma_t X_t \right) = \Gamma_t\left[b^1_t X_t + b^0_t + \alpha_t X_t + \beta_t \cdot [\sigma^1_t X_t + \sigma^0_t] \right] dt + \Gamma_t\left[\sigma^1_t X_t + \sigma^0_t + \beta_t X_t \right] \cdot dB_t$$

$$= \Gamma_t\left[[b^1_t + \alpha_t + \beta_t \cdot \sigma^1_t] X_t + [b^0_t + \beta_t \cdot \sigma^0_t] \right] dt + \Gamma_t\left[[\sigma^1_t + \beta_t] X_t + \sigma^0_t \right] \cdot dB_t.$$

Set

$$b^1_t + \alpha_t + \beta_t \cdot \sigma^1_t = 0, \ \ \ \sigma^1_t + \beta_t = 0.$$

That is,

$$\beta := -\sigma^1, \ \alpha := -b^1 + |\sigma^1|^2, \text{ and thus } \Gamma_t := \exp\left(-\int_0^t \sigma^1_s \cdot dB_s - \int_0^t [b^1_s - \frac{1}{2}|\sigma^1_s|^2] ds \right). \quad (3.1.8)$$

Then we have

$$d\left(\Gamma_t X_t \right) = \Gamma_t[b^0_t - \sigma^0_t \cdot \sigma^1_t] dt + \Gamma_t \sigma^0_t \cdot dB_t.$$

Notice that $\Gamma > 0$ is continuous and thus $\Gamma_T^* < \infty$, a.s. Then, under our conditions, $\Gamma[b^0 - \sigma^0 \cdot \sigma^1] \in \mathbb{L}_{loc}^1(\mathbb{F})$ and $\Gamma\sigma^0 \in \mathbb{L}_{loc}^2(\mathbb{F}, \mathbb{R}^d)$. Moreover, note that the above leads to

$$X_t = (\Gamma_t)^{-1}\Big[\eta + \int_0^t \Gamma_s[b_s^0 - \sigma_s^0 \cdot \sigma_s^1]ds + \int_0^t \Gamma_s\sigma_s^0 \cdot dB_s\Big]. \tag{3.1.9}$$

This implies that (3.1.1) has a unique solution. ∎

Remark 3.1.1 When X is multidimensional, following the general well-posedness result in the remaining sections of this chapter, under appropriate conditions the linear SDE (3.1.1) is still well-posed. However, we do not have the explicit formulae (3.1.8) and (3.1.9) due to the noncommutativity of matrices. By using some dimension reduction technique, or say decoupling strategy, one may derive some semi-explicit representation formula for multidimensional SDEs. The idea will be illustrated briefly in Problem 3.7.2. ∎

3.2 A Priori Estimates for SDEs

We now study the general nonlinear SDE (3.0.1). Recall that for notational simplicity we omit the ω in the coefficients. We shall assume

Assumption 3.2.1

  (i) b, σ are $\mathbb{F}$-measurable with appropriate dimensions;
  (ii) b and σ are uniformly Lipschitz continuous in x with a Lipschitz constant L. To be precise, there exists a constant $L \geq 0$ such that, for $dt \times dP$-a.s. (t, ω),

$$|b_t(x_1) - b_t(x_2)| + |\sigma_t(x_1) - \sigma_t(x_2)| \leq L|x_1 - x_2|, \quad \text{for all} \quad x_1, x_2 \in \mathbb{R}^{d_1}. \tag{3.2.1}$$

  (iii) $\eta \in \mathbb{L}^2(\mathscr{F}_0, \mathbb{R}^{d_1})$, $b^0 := b(0) \in \mathbb{L}^{1,2}(\mathbb{F}, \mathbb{R}^{d_1})$, and $\sigma^0 := \sigma(0) \in \mathbb{L}^2(\mathbb{F}, \mathbb{R}^{d_1 \times d})$.

Notice that Assumption 3.2.1 clearly implies the following linear growth condition:

$$|b_t(x)| \leq |b_t^0| + L|x|, \quad |\sigma_t(x)| \leq |\sigma_t^0| + L|x|. \tag{3.2.2}$$

Our first result is:

Theorem 3.2.2 *Let Assumption 3.2.1 hold and $X \in \mathbb{L}^2(\mathbb{F}, \mathbb{R}^{d_1})$ be a solution to SDE (3.0.1). Then $X \in \mathbb{S}^2(\mathbb{F})$ and there exists a constant C, which depends only on T, L, and d, d_1, such*

$$\mathbb{E}[|X_T^*|^2] \leq CI_0^2, \quad \text{where} \quad I_0^2 := \mathbb{E}\Big[|\eta|^2 + \Big(\int_0^T |b_t^0|dt\Big)^2 + \int_0^T |\sigma_t^0|^2 dt\Big]. \tag{3.2.3}$$

In this proof and in the sequel, we shall denote by C a generic constant, which depends only on T, L, and d, d_1, and may vary from line to line.

Proof For notational simplicity, we assume $d = d_1 = 1$. We proceed in several steps.

Step 1. We first show that

$$\mathbb{E}[|X_T^*|^2] \le CI_0^2 + C\mathbb{E}\Big[\int_0^T |X_t|^2 dt\Big] < \infty. \tag{3.2.4}$$

Indeed, note that

$$X_T^* \le |\eta| + \int_0^T |b_t(X_t)| dt + \sup_{0 \le t \le T} \Big| \int_0^t \sigma_s(X_s) dB_s \Big|.$$

Square both sides, take the expectation, and apply the Burkholder-Davis-Gundy Inequality, we obtain

$$\mathbb{E}[|X_T^*|^2] \le C\mathbb{E}\Big[|\eta|^2 + \Big(\int_0^T |b_t(X_t)| dt\Big)^2 + \sup_{0 \le t \le T} \Big(\int_0^t \sigma_s(X_s) dB_s\Big)^2\Big]$$

$$\le C\mathbb{E}\Big[|\eta|^2 + \Big(\int_0^T [|b_t^0| + |X_t|] dt\Big)^2 + \int_0^T |\sigma_t(X_t)|^2 dt\Big]$$

$$\le C\mathbb{E}\Big[|\eta|^2 + \Big(\int_0^T |b_t^0| dt\Big)^2 + \int_0^T \big[|\sigma_t^0|^2 + |X_t|^2\big] dt\Big],$$

which implies (3.2.4) immediately.

Step 2. We next show that, for any $\varepsilon > 0$, there exists a constant $C_\varepsilon > 0$, which may depend on ε as well, such that

$$\sup_{0 \le t \le T} \mathbb{E}[|X_t|^2] \le \varepsilon \mathbb{E}[|X_T^*|^2] + C_\varepsilon I_0^2. \tag{3.2.5}$$

Indeed, applying Itô formula we have

$$d|X_t|^2 = \Big[2X_t b_t(X_t) + |\sigma_t(X_t)|^2\Big] dt + 2X_t \sigma_t(X_t) dB_t. \tag{3.2.6}$$

Note that

$$\mathbb{E}[|X_T^*|^2] < \infty, \quad \mathbb{E}\Big[\int_0^T |2\sigma_t(X_t)|^2 dt\Big] \le C\mathbb{E}\Big[\int_0^T [|\sigma_t^0|^2 + |X_t|^2] dt\Big] < \infty.$$

By Problem 2.10.7 (i), $\int_0^t 2X_s \sigma_s(X_s) dB_s$ is a true martingale. Take expectation on both sides of (3.2.6), we have

$$\mathbb{E}[|X_t|^2] = \mathbb{E}\Big[|\eta|^2 + \int_0^t \big[2X_s b_s(X_s) + |\sigma_s(X_s)|^2\big]ds\Big]$$

$$\leq \mathbb{E}\Big[|\eta|^2 + \int_0^t \big[C|X_s|^2 + 2|X_s||b_s^0| + C|\sigma_s^0|^2\big]ds\Big]$$

$$\leq C\int_0^t \mathbb{E}[|X_s|^2]ds + 2\mathbb{E}\Big[X_T^* \int_0^T |b_s^0|ds\Big] + CI_0^2.$$

Apply Gronwall's inequality and note that $2ab \leq \varepsilon a^2 + \varepsilon^{-1}b^2$, we get

$$\mathbb{E}[|X_t|^2] \leq C\mathbb{E}\Big[X_T^* \int_0^T |b_s^0|ds\Big] + CI_0^2 \leq \varepsilon\mathbb{E}[|X_T^*|^2] + C\varepsilon^{-1}\mathbb{E}\Big[\big(\int_0^T |b_s^0|ds\big)^2\Big] + CI_0^2.$$

This implies (3.2.5) immediately.

Step 3. Plug (3.2.5) into (3.2.4), we get

$$\mathbb{E}[|X_T^*|^2] \leq C\varepsilon\mathbb{E}[|X_T^*|^2] + C\varepsilon^{-1}I_0^2.$$

Set $\varepsilon := \frac{1}{2C}$ for the above constant C and recall (3.2.4), we prove (3.2.3). ∎

Remark 3.2.3 From the above proof one can easily see that in Theorem 3.2.2 the Lipschitz condition (3.2.1) can be replaced with the linear growth condition (3.2.2). ∎

Theorem 3.2.4 *For $i = 1, 2$, assume (η_i, b^i, σ^i) satisfy Assumption 3.2.1, and $X^i \in \mathbb{L}^2(\mathbb{F}, \mathbb{R}^{d_1})$ is a solution to SDE (3.0.1) with coefficients (η_i, b^i, σ^i). Then*

$$\mathbb{E}\Big[|(\Delta X)_T^*|^2\Big] \leq C\mathbb{E}\Big[|\Delta\eta|^2 + \big(\int_0^T |\Delta b_t(X_t^1)|dt\big)^2 + |\Delta\sigma_t(X_t^1)|^2 dt\Big], \quad (3.2.7)$$

where

$$\Delta X := X^1 - X^2, \quad \Delta\eta := \eta_1 - \eta_2, \quad \Delta b := b^1 - b^2, \quad \Delta\sigma := \sigma^1 - \sigma^2. \quad (3.2.8)$$

Proof Again we assume $d = d_1 = 1$. Notice that

$$\Delta X_t = \Delta\eta + \int_0^t [\Delta b_s(X_s^1) + \alpha_s \Delta X_s]ds + \int_0^t [\Delta\sigma_s(X_s^1) + \beta_s \Delta X_s]dB_s, \quad (3.2.9)$$

where

$$\alpha_t := \frac{b_t^2(X_t^1) - b_t^2(X_t^2)}{\Delta X_t}\mathbf{1}_{\{\Delta X_t \neq 0\}}, \quad \beta_t := \frac{\sigma_t^2(X_t^1) - \sigma_t^2(t, X_t^2)}{\Delta X_t}\mathbf{1}_{\{\Delta X_t \neq 0\}} \quad (3.2.10)$$

are bounded by L, thanks to (3.2.1). Now we may view $\Delta X \in \mathbb{L}^2(\mathbb{F})$ as the solution to the linear SDE (3.2.9) with coefficients:

$$\eta := \Delta\eta, \quad b_t(\omega, x) := \alpha_t(\omega)x + \Delta b_t(\omega, X_t^1(\omega)), \quad \sigma_t(\omega, x) := \beta_t(\omega)x + \Delta\sigma_t(\omega, X_t^1(\omega)).$$

One may easily check that (η, b, σ) satisfy Assumption 3.2.1. Then applying Theorem 3.2.2 on SDE (3.2.9) we obtain (3.2.7).   ∎

3.3   Well-Posedness of SDEs

We now establish the existence and uniqueness of solutions to SDE (3.0.1).

Theorem 3.3.1 *Under Assumption 3.2.1 hold, SDE (3.0.1) admits a unique solution $X \in \mathbb{L}^2(\mathbb{F}, \mathbb{R}^{d_1})$.*
We remark that, by Theorem 3.2.2, any $\mathbb{L}^2(\mathbb{F})$ solution is actually in $\mathbb{S}^2(\mathbb{F})$. We still use $\mathbb{L}^2(\mathbb{F})$ in the statement of this theorem for a more general uniqueness result.

Proof The uniqueness follows immediately from Theorem 3.2.4. We prove the existence using two methods. The first method is neater; however, the second method works better for forward-backward SDEs, see Chapter 8 below. Again, we assume $d = d_1 = 1$.

Method 1. (Global Approach) We construct the solution via Picard Iteration. Let $X_t^0 := \eta, 0 \le t \le T$, and for $n = 0, 1, \cdots$, define

$$X_t^{n+1} := \eta + \int_0^t b_s(X_s^n)ds + \int_0^t \sigma_s(X_s^n)dB_s. \tag{3.3.1}$$

We note that $X^n \in \mathbb{S}^2(\mathbb{F})$ for all $n \ge 0$. Indeed, it is clear that $X^0 \in \mathbb{S}^2(\mathbb{F})$. Assume $X^n \in \mathbb{S}^2(\mathbb{F})$. By Assumption 3.2.1 (iii) and (3.2.2) one can easily see that $b(X^n) \in \mathbb{L}^{1,2}(\mathbb{F})$ and $\sigma(X^n) \in \mathbb{L}^2(\mathbb{F})$, and thus it follows from Theorem 3.2.2 (or more directly from the Burkholder-Davis-Gundy inequality) that $X^{n+1} \in \mathbb{S}^2(\mathbb{F})$.

Now denote $\Delta X^n := X^n - X^{n-1}$. Then

$$\Delta X_t^{n+1} = \int_0^t [b_s(X_s^n) - b_s(X_s^{n-1})]ds + \int_0^t [\sigma_s(X_s^n) - \sigma_s(X_s^{n-1})]dB_s$$

$$= \int_0^t \alpha_s^n \Delta X_s^n ds + \int_0^t \beta_s^n \Delta X_s^n dB_s, \tag{3.3.2}$$

where α^n, β^n are defined in the spirit of (3.2.10) and are bounded by L. Let $\lambda > 0$ be a constant which will be determined later. Apply Itô formula on $e^{-\lambda t}|\Delta X_t^{n+1}|^2$ we have

$$d\big(e^{-\lambda t}|\Delta X_t^{n+1}|^2\big) = e^{-\lambda t}\Big[2\Delta X_t^{n+1}\alpha_t^n \Delta X_t^n + |\beta_s^n \Delta X_s^n|^2 - \lambda|\Delta X_t^{n+1}|^2\Big]dt$$

$$+ e^{-\lambda t}2\Delta X_t^{n+1}\beta_t^n \Delta X_t^n dB_t.$$

By Problem 2.10.7, $e^{-\lambda t}2\Delta X_t^{n+1}\beta_t^n\Delta X_t^n dB_t$ is a true martingale. Note that $\Delta X_0^n = 0$ for all n. Then taking expectation in the above equation we obtain

$$\lambda\mathbb{E}\left[\int_0^T e^{-\lambda t}|\Delta X_t^{n+1}|^2 dt\right] \leq \mathbb{E}\left[e^{-\lambda T}|\Delta X_T^{n+1}|^2 + \lambda\int_0^T e^{-\lambda t}|\Delta X_t^{n+1}|^2 dt\right]$$

$$= \mathbb{E}\left[\int_0^T e^{-\lambda t}\left[2\Delta X_t^{n+1}\alpha_t^n\Delta X_t^n + |\beta_t^n\Delta X_t^n|^2\right]dt\right]$$

$$\leq \mathbb{E}\left[\int_0^T e^{-\lambda t}\left[|\Delta X_t^{n+1}|^2 + 2L^2|\Delta X_t^n|^2\right]dt\right]$$

Set $\lambda := 1 + 8L^2$, we obtain:

$$\mathbb{E}\left[\int_0^T e^{-\lambda t}|\Delta X_t^{n+1}|^2 dt\right] \leq \frac{1}{4}\mathbb{E}\left[\int_0^T e^{-\lambda t}|\Delta X_t^n|^2 dt\right].$$

By induction we have, for all $n \geq 1$,

$$\mathbb{E}\left[\int_0^T e^{-\lambda t}|\Delta X_t^n|^2 dt\right] \leq \frac{1}{4^{n-1}}\mathbb{E}\left[\int_0^T e^{-\lambda t}|\Delta X_t^1|^2 dt\right] = \frac{C}{4^n}.$$

Recall the norms $\|\cdot\|_2$ and $\|\cdot\|_{\infty,2}$ in Section 1.3.1. This implies

$$\|\Delta X^n\|_2^2 \leq C\mathbb{E}\left[\int_0^T e^{-\lambda t}|\Delta X_t^n|^2 dt\right] \leq \frac{C}{4^n}.$$

Applying the Burkholder-Davis-Gundy inequality, it follows from (3.3.2) that

$$\|\Delta X^n\|_{\infty,2}^2 \leq \frac{C}{4^n}.$$

Then, for $m > n$,

$$\|X^m - X^n\|_{\infty,2} \leq \sum_{k=n+1}^m \|\Delta X^{k+1}\|_{\infty,2} \leq \sum_{k=n+1}^m \frac{C}{2^k} \leq \frac{C}{2^n} \to 0, \quad \text{as} \quad n \to \infty.$$

Therefore, there exists $X \in \mathbb{S}^2(\mathbb{F})$ such that

$$\|X^n - X\|_{\infty,2} \leq \frac{C}{2^n} \to 0, \quad \text{as } n \to \infty. \tag{3.3.3}$$

Now send $n \to \infty$ in (3.3.1), one can easily see that X satisfies (3.0.1).

Method 2. (Local Approach) Let $\delta > 0$ be a small number which depends only on L (and the dimensions d, d_1) and will be specified later.

Step 1. We first assume $T \leq \delta$. Define X^n and ΔX^n as in Method 1. Square both sides of (3.3.2) and take expectations, we get

$$\mathbb{E}[|\Delta X_t^{n+1}|^2] \leq 2\mathbb{E}\Big[\Big|\int_0^t \alpha_s^n \Delta X_s^n ds\Big|^2 + \int_0^t |\beta_s^n \Delta X_s^n|^2 ds\Big]$$

$$\leq 2L^2 \mathbb{E}\Big[\Big(\int_0^t |\Delta X_s^n| ds\Big)^2 + \int_0^t |\Delta X_s^n|^2 ds\Big] \leq 2L^2(1+\delta)\mathbb{E}\Big[\int_0^T |\Delta X_s^n|^2 ds\Big].$$

This implies that

$$\mathbb{E}\Big[\int_0^T |\Delta X_s^{n+1}|^2 ds\Big] \leq 2L^2(1+\delta)\delta \mathbb{E}\Big[\int_0^T |\Delta X_s^n|^2 ds\Big].$$

Choose δ small enough that $2L^2(1+\delta)\delta \leq \frac{1}{4}$, then

$$\|\Delta X^{n+1}\|_2 \leq \frac{1}{2}\|\Delta X^n\|_2, \quad \text{for all } n \geq 1.$$

We emphasize again that δ depends only on L (and the dimensions), but it does not depend on η. Now following similar arguments as in Method 1 we obtain the solution X.

Step 2. We next prove the existence for arbitrary T. Let $\delta > 0$ be the constant above, and consider a time partition $0 = t_0 < \cdots < t_n = T$ such that $t_{i+1} - t_i \leq \delta$ for $i = 0, \cdots, n-1$. We now define $X_{t_0} := \eta$, and then define X recursively on $(t_i, t_{i+1}]$ for $i = 0, \cdots, n-1$:

$$X_t = X_{t_i} + \int_{t_i}^t b_s(X_s)ds + \int_{t_i}^t \sigma_s(X_s)dB_s, \quad t \in [t_i, t_{i+1}].$$

The existence of the above SDE is guaranteed by Step 1 since $t_{i+1} - t_i \leq \delta$. Since n is finite, we see that $X \in \mathbb{L}^2(\mathbb{F})$, and thus we obtain the solution over the whole interval $[0, T]$. ∎

3.4   Some Properties of SDEs

We first establish the comparison theorem. We emphasize that X is 1-dimensional and the two systems have the same σ.

Theorem 3.4.1 *Assume $d_1 = 1$ (but B can be multidimensional in general). Let (η_i, b^i, σ), $i = 1, 2$, satisfy Assumption 3.2.1 and $X^i \in \mathbb{S}^2(\mathbb{F})$ be the solution to the corresponding SDE (3.0.1). If $\eta_1 \leq \eta_2$, a.s. and $b^1(x) \leq b^2(x)$, $dt \times d\mathbb{P}$-a.s. for all $x \in \mathbb{R}$, then $X_t^1 \leq X_t^2$, $0 \leq t \leq T$, a.s.*

Proof Assume $d = 1$ for simplicity. First, by Problem 1.4.6 (ii), we see that

$$b^1(x) \le b^2(x), \quad \text{for all } x \in \mathbb{R}, \quad dt \times d\mathbb{P} - \text{a.s.} \tag{3.4.1}$$

Recall the notations ΔX, $\Delta \eta$, and Δb in (3.2.8). By (3.2.9) we have

$$\Delta X_t = \Delta \eta + \int_0^t [\Delta b_s(X_s^1) + \alpha_s \Delta X_s] ds + \int_0^t \beta_s \Delta X_s dB_s,$$

where α, β are bounded by L. Follow the arguments in Section 3.1 Case 3, we see that (3.1.6) and (3.1.9) become

$$\Gamma_t := \exp\left(-\int_0^t \beta_s dB_s - \int_0^t [\alpha_s - \frac{1}{2}|\beta_s|^2] ds \right),$$

$$\Delta X_t = (\Gamma_t)^{-1}\left[\Delta \eta + \int_0^t \Gamma_s \Delta b(s, X_s^1) ds \right]. \tag{3.4.2}$$

Since $\Gamma > 0$, $\Delta \eta \le 0$, a.s. and by (3.4.1), $\Delta b(t, X_t^1) \le 0$, $dt \times d\mathbb{P}$-a.s. Then it follows from Problem 1.4.6 (i) that $\Delta X_t \le 0$, $0 \le t \le T$, a.s. ∎

We remark that the comparison principle typically does not hold true in multidimensional case. We have some brief discussion on this in Problem 3.7.6.

We next establish the stability result.

Theorem 3.4.2 *Let (η, b, σ) and (η_n, b^n, σ^n), $n = 1, 2, \cdots$, satisfy Assumption 3.2.1 with the same constant L, and $X, X^n \in \mathbb{S}^2(\mathbb{F})$ be the solution to the corresponding SDE (3.0.1). Denote*

$$\Delta X^n := X^n - X, \quad \Delta \eta_n := \eta_n - \eta, \quad \Delta b^n := b^n - b, \quad \Delta \sigma^n := \sigma^n - \sigma.$$

Assume

$$\lim_{n \to \infty} \mathbb{E}\left[|\Delta \eta_n|^2 + \left(\int_0^T |\Delta b_s^n(0)| ds \right)^2 + \int_0^T |\Delta \sigma_s^n(0)|^2 ds \right] = 0, \tag{3.4.3}$$

and $\Delta b^n(x) \to 0, \Delta \sigma^n(x) \to 0$, in measure $dt \times d\mathbb{P}$, for all $x \in \mathbb{R}^{d_1}$. Then

$$\lim_{n \to \infty} \mathbb{E}\left[|(\Delta X^n)_T^*|^2 \right] = 0.$$

Proof First, applying Theorem 3.2.4 we have

$$\mathbb{E}\left[|(\Delta X^n)_T^*|^2 \right] \le C\mathbb{E}\left[|\Delta \eta_n|^2 + \left(\int_0^T |\Delta b_t^n(X_t)| dt \right)^2 |\Delta \sigma_t^n(X_t)|^2 dt \right]$$

$$\le C\mathbb{E}\left[|\Delta \eta_n|^2 + \left(\int_0^T |\Delta b_t^n(0)| dt \right)^2 + \int_0^T |\Delta \sigma_t^n(0)|^2 dt \right. \tag{3.4.4}$$

$$\left. + \left(\int_0^T |\Delta b_t^n(X_t) - \Delta b_t^n(0)| dt \right)^2 + \int_0^T |\Delta \sigma_t^n(X_t) - \Delta \sigma_t^n(0)|^2 dt \right].$$

By Problem 1.4.6 (iii), $\Delta b^n(X) \to 0$ and $\Delta \sigma^n(X) \to 0$, in measure $dt \times d\mathbb{P}$. Note that

$$|\Delta b_t^n(X_t) - \Delta b_t^n(0)| \le 2L|X_t|, \quad |\Delta \sigma_t^n(X_t) - \Delta \sigma_t^n(0)| \le 2L|X_t|.$$

Applying the dominated convergence Theorem we have

$$\lim_{n \to \infty} \mathbb{E}\Big[\Big(\int_0^T |\Delta b_t^n(X_t) - \Delta b_t^n(0)|dt\Big)^2 + \int_0^T |\Delta \sigma_t^n(X_t) - \Delta \sigma_t^n(0)|^2 dt\Big] = 0.$$

This, together with (3.4.3) and (3.4.4), leads to the result. ∎

We conclude this section by extending the well-posedness result to $\mathbb{L}^p(\mathbb{F})$ for $p \ge 2$.

Theorem 3.4.3 *Let Assumption (3.2.1) hold and $X \in \mathbb{S}^2(\mathbb{F}, \mathbb{R}^{d_1})$ be the unique solution to SDE (3.0.1). Assume $\eta \in \mathbb{L}^p(\mathscr{F}_0, \mathbb{R}^{d_1})$, $b^0 \in \mathbb{L}^{1,p}(\mathbb{F}, \mathbb{R}^{d_1})$, $\sigma^0 \in \mathbb{L}^{2,p}(\mathbb{F}, \mathbb{R}^{d_1 \times d})$ for some $p \ge 2$, then*

$$\mathbb{E}\Big[|X_T^*|^p\Big] \le C_p I_p^p \quad \text{where} \quad I_p^p := \mathbb{E}\Big[|\eta|^p + \Big(\int_0^T |b_t^0|dt\Big)^p + \Big(\int_0^T |\sigma_t^0|^2 dt\Big)^{\frac{p}{2}}\Big]. \quad (3.4.5)$$

Proof Again we assume $d = d_1 = 1$. We proceed in two steps.

Step 1. We first assume $X \in \mathbb{L}^p(\mathbb{F})$ and prove (3.4.5). We shall follow the arguments in Theorem 3.2.2. First, analogous to (3.2.4) one can easily show that

$$\mathbb{E}\Big[|X_T^*|^p\Big] \le C_p I_p^p + C_p \mathbb{E}\Big[\int_0^T |X_t|^p dt\Big] < \infty. \quad (3.4.6)$$

Next, applying Itô formula we have

$$d(|X_t|^2) = [2X_t b_t(X_t) + |\sigma_t(X_t)|^2]dt + 2X_t \sigma_t(X_t)dB_t;$$

$$d(|X_t|^p) = d([|X_t|^2]^{\frac{p}{2}}) = |X_t|^{p-2}\Big[pX_t b_t(X_t) + \frac{p(p-1)}{2}|\sigma_t(X_t)|^2\Big]dt + p|X_t|^{p-2}X_t \sigma_t(X_t)dB_t.$$

Let $q := \frac{p}{p-1}$ be the conjugate of p. By (3.4.6) we have $|X_T^*|^{p-1} \in \mathbb{L}^q(\mathscr{F}_T)$. By our conditions it is clear that $\sigma(X) \in \mathbb{L}^{2,p}(\mathbb{F})$. Then Problem 2.10.7 (i) implies that $\int_0^t p|X_s|^{p-2}X_s \sigma_s(X_s)dB_s$ is a true martingale. Thus following the arguments for (3.2.5) we obtain, for any $\varepsilon > 0$,

$$\sup_{0 \le t \le T} \mathbb{E}[|X_t|^p] \le \varepsilon \mathbb{E}\Big[|X_T^*|^p\Big] + C_p \varepsilon^{-1} I_p^p.$$

This, together with (3.4.6) and by choosing ε small, proves (3.4.5).

Step 2. In the general case, we shall use truncation arguments. We provide two types of truncations.

Method 1. (Time Truncation) For each $n \ge 1$, denote

$$\tau_n := \inf\Big\{t \ge 0 : |X_t| \ge n\Big\} \wedge T.$$

Then $\tau_n \in \mathcal{T}$, $\tau_n \uparrow T$, $\tau_n = T$ when n is large enough, and $|X_t| \le n$ for $t \in [0, \tau_n]$. Denote

$$X_t^n := X_{\tau_n \wedge t}, \quad b_t^n(x) := b_t(x)\mathbf{1}_{[0,\tau_n]}(t), \quad \sigma_t^n(x) := \sigma_t(x)\mathbf{1}_{[0,\tau_n]}(t).$$

Then (η, b^n, σ^n) satisfy all the conditions of this theorem with the same Lipschitz constant L, and $X^n \in \mathbb{L}^p(\mathbb{F})$ is the solution to SDE (3.0.1) with coefficients (η, b^n, σ^n). By Step 1 we have

$$\mathbb{E}\Big[|X_{\tau_n}^*|^p\Big] = \mathbb{E}\Big[|(X^n)_T^*|^p\Big] \le C_p \mathbb{E}\Big[|\eta|^p + \Big(\int_0^T |b_t^n(0)|dt\Big)^p + \Big(\int_0^T |\sigma_t^n(0)|^2 dt\Big)^{\frac{p}{2}}\Big]$$

$$= C_p \mathbb{E}\Big[|\eta|^p + \Big(\int_0^{\tau_n} |b_t^0|dt\Big)^p + \Big(\int_0^{\tau_n} |\sigma_t^0|^2 dt\Big)^{\frac{p}{2}}\Big] \le C_p I_p^p.$$

Now send $n \to \infty$, applying the Monotone Convergence Theorem we obtain (3.4.5).

Method 2. (Space Truncation) For each $n \ge 1$, denote $b^n := (-n) \vee b \wedge n$, $\sigma^n := (-n) \vee \sigma \wedge n$. Clearly (η, b^n, σ^n) satisfy all the conditions of this theorem with the same Lipschitz constant L, and

$$(b^n, \sigma^n) \to (b, \sigma), \quad |b^n| \le |b|, |\sigma_n| \le |\sigma|, \quad |b^n| \le n, |\sigma_n| \le n, \quad \text{for all } (t, \omega, x).$$

Let $X^n \in \mathbb{L}^2(\mathbb{F})$ be the unique solution to SDE (3.0.1) with coefficients (η, b^n, σ^n). Applying Burkholder-Davis-Gundy Inequality we have $X^n \in \mathbb{L}^p(\mathbb{F})$. Then it follows from Step 1 that

$$\mathbb{E}\Big[|(X^n)_T^*|^p\Big] \le C_p \mathbb{E}\Big[|\eta|^p + \Big(\int_0^T |b_t^n(0)|dt\Big)^p + \Big(\int_0^T |\sigma_t^n(0)|^2 dt\Big)^{\frac{p}{2}}\Big] \le C_p I_p^p.$$

Moreover, by Theorem 3.4.2, $\mathbb{E}\Big[|(X^n - X)_T^*|^2\Big] \to 0$. This implies $(X^n)_T^* \to X_T^*$, in measure $\mathbb{P}$. Then (3.4.5) follows from the Fatou Lemma. ∎

3.5 Weak Solutions of SDEs

In this section we introduce weak solutions for SDEs, which does not involve the estimates in the previous sections. The materials are more relevant to the weak formulation which will be the main setting for the fully nonlinear theory in Part III.

Let us start with the following SDE with deterministic coefficients b, σ:

$$X_t = x + \int_0^t b(s, X_s)ds + \int_0^t \sigma(s, X_s)dB_s, \quad 0 \le t \le T. \tag{3.5.1}$$

When b and σ satisfy Assumption 3.2.1, the above SDE has a unique solution X, which we call a strong solution because X is $\mathbb{F}^B$-measurable. However, for more general b and σ, the SDE may not have a strong solution, instead we can look for weak solutions which may not be $\mathbb{F}^B$-measurable. For our purpose, we are particularly interested in the following path dependent SDE:

$$X_t = x + \int_0^t b(s, X.)ds + \int_0^t \sigma(s, X.)dB_s, \quad 0 \le t \le T. \tag{3.5.2}$$

Here $X.$ denotes the path of X. We require b and σ to be progressively measurable in the sense that, for $\varphi = b, \sigma$, $\varphi(t, X.) = \varphi(t, X._{\wedge t})$ depends only on the path of X up to time t.

Definition 3.5.1

(i) We call $(\Omega, \mathscr{F}, \mathbb{P}, B, X)$ a weak solution to SDE (3.5.2) if $(\Omega, \mathscr{F}, \mathbb{P})$ is a probability space, (B, X) are two processes on it, B is a $(\mathbb{F}^{B,X}, \mathbb{P})$-Brownian motion, and (3.5.2) hold $\mathbb{P}$-a.s.

(ii) We may also call $(\Omega, \mathscr{F}, \mathbb{P}, X)$ a weak solution of SDE (3.5.2) if there exists a weak solution $(\tilde{\Omega}, \tilde{\mathscr{F}}, \tilde{\mathbb{P}}, \tilde{B}, \tilde{X})$ such that the $\mathbb{P}$-distribution of X is equal to the $\tilde{\mathbb{P}}$-distribution of $\tilde{X}$.

Remark 3.5.2 Let $(\Omega, \mathscr{F}, \mathbb{P}, B, X)$ be a weak solution.

(i) We say it is a strong solution if X is $\mathbb{F}^B$-measurable, or more precisely, if X has a $\mathbb{P}$-modification $\tilde{X} \in \mathbb{L}^0(\mathbb{F}^B)$. That is, $\overline{\mathbb{F}^X}^{\mathbb{P}} \subset \overline{\mathbb{F}^B}^{\mathbb{P}}$.

(ii) When σ is invertible, we see that $B \in \mathbb{L}^0(\mathbb{F}^X)$, namely $\overline{\mathbb{F}^B}^{\mathbb{P}} \subset \overline{\mathbb{F}^X}^{\mathbb{P}}$. However, in general it is possible that $\overline{\mathbb{F}^X}^{\mathbb{P}}$ and $\overline{\mathbb{F}^B}^{\mathbb{P}}$ do not include each other.

(iii) We say the weak solution is unique (in law) if the $\mathbb{P}$-distribution of X is unique.

∎

Recall the Girsanov Theorem 2.6.4 in Section 2.6, we have

Example 3.5.3 Let $(\Omega, \mathscr{F}, \mathbb{P})$ be a probability space, B a d-dimensional $\mathbb{P}$-Brownian motion, and $\theta \in \mathbb{L}^\infty(\mathbb{F}^B, \mathbb{R}^d)$. Then $(\Omega, \mathscr{F}, \mathbb{P}^\theta, B^\theta, B)$ is the unique weak solution to the SDE:

$$X_t = \int_0^t \theta_s(X.)ds + B_t. \tag{3.5.3}$$

We remark that, since $\theta \in \mathbb{L}^\infty(\mathbb{F}^B, \mathbb{R}^d)$, we may view $\theta = \theta(B.)$ as a mapping on continuous paths, and thus $\theta(X.)$ in the right side of (3.5.3) makes sense.

Proof First, by Girsanov Theorem 2.6.4 it is clear that $(\Omega, \mathscr{F}, \mathbb{P}^\theta, B^\theta, B)$ is a weak solution to SDE (3.5.3). We now prove the uniqueness. For this purpose, let $(\tilde{\Omega}, \tilde{\mathscr{F}}, \tilde{\mathbb{P}}, \tilde{B}, \tilde{X})$ be an arbitrary weak solution. Denote

$$d\tilde{\mathbb{P}}^\theta := \tilde{M}_T^\theta d\tilde{\mathbb{P}} := \exp\left(-\int_0^T \theta_t(\tilde{X}.) \cdot d\tilde{B}_t - \frac{1}{2}\int_0^T |\theta_t(\tilde{X}.)|^2 dt\right)d\tilde{\mathbb{P}}.$$

By Girsanov Theorem 2.6.4 again, $\tilde{X}$ is a $\tilde{\mathbb{P}}^\theta$-Brownian motion, and consequently, the $\tilde{\mathbb{P}}^\theta$-distribution of $\tilde{X}$ is equal to the $\mathbb{P}$-distribution of B. This implies that the $\tilde{\mathbb{P}}^\theta$-distribution of $(\tilde{X}, \tilde{M}^\theta)$ is equal to the $\mathbb{P}$-distribution of (B, M^θ). Therefore, it follows from $d\tilde{\mathbb{P}} = [\tilde{M}_T^\theta]^{-1} d\tilde{\mathbb{P}}^\theta$ and $d\mathbb{P} = [M_T^\theta]^{-1} d\mathbb{P}^\theta$ that the $\tilde{\mathbb{P}}$-distribution of $\tilde{X}$ is equal to the $\mathbb{P}^\theta$-distribution of B.   ∎

We note that a SDE may have a weak solution but without strong solution.

Example 3.5.4 (Tanaka's Example) *Let $d = d_1 = 1$. The following SDE has a unique weak solution but has no strong solution: denoting* $sign(x) := \mathbf{1}_{\{x \geq 0\}} - \mathbf{1}_{\{x < 0\}}$,

$$X_t = \int_0^t sign(X_s) dB_s. \tag{3.5.4}$$

Proof First, let B be a $\mathbb{P}$-Brownian motion and denote $\tilde{B}_t := \int_0^t sign(B_s) dB_s$, $\mathbb{P}$-a.s. By Problem 3.7.1, which follows the arguments in Theorem 2.6.4, or applying Levy's martingale characterization of Brownian motion in Remark 2.6.5 directly, $\tilde{B}$ is also a $\mathbb{P}$-Brownian motion. Then one can easily see that $(\Omega, \mathscr{F}, \mathbb{P}, \tilde{B}, B)$ is a weak solution to SDE (3.5.4). Moreover, for any weak solution $(\Omega, \mathscr{F}, \mathbb{P}, B, X)$, by Problem 3.7.1 again, X has to be a $\mathbb{P}$-Brownian motion, and thus the weak solution is unique (in law).

It remains to show that SDE (3.5.4) has no strong solution. Assume to the contrary that the SDE has a strong solution $X \in \mathbb{L}^0(\mathbb{F}^B)$. By Problem 3.7.1, X is also $\mathbb{P}$-Brownian motion, and thus $\mathbb{P}(X_t = 0) = 0$, for any $0 < t \leq T$. Then, applying Itô formula we have

$$d|X_t|^2 = 2X_t sign(X_t) dB_t + dt = 2|X_t| dB_t + dt = [2|X_t| + \mathbf{1}_{\{|X_t|=0\}}] dB_t + dt, \quad \mathbb{P}\text{-a.s.}$$

This implies that

$$dB_t = \frac{1}{2|X_t| + \mathbf{1}_{\{|X_t|=0\}}} d(|X_t|^2 - t).$$

Here, noting that $M_t := |X_t|^2 - t$ is a $\mathbb{P}$-martingale with quadratic variation process $\langle X \rangle_t = 4 \int_0^t |X_s|^2 ds$, the stochastic integral in the right side should be understood in the sense of Problem 2.10.9. Then clearly B is $\mathbb{F}^{|X|}$-measurable. This implies further that X is $\mathbb{F}^{|X|}$-measurable, which is impossible.   ∎

We remark that in the literature of weak solutions, the central problem is the existence and uniqueness of weak solutions. In Part III of this book, we shall use the weak formulation, which can be viewed as weak solutions in canonical space. However, our focus is somewhat different, and in particular the uniqueness of weak solutions will not be relevant. Since the emphasis of this chapter (and of Parts I and II) is the a priori estimates for strong solutions, we do not get into the details of weak solutions in this section.

We also remark that in the literature weak solution is equivalent to the so-called martingale problem, see Remark 9.2.11 in Part III.

3.6   Bibliographical Notes

The materials of the first four sections are very standard. We refer to the books
Karatzas & Shreve [117], Revuz & Yor [206], Protter [196], and the references
therein. For the dimension reduction technique for multidimensional linear SDEs
we refer to Keller & Zhang [123] (in the form of rough differential equations)

For weak solutions of Markov SDE (3.5.1), one can find general existence result
in Krylov [129], and very deep uniqueness results in Stroock & Varadhan [218]
(in the equivalent form of the so-called martingale problem), as well as Krylov
[128] for some uniqueness results. The recent work Costantini & Kurtz [41] studies
the problem in an abstract framework by using viscosity methods. We also refer
to Barlow [6], Fabes & Kenig [87], and Tsirel'son [229] for some interesting
counterexamples, and the monograph Cherny & Engelbert [33] for a thorough study
on SDEs with irregular coefficients.

3.7   Exercises

Problem 3.7.1   Let $(\Omega, \mathscr{F}, \mathbb{F}, \mathbb{P})$ be a filtered probability space, B a d-dimensional
$(\mathbb{P}, \mathbb{F})$-Brownian motion, $\sigma \in \mathbb{L}^0(\mathbb{F}, \mathbb{R}^d)$ such that $|\sigma| = 1$. Show that $X_t := \int_0^t \sigma_s \cdot dB_s$ is also a $(\mathbb{P}, \mathbb{F})$-Brownian motion.   ∎

Problem 3.7.2   In this problem we illustrate the idea of dimension reduction, or say
decoupling strategy, for multidimensional linear SDE. For simplicity, we consider
the following linear SDE with $d = 1$ and $d_1 = 2$:

$$X_t^i = \eta_i + \int_0^t [\sum_{j=1}^2 b_s^{ij} X_s^j + b_s^{0i}]ds + \int_0^t [\sum_{j=1}^2 \sigma_s^{ij} X_s^j + \sigma_s^{0i}]dB_s, \quad i = 1, 2. \quad (3.7.1)$$

Here $\eta_i \in \mathbb{L}^2(\mathscr{F}_0, \mathbb{R})$, $b^{ij}, \sigma^{ij} \in \mathbb{L}^\infty(\mathbb{F}, \mathbb{R})$, and $b^{0,i} \in \mathbb{L}^{1,2}(\mathbb{F}, \mathbb{R})$, $\sigma^{0i} \in \mathbb{L}^2(\mathbb{F}, \mathbb{R})$.

(i) Let $\overline{X}_t := X_t^1 + \Gamma_t X_t^2$, where Γ is an adjoint process taking the form:

$$\Gamma_t := \int_0^t \alpha_s ds + \int_0^t \beta_s dB_s. \quad (3.7.2)$$

Show that, by choosing

$$\beta := \Gamma[\sigma^{11} + \Gamma\sigma^{12}] - [\sigma^{12} + \Gamma\sigma^{22}], \quad \alpha := \Gamma[b^{11} + \Gamma b^{21} + \beta\sigma^{21}]$$
$$-[b^{12} + \Gamma b^{22} + \beta\sigma^{22}], \quad (3.7.3)$$

we have

$$\overline{X}_t = \eta_1 + \int_0^t \Big[b_s^{11} + \Gamma_s b_s^{21} + \beta_s \sigma_s^{21}] \overline{X}_s + [b_s^{01} + \Gamma_s b_s^{02} + \beta_s \sigma_s^{02}] \Big] ds$$

$$+ \int_0^t \Big[[\sigma_s^{11} + \Gamma_s \sigma_s^{21}] \overline{X}_s + [\sigma_s^{01} + \Gamma_s \sigma_s^{02}] \Big] dB_s. \qquad (3.7.4)$$

(ii) Note that the SDE (3.7.2)–(3.7.3) involves $|\Gamma|^2$ and $|\Gamma|^3$ and thus is a Riccati type of equation. Show that there exists $\tau \in \mathscr{T}(\mathbb{F})$ such that the SDE (3.7.2)–(3.7.3) has a unique bounded solution on $[0, \tau]$.

(iii) Show that there exist a sequence of stopping times $\tau_n \in \mathscr{T}(\mathbb{F})$, $n \geq 1$, such that

 - $\tau_n \uparrow T$ and $\tau_n = T$ when n is large enough;
 - For each n, the SDE (3.7.2)–(3.7.3) on $[\tau_n, \tau_{n+1}]$ with initial condition $\Gamma_{\tau_n} = 0$ has a bounded solution.

(iv) By using the Γ constructed in (iii), solve the system (3.7.1) by solving two 1-dimensional SDEs: first for $\overline{X}$ and then for X^2.

We remark that the SDEs for $\overline{X}$ and X^2 are linear and thus have explicit formulae as in (3.1.9). However, these formulae depend on Γ which is a solution to a nonlinear SDE and does not have an explicit formula. In this sense we say we provide a semi-explicit formula for the system (3.7.1). ∎

Problem 3.7.3 Provide an alternative proof for Theorem 3.2.2 by using the global approach in Theorem 3.3.1. That is, under the conditions of Theorem 3.2.2, first prove $\mathbb{E}\Big[\int_0^T e^{-\lambda t} |X_t|^2 dt \Big] \leq C I_0^2$ for $\lambda > 0$ large enough, and then prove (3.2.3). ∎

Problem 3.7.4 Let $d_1 = 1$, (η, b, σ) satisfy Assumption 3.2.1 (i), (iii), and assume (b, ξ) are continuous in y and satisfy the linear growth condition (3.2.2). Show that SDE (3.0.1) has a solution $X \in \mathbb{S}^2(\mathbb{F})$ satisfying (3.2.3). (Hint: mollify the coefficients and apply the comparison Theorem 3.4.1.) ∎

Problem 3.7.5 Provide an alternative proof for Theorem 3.3.1 by using contraction mapping. That is, define a mapping $F : \mathbb{L}^2(\mathbb{F}, \mathbb{R}^{d_1}) \to \mathbb{L}^2(\mathbb{F}, \mathbb{R}^{d_1})$ by $F(X) := \tilde{X}$, where

$$\tilde{X}_t := x + \int_0^t b_s(X_s) ds + \int_0^t \sigma_s(X_s) dB_s.$$

Show that F is a contraction mapping under the norm $\|X\|_\lambda^2 := \mathbb{E}\Big[\int_0^T e^{-\lambda t} |X_t|^2 dt \Big]$ for $\lambda > 0$ large enough. ∎

Problem 3.7.6 This problem concerns comparison principle.

(i) Find a counterexample for the case $\sigma^1 \neq \sigma^2$ in Theorem 3.4.1. That is, $d_1 = 1$, $\eta_1 \leq \eta_2$, $b^1 \leq b^2$, $\sigma^1 \leq \sigma^2$, but it does not hold that $X^1 \leq X^2$.

(ii) Find a counterexample for multidimensional case. To be precise, let $d_1 = 2, d = 1$, (X, η, b, σ) be as in (3.0.1), and $(\tilde{X}, \tilde{\eta}, \tilde{b}, \tilde{\sigma})$ be another system. We want $\eta_i \leq \tilde{\eta}_i$, $b^i \leq \tilde{b}^i$, $\sigma^i = \tilde{\sigma}^i$, $i = 1, 2$, but it does not hold that $X^i \leq \tilde{X}^i$, $i = 1, 2$.

(iii) We now provide a special multidimensional SDE for which comparison holds. Let (η, b, σ) and $(\tilde{\eta}, \tilde{b}, \tilde{\sigma})$ satisfy Assumption 3.2.1, and X, $\tilde{X}$ be the corresponding solution to SDE (3.0.1). Assume

$$\eta_i \leq \tilde{\eta}_i, \quad b^i \leq \tilde{b}^i, \quad \sigma^i_t(x) = \tilde{\sigma}^i(x) = a^i(x_i), \quad i = 1, \cdots, d_1,$$

for some $a^i : [0, T] \times \Omega \times \mathbb{R} \to \mathbb{R}^{1 \times d}$. Here σ^i, $\tilde{\sigma}^i$ denote the i-th row of the matrices, and the last condition implies that σ^i depends only on X^i, not on other X^j. Moreover, for $i = 1, \cdots, d_1$, assume b^i is increasing in x_j for all $j \neq i$. Show that

$$X^i_t \leq \tilde{X}^i_t, \quad 0 \leq t \leq T, \quad \mathbb{P}\text{-a.s.}, \quad i = 1, \cdots, d_1.$$

∎

Problem 3.7.7 Extend Theorem 3.4.3 to the case $1 \leq p < 2$. That is, assume Assumption 3.2.1 (i) and (ii), but replace (iii) with

$$\eta \in \mathbb{L}^p(\mathcal{F}_0, \mathbb{R}^{d_1}), \quad b \in \mathbb{L}^{1,p}(\mathbb{F}, \mathbb{R}^{d_1}), \quad \sigma \in \mathbb{L}^{2,p}(\mathbb{F}, \mathbb{R}^{d_1 \times d}).$$

Show that SDE (3.0.1) has a unique solution $X \in \mathbb{S}^p(\mathbb{F}, \mathbb{R}^{d_1})$ and the estimate (3.4.5) still holds. ∎

Problem 3.7.8 Consider the following path dependent SDE:

$$X_t = \eta + \int_0^t b_s(X_.)ds + \sigma_s(X_.)dB_s, \quad \mathbb{P}\text{-a.s.} \tag{3.7.5}$$

where $b : [0, T] \times \Omega \times C([0, T], \mathbb{R}^{d_1}) \to \mathbb{R}^{d_1}$, $\sigma : [0, T] \times \Omega \times C([0, T], \mathbb{R}^{d_1}) \to \mathbb{R}^{d_1 \times d}$. Assume

(i) b, σ are $\mathbb{F}$-measurable and are adapted to X: $b_t(X_.) = b_t(X_{. \wedge t}), \sigma_t(X_.) = \sigma_t(X_.)$;

(ii) b, σ are uniformly Lipschitz continuous in X:

$$|b_t(X_.) - b_t(\tilde{X}_.)| + |\sigma_t(X_.) - \sigma_t(\tilde{X}_.)| \leq L(X - \tilde{X})^*_t; \tag{3.7.6}$$

(iii) $\eta \in \mathbb{L}^0(\mathbb{F}, \mathbb{R}^{d_1})$, $b^0 := b(0) \in \mathbb{L}^{1,2}(\mathbb{F}, \mathbb{R}^{d_1})$, and $\sigma^0 := \sigma(0) \in \mathbb{L}^2(\mathbb{F}, \mathbb{R}^{d_1 \times d})$.

Show that SDE (3.7.5) admits a unique solution $X \in \mathbb{S}^2(\mathbb{F}, \mathbb{R}^{d_1})$. ∎

Chapter 4
Backward Stochastic Differential Equations

Let $(\Omega, \mathscr{F}, \mathbb{F}, \mathbb{P})$ be given, and B a d-dimensional Brownian motion. In order to apply the martingale representation theorem, in this chapter we shall always assume

$$\mathbb{F} = \mathbb{F}^B. \tag{4.0.1}$$

While SDE is a nonlinear extension of the stochastic integration, Backward SDE is a nonlinear version of the martingale representation theorem. In fact, both the results and the arguments in this chapter are analogous to those for SDEs, combined with the martingale representation theorem.

Given $\xi \in \mathbb{L}^2(\mathbb{F})$, it induces naturally a martingale $Y_t := \mathbb{E}[\xi|\mathscr{F}_t]$. By the martingale representation theorem, there exists unique $Z \in \mathbb{L}^2(\mathbb{F})$ such that

$$dY_t = Z_t dB_t, \quad \text{or equivalently,} \quad Y_t = \xi - \int_t^T Z_s dB_s. \tag{4.0.2}$$

This is a linear SDE with terminal condition $Y_T = \xi$, and thus is called a *Backward SDE* (BSDE, for short). We emphasize that the solution to a BSDE is a *pair* of $\mathbb{F}$-measurable processes (Y, Z). As we will see more clearly in Section 9.4, the component Z is essentially the derivative of Y with respect to B and thus is uniquely determined by Y (and B). We also emphasize that the presence of Z is crucial to ensure the $\mathbb{F}$-measurability of Y. Indeed, if we consider a SDE with terminal condition in the following form:

$$dY_t = \sigma_t(Y_t) dB_t, \quad Y_T = \xi.$$

Then typically the equation has no $\mathbb{F}$-measurable solution Y. For example, if $\sigma = 0$, then the candidate solution has to be $Y_t = \xi$ for all t, which is not $\mathbb{F}$-measurable unless $\xi \in \mathscr{F}_0$.

© Springer Science+Business Media LLC 2017
J. Zhang, *Backward Stochastic Differential Equations*, Probability Theory and Stochastic Modelling 86, DOI 10.1007/978-1-4939-7256-2_4

In this chapter we consider the following nonlinear BSDE:

$$Y_t = \xi + \int_t^T f_s(Y_s, Z_s)ds - \int_t^T Z_s dB_s, \quad 0 \le t \le T, \ \mathbb{P}\text{-a.s.} \qquad (4.0.3)$$

where $Y \in \mathbb{L}^2(\mathbb{F}, \mathbb{R}^{d_2})$, $Z \in \mathbb{L}^2(\mathbb{F}, \mathbb{R}^{d_2 \times d})$ for some dimension d_2. We call f the (nonlinear) generator and ξ the terminal condition of the BSDE. We shall always assume

Assumption 4.0.1

(i) (4.0.1) *holds;*
(ii) $f : [0, T] \times \Omega \times \mathbb{R}^{d_2} \times \mathbb{R}^{d_2 \times d} \to \mathbb{R}^{d_2}$ *is* $\mathbb{F}$-*measurable in all variables;*
(iii) f *is uniformly Lipschitz continuous in* (y, z) *with a Lipschitz constant L;*
(iv) $\xi \in \mathbb{L}^2(\mathscr{F}_T, \mathbb{R}^{d_2})$ *and* $f^0 := f(0,0) \in \mathbb{L}^{1,2}(\mathbb{F}, \mathbb{R}^{d_2})$.

As in Chapter 3, for notational simplicity we shall assume $d_2 = d = 1$ in most proofs. We remark that, in the standard literature, it is required that $f^0 \in \mathbb{L}^2(\mathbb{F})$. Our condition here is slightly weaker.

4.1 Linear Backward Stochastic Differential Equations

In this section we study the case when f is linear. We first have the following simple result.

Proposition 4.1.1 *Let* $\xi \in \mathbb{L}^2(\mathscr{F}_T, \mathbb{R}^{d_2})$ *and* $f^0 \in \mathbb{L}^{1,2}(\mathbb{F}, \mathbb{R}^{d_2})$. *Then, the following linear BSDE has a unique solution* $(Y, Z) \in \mathbb{S}^2(\mathbb{F}, \mathbb{R}^{d_2}) \times \mathbb{L}^2(\mathbb{F}, \mathbb{R}^{d_2 \times d})$:

$$Y_t = \xi + \int_t^T f_s^0 ds - \int_t^T Z_s dB_s \qquad (4.1.1)$$

Proof It is obvious that

$$Y_t = \mathbb{E}\Big[\xi + \int_t^T f_s^0 ds \Big| \mathscr{F}_t\Big].$$

Note that

$$\tilde{Y}_t := Y_t + \int_0^t f_s^0 ds = \mathbb{E}\Big[\xi + \int_0^T f_s^0 ds \Big| \mathscr{F}_t\Big]$$

is a square integrable martingale. By the martingale representation theorem, there exists unique $Z \in \mathbb{L}^2(\mathbb{F}, \mathbb{R}^{d_2 \times d})$ such that

$$d\tilde{Y}_t = Z_t dB_t.$$

One can check straightforwardly that the above pair (Y, Z) satisfies (4.1.1), and, from the above derivation, it is the unique solution. ∎

We next consider the general linear BSDE with $d_2 = 1$:

$$Y_t = \xi + \int_t^T [\alpha_s Y_s + Z_s \beta_s + f_s^0] ds - \int_t^T Z_s dB_s. \tag{4.1.2}$$

The well-posedness of this BSDE will follow from the general theory. Here we provide a representation formula for its solution.

Proposition 4.1.2 *Let* $d_2 = 1$, $\xi \in \mathbb{L}^2(\mathscr{F}_T, \mathbb{R})$, $\alpha \in \mathbb{L}^\infty(\mathbb{F}, \mathbb{R})$, $\beta \in \mathbb{L}^\infty(\mathbb{F}, \mathbb{R}^d)$, *and* $f^0 \in \mathbb{L}^{1,2}(\mathbb{F}, \mathbb{R})$. *If* $(Y, Z) \in \mathbb{L}^2(\mathbb{F}, \mathbb{R}) \times \mathbb{L}^2(\mathbb{F}, \mathbb{R}^{1 \times d})$ *satisfies the linear BSDE* (4.1.2), *then*

$$Y_t = \Gamma_t^{-1} \mathbb{E} \Big[\Gamma_T \xi + \int_t^T \Gamma_s f_s^0 ds \Big| \mathscr{F}_t \Big], \tag{4.1.3}$$

where

$$\Gamma_t = 1 + \int_0^t \Gamma_s [\alpha_s dt + \beta_s \cdot dB_s], \ or \ say, \ \Gamma_t := \exp \Big(\int_0^t \beta_s \cdot dB_s + \int_0^t [\alpha_s - \frac{1}{2} |\beta_s|^2] ds \Big). \tag{4.1.4}$$

Proof Applying Itô formula we have

$$d(\Gamma_t Y_t) = -\Gamma_t f_t^0 dt + \Gamma_t [Y_t \beta_t^\top + Z_t] dB_t.$$

Denote

$$\hat{Y}_t := \Gamma_t Y_t; \quad \hat{Z}_t := \Gamma_t [Y_t \beta_t^\top + Z_t]; \quad \hat{\xi} := \Gamma_T \xi; \quad \hat{f}_t^0 := \Gamma_t f_t^0. \tag{4.1.5}$$

Then, one may rewrite (4.1.2) as

$$\hat{Y}_t = \hat{\xi} + \int_t^T \hat{f}_s^0 ds - \int_t^T \hat{Z}_s dB_s.$$

This is a linear BSDE in the form (4.1.1). By Lemma 2.6.1 and Problem 2.10.7 (i) we see that $\int_0^t \hat{Z}_s dB_s$ is a martingale. Then

$$\hat{Y}_t := \mathbb{E} \Big[\hat{\xi} + \int_t^T \hat{f}_s^0 ds \Big| \mathscr{F}_t \Big],$$

which implies (4.1.3) immediately. ∎

4.2   A Priori Estimates for BSDEs

We now investigate the nonlinear BSDE (4.0.3).

Theorem 4.2.1 *Let Assumption 4.0.1 hold and* $(Y, Z) \in \mathbb{L}^2(\mathbb{F}, \mathbb{R}^{d_2}) \times \mathbb{L}^2(\mathbb{F}, \mathbb{R}^{d_2 \times d})$ *be a solution to BSDE (4.0.3). Then* $Y \in \mathbb{S}^2(\mathbb{F}, \mathbb{R}^{d_2})$ *and there exists a constant C, depending only on T, L, and* d, d_2, *such that*

$$\|(Y, Z)\|^2 := \mathbb{E}\Big[|Y_T^*|^2 + \int_0^T |Z_t|^2 dt \Big] \le C I_0^2, \text{ where } I_0^2 := \mathbb{E}\Big[|\xi|^2 + \big(\int_0^T |f_t^0| dt \big)^2 \Big].$$

(4.2.1)

Proof Foy simplicity, we assume $d = d_2 = 1$. We proceed in several steps.

   Step 1. We first show that

$$\mathbb{E}[|Y_T^*|^2] \le C\mathbb{E}\Big[\int_0^T [|Y_t|^2 + |Z_t|^2] dt \Big] + C I_0^2 < \infty.$$

(4.2.2)

Indeed, note that

$$|Y_t| \le |\xi| + \int_t^T [|f_s^0| + C|Y_s| + C|Z_s|] ds + \Big| \int_t^T Z_s dB_s \Big|.$$

Then,

$$Y_T^* \le C\Big[|\xi| + \int_0^T [|f_t^0| + |Y_t| + |Z_t|] dt + \sup_{0 \le t \le T} \Big| \int_0^t Z_s dB_s \Big| \Big].$$

Applying Burkholder-Davis-Gundy inequality we have

$$\mathbb{E}[|Y_T^*|^2] \le C\mathbb{E}\Big[|\xi|^2 + \Big(\int_0^T |f_t^0| dt \Big)^2 + \int_0^T [|Y_t|^2 + |Z_t|^2] dt \Big],$$

which implies (4.2.2) immediately.

   Step 2. We next show that, for any $\varepsilon > 0$,

$$\sup_{0 \le t \le T} \mathbb{E}[|Y_t|^2] + \mathbb{E}\Big[\int_0^T |Z_t|^2 dt \Big] \le \varepsilon [|Y_T^*|^2] + C\varepsilon^{-1} I_0^2.$$

(4.2.3)

Indeed, by Itô formula,

$$d|Y_t|^2 = 2Y_t dY_t + |Z_t|^2 dt = -2Y_t f_t(Y_t, Z_t) dt + 2Y_t Z_t dB_t + |Z_t|^2 dt.$$

(4.2.4)

Thus,

$$|Y_t|^2 + \int_t^T |Z_s|^2 ds = |\xi|^2 + 2\int_t^T Y_s f_s(Y_s, Z_s) ds + 2\int_t^T Y_s Z_s dB_s.$$

(4.2.5)

By (4.2.2) and Problem 2.10.7 (i) we know $\int_0^t Y_s Z_s dB_s$ is a true martingale. Now, taking expectation on both sides of (4.2.5) and noting that $ab \leq \frac{1}{2}a^2 + \frac{1}{2}b^2$, we have

$$
\begin{aligned}
\mathbb{E}\left[|Y_t|^2 + \int_t^T |Z_s|^2 ds\right] &= \mathbb{E}\left[|\xi|^2 + 2\int_t^T Y_s f_s(Y_s, Z_s) ds\right] \\
&\leq \mathbb{E}\left[|\xi|^2 + C\int_t^T |Y_s|[|f_s^0| + |Y_s| + |Z_s|]ds\right] \\
&\leq \mathbb{E}\left[|\xi|^2 + CY_T^* \int_0^T |f_s^0|ds + C\int_0^T [|Y_s|^2 + |Y_s Z_s|]ds\right] \\
&\leq \mathbb{E}\left[|\xi|^2 + CY_T^* \int_0^T |f_s^0|ds + C\int_t^T |Y_s|^2 ds + \frac{1}{2}\int_t^T |Z_s|^2 ds\right].
\end{aligned}
$$

This leads to

$$
\mathbb{E}\left[|Y_t|^2 + \frac{1}{2}\int_t^T |Z_s|^2 ds\right] \leq \mathbb{E}\left[C\int_t^T |Y_s|^2 ds + |\xi|^2 + CY_T^* \int_0^T |f_s^0|ds\right], \quad (4.2.6)
$$

which, together with Fubini Theorem, implies that

$$
\mathbb{E}[|Y_t|^2] \leq \mathbb{E}\left[|\xi|^2 + CY_T^* \int_0^T |f_s^0|ds\right] + C\int_t^T E[|Y_s|^2]ds.
$$

Applying (backward) Gronwall inequality, we get

$$
\mathbb{E}[|Y_t|^2] \leq C\mathbb{E}\left[|\xi|^2 + Y_T^* \int_0^T |f_s^0|ds\right], \quad \forall t \in [0, T]. \quad (4.2.7)
$$

Then, by letting $t = 0$ and plug (4.2.7) into (4.2.6) we have

$$
\mathbb{E}\left[\int_0^T |Z_s|^2 ds\right] \leq C\mathbb{E}\left[|\xi|^2 + Y_T^* \int_0^T |f_s^0|ds\right]. \quad (4.2.8)
$$

By (4.2.7) and (4.2.8) and noting that $2ab \leq \varepsilon a^2 + \varepsilon^{-1}b^2$, we obtain (4.2.3) immediately.

Step 3. Plug (4.2.3) into (4.2.2), we get

$$
\mathbb{E}[|Y_T^*|^2] \leq C\varepsilon\mathbb{E}[|Y_T^*|^2] + C\varepsilon^{-1}I_0^2.
$$

By choosing $\varepsilon = \frac{1}{2C}$ for the constant C above, we obtain

$$
\mathbb{E}[|Y_T^*|^2] \leq CI_0^2.
$$

This, together with (4.2.3), proves (4.2.1). ∎

Remark 4.2.2 Similar to Remark 3.2.3, Theorem 4.2.1 remains true if we weaken the Lipschitz condition of Assumption 4.0.1 (iii) to the linear growth condition:

$$|f_t(y,z)| \le |f_t^0| + L[|y| + |z|]. \tag{4.2.9}$$

■

Theorem 4.2.3 *For* $i = 1, 2$, *assume* (ξ_i, f^i) *satisfy Assumption 4.0.1 and* $(Y^i, Z^i) \in \mathbb{L}^2(\mathbb{F}, \mathbb{R}^{d_2}) \times \mathbb{L}^2(\mathbb{F}, \mathbb{R}^{d_2 \times d})$ *is a solution to BSDE (4.0.3) with coefficients* (ξ_i, f^i). *Then*

$$\|(\Delta Y, \Delta Z)\|^2 \le C\mathbb{E}\Big[|\Delta\xi|^2 + \Big(\int_0^T |\Delta f_t(Y_t^1, Z_t^1)|dt\Big)^2\Big], \tag{4.2.10}$$

where

$$\Delta Y := Y^1 - Y^2, \quad \Delta Z := Z^1 - Z^2, \quad \Delta\xi := \xi_1 - \xi_2, \quad \Delta f := f^1 - f^2.$$

Proof Again assume $d = d_2 = 1$. Note that

$$\Delta Y_t = \Delta\xi + \int_t^T [f_s^1(Y_s^1, Z_s^1) - f_s^2(Y_s^2, Z_s^2)]ds - \int_t^T \Delta Z_s dB_s$$

$$= \Delta\xi + \int_t^T [\Delta f_s(Y_s^1, Z_s^1) + \alpha_s \Delta Y_s + \beta_s \Delta Z_s]ds - \int_t^T \Delta Z_s dB_s,$$

where, similar to (3.2.10)

$$\alpha_t := \frac{f_t^2(Y_t^1, Z_t^1) - f_t^2(Y_t^2, Z_t^1)}{\Delta Y_t} \mathbf{1}_{\{\Delta Y_t \ne 0\}}, \quad \beta_t := \frac{f_t^2(Y_t^2, Z_t^1) - f_t^2(Y_t^2, Z_t^2)}{\Delta Z_t} \mathbf{1}_{\{\Delta Z_t \ne 0\}}$$

$$\tag{4.2.11}$$

are bounded by L. Then, by Theorem 4.2.1 we obtain the result immediately. ■

4.3 Well-Posedness of BSDEs

We now establish the well-posedness of BSDE (4.0.3).

Theorem 4.3.1 *Under Assumption 4.0.1, BSDE (4.0.3) has a unique solution* $(Y, Z) \in \mathbb{L}^2(\mathbb{F}, \mathbb{R}^{d_2}) \times \mathbb{L}^2(\mathbb{F}, \mathbb{R}^{d_2 \times d})$.

Proof Uniqueness follows directly from Theorem 4.2.3. In particular, the uniqueness means

$$Y_t^1 = Y_t^2 \text{ for all } t \in [0, T], \ \mathbb{P}\text{-a.s. and } Z_t^1 = Z_t^2, \ dt \times d\mathbb{P}\text{-a.s.} \tag{4.3.1}$$

We now prove the existence by using the Picard iteration. We shall use the local approach similar to that used in the proof of Theorem 3.3.1 and leave the global approach to Exercise. For simplicity we assume $d = d_2 = 1$.

Step 1. Let $\delta > 0$ be a constant which will be specified later, and assume $T \le \delta$. We emphasize that δ will depend only on the Lipschitz constant L (and the dimensions). In particular, it does not depend on the terminal condition ξ.

Denote $Y_t^0 := 0, Z_t^0 := 0$. For $n = 1, 2, \cdots$, let

$$Y_t^n = \xi + \int_t^T f_s(Y_s^{n-1}, Z_s^{n-1})ds - \int_t^T Z_s^n dB_s. \tag{4.3.2}$$

Assume $(Y^{n-1}, Z^{n-1}) \in \mathbb{L}^2(\mathbb{F}) \times \mathbb{L}^2(\mathbb{F})$. Note that

$$|f_t(Y_t^{n-1}, Z_t^{n-1})| \le C\Big[|f_t^0| + |Y_t^{n-1}| + |Z_t^{n-1}|\Big].$$

Then, $f_t(Y_t^{n-1}, Z_t^{n-1}) \in \mathbb{L}^{1,2}(\mathbb{F})$. By Proposition 4.1.1, the linear BSDE (4.3.2) uniquely determines $(Y^n, Z^n) \in \mathbb{L}^2(\mathbb{F}) \times \mathbb{L}^2(\mathbb{F})$, and then Theorem 4.2.1 implies further that $(Y^n, Z^n) \in \mathbb{S}^2(\mathbb{F}) \times \mathbb{L}^2(\mathbb{F})$. By induction we have $(Y^n, Z^n) \in \mathbb{S}^2(\mathbb{F}) \times \mathbb{L}^2(\mathbb{F})$ for all $n \ge 0$.

Denote $\Delta Y_t^n := Y_t^n - Y_t^{n-1}, \Delta Z_t^n := Z_t^n - Z_t^{n-1}$. Then,

$$\Delta Y_t^n = \int_t^T [\alpha_s^{n-1} \Delta Y_s^{n-1} + \beta_s^{n-1} \Delta Z_s^{n-1}]ds - \int_t^T \Delta Z_s^n dB_s,$$

where α^n, β^n are defined in a similar way as in (4.2.11) and are bounded by L. Applying Itô formula we have

$$d(|\Delta Y_t^n|^2) = -2\Delta Y_t^n[\alpha_t^{n-1}\Delta Y_t^{n-1} + \beta_t^{n-1}\Delta Z_t^{n-1}]dt + 2\Delta Y_t^n \Delta Z_t^n dB_t + |\Delta Z_t^n|^2 dt.$$

By Problem 2.10.7 (i), $\int_0^t \Delta Y_s^n \Delta Z_s^n dB_s$ is a true martingale. Noting that $\Delta Y_T^n = 0$, we get

$$\mathbb{E}\Big[|\Delta Y_t^n|^2 + \int_t^T |\Delta Z_s^n|^2 ds\Big] = \mathbb{E}\Big[2\int_t^T [\Delta Y_s^n[\alpha_s^{n-1}\Delta Y_s^{n-1} + \beta_s^{n-1}\Delta Z_s^{n-1}]ds\Big]$$

$$\le C\mathbb{E}\Big[\int_0^T |\Delta Y_s^n|[|\Delta Y_s^{n-1}| + |\Delta Z_s^{n-1}|]ds\Big]. \tag{4.3.3}$$

Thus

$$\mathbb{E}\Big[\int_0^T |\Delta Y_t^n|^2 dt\Big] \le C\delta\mathbb{E}\Big[\int_0^T |\Delta Y_s^n|[|\Delta Y_s^{n-1}| + |\Delta Z_s^{n-1}|]ds\Big]$$

$$\le C\delta\mathbb{E}\Big[\int_0^T [|\Delta Y_t^n|^2 + |\Delta Y_t^{n-1}|^2 + |\Delta Z_t^{n-1}|^2]dt\Big]$$

Assume $\delta < \frac{1}{2C}$ for the above constant C and thus $1 - C\delta \le \frac{1}{2}$, then,

$$\mathbb{E}\Big[\int_0^T |\Delta Y_t^n|^2 dt \Big] \le C\delta \mathbb{E}\Big[\int_0^T [|\Delta Y_t^{n-1}|^2 + |\Delta Z_t^{n-1}|^2] dt \Big].$$

Moreover, by setting $t = 0$ in (4.3.3), we have

$$\mathbb{E}\Big[\int_0^T |\Delta Z_t^n|^2 dt \Big] \le C\mathbb{E}\Big[\int_0^T |\Delta Y_s^n|^2 dt \Big] + \frac{1}{8}\mathbb{E}\Big[\int_0^T [|\Delta Y_t^{n-1}|^2 + |\Delta Z_t^{n-1}|^2] dt \Big]$$

$$\le \Big[C\delta + \frac{1}{8} \Big] \mathbb{E}\Big[\int_0^T [|\Delta Y_t^{n-1}|^2 + |\Delta Z_t^{n-1}|^2] dt \Big].$$

Thus

$$\mathbb{E}\Big[\int_0^T [|\Delta Y_t^n|^2 + |\Delta Z_t^n|^2] dt \Big] \le \Big[C\delta + \frac{1}{8} \Big] \mathbb{E}\Big[\int_0^T [|\Delta Y_t^{n-1}|^2 + |\Delta Z_t^{n-1}|^2] dt \Big].$$

Set $\delta := \frac{1}{8C}$ for the above C. Then

$$\mathbb{E}\Big[\int_0^T [|\Delta Y_t^n|^2 + |\Delta Z_t^n|^2] dt \Big] \le \frac{1}{4} \mathbb{E}\Big[\int_0^T [|\Delta Y_t^{n-1}|^2 + |\Delta Z_t^{n-1}|^2] dt \Big].$$

By induction we have

$$\mathbb{E}\Big[\int_0^T [|\Delta Y_t^n|^2 + |\Delta Z_t^n|^2] dt \Big] \le \frac{C}{4^n}, \quad \forall n \ge 1.$$

Now following the arguments in Theorem 3.3.1 one can easily see that there exists $(Y, Z) \in \mathbb{S}^2(\mathbb{F}) \times \mathbb{L}^2(\mathbb{F})$ such that

$$\lim_{n \to \infty} \|(Y_t^n - Y_t, Z_t^n - Z_t)\| = 0.$$

Therefore, by letting $n \to \infty$ in BSDE (4.3.2) we know that (Y, Z) satisfies BSDE (4.0.3).

   Step 2. We now prove the existence for arbitrary T. Let $\delta > 0$ be the constant in Step 1. Consider a partition $0 = t_0 < \cdots < t_n = T$ such that $t_{i+1} - t_i \le \delta$, $= 0, \cdots, n - 1$. Define $Y_{t_n} := \xi$, and for $i = n - 1, \cdots, 0$ and $t \in [t_i, t_{i+1})$, let (Y_t, Z_t) be the solution to the following BSDE on $[t_i, t_{i+1}]$:

$$Y_t = Y_{t_{i+1}} + \int_t^{t_{i+1}} f_s(Y_s, Z_s) ds - \int_t^{t_{i+1}} Z_s dB_s, \quad t \in [t_i, t_{i+1}].$$

Since $t_{i+1} - t_i \leq \delta$, by Step 1 the above BSDE is well posed. Moreover, since n is finite here, we see that $(Y, Z) \in \mathbb{L}^2(\mathbb{F}) \times \mathbb{L}^2(\mathbb{F})$, and thus they are a global solution on the whole interval $[0, T]$. ∎

Remark 4.3.2 Assume f satisfies Assumption 4.0.1, $\tau \in \mathscr{T}(\mathbb{F})$, and $\xi \in \mathbb{L}^2(\mathscr{F}_\tau)$. Consider the following BSDE

$$Y_t = \xi + \int_t^T \tilde{f}_s(Y_s, Z_s) ds - \int_t^T Z_s dB_s, \quad \text{where} \quad \tilde{f}_s(y, z) := f_s(y, z) \mathbf{1}_{[0, \tau]}(s).$$

$$(4.3.4)$$

One can easily see that $\tilde{f}$ also satisfies Assumption 4.0.1, and thus the above BSDE has a unique solution. Since $\xi \in \mathscr{F}_\tau$, we see immediately that $Y_s := \xi, Z_s := 0$ satisfy (4.3.4) for $s \in [\tau, T]$. Therefore, we may rewrite (4.3.4) as

$$Y_t = \xi + \int_t^\tau f_s(Y_s, Z_s) ds - \int_t^\tau Z_s dB_s, \quad 0 \leq t \leq \tau, \qquad (4.3.5)$$

and it is also well posed. ∎

4.4 Basic Properties of BSDEs

As in Section 3.4, we start with the comparison result, in the case $d_2 = 1$.

Theorem 4.4.1 (Comparison Theorem) *Let $d_2 = 1$. Assume, for $i = 1, 2$, (ξ_i, f^i) satisfies Assumption 4.0.1 and $(Y^i, Z^i) \in \mathbb{S}^2(\mathbb{F}, \mathbb{R}) \times \mathbb{L}^2(\mathbb{F}, \mathbb{R}^{1 \times d})$ is the unique solution to the following BSDE:*

$$Y_t^i = \xi_i + \int_t^T f_s^i(Y_s^i, Z_s^i) ds - \int_t^T Z_s^i dB_s. \qquad (4.4.1)$$

Assume further that $\xi_1 \leq \xi_2$, $\mathbb{P}$-a.s., and $f^1(y, z) \leq f^2(y, z)$, $dt \times d\mathbb{P}$-a.s. that for any (y, z). Then,

$$Y_t^1 \leq Y_t^2, \quad 0 \leq t \leq T, \quad \mathbb{P}\text{-a.s.} \qquad (4.4.2)$$

Proof Denote

$$\Delta Y_t := Y_t^1 - Y_t^2; \quad \Delta Z_t := Z_t^1 - Z_t^2; \quad \Delta \xi := \xi_1 - \xi_2, \quad \Delta f := f^1 - f^2.$$

Then,

$$\Delta Y_t = \Delta \xi + \int_t^T [f_s^1(Y_s^1, Z_s^1) - f_s^2(Y_s^2, Z_s^2)] ds - \int_t^T \Delta Z_s dB_s$$

$$= \Delta \xi + \int_t^T [\alpha_s \Delta Y_s + \Delta Z_s \beta_s + \Delta f_s(Y_s^2, Z_s^2)] ds - \int_t^T \Delta Z_s dB_s,$$

where α and β are bounded. Define Γ by (4.1.4). By (4.1.3) we have

$$\Delta Y_t = \Gamma_t^{-1} \mathbb{E}\Big[\Gamma_T \Delta \xi + \int_t^T \Gamma_s \Delta f_s(Y_s^2, Z_s^2) ds \Big| \mathscr{F}_t\Big]. \tag{4.4.3}$$

Similar to (3.4.1), by Problem 1.4.6 (ii) we have

$$f^1(y, z) \le f^2(y, z) \text{ for all } (y, z), \; dt \times d\mathbb{P}\text{-a.s.}$$

This implies that $\Delta f(Y^2, Z^2) \le 0$, $dt \times d\mathbb{P}$-a.s. Since $\Gamma \ge 0$ and $\Delta \xi \le 0$, then (4.4.2) follows from (4.4.3) immediately.   ∎

Remark 4.4.2 In the Comparison Theorem we require the process Y to be scalar. The comparison principle for general multidimensional BSDEs is an important but very challenging subject. See Problem 4.7.5 for some simple result.   ∎

We next establish the stability result.

Theorem 4.4.3 (Stability) *Let* (ξ, f) *and* (ξ_n, f^n), $n = 1, 2, \cdots$, *satisfy Assumption 4.0.1 with the same Lipschitz constant L, and* $(Y, Z), (Y^n, Z^n) \in \mathbb{S}^2(\mathbb{F}, \mathbb{R}^{d_2}) \times \mathbb{L}^2(\mathbb{F}, \mathbb{R}^{d_2 \times d})$ *be the solution to the corresponding BSDE (4.0.3). Denote*

$$\Delta Y^n := Y^n - Y, \quad \Delta Z^n := Z^n - Z; \quad \Delta \xi_n := \xi_n - \xi, \quad \Delta f^n := f^n - f.$$

Assume

$$\lim_{n \to \infty} \mathbb{E}\Big[|\Delta \xi_n|^2 + \Big(\int_0^T |\Delta f_t^n(0, 0)| dt\Big)^2\Big] = 0, \tag{4.4.4}$$

and that $\Delta f^n(y, z) \to 0$ *in measure* $dt \times d\mathbb{P}$, *for all* (y, z). *Then,*

$$\lim_{n \to \infty} \|(\Delta Y^n, \Delta Z^n)\| = 0. \tag{4.4.5}$$

Proof First, by (4.2.10) we have

$$\|(\Delta Y^n, \Delta Z^n)\|^2 \le C\mathbb{E}\Big[|\Delta \xi_n|^2 + \Big(\int_0^T |\Delta f_t^n(Y_t, Z_t)| dt\Big)^2\Big]$$

$$\le C\mathbb{E}\Big[|\Delta \xi_n|^2 + \Big(\int_0^T |\Delta f_t^n(0, 0)| dt\Big)^2 + \Big(\int_0^T |\Delta f_t^n(Y_t, Z_t) - \Delta f_t^n(0, 0)| dt\Big)^2\Big]. \tag{4.4.6}$$

By Problem 1.4.6 (iii), $\Delta f^n(Y, Z) \to 0$, in measure $dt \times d\mathbb{P}$. Note that

$$|\Delta f_n(t, Y_t, Z_t) - \Delta f_n(t, 0, 0)| \le C[|Y_t| + |Z_t|].$$

Applying the dominated convergence Theorem we have

$$\lim_{n \to \infty} \mathbb{E}\Big[\Big(\int_0^T |\Delta f_t^n(Y_t, Z_t) - \Delta f_t^n(0, 0)| dt\Big)^2\Big] = 0.$$

This, together with (4.4.4) and (4.4.6), leads to the result.   ∎

We conclude this section by extending the well-posedness result to $\mathbb{L}^p(\mathbb{F})$ for $p \geq 2$.

Theorem 4.4.4 *Assume Assumption 4.0.1 holds and* $\xi \in \mathbb{L}^p(\mathscr{F}_T, \mathbb{R}^{d_2})$, $f^0 \in \mathbb{L}^{1,p}(\mathbb{F}, \mathbb{R}^{d_2})$ *for some* $p \geq 2$. *Let* $(Y, Z) \in \mathbb{S}^2(\mathbb{F}, \mathbb{R}^{d_2}) \times \mathbb{L}^2(\mathbb{F}, \mathbb{R}^{d_2 \times d})$ *be the unique solution to BSDE* (4.0.3). *Then,*

$$\mathbb{E}\Big[|Y_T^*|^p + \Big(\int_0^T |Z_t|^2 dt\Big)^{\frac{p}{2}}\Big] \leq C_p I_p^p, \text{ where } I_p^p := \mathbb{E}\Big[|\xi|^p + \Big(\int_0^T |f_t^0| dt\Big)^p\Big].$$

$$(4.4.7)$$

Proof As in Theorem 3.4.3 we proceed in two steps. Again assume $d = d_2 = 1$ for simplicity.

Step 1. We first assume $Y \in \mathbb{L}^{\infty,p}(\mathbb{F})$, $Z \in \mathbb{L}^{2,p}(\mathbb{F})$ and prove (4.4.7). Applying Itô formula we have

$$d|Y_t|^2 = -2Y_t f_t(Y_t, Z_t) dt + |Z_t|^2 dt + 2Y_t Z_t dB_t;$$

$$d(|Y_t|^p) = d(|Y_t|^2)^{\frac{p}{2}} = -p|Y_t|^{p-2} Y_t f_t(Y_t, Z_t) dt + \frac{1}{2} p(p-1)|Y_t|^{p-2}|Z_t|^2 dt + p|Y_t|^{p-2} Y_t Z_t dB_t.$$

$$(4.4.8)$$

Following the arguments in Theorem 4.2.1 Steps 1 and 2 one can easily show that, for any $\varepsilon > 0$,

$$\mathbb{E}\Big[|Y_T^*|^p\Big] \leq C_p \sup_{0 \leq t \leq T} \mathbb{E}[|Y_t|^p] + C_p \mathbb{E}\Big[\int_0^T |Y_t|^{p-2}|Z_t|^2 dt\Big] + C_p I_p^p;$$

$$\sup_{0 \leq t \leq T} \mathbb{E}[|Y_t|^p] + \mathbb{E}\Big[\int_0^T |Y_t|^{p-2}|Z_t|^2 dt\Big] \leq \varepsilon \mathbb{E}\Big[|Y_T^*|^p\Big] + C_p \varepsilon^{-1} I_p^p.$$

Then, by choosing $\varepsilon > 0$ small enough we obtain

$$\mathbb{E}\Big[|Y_T^*|^p\Big] \leq C_p I_p^p. \qquad (4.4.9)$$

Next, by (4.4.8) we see that

$$\int_0^T |Z_t|^2 dt = |\xi|^2 - |Y_0|^2 + 2\int_0^T Y_t f_t(Y_t, Z_t) dt - 2\int_0^T Y_t Z_t dB_t$$

$$\leq C|Y_T^*|^2 + C\int_0^T |Y_t|[|f_t^0| + |Y_t| + |Z_t|] dt + C\Big|\int_0^T Y_t Z_t dB_t\Big|$$

$$\leq C|Y_T^*|^2 + C\Big(\int_0^T |f_t^0| dt\Big)^2 + \frac{1}{2}\int_0^T |Z_t|^2 dt + C\Big|\int_0^T Y_t Z_t dB_t\Big|.$$

Then, by (4.4.9) and Burkholder-Davis-Gundy inequality,

$$\mathbb{E}\Big[\Big(\int_0^T |Z_t|^2 dt\Big)^{\frac{p}{2}}\Big] \le C_p I_p^2 + C_p \mathbb{E}\Big[|Y_T^*|^p + \big|\int_0^T Y_t Z_t dB_t\big|^{\frac{p}{2}}\Big]$$

$$\le C_p I_p^2 + C_p \mathbb{E}\Big[\Big(\int_0^T |Y_t Z_t|^2 dt\Big)^{\frac{p}{4}}\Big] \le C_p I_p^2 + C_p \mathbb{E}\Big[|Y_T^*|^{\frac{p}{2}} \Big(\int_0^T |Z_t|^2 dt\Big)^{\frac{p}{4}}\Big]$$

$$\le C_p I_p^2 + C_p \mathbb{E}[|Y_T^*|^p] + \frac{1}{2}\mathbb{E}\Big[\Big(\int_0^T |Z_t|^2 dt\Big)^{\frac{p}{2}}\Big] \le C_p I_p^2 + \frac{1}{2}\mathbb{E}\Big[\Big(\int_0^T |Z_t|^2 dt\Big)^{\frac{p}{2}}\Big].$$

This leads to the desired estimate for Z, and together with (4.4.9), proves further (4.4.7).

Step 2. In the general case, we shall use the space truncation arguments in Theorem 3.4.3. We note that the time truncation does not work well here because it will involve Y_{τ_n} which still lacks desired integrability. For each $n \ge 1$, denote $\xi_n := (-n) \vee \xi \wedge n, f_n := (-n) \vee f \wedge n$. Clearly (ξ_n, f^n) satisfy all the conditions of this theorem with the same Lipschitz constant L, and

$$(\xi_n, f^n) \to (\xi, f), \quad |\xi_n| \le |\xi|, |f^n| \le |f|, \quad |\xi_n| \le n, |f^n| \le n, \quad \text{for all } (t, \omega, y, z).$$

Let $(Y^n, Z^n) \in \mathbb{S}^2(\mathbb{F}) \times \mathbb{L}^2(\mathbb{F})$ be the unique solution to BSDE (4.0.3) with coefficients (ξ_n, f^n). Then

$$Y_t^n = \mathbb{E}\Big[\xi_n + \int_t^T f_s^n(Y_s^n, Z_s^n) ds \Big| \mathscr{F}_t\Big], \quad \int_0^t Z_s^n dB_s = Y_t^n - Y_0^n + \int_0^t f_s^n(Y_s^n, Z_s^n) ds$$

are bounded. By the Burkholder-Davis-Gundy inequality, this implies further that $Z^n \in \mathbb{L}^{2,p}(\mathbb{F})$. Then it follows from Step 1 that

$$\mathbb{E}\Big[|(Y^n)_T^*|^p + \Big(\int_0^T |Z_t^n|^2 dt\Big)^{\frac{p}{2}}\Big] \le C_p \mathbb{E}\Big[|\xi_n|^p + \Big(\int_0^T |f_t^n(0,0)| dt\Big)^p\Big] \le C_p I_p^p.$$

Now similar to the arguments in Theorem 4.2.1, (4.4.7) follows from Theorem 4.4.3 and Fatou lemma. ∎

4.5 Some Applications of BSDEs

The theory of BSDEs has wide applications in many fields, most notably in mathematical finance, stochastic control theory, and probabilistic numerical methods for nonlinear PDEs. We shall discuss its connection with PDE rigorously in the next chapter. In this section we present the first two types of applications in very simple settings and in a heuristic way, just to illustrate the idea.

4.5.1 Application in Asset Pricing and Hedging Theory

Consider the Black-Scholes model in Section 2.8. Assume a self-financing portfolio (λ, h) hedges ξ. By (2.8.8) and (2.8.7) we have:

$$dV_t = \left[\lambda_t r e^{rt} + h_t S_t \mu\right] dt + h_t S_t \sigma dB_t$$

$$= \left[r(V_t - h_s S_t) + h_t S_t \mu\right] dt + h_t S_t \sigma dB_t. \tag{4.5.1}$$

Denote

$$Y_t := V_t, \quad Z_t := \sigma S_t h_t. \tag{4.5.2}$$

Then (4.5.1) leads to

$$dY_t = \left[r[Y_t - \frac{Z_t}{\sigma S_t}] + \frac{\mu Z_t}{\sigma S_t}\right] dt + Z_t dB_t, \quad Y_T = \xi, \quad \mathbb{P}\text{-a.s.} \tag{4.5.3}$$

This is a linear BSDE. Once we solve it, we obtain that:

Y is the price of the option ξ and Z induces the hedging portfolio: $h_t = \dfrac{Z_t}{\sigma S_t}$.

$$\tag{4.5.4}$$

We remark that BSDE (4.5.3) is under the market measure $\mathbb{P}$. In this approach, there is no need to talk about the risk neutral measure.

Note that BSDE (4.5.3) is linear, which can be solved explicitly. In particular, for the special example we are presenting, Y_0 can be computed via the well-known Black-Scholes formula. To motivate nonlinear BSDEs, let us assume in a more practical manner that the lending interest rate r_1 is less than the borrowing interest rate r_2. That is, the self-financing condition (4.5.1) should be replaced by

$$dV_t := \left[r_1(V_t - h_t S_t)^+ - r_2(V_t - h_t S_t)^-\right] dt + h_t dS_t, \tag{4.5.5}$$

and therefore, BSDE (4.5.3) becomes a nonlinear one:

$$dY_t = \left[r_1(Y_t - \frac{Z_t}{\sigma S_t})^+ - r_2(Y_t - \frac{Z_t}{\sigma S_t})^- + \frac{\mu Z_t}{\sigma S_t}\right] dt + Z_t dB_t, \quad Y_T = \xi. \tag{4.5.6}$$

Nonlinear BSDEs typically do not have explicit formula. We shall discuss its numerical method in the next chapter.

4.5.2 Applications in Stochastic Control

Consider a controlled SDE:

$$X_t^k = x + \int_0^t b(s, X_s^k, k_s) ds + \int_0^t \sigma(s, X_s^k, k_s) dB_s, \quad 0 \le t \le T, \ \mathbb{P}\text{-a.s. (4.5.7)}$$

Here B, X, b, σ take values in $\mathbb{R}^d$, $\mathbb{R}^{d_1}$, $\mathbb{R}^{d_1}$, and $\mathbb{R}^{d_1 \times d}$, respectively, and $k \in \mathscr{K}$ are admissible controls. We assume k takes values in certain Polish space $\mathbb{K}$ and is $\mathbb{F}$-measurable. Our goal is the following stochastic optimization problem (with superscript ^{S} indicating strong formulation in contrast to the weak formulation in (4.5.12) below):

$$V_0^S := \sup_{k \in \mathscr{K}} J_S(k) \quad \text{where} \quad J_S(k) := \mathbb{E}^{\mathbb{P}} \Big[g(X_T^k) + \int_0^T f(t, X_t^k, k_t) dt \Big], \ (4.5.8)$$

where f and g are 1-dimensional and thus J_S and V_0^S are scalars.

If we follow the standard stochastic maximum principle, the above problem will lead to a forward-backward SDE, which is the main subject of Chapter 8 and is in general not solvable. We thus transform the problem to weak formulation as follows. We remark that the weak formulation, especially when there is diffusion control (namely σ depends on k), will be our main formulation for stochastic control problems and will be explored in details in Part III. Here we just present some very basic ideas. For this purpose, we assume

Assumption 4.5.1

  (i) *b, σ, f, g are deterministic, Borel measurable in all variables, and bounded (for simplicity);*
  (ii) *$\sigma = \sigma(t, x)$ does not contain the control k, and is uniformly Lipschitz in x;*
  (iii) *There exists a bounded $\mathbb{R}^d$-valued function $\theta(t, x, k)$ such that $b(t, x, k) = \sigma(t, x) \theta(t, x, k)$.*

We note that, when $d = d_1$ and $\sigma \in \mathbb{S}^d$ is invertible, it is clear that $\theta(t, x, k) = \sigma^{-1}(t, x) b(t, x, k)$ and is unique.

Let X be the unique solution to the following SDE (without control):

$$X_t = x + \int_0^t \sigma(s, X_s) dB_s, \quad 0 \le t \le T, \ \mathbb{P}\text{-a.s.} \tag{4.5.9}$$

For each $k \in \mathscr{K}$, recall the notations in Section 2.6 and denote

$$\theta_t^k := \theta(t, X_t, k_t), \quad B_t^k := B_t - \int_0^t \theta_s^k ds, \quad M^k := M^{\theta^k}, \quad \mathbb{P}^k := \mathbb{P}^{\theta^k}. (4.5.10)$$

Under Assumption 4.5.1 (iii), θ^k is bounded and thus it follows from the Girsanov Theorem that B^k is a $\mathbb{P}^k$-Brownian motion. Since $\mathbb{P}^k$ is equivalent to $\mathbb{P}$, then (4.5.9) leads to

$$X_t = x + \int_0^t b(s, X_s, k_s)ds + \int_0^t \sigma(s, X_s)dB_s^k, \quad 0 \le t \le T, \ \mathbb{P}^k\text{-a.s. (4.5.11)}$$

Compare (4.5.11) with (4.5.7), we modify (4.5.8) as

$$V_0 := \sup_{k \in \mathcal{K}} J(k), \quad \text{where} \quad J(k) := \mathbb{E}^{\mathbb{P}^k}\left[g(X_T) + \int_0^T f(t, X_t, k_t)dt\right]. (4.5.12)$$

This is the stochastic optimization problem under weak formulation (with drift control only).

Remark 4.5.2

(i) In strong formulation (4.5.8), $\mathbb{P}$ is fixed and one controls the state process X^k, while in weak formulation (4.5.8), the state process X is fixed and one controls the probability $\mathbb{P}^k$, or more precisely controls the distribution of X.

(ii) Although formally (4.5.11) looks very much like (4.5.7), the $\mathbb{P}^k$-distribution of k is different from the $\mathbb{P}$-distribution of k, then the joint $\mathbb{P}^k$-distribution of (B^k, k, X) is different from the joint $\mathbb{P}$-distribution of (B, k, X^k). Consequently, for given $k \in \mathcal{K}$, typically $J(k) \ne J_S(k)$.

(iii) In most interesting applications, it holds that $V_0^S = V_0$. However, in general it is possible that they are not equal. Nevertheless, in this section we investigate V_0. This is partially because the optimization problem (4.5.12) is technically easier, and more importantly because the weak formulation is more appropriate in many applications, as we discuss next.

(iv) As discussed in Section 2.8.3, in many applications one can actually observe the state process X, rather than the noise B. So it makes more sense to assume the control k depends on X, instead of on B (or ω). That is, weak formulation is more appropriate than strong formulation in many applications, based on the information one observes. In this case, of course, we shall either restrict $\mathcal{K}$ to $\mathbb{F}^X$-measurable processes or assume $\mathbb{F}^X = \mathbb{F}^B$ (e.g., when $d = d_1$ and $\sigma > 0$).

(v) Even when $V_0^S = V_0$, it is much more likely to have the existence of optimal control in weak formulation than in strong formulation. See Remark 4.5.4 below. ∎

We now solve (4.5.12). For each $k \in \mathcal{K}$, applying Theorem 2.6.6, the martingale representation theorem under Girsanov setting, one can easily see that the following linear BSDE under $\mathbb{P}^k$ has a unique solution (Y^k, Z^k):

$$Y_t^k = g(X_T) + \int_t^T f(s, X_s, k_s)ds - \int_t^T Z_s^k dB_s^k, \quad \mathbb{P}^k\text{-a.s.} \qquad (4.5.13)$$

Clearly $J(k) = Y_0^k$. By (4.5.10) and noting that $\mathbb{P}^k$ and $\mathbb{P}$ are equivalent, we may rewrite (4.5.13) as

$$Y_t^k = g(X_T) + \int_t^T \Big[f(s, X_s, k_s) + Z_s^k \theta(s, X_s, k_s) \Big] ds - \int_t^T Z_s^k dB_s, \quad \mathbb{P}\text{-a.s.}$$

$$(4.5.14)$$

Define the Hamiltonians:

$$H^*(t, x, z) := \sup_{k \in \mathbb{K}} H(t, x, z, k), \quad \text{where} \quad H(t, x, z, k) := f(t, x, k) + z\theta(t, x, k).$$

$$(4.5.15)$$

By Assumption 4.5.1 (iii) and (i), H^* is uniformly Lipschitz continuous in z and $H^*(t, x, 0)$ is bounded. Then the following BSDE has a unique solution (Y^*, Z^*):

$$Y_t^* = g(X_T) + \int_t^T H^*(s, X_s, Z_s^*) ds - \int_t^T Z_s^* dB_s, \quad \mathbb{P}\text{-a.s.} \qquad (4.5.16)$$

We have the following main result for this subsection.

Theorem 4.5.3 *Under Assumption 4.5.1, we have*

$$V_0 = Y_0^*. \qquad (4.5.17)$$

Moreover, if there exists a Borel measurable function $I : [0, T] \times \mathbb{R}^{d_1} \times \mathbb{R}^d \to \mathbb{K}$ *such that*

$$H^*(t, x, z) = H(t, x, z, I(t, x, z)). \qquad (4.5.18)$$

Then

$$k_t^* := I(t, X_t, Z_t^*) \text{ is an optimal control.} \qquad (4.5.19)$$

Proof First, applying comparison theorem, we have $Y_0^k \leq Y_0^*$ for all $k \in \mathscr{K}$, and thus $V_0 \leq Y_0^*$. On the other hand, for any $\varepsilon > 0$, by standard measurable selection there exists a Borel measurable function $I^\varepsilon : [0, T] \times \mathbb{R}^{d_1} \times \mathbb{R}^d \to \mathbb{K}$ such that

$$H^*(t, x, z) \leq H(t, x, z, I^\varepsilon(t, x, z)) + \varepsilon.$$

Denote $k_t^\varepsilon := I^\varepsilon(t, X_t, Z_t^*)$, and thus $H^*(t, X_t, Z_t^*) \leq H(t, X_t, Z_t^*, k_t^\varepsilon) + \varepsilon$. Note that

$$Y_t^{k^\varepsilon} = g(X_T) + \int_t^T H(s, X_s, Z_s^{k^\varepsilon}, k_s^\varepsilon) ds - \int_t^T Z_s^{k^\varepsilon} dB_s.$$

Denote $\Delta Y^\varepsilon := Y^* - Y^{k^\varepsilon}$, $\Delta Z^\varepsilon := Z^* - Z^{k^\varepsilon}$. Then

$$\Delta Y_t^\varepsilon = \int_t^T \Big[H^*(s,X_s,Z_s^*) - H(s,X_s,Z_s^*,k_s^\varepsilon) + \Delta Z_s^\varepsilon \theta(s,X_s,k_s^\varepsilon) \Big] ds - \int_t^T \Delta Z_s^\varepsilon dB_s$$

$$= \int_t^T \Big[H^*(s,X_s,Z_s^*) - H(s,X_s,Z_s^*,k_s^\varepsilon) \Big] ds - \int_t^T \Delta Z_s^\varepsilon dB_s^{k^\varepsilon} \le \varepsilon(T-t) - \int_t^T \Delta Z_s^\varepsilon dB_s^{k^\varepsilon}.$$

This implies that $\Delta Y_0^\varepsilon \le T\varepsilon$. Since $\varepsilon > 0$ is arbitrary, we obtain $Y_0^* \le V_0$, and hence the equality holds.

Finally, under (4.5.18) it is clear that $Y^* = Y^{k^*}$, which implies (4.5.19) immediately. ∎

Remark 4.5.4 We emphasize that the optimal control k^* in (4.5.19) is optimal in weak formulation, but not necessarily in strong formulation. To illustrate the main idea, let us consider a special case: $d = d_1 = 1$, $\sigma = 1$, $x = 0$, and then $X = B$. Since k^* is $\mathbb{F}^B$-measurable, so we may write $k^* = k^*(B) = k^*(X)$. Assume $V_0^S = V_0$, then the above k^* provides an optimal control in strong formulation amounts to say the following SDE admits a strong solution:

$$X_t = x + \int_0^t b(s,X_s,k_s^*(X.))ds + \int_0^t \sigma(s,X_s)dB_s, \quad \mathbb{P}\text{-a.s.} \quad (4.5.20)$$

We remark that, in this special case here, actually one can show that $k_t^* = k^*(t,X_t^*)$ depends only on X_t^*. However, k^* may be discontinuous in X, and thus it is difficult to establish a general theory for the strong solvability of SDE (4.5.20). Moreover, one may easily extend Theorem 4.5.3 to the path dependent case, namely b, f, and/or g depend on the paths of X. In this case k^* may also depend on the paths of X^* and thus (4.5.20) becomes path dependent. Typically this SDE does not have a strong solution, see a counterexample in Wang & Zhang [231] which is based on Tsirelson's [229] counterexample. Consequently, the optimization problem (4.5.8) (or its extension to path dependent case) in strong formulation may not have an optimal control. ∎

4.6 Bibliographical Notes

The linear BSDE was first proposed by Bismut [16], motivated from applications in stochastic control, and the well-posedness of nonlinear BSDEs was established by the seminal paper Paradox & Peng [167]. There is an excellent exposition on the basic theory and applications of BSDEs in El Karoui, Peng, & Quenez [81], and Peng [182] provides a detailed survey on the theory and its further developments. Another application which independently leads to the connection with BSDE is the recursive utility proposed by Duffie and Epstein [69, 70]. We also refer to some book chapters El Karoui & Mazliak [80], Peng [175], Yong & Zhou [242], Pham

[190], Cvitanic & Zhang [52], Touzi [227], as well as the recent book Pardoux & Rascanu [170] on theory and applications of BSDEs. In particular, many materials of this and the next chapter follow from the presentation in [52].

We note that the materials in this chapter are very basic. There have been various extensions of the theory, with some important ones presented in the next chapter and Part II. The further extension to fully nonlinear situation is the subject of Part III. Besides those and among many others, we note that Lepeltier & San Martin [135] studied BSDEs with non-Lipschitz continuous generators, Tang & Li [223] studied BSDEs driven by jump processes, Fuhrman & Tessitore [95] studied BSDEs in infinite dimensional spaces, Darling & Pardoux [54] studied BSDEs with random terminal time, Buckdahn, Engelbert, & Rascanu [24] studied weak solutions of BSDEs, and Pardoux & Peng [169] studied backward doubly SDEs which provides a representation for solutions to (forward) stochastic PDEs. Moreover, we note that Hu & Peng [110] provided some general result concerning comparison principle for multidimensional BSDEs, and Hamadene & Lepeltier [103] extended the stochastic optimization problem to a zero-sum stochastic differential game problem, again in weak formulation. Another closely related concept is the g-expectation developed by Peng [176, 179], see also Coquet, Hu, Memin, & Peng [38], Chen & Epstein [30], and Delbaen, Peng, & Rosazza Gianin [53]. This is a special type of the nonlinear expectation which we will introduce in Chapter 10.

4.7 Exercises

Problem 4.7.1 Similar to Problem 3.7.2, this problem consider the decoupling strategy for multidimensional linear BSDE. For simplicity, we consider the following linear BSDE with $d = 1$ and $d_2 = 2$:

$$Y_t^i = \xi_i + \int_0^t \Big[\sum_{j=1}^2 [\alpha_s^{ij} Y_s^j + \beta_s^{ij} Z_s^j] + \gamma_s^i \Big] ds + \int_0^t Z_s^i dB_s, \quad i = 1, 2. \quad (4.7.1)$$

Here $\xi_i \in \mathbb{L}^2(\mathscr{F}_T, \mathbb{R})$, $\alpha^{ij}, \beta^{ij} \in \mathbb{L}^\infty(\mathbb{F}, \mathbb{R})$, and $\gamma^i \in \mathbb{L}^{1,2}(\mathbb{F}, \mathbb{R})$. Show that there exists a process Γ such that $\overline{Y} := Y^1 + \Gamma Y^2$ solves a one-dimensional BSDE, whose coefficients may depend on Γ. ∎

Problem 4.7.2

(i) Provide an alternative proof for Theorem 4.3.1 by using the global approach similar to that used in the proof of Theorem 3.3.1. (Hint: first provide a priori estimate for $\|(Y, Z)\|_\lambda^2 := \sup_{0 \le t \le T} \mathbb{E}[e^{\lambda t}|Y_t|^2] + \mathbb{E}\Big[\int_0^T e^{\lambda t}|Z_t|^2 dt \Big]$ for some $\lambda > 0$ large enough.)

(ii) Provide another proof for Theorem 4.3.1 by using contraction mapping. That is, define a mapping $F : \mathbb{L}^2(\mathbb{F}, \mathbb{R}^{d_2}) \times \mathbb{L}^2(\mathbb{F}, \mathbb{R}^{d_2 \times d}) \to \mathbb{L}^2(\mathbb{F}, \mathbb{R}^{d_2}) \times \mathbb{L}^2(\mathbb{F}, \mathbb{R}^{d_2 \times d})$

by $F(Y,Z) := (\tilde{Y}, \tilde{Z})$, where

$$\tilde{Y}_t := \xi + \int_t^T f_s(Y_s, Z_s)ds - \int_t^T Z_s dB_s.$$

Show that F is a contraction mapping under the norm $\|(Y,Z)\|_\lambda^2$ for $\lambda > 0$ large enough.   ∎

Problem 4.7.3 Show that the result of Theorem 4.3.1 still holds if, in Assumption 4.0.1, the Lipschitz continuity of f in y is replaced with the following slightly weaker monotonicity condition:

$$[f_t(y_1, z) - f_t(y_2, z)] \cdot [y_1 - y_2] \le L|y_1 - y_2|^2, \quad \forall (t, \omega), y_1, y_2, z.$$

∎

Problem 4.7.4 Let f satisfy Assumption 4.0.1 (i), (ii), (iv), and the linear growth condition (4.2.9).

(i) If $\xi \in \mathbb{L}^2(\mathscr{F}_T, \mathbb{R}^{d_2})$ and $(Y,Z) \in \mathbb{L}^2(\mathbb{F}, \mathbb{R}^{d_2}) \times \mathbb{L}^2(\mathbb{F}, \mathbb{R}^{d_2 \times d})$ is a solution to BSDE (4.0.3). Show that (Y,Z) satisfies the a priori estimate (4.2.1).

(ii) Let $(Y^n, Z^n) \in \mathbb{L}^2(\mathbb{F}, \mathbb{R}^{d_2}) \times \mathbb{L}^2(\mathbb{F}, \mathbb{R}^{d_2 \times d})$ be a solution to BSDE (4.0.3) with terminal condition $\xi_n \in \mathbb{L}^2(\mathscr{F}_T, \mathbb{R}^{d_2})$. Assume $\lim_{n \to \infty} \mathbb{E}\left[|\xi_n - \xi|^2 + \int_0^T |Y_t^n - Y_t|^2 dt\right] = 0$ for some $\xi \in \mathbb{L}^2(\mathscr{F}_T, \mathbb{R}^{d_2})$ and $Y \in \mathbb{L}^2(\mathbb{F}, \mathbb{R}^{d_2})$. Show that there exists $Z \in \mathbb{L}^2(\mathbb{F}, \mathbb{R}^{d_2 \times d})$ such that $\lim_{n \to \infty} \mathbb{E}\left[\int_0^T |Z_t^n - Z_t|^2 dt\right] = 0$ and (Y,Z) is a solution to BSDE (4.0.3) with terminal condition ξ.

(iii) Assume $d_2 = 1$, $\xi \in \mathbb{L}^2(\mathscr{F}_T, \mathbb{R})$, and f is continuous in (y,z). Show that BSDE (4.0.3) has a solution $(Y,Z) \in \mathbb{S}^2(\mathbb{F}, \mathbb{R}) \times \mathbb{L}^2(\mathbb{F}, \mathbb{R}^{1 \times d})$.

(iv) Under the conditions in (iii), find a counterexample such that the BSDE has multiple solutions.   ∎

Problem 4.7.5

(i) Find a counterexample for comparison principle of multidimensional BSDEs. To be precise, let $d_2 = 2, d = 1$, (ξ, f, Y, Z) be as in (4.0.3), and $(\tilde{\xi}, \tilde{f}, \tilde{Y}, \tilde{Z})$ be another system. We want $\xi_i \le \tilde{\xi}_i$ and $f^i \le \tilde{f}^i$, $i = 1, 2$, but it does not hold that $Y^i \le \tilde{Y}^i$, $i = 1, 2$.

(ii) Prove the comparison for the following special multidimensional BSDE. Let (ξ, f) and $(\tilde{\xi}, \tilde{f})$ satisfy Assumption 4.0.1, and (Y,Z), $(\tilde{Y}, \tilde{Z})$ be the corresponding solution to BSDE (4.0.3). Assume

$$\xi_i \le \tilde{\xi}_i, \quad f^i \le \tilde{f}^i, \quad i = 1, \cdots, d_2.$$

Moreover, for $i = 1, \cdots, d_2$, assume f^i does not depend on z_j and is increasing in y_j for all $j \neq i$. Show that

$$Y_t^i \leq \tilde{Y}_t^i, \quad 0 \leq t \leq T, \quad \mathbb{P}\text{-a.s.}, \quad i = 1, \cdots, d_2.$$

∎

Problem 4.7.6 This problem extends the optimization problem in Subsection 4.5.2 to a game problem, still in weak formulation. Assume $\mathbb{K} = \mathbb{K}_1 \times \mathbb{K}_2$, its elements are denoted as $k = (k_1, k_2)$, and denote $\mathscr{K}_1, \mathscr{K}_2$ in obvious sense. Assume Assumption 4.5.1 holds true. Denote

$$\overline{H}(t,x,z) := \inf_{k_1 \in \mathbb{K}_1} \sup_{k_2 \in \mathbb{K}_2} H(t,x,z,u), \quad \underline{H}(t,x,z) := \sup_{k_2 \in \mathbb{K}_2} \inf_{k_1 \in \mathbb{K}_1} H(t,x,z,u),$$

(4.7.2)

and let $(\overline{Y}, \overline{Z})$, $(\underline{Y}, \underline{Z})$ denote the solution to the following BSDEs:

$$\overline{Y}_t = g(X_T) + \int_t^T \overline{H}(s, X_s, \overline{Z}_s) ds - \int_t^T \overline{Z}_s dB_s,$$
$$\underline{Y}_t = g(X_T) + \int_t^T \underline{H}(s, X_s, \underline{Z}_s) ds - \int_t^T \underline{Z}_s dB_s, \qquad \mathbb{P}\text{-a.s.} \quad (4.7.3)$$

(i) Show that

$$\overline{Y}_0 = \inf_{k_1 \in \mathscr{K}_1} \sup_{k_2 \in \mathscr{K}_2} Y_0^{k_1,k_2}, \quad \underline{Y}_0 = \sup_{k_2 \in \mathscr{K}_2} \inf_{k_1 \in \mathscr{K}_1} Y_0^{k_1,k_2} \quad (4.7.4)$$

Moreover, if the following Isaacs condition holds:

$$\overline{H} = \underline{H} =: H^*, \quad (4.7.5)$$

then the game value exists, namely

$$\inf_{k_1 \in \mathscr{K}_1} \sup_{k_2 \in \mathscr{K}_2} Y_0^{k_1,k_2} = \sup_{k_2 \in \mathscr{K}_2} \inf_{k_1 \in \mathscr{K}_1} Y_0^{k_1,k_2} = Y_0^*, \quad (4.7.6)$$

where Y^* is the solution to BSDE (4.5.16) with the generator H^* defined by (4.7.5).

(ii) Assume further that there exists Borel measurable functions $I_1(t,x,z) \in \mathbb{K}_1$ and $I_2(t,x,z) \in \mathbb{K}_2$ such that, for all (t,x,z) and all $(k_1, k_2) \in \mathbb{K}_1 \times \mathbb{K}_2$,

$$H(t,x,z,k_1,I_2(t,x,z)) \geq H(t,x,z,I_1(t,x,z),I_2(t,x,z)) \geq H(t,x,z,I_1(t,x,z),k_2).$$

(4.7.7)

Then Isaacs condition (4.7.5) holds and the game has a saddle point:

$$k_t^{1,*} := I_1(t, X_t, Z_t^*), \quad k_t^{2,*} := I_2(t, X_t, Z_t^*), \tag{4.7.8}$$

where H^*, Y^*, Z^* are as in (i). Here the saddle point, also called equilibrium, means:

$$Y_0^{k^1, k^{2,*}} \geq Y_0^* \geq Y_0^{k^{1,*}, k^2}, \quad \forall (k^1, k^2) \in \mathcal{K}_1 \times \mathcal{K}_2. \tag{4.7.9}$$

∎

Chapter 5
Markov BSDEs and PDEs

Again we are given $(\Omega, \mathcal{F}, \mathbb{F}, \mathbb{P})$ and a d-dimensional Brownian motion B. The filtration assumption (4.0.1) is irrelevant in this chapter, because the involved processes will automatically be $\mathbb{F}^B$-measurable. However, for notational simplicity we nevertheless still assume $\mathbb{F} := \mathbb{F}^B$. Moreover, for $0 \leq t \leq T$, denote

$$\mathbb{F}^t := \{\mathcal{F}_s^t\}_{t \leq s \leq T} \quad \text{where} \quad \mathcal{F}_s^t := \sigma(B_{t,r}, t \leq r \leq s). \tag{5.0.1}$$

Note that $\mathbb{F}^t$ and $\mathcal{F}_t$ are independent.

Our main objective of this chapter is to study the following decoupled Forward-Backward SDE with deterministic coefficients:

$$\begin{cases} X_t = x + \displaystyle\int_0^t b(s, X_s)ds + \int_0^t \sigma(s, X_s)dB_s; \\ Y_t = g(X_T) + \displaystyle\int_t^T f(s, X_s, Y_s, Z_s)ds - \int_t^T Z_s dB_s. \end{cases} \tag{5.0.2}$$

Throughout the chapter, unless otherwise stated, the processes B, X, Y, and Z take values in $\mathbb{R}^d$, $\mathbb{R}^{d_1}$, $\mathbb{R}^{d_2}$, $\mathbb{R}^{d_2 \times d}$, respectively. As before, for notational simplicity we may carry out many proofs only for $d = d_1 = d_2 = 1$. We shall always assume

Assumption 5.0.1

(i) b, σ, f, g are deterministic taking values in $\mathbb{R}^{d_1}$, $\mathbb{R}^{d_1 \times d}$, $\mathbb{R}^{d_2}$, $\mathbb{R}^{d_2}$, respectively; and $b(\cdot, 0), \sigma(\cdot, 0), f(\cdot, 0, 0, 0)$ and $g(0)$ are bounded.
(ii) b, σ, f, g are uniformly Lipschitz continuous in (x, y, z) with Lipschitz constant L.
(iii) b, σ, f are uniformly Hölder-$\frac{1}{2}$ continuous in t with Hölder constant L.

© Springer Science+Business Media LLC 2017
J. Zhang, *Backward Stochastic Differential Equations*, Probability Theory and Stochastic Modelling 86, DOI 10.1007/978-1-4939-7256-2_5

Under the above conditions, it is clear that FBSDE (5.0.2) is well posed. We remark that the Holder continuity in (iii) above is mainly for the regularity and numerical method later, not for the well-posedness. In this chapter, the generic constant C will depend on T, the dimensions d, d_1, d_2, and the bounds in Assumption 5.0.1.

5.1 Markov Property and Nonlinear Feynman-Kac Formula

5.1.1 Markov SDEs

To investigate the Markov property of X, we first introduce two notations. For any $(t,x) \in [0,T] \times \mathbb{R}^{d_1}$, and $\eta \in \mathbb{L}^2(\mathscr{F}_t, \mathbb{R}^{d_1})$, let $X^{t,x}$ and $\mathscr{X}^{t,\eta}$ denote the unique solution to the following SDEs on $[t,T]$, respectively:

$$
\begin{aligned}
X_s^{t,x} &= x + \int_t^s b(r, X_r^{t,x})dr + \int_t^s \sigma(r, X_r^{t,x})dB_r; \\
\mathscr{X}_s^{t,\eta} &= \eta + \int_t^s b(r, \mathscr{X}_r^{t,\eta})dr + \int_t^s \sigma(r, \mathscr{X}_r^{t,\eta})dB_r
\end{aligned}
\tag{5.1.1}
$$

It is clear that $X^{0,x} = X$, $\mathscr{X}^{t,x} = X^{t,x}$, and by the uniqueness, $X_s = \mathscr{X}_s^{t,X_t}$ for $t \le s \le T$. Moreover,

Lemma 5.1.1 *Let Assumption 5.0.1 hold, and fix $0 \le t \le T$.*

(i) The mapping $x \mapsto X^{t,x}$ is Lipschitz continuous in the following sense:

$$
\mathbb{E}\Big[\big|(X^{t,x_1} - X^{t,x_2})_T^*\big|^2\Big] \le C|x_1 - x_2|^2.
\tag{5.1.2}
$$

(ii) There exists a version of $X^{t,x}$ for each x such that the mapping $(x,s,\omega) \mapsto X_s^{t,x}(\omega)$ is $\mathbb{F}^t$-progressively measurable. In particular, $X^{t,x}$ is independent of $\mathscr{F}_t$.

Proof (i) is a direct consequence of Theorem 3.2.4. To see (ii), denote $X_s^{t,x,0} := x$, $t \le s \le T$, and

$$
X_s^{t,x,n+1} := x + \int_t^s b(r, X_r^{t,x,n})dr + \int_t^s \sigma(r, X_r^{t,x,n})dB_r, \quad t \le s \le T, \quad n = 0, 1, \cdots
$$

Clearly $X^{t,x,0}$ satisfies all the properties in (ii). By induction one can easily show that $X^{t,x,n}$ also satisfies all the properties in (ii), then so does $\bar{X}^{t,x} := \limsup_{n\to\infty} X^{t,x,n}$. Now by the proof of Theorem 3.3.1, in particular by (3.3.3), we see that

$$
\mathbb{E}\Big[\sum_{n=1}^\infty \sup_{t \le s \le T} |X_s^{t,x,n} - X_s^{t,x}|^2\Big] < \infty.
$$

This implies that $X^{t,x,n}$ converges to $X^{t,x}$, a.s. Then $X^{t,x} = \bar{X}^{t,x}$, a.s. for all x. Therefore, $\bar{X}^{t,x}$ is a desired version of $X^{t,x}$. ∎

Our main result of this section is:

Theorem 5.1.2 *Let Assumption 5.0.1 hold. For any t and $\eta \in \mathbb{L}^2(\mathscr{F}_t, \mathbb{R}^{d_1})$, we have*

$$\mathscr{X}_s^{t,\eta}(\omega) = X_s^{t,\eta(\omega)}(\omega), \quad t \le s \le T, \quad \text{for } \mathbb{P}\text{-a.e. } \omega. \tag{5.1.3}$$

Consequently, X is Markov.

Proof We first show that (5.1.3) implies the Markov property. Indeed, for any $t < s$ and any bounded Borel measurable function φ, by (5.1.3) we have

$$\mathbb{E}\Big[\varphi(X_s)\Big|\mathscr{F}_t\Big] = \mathbb{E}\Big[\varphi(\mathscr{X}_s^{t,X_t})\Big|\mathscr{F}_t\Big] = \mathbb{E}\Big[\varphi(X_s^{t,X_t})\Big|\mathscr{F}_t\Big].$$

Now by Lemma 5.1.1 (ii) and Problem 5.7.1 we see that

$$\mathbb{E}\Big[\varphi(X_s)\Big|\mathscr{F}_t\Big] = \psi(X_t) \quad \text{where} \quad \psi(x) := \mathbb{E}[\varphi(X_s^{t,x})] \text{ is deterministic.}$$

This means that X is Markov.

We now prove (5.1.3) in two steps.

Step 1. We first assume $\eta = \sum_{i=1}^{\infty} x_i \mathbf{1}_{A_i}$, where $x_i \in \mathbb{R}^{d_1}$ are constants and $\{A_i, i \ge 1\} \subset \mathscr{F}_t$ is a partition of Ω. Define $\tilde{X}_s(\omega) := \sum_{i=1}^{\infty} X_s^{t,x_i}(\omega)\mathbf{1}_{A_i}(\omega)$. Note that, for any function $\varphi(s, \omega, x)$, we have

$$\varphi(s, \omega, \eta(\omega)) = \sum_{i=1}^{\infty} \varphi(s, \omega, x_i)\mathbf{1}_{A_i}(\omega).$$

Then,

$$\tilde{X}_s = \sum_{i=1}^{\infty} X_s^{t,x_i} \mathbf{1}_{A_i} = \sum_{i=1}^{\infty} \Big[x_i + \int_t^s b(r, X_r^{t,x_i})dr + \int_t^s \sigma(r, X_r^{t,x_i})dB_r\Big]\mathbf{1}_{A_i}$$

$$= \sum_{i=1}^{\infty} x_i \mathbf{1}_{A_i} + \int_t^s \sum_{i=1}^{\infty} b(r, X_r^{t,x_i})\mathbf{1}_{A_i}dr + \int_t^s \sum_{i=1}^{\infty} \sigma(r, X_r^{t,x_i})\mathbf{1}_{A_i}dB_r$$

$$= \eta + \int_t^s b(r, \tilde{X}_r)dr + \int_t^s \sigma(r, \tilde{X}_r)dB_r,$$

where the second line used the fact that $A_i \in \mathscr{F}_t$ and Lemma 2.2.3 (iii). Now by the uniqueness of solutions to SDEs, we obtain $\tilde{X} = \mathscr{X}^{t,\eta}$, which is (5.1.3).

Step 2. In the general case, let $\eta_n \in \mathbb{L}^2(\mathscr{F}_t, \mathbb{R}^{d_1})$ be a discrete approximation of η such that each η_n takes countably many values and $|\eta_n - \eta| \le \frac{1}{n}$. By Step 1,

$$\mathscr{X}_s^{t,\eta_n}(\omega) = X_s^{t,\eta_n(\omega)}(\omega), \quad t \le s \le T, \quad \text{for } \mathbb{P}\text{-a.e. } \omega. \tag{5.1.4}$$

Applying Lemma 5.1.1 (ii) and Problem 5.7.1, we have

$$\mathbb{E}\Big[|(X^{t,\eta(\omega)} - X^{t,\eta_n(\omega)})^*_T|^2\Big|\mathcal{F}_t\Big] = \mathbb{E}\big[|(X^{t,x} - X^{t,x_n})^*_T|^2\big]\Big|_{x=\eta(\omega),x_n=\eta_n(\omega)}.$$

Then it follows from Lemma 5.1.1 (i) that

$$\mathbb{E}\Big[|(X^{t,\eta(\omega)} - X^{t,\eta_n(\omega)})^*_T|^2\Big] = \mathbb{E}\Big[\mathbb{E}\big[|(X^{t,\eta(\omega)} - X^{t,\eta_n(\omega)})^*_T|^2\big|\mathcal{F}_t\big]\Big] \le C\mathbb{E}[|\eta_n - \eta|^2] \le \frac{C}{n}.$$

On the other hand, applying Theorem 3.2.4 it is clear that

$$\mathbb{E}\Big[|(\mathscr{X}^{t,\eta} - \mathscr{X}^{t,\eta_n})^*_T|^2\Big] \le C\mathbb{E}[|\eta_n - \eta|^2] \le \frac{C}{n}.$$

Then by (5.1.4) we have

$$\mathbb{E}\Big[|(X^{t,\eta(\omega)} - \mathscr{X}^{t,\eta})^*_T|^2\Big] \le \frac{C}{n}.$$

Send $n \to \infty$, we obtain (5.1.3) immediately. ∎

5.1.2   Markov BSDEs

Similar to (5.1.1), for any t, x, and $\eta \in \mathbb{L}^2(\mathcal{F}_t)$, let $(Y^{t,x}, Z^{t,x})$ and $(\mathscr{Y}^{t,\eta}, \mathscr{Z}^{t,\eta})$ denote the unique solution to the following BSDEs on $[t, T]$, respectively:

$$
\begin{aligned}
Y^{t,x}_s &= g(X^{t,x}_T) - \int_s^T f(r, X^{t,x}_r, Y^{t,x}_r, Z^{t,x}_r)dr + \int_s^T Z^{t,x}_r dB_r; \\
\mathscr{Y}^{t,\eta}_s &= g(\mathscr{X}^{t,\eta}_T) - \int_s^T f(r, \mathscr{X}^{t,\eta}_r, \mathscr{Y}^{t,\eta}_r, \mathscr{Z}^{t,\eta}_r)dr + \int_s^T \mathscr{Z}^{t,\eta}_r dB_r.
\end{aligned}
\tag{5.1.5}
$$

By the uniqueness of solutions, we have $(Y_s, Z_s) = (\mathscr{Y}^{t,X_t}_s, \mathscr{Z}^{t,X_t}_s)$. Similar to Lemma 5.1.1 and Theorem 5.1.2, one may easily prove

Theorem 5.1.3 *Let Assumption 5.0.1 hold and fix $0 \le t \le T$.*

- (i) *There exists a version of $(Y^{t,x}, Z^{t,x})$ for each x such that the mapping $(x, s, \omega) \mapsto (Y^{t,x}_s(\omega), Z^{t,x}_s(\omega))$ is $\mathbb{F}^t$-progressively measurable. In particular, $(Y^{t,x}, Z^{t,x})$ is independent of $\mathcal{F}_t$.*
- (ii) *For any $\eta \in \mathbb{L}^2(\mathcal{F}_t)$, we have*

$$(\mathscr{Y}^{t,\eta}_s, \mathscr{Z}^{t,\eta}_s(\omega)) = (Y^{t,\eta(\omega)}_s(\omega), Z^{t,\eta(\omega)}_s(\omega)), \quad for\ ds \times d\mathbb{P}\text{-}a.e.\ (s, \omega). \tag{5.1.6}$$

- (iii) *Consequently, (X, Y, Z) is Markov.*

From now on, we shall always use the version of $(X^{t,x}, Y^{t,x}, Z^{t,x})$ as in Lemma 5.1.2 (ii) and Theorem 5.1.3 (i).

Now define

$$u(t, x) := Y_t^{t,x}. \tag{5.1.7}$$

Then $u(t, x)$ is both $\mathscr{F}_t$-measurable and independent of $\mathscr{F}_t$, and thus is deterministic. Moreover, since $Y_t = \mathscr{Y}_t^{t,X_t} = Y_t^{t,X_t}$, we have

$$Y_t = u(t, X_t), \quad 0 \le t \le T. \tag{5.1.8}$$

5.1.3 Nonlinear Feynman-Kac Formula

We now derive the PDE which the above function u should satisfy. First we assume $u \in C^{1,2}([0, T] \times \mathbb{R}^{d_1}, \mathbb{R}^{d_2})$ and for notational simplicity assume $d_2 = 1$. Applying Itô formula we have

$$du(t, X_t) = [\partial_t u + \partial_x u b + \frac{1}{2} \partial_{xx} u : (\sigma \sigma^\top)](t, X_t) dt + \partial_x u \sigma(t, X_t) dB_t.$$

Compare this with

$$dY_t = -f(t, X_t, Y_t, Z_t) dt + Z_t dB_t.$$

We obtain

$$Z_t = \partial_x u \sigma(t, X_t), \quad [\partial_t u + \partial_x u b + \frac{1}{2} \partial_{xx} u : (\sigma \sigma^\top)](t, X_t) + f(t, X_t, Y_z, Z_t) = 0,$$

and thus

$$\mathbb{L}u(t, x) := \partial_t u + \frac{1}{2} \partial_{xx} u : (\sigma \sigma^\top) + \partial_x u b + f(t, x, u, \partial_x u \sigma) = 0; \quad u(T, x) = g(x). \tag{5.1.9}$$

We now state this result in multidimensional case, whose proof is obvious.

Theorem 5.1.4 (Nonlinear Feynman-Kac Formula) *Assume Assumption 5.0.1 holds true, and $u \in C^{1,2}([0, T] \times \mathbb{R}^{d_1}, \mathbb{R}^{d_2})$ is a classical solution to the following system of PDEs:*

$$\partial_t u^i + \partial_x u^i b + \frac{1}{2} \partial_{xx} u^i : (\sigma \sigma^\top) + f^i(t, x, u, \partial_x u \sigma) = 0, \, i = 1, \cdots, d_2; \, u(T, x) = g(x). \tag{5.1.10}$$

Then

$$Y_t = u(t, X_t), \quad Z_t = \partial_x u \sigma(t, X_t). \tag{5.1.11}$$

Remark 5.1.5 For the European call option in the Black-Scholes model, as in Sections 2.8 and 4.5.1, the PDE (5.1.9) is linear and the Black-Scholes formula is, in fact, obtained via the solution to (5.1.9), that is, $Y_0 = u(0, x)$ gives the option price. Moreover, recall that $Z_t\sigma^{-1}$ represents the hedging portfolio. By (5.1.11), $Z_t\sigma^{-1} = \partial_x u(t, X_t)$ is the sensitivity of the option price Y_t with respect to the stock price X_t. This is exactly the so-called Δ-hedging in the option pricing theory. ∎

In general, the function u defined by (5.1.7) is not smooth. In the case $d_2 = 1$, we will show in Section 5.5 below that u is the unique viscosity solution of PDE (5.1.9).

5.2 Regularity of Solutions

In this section we prove some regularities of (X, Y, Z) and u, which will be crucial for the convergence of our numerical algorithms in the next section. We start with some simple properties.

Theorem 5.2.1 *Let Assumption 5.0.1 hold. Then, for any t, x, x_1, x_2,*

(i) $\mathbb{E}\Big[|X_T^*|^2 + |Y_T^*|^2\Big] \le C(1 + |x|^2);$

(ii) $\mathbb{E}\Big[|(X^{0,x_1} - X^{0,x_2})_T^*|^2 + |(Y^{0,x_1} - Y^{0,x_2})_T^*|^2\Big] + \int_0^T |Z_t^{0,x_1} - Z_t^{0,x_2}|^2 dt\Big] \le C|x_1 - x_2|^2;$

(iii) $|u(t, x)| \le C(1 + |x|)$ *and* $|u(t, x_1) - u(t, x_2)| \le C|x_1 - x_2|;$

Proof (i) follows from Theorems 3.2.2 and 4.2.1; (ii) follows from Theorems 3.2.4 and 4.2.3; and (iii) follows from (i) and (ii). ∎

We next provide the time regularity of X, Y, and u, as well as some basic properties of Z.

Theorem 5.2.2 *Let Assumption 5.0.1 hold. Then, for any t, x and $t_1 < t_2$,*

(i) $|Z_t| \le C|\sigma(t, X_t)| \le C[1 + |X_t|]$, $dt \times d\mathbb{P}$-*a.s.*

(ii) $\mathbb{E}\Big[\sup_{t_1 \le t \le t_2} [|X_{t_1,t}|^2 + |Y_{t_1,t}|^2]\Big] \le C(1 + |x|^2)(t_2 - t_1);$

(iii) $|u(t_1, x) - u(t_2, x)| \le C(1 + |x|)\sqrt{t_2 - t_1};$

Proof For simplicity, we assume all processes are one dimensional.

(i) First, when u is smooth, by Theorem 5.2.1 (iii) we see that $\partial_x u$ is bounded, where the bound depends only on the parameters in Assumption 5.0.1. Then the result follows from (5.1.11).

   In the general case, let b_n, σ_n, f_n, g_n be a smooth mollifier of (b, σ, f, g) with all variables, see Problem 1.4.14, and define (X^n, Y^n, Z^n) and u_n correspondingly. Then by the standard PDE literature (see, e.g., Lieberman [136], or see Problem 5.7.4 for a purely probabilistic arguments), we see that u_n is smooth, and thus $|Z^n| \le C|\sigma_n(t, X_t^n)|$, where the constant C is independent of n. Note that $(X^n, Z^n) \to (X, Z)$ in $\mathbb{L}^2$, and $\sigma_n \to \sigma$ uniformly. Then we prove the result immediately.

(ii) Note that

$$X_{t,s} = \int_t^s b(r,X_r)dr + \int_t^s \sigma(r,X_r)dB_r, \quad Y_{t,s} = -\int_t^s f(r,X_r,Y_r,Z_r)dr + \int_t^s Z_r dB_r.$$

By (i) and Theorem 5.2.1 (i) we obtain the estimates immediately.

(iii) Note that

$$u(t_1,x) - u(t_2,x) = \mathbb{E}\Big[Y_{t_1}^{t_1,x} - Y_{t_2}^{t_1,x} + u(t_2,X_{t_2}^{t_1,x}) - u(t_2,x) \Big]. \quad (5.2.1)$$

Then, by Theorem 5.2.1 (iii) and then by (ii) we have

$$|u(t_1,x) - u(t_2,x)| \leq \mathbb{E}\Big[\int_{t_1}^{t_2} |f(t,X_t^{t_1,x},Y_t^{t_1,x},Z_t^{t_1,x})|dt + C|X_{t_2}^{t_1,x} - x| \Big]$$

$$\leq C(1 + |x|)\sqrt{t_2 - t_1},$$

completing the proof. ∎

In order to obtain the regularity of Z, we first establish a representation formula for Z.

Lemma 5.2.3 *Assume Assumption 5.0.1 holds true, and b, σ, f and g are continuously differentiable in (x,y,z). Then u is continuously differentiable in x with bounded derivatives and*

$$\partial_x u(t,X_t) = \nabla Y_t(\nabla X_t)^{-1}, \quad Z_t = \nabla Y_t(\nabla X_t)^{-1}\sigma(t,X_t), \quad (5.2.2)$$

where $(\nabla X, \nabla Y) \in \mathbb{L}^2(\mathbb{F}, \mathbb{R}^{d_1 \times d_1} \times \mathbb{R}^{d_2 \times d_1})$ and $\nabla Z^k \in \mathbb{L}^2(\mathbb{F}, \mathbb{R}^{d_2 \times d_1})$, $k = 1, \cdots, d$, satisfy the following decoupled linear FBSDE (with random coefficients):

$$\nabla X_t = I_{d_1} + \int_0^t \partial_x b(s,X_s)\nabla X_s ds + \sum_{k=1}^d \int_0^t \partial_x \sigma^k(s,X_s)\nabla X_s dB_s^k;$$

$$\nabla Y_t = \partial_x g(X_T)\nabla X_T + \int_t^T \Big[\partial_x f \nabla X_s + \partial_y f \nabla Y_s + \sum_{k=1}^d \partial_{z_k} f \nabla Z_s^k \Big](s,X_s,Y_s,Z_s)ds$$

$$- \sum_{k=1}^d \int_t^T \nabla Z_s^k dB_s^k;$$

$$(5.2.3)$$

Here σ^k is the k-th column of σ, z_k is the k-th column of $z \in \mathbb{R}^{d_2 \times d}$, $\partial_x b$ takes values in $\mathbb{R}^{d_1 \times d_1}$ with its (i,j)-th component $\partial_{x_j} b^i$, and similarly for the other matrix valued derivatives.

Proof The boundedness of $\partial_x u$ follows from Theorem 5.2.1 (iii). In the remaining of the proof, for simplicity we shall assume $d = d_1 = d_2 = 1$. First, applying (3.1.5) on the first equation in (5.2.3) we see that $\nabla X_t > 0$ and thus $(\nabla X_t)^{-1}$ makes sense.

We first prove that $u(0, \cdot)$ is continuous differentiable in x. Fix x and denote, for any $\varepsilon > 0$,

$$x_\varepsilon := x + \varepsilon, \quad \Delta X^\varepsilon := X^{0,x_\varepsilon} - X, \quad \Delta Y^\varepsilon := Y^{0,x_\varepsilon} - Y, \quad \Delta Z^\varepsilon := Z^{0,x_\varepsilon} - Z.$$

By Theorem 5.2.1 we have

$$\mathbb{E}\left[|(\Delta X^\varepsilon)^*_T|^2 + |(\Delta Y^\varepsilon)^*_T|^2 + \int_0^T |\Delta Z^\varepsilon_t|^2 dt\right] \le C\varepsilon^2 \to 0, \qquad \text{as } \varepsilon \to 0. \quad (5.2.4)$$

Denote

$$\nabla X^\varepsilon := \frac{\Delta X^\varepsilon}{\varepsilon}, \quad \nabla Y^\varepsilon := \frac{\Delta Y^\varepsilon}{\varepsilon}, \quad \nabla Z^\varepsilon := \frac{\Delta Z^\varepsilon}{\varepsilon}. \quad (5.2.5)$$

We have

$$\nabla X^\varepsilon_t = 1 + \int_0^t b^\varepsilon_x(s, X_s) \nabla X^\varepsilon_t ds + \int_0^t \sigma^\varepsilon_x(s, X_s) \nabla X^\varepsilon_t dB_s;$$

$$\nabla Y^\varepsilon_t = g^\varepsilon_x(X_T) \nabla X^\varepsilon_T + \int_t^T \left[f^\varepsilon_x \nabla X^\varepsilon_s + f^\varepsilon_y \nabla Y^\varepsilon_s + f^\varepsilon_z \nabla Z^\varepsilon_s \right] ds - \int_t^T \nabla Z^\varepsilon_s dB_s;$$

where

$$b^\varepsilon_x(s, X_s) := \int_0^1 \partial_x b(s, X_s + \theta \Delta X^\varepsilon_s) d\theta; \quad g^\varepsilon_x(X_T) := \int_0^1 \partial_x g(X_T + \theta \Delta X^\varepsilon_T) d\theta;$$

$$f^\varepsilon_x(s, X_s, Y_s, Z_s) := \int_0^1 \partial_x f(s, X_s + \theta \Delta X^\varepsilon_s, Y_s + \theta \Delta Y^\varepsilon_s, Z_s + \theta \Delta Z^\varepsilon_s) d\theta;$$

and $\sigma^\varepsilon_x, f^\varepsilon_y, f^\varepsilon_z$ are defined similarly. Now applying Theorems 3.4.2 and 4.4.3 and by (5.2.4) we have

$$\lim_{\varepsilon \to 0} \mathbb{E}\left[|(\nabla X^\varepsilon - \nabla X)^*_T|^2 + |(\nabla Y^\varepsilon - \nabla Y)^*_T|^2 + \int_0^T |\nabla Z^\varepsilon_t - \nabla Z_t|^2 dt\right] = 0. \quad (5.2.6)$$

Note that

$$\nabla Y^\varepsilon_0 = \frac{1}{\varepsilon}[Y^\varepsilon_0 - Y_0] = \frac{1}{\varepsilon}[u(0, x + \varepsilon) - u(x)].$$

Then we see that $\partial_x u(0, x) = \nabla Y_0$ exists, and by applying Theorems 3.4.2 and 4.4.3 first on (5.0.2) and then on (5.2.3), one can easily see that $\partial_x u(0, x)$ is continuous in x.

Similarly, one can show that $u(t, \cdot)$ is continuously differentiable for all t. Moreover, noticing that

$$\nabla Y_t^\varepsilon = \frac{1}{\varepsilon}[Y_t^{0,x_\varepsilon} - Y_t] = \frac{1}{\varepsilon}[u(t, X_t^{0,x_\varepsilon}) - u(t, X_t)] = \frac{u(t, X_t^{0,x_\varepsilon}) - u(t, X_t)}{X_t^{0,x_\varepsilon} - X_t}\nabla X_t^\varepsilon,$$

by sending $\varepsilon \to 0$ we obtain $\nabla Y_t = \partial_x u(t, X_t)\nabla X_t$, which is the first equality in (5.2.2).

Finally, when u is smooth, the second equality of (5.2.2) follows from (5.1.11). In the general case it follows from the approximating arguments in Theorem 5.2.2 (i). ∎

We end this section with the so-called $\mathbb{L}^2$-modulus regularity for the process Z, which plays a crucial role for the discretization of BSDEs in the next section. Fix n and denote

$$h := h_n := \frac{T}{n}, \quad t_i := t_i^n := ih, \ i = 0, \cdots, n. \qquad (5.2.7)$$

Theorem 5.2.4 *Assume Assumption 5.0.1. Then*

$$\sum_{i=0}^{n-1} \mathbb{E}\left[\int_{t_i}^{t_{i+1}} |Z_t - \hat{Z}_{t_i}^n|^2 dt\right] \leq C(1 + |x|^2)h \ \text{where} \ \hat{Z}_{t_i}^n := \frac{1}{h}\mathbb{E}\left[\int_{t_i}^{t_{i+1}} Z_t dt \,\Big|\, \mathscr{F}_{t_i}\right].$$

$$(5.2.8)$$

Remark 5.2.5
(i) We note that $\hat{Z}_{t_i}^n$ is the best approximation for Z on $[t_i, t_{i+1}]$ in the following sense:

$$\mathbb{E}\left[\int_{t_i}^{t_{i+1}} |Z_t - \hat{Z}_{t_i}^n|^2 dt\right] \leq \mathbb{E}\left[\int_{t_i}^{t_{i+1}} |Z_t - \eta|^2 dt\right], \quad \text{for any } \eta \in \mathbb{L}^2(\mathscr{F}_{t_i}).$$

$$(5.2.9)$$

In the next section, we shall construct certain $Z_{t_i}^n \in \mathbb{L}^2(\mathscr{F}_{t_i})$ and use piecewise constant process $Z_t^n := \sum_{i=0}^{n-1} Z_{t_i}^n \mathbf{1}_{[t_i, t_{i+1})}(t)$ to approximate Z. Then we see that

$$\sum_{i=0}^{n-1} \mathbb{E}\left[\int_{t_i}^{t_{i+1}} |Z_t - \hat{Z}_{t_i}^n|^2 dt\right] \leq \mathbb{E}\left[\int_0^T |Z_t - Z_t^n|^2 dt\right]. \qquad (5.2.10)$$

In other words, the estimate in this theorem provides a benchmark for the error of all possible discretizations.
(ii) Although depending on n, the process $\hat{Z}_{t_i}^n$ is defined through the true solution Z, not through some approximations. Therefore, (5.2.8) is a type of regularity of Z.

(iii) We shall actually prove the following result, which is slightly stronger than (5.2.8) by setting $\eta = Z_{t_i}$ in (5.2.9) for each i:

$$\sum_{i=0}^{n-1} \mathbb{E}\Big[\int_{t_i}^{t_{i+1}} |Z_t - Z_{t_i}|^2 dt \Big] \le C(1 + |x|^2)h. \qquad (5.2.11)$$

However, we prefer to state the result in the form of (5.2.8) for three reasons. Firstly, this type of regularity is in $\mathbb{L}^2$-sense rather than point wise sense (in terms of time t), so $\hat{Z}_t^n$ is a more natural approximation of Z on $[t_i, t_{i+1}]$ than Z_{t_i}; Secondly, as mentioned in (i), the estimate (5.2.8) provides the benchmark for the error of all possible discretizations, but (5.2.11) does not serve for this purpose; Thirdly, (5.2.11) requires the time regularity of Z, but (5.2.8) does not, so there is more hope to extend (5.2.8) to more general situations. ∎

Proof of Theorem 5.2.4 Again we shall assume $d = d_1 = d_2 = 1$ for simplicity. First, following the approximating arguments in Theorem 5.2.2 (i), without loss of generality we may assume that b, σ, f, g are continuously differentiable in (x, y, z) and thus we may apply Lemma 5.2.3. By Remark 5.2.5 (iii), we shall actually prove (5.2.11).

Apply Itô formula on the ∇X in (5.2.3), we have

$$(\nabla X_t)^{-1} = 1 - \int_0^t [\partial_x b - \frac{1}{2}|\partial_x \sigma|^2](s, X_s)(\nabla X_s)^{-1} ds - \int_0^t \partial_x \sigma(s, X_s)(\nabla X_s)^{-1} dB_s.$$
$$(5.2.12)$$

By Theorem 4.4.4, for any $p \ge 2$ we have

$$\mathbb{E}[|X_T^*|^p] \le C_p(1 + |x|^p), \quad \mathbb{E}\Big[|(\nabla X^{-1})_T^*|^p + |(\nabla Y)_T^*|^p + \big(\int_0^T |\nabla Z_t|^2 dt \big)^{\frac{p}{2}} \Big] \le C_p;$$
$$\mathbb{E}\Big[|X_{t_1, t_2}|^p \Big] \le C_p(1 + |x|^p)|t_2 - t_1|^{\frac{p}{2}}; \quad \mathbb{E}\Big[|(\nabla X_{t_1})^{-1} - (\nabla X_{t_2})^{-1}|^p \Big] \le C_p|t_2 - t_1|^{\frac{p}{2}}.$$
$$(5.2.13)$$

Apply Lemma 5.2.3, for $t \in [t_i, t_{i+1}]$ we have

$$|Z_{t_i, t}| = |\nabla Y_t(\nabla X_t)^{-1}\sigma(t, X_t) - \nabla Y_{t_i}(\nabla X_{t_i})^{-1}\sigma(t_i, X_{t_i})| \le I_1(t) + I_2(t), \text{ where}$$
$$I_1(t) := |\nabla Y_t(\nabla X_t)^{-1}||\sigma(t, X_t) - \sigma(t_i, X_{t_i})| + |\nabla Y_t||(\nabla X_t)^{-1} - (\nabla X_{t_i})^{-1}||\sigma(t_i, X_{t_i})|,$$
$$I_2(t) := |\nabla Y_t - \nabla Y_{t_i}||(\nabla X_{t_i})^{-1}\sigma(t_i, X_{t_i})|.$$
$$(5.2.14)$$

By (5.2.13) and Assumption 5.0.1, in particular Assumption 5.0.1 (iii), we have

$$\mathbb{E}[|I_1(t)|^2] \le C\mathbb{E}\Big[|(\nabla Y)_T^*|^2 |(\nabla X^{-1})_T^*|^2 [h + |X_{t_i, t}|^2]$$

$$+ |(\nabla Y)_T^*|^2 [1 + |X_T^*|^2] |(\nabla X_t)^{-1} - (\nabla X_{t_i})^{-1}|^2 \Big]$$

$$\leq C\Big(\mathbb{E}\big[|(\nabla Y)_T^*|^6\big]\Big)^{\frac{1}{3}}\Big(\mathbb{E}\big[|(\nabla X^{-1})_T^*|^6\big]\Big)^{\frac{1}{3}}\Big(\mathbb{E}\big[h^3 + |X_{t_i,t}|^6\big]\Big)^{\frac{1}{3}}$$

$$+C\Big(\mathbb{E}\big[|(\nabla Y)_T^*|^6\big]\Big)^{\frac{1}{3}}\Big(\mathbb{E}\big[1 + |X_T^*|^6\big]\Big)^{\frac{1}{3}}\Big(\mathbb{E}\big[|(\nabla X_t)^{-1} - (\nabla X_{t_i})^{-1}|^6\big]\Big)^{\frac{1}{3}}$$

$$\leq C(1 + |x|^2)h. \tag{5.2.15}$$

Moreover, by (5.2.3) we have

$$\mathbb{E}[|I_2(t)|^2] = \mathbb{E}\Big[|(\nabla X_{t_i})^{-1}\sigma(t_i, X_{t_i})|^2\Big(\int_{t_i}^t [\partial_x f \nabla X_s + \partial_y f \nabla Y_s]$$

$$+\partial_x f \nabla Z_s]ds - \int_{t_i}^t \nabla Z_s dB_s\Big)^2\Big]$$

$$\leq C\mathbb{E}\Big[|(\nabla X^{-1})_T^*|^2[1 + |X_T^*|^2]\int_{t_i}^t [|\nabla X_s|^2 + |\nabla Y_s|^2 + |\nabla Z_s|^2]ds\Big]h$$

$$+C\mathbb{E}\Big[|(\nabla X_{t_i})^{-1}\sigma(t_i, X_{t_i})|^2\mathbb{E}_{t_i}\big[\big(\int_{t_i}^t \nabla Z_s dB_s\big)^2\big]\Big]$$

$$\leq C\Big(\mathbb{E}\big[|(\nabla X^{-1})_T^*|^6\big]\Big)^{\frac{1}{3}}\Big(\mathbb{E}\big[1 + |X_T^*|^6\big]\Big)^{\frac{1}{3}}\Big(\mathbb{E}\big[\big(\int_0^T [|\nabla X_s|^2$$

$$+|\nabla Y_s|^2 + |\nabla Z_s|^2]ds\big)^3\big]\Big)^{\frac{1}{3}}h + C\mathbb{E}\Big[|(\nabla X_{t_i})^{-1}\sigma(t_i, X_{t_i})|^2\int_{t_i}^t |\nabla Z_s|^2 ds\Big]$$

$$\leq C(1 + |x|^2)h + C\mathbb{E}\Big[|(\nabla X^{-1})_T^*|^2(1 + |X_T^*|^2)\int_{t_i}^{t_{i+1}} |\nabla Z_s|^2 ds\Big].$$

Plug this and (5.2.15) into (5.2.14), we obtain

$$\mathbb{E}[|Z_{t_i,t}|^2] \leq C(1 + |x|^2)h + C\mathbb{E}\Big[|(\nabla X^{-1})_T^*|^2(1 + |X_T^*|^2)\int_{t_i}^{t_{i+1}} |\nabla Z_s|^2 ds\Big], \quad t \in [t_i, t_{i+1}].$$

Then

$$\mathbb{E}\Big[\sum_{i=0}^{n-1}\int_{t_i}^{t_{i+1}} |Z_{t_i,t}|^2 dt\Big]$$

$$\leq C\sum_{i=0}^{n-1}(1 + |x|^2)h^2 + Ch\sum_{i=0}^{n-1}\mathbb{E}\Big[|(\nabla X^{-1})_T^*|^2(1 + |X_T^*|^2)\int_{t_i}^{t_{i+1}} |\nabla Z_s|^2 ds\Big]$$

$$= C(1 + |x|^2)h + Ch\mathbb{E}\Big[|(\nabla X^{-1})_T^*|^2(1 + |X_T^*|^2)\int_0^T |\nabla Z_s|^2 ds\Big]$$

$$\leq C(1 + |x|^2)h + Ch\Big(\mathbb{E}\big[|(\nabla X^{-1})_T^*|^6\big]\Big)^{\frac{1}{3}}\Big(\mathbb{E}\big[1 + |X_T^*|^6\big]\Big)^{\frac{1}{3}}\Big(\mathbb{E}\big[\big(\int_0^T |\nabla Z_s|^2 ds\big)^3\big]\Big)^{\frac{1}{3}}$$

$$\leq C(1 + |x|^2)h.$$

This completes the proof. ∎

Remark 5.2.6 Theorem 5.2.4 holds true for nonuniform partition $\pi : 0 = t_0 < \cdots < t_n = T$ as well, with (5.2.8) replaced with the following estimate:

$$\sum_{i=0}^{n-1} \mathbb{E}\Big[\int_{t_i}^{t_{i+1}} |Z_t - \hat{Z}_{t_i}^n|^2 dt \Big] \le C(1 + |x|^2)|\pi|.$$

∎

5.3   Time Discretization of SDEs and BSDEs

In this section we fix n and recall the time partition (5.2.7).

5.3.1   Euler Scheme for SDEs

We first introduce the Euler scheme for the forward SDE in (5.0.2). We remark that in this subsection we can actually allow the coefficients to be random. Define $X_{t_0}^n := x$, and for $i = 0, \cdots, n-1$,

$$X_{t_{i+1}}^n := X_{t_i}^n + b(t_i, X_{t_i}^n)h + \sigma(t_i, X_{t_i}^n)B_{t_i,t_{i+1}}. \tag{5.3.1}$$

Theorem 5.3.1 *Let Assumption 5.0.1 hold. Then*

$$\max_{0 \le i \le n-1} \mathbb{E}\Big[\sup_{t_i \le t \le t_{i+1}} |X_t - X_{t_i}^n|^2 \Big] \le C[1 + |x|^2]h.$$

Proof We first claim that, denoting $\Delta X_{t_i}^n := X_{t_i} - X_{t_i}^n$,

$$\mathbb{E}[|\Delta X_{t_{i+1}}^n|^2] \le (1 + Ch)\mathbb{E}[|\Delta X_{t_i}^n|^2] + C(1 + |x|^2)h^2. \tag{5.3.2}$$

Then, noting that $\Delta X_{t_0}^n = 0$, it follows from the discrete Gronwall Inequality that

$$\max_{0 \le i \le n} \mathbb{E}[|\Delta X_{t_i}^n|^2] \le C(1 + |x|^2)h.$$

This, together with Theorem 5.2.2 (ii), proves the theorem.

We now prove (5.3.2). Note that

$$\Delta X_{t_{i+1}}^n = \Delta X_{t_i}^n + [b(t_i, X_{t_i}) - b(t_i, X_{t_i}^n)]h + \int_{t_i}^{t_{i+1}} [\sigma(t, X_t) - \sigma(t_i, X_{t_i}^n)]dB_t$$

$$+ \int_{t_i}^{t_{i+1}} [b(t, X_t) - b(t_i, X_{t_i})]dt.$$

Square both sides and take the expectation, and note that $(a + b)^2 \leq (1+h)a^2 + (1 + h^{-1})b^2$. Then it follows from the regularity conditions of b, σ in Assumption 5.0.1 that

$$
\begin{aligned}
\mathbb{E}[|\Delta X_{t_{i+1}}^n|^2] &\leq (1 + h)\mathbb{E}\Big[\Big(\Delta X_{t_i}^n + [b(t_i, X_{t_i}) - b(t_i, X_{t_i}^n)]h \\
&\quad + \int_{t_i}^{t_{i+1}} [\sigma(t, X_t) - \sigma(t_i, X_{t_i}^n)]dB_t\Big)^2\Big] \\
&\quad + (1 + h^{-1})\mathbb{E}\Big[\Big(\int_{t_i}^{t_{i+1}} [b(t, X_t) - b(t_i, X_{t_i})]dt\Big)^2\Big] \\
&\leq (1 + h)\mathbb{E}\Big[\Big(\Delta X_{t_i}^n + [b(t_i, X_{t_i}) - b(t_i, X_{t_i}^n)]h\Big)^2 \\
&\quad + \int_{t_i}^{t_{i+1}} [\sigma(t, X_t) - \sigma(t_i, X_{t_i}^n)]^2 dt\Big] \\
&\quad + (1 + h^{-1})\mathbb{E}\Big[h\int_{t_i}^{t_{i+1}} [b(t, X_t) - b(t_i, X_{t_i})]^2 dt\Big] \\
&\leq (1 + Ch)\mathbb{E}[|\Delta X_{t_i}^n|^2] + C\mathbb{E}\Big[\int_{t_i}^{t_{i+1}} [\sigma(t, X_t) - \sigma(t_i, X_{t_i})]^2 dt\Big] \\
&\quad + C(1 + h)\mathbb{E}\Big[\int_{t_i}^{t_{i+1}} [b(t, X_t) - b(t_i, X_{t_i})]^2 dt\Big] \\
&\leq (1 + Ch)\mathbb{E}[|\Delta X_{t_i}^n|^2] + C\mathbb{E}\Big[\int_{t_i}^{t_{i+1}} [h + |X_{t_i,t}|^2]dt\Big].
\end{aligned}
$$

This, together with Theorem 5.2.2 (ii), implies (5.3.2) immediately. ∎

5.3.2 Backward Euler Scheme for BSDEs

We now propose the Backward Euler Scheme for the BSDE in (5.0.2): $Y_{t_n}^n := g(X_{t_n}^n)$, and

$$
Z_{t_i}^n := \frac{1}{h}\mathbb{E}_{t_i}\Big[Y_{t_{i+1}}^n B_{t_i, t_{i+1}}^\top\Big]; \quad Y_{t_i}^n := \mathbb{E}_{t_i}\Big[Y_{t_{i+1}}^n + f(t_i, X_{t_i}^n, Y_{t_{i+1}}^n, Z_{t_i}^n)h\Big], \quad i = n-1, \cdots, 0.
$$

$$(5.3.3)$$

Remark 5.3.2

(i) To motivate the backward Euler scheme, we note that

$$Y_{t_i} = Y_{t_{i+1}} + \int_{t_i}^{t_{i+1}} f(t, X_t, Y_t, Z_t) dt - \int_{t_i}^{t_{i+1}} Z_t dB_t$$

$$\approx Y_{t_{i+1}} + f(t_i, X_{t_i}, Y_{t_i}, Z_{t_i}) h - Z_{t_i} B_{t_i,t_{i+1}}.$$

Multiply both sides by $B_{t_i,t_{i+1}}^\top$ and then take conditional expectation $\mathbb{E}_{t_i}$, we get

$$0 \approx \mathbb{E}_{t_i}[Y_{t_{i+1}} B_{t_i,t_{i+1}}^\top] - Z_{t_i} h \text{ and thus } Z_{t_i} \approx \frac{1}{h} \mathbb{E}_{t_i}[Y_{t_{i+1}} B_{t_i,t_{i+1}}^\top].$$

(ii) The scheme (5.3.3) is called *explicit scheme*, because $Y_{t_i}^n$ is defined explicitly. One may also consider the following *implicit scheme*: $\tilde{Y}_{t_n}^n := g(X_{t_n}^n)$, and for $i = n - 1, \cdots, 0$,

$$\tilde{Z}_{t_i}^n := \frac{1}{h} \mathbb{E}_{t_i}[\tilde{Y}_{t_{i+1}}^n B_{t_i,t_{i+1}}^\top]; \quad \tilde{Y}_{t_i}^n = \mathbb{E}_{t_i}[\tilde{Y}_{t_{i+1}}^n] + f(t_i, X_{t_i}^n, \tilde{Y}_{t_i}^n, \tilde{Z}_{t_i}^n) h. \quad (5.3.4)$$

This scheme will yield the same rate of convergence in Theorem 5.3.3 below. We note that, in (5.3.4) $\tilde{Y}_{t_i}^n$ is determined through an equation. Since the mapping $y \mapsto f(t, x, y, z) h$ has Lipschitz constant $Lh < 1$ for h small enough, the second equation in (5.3.4) has a unique solution and thus $\tilde{Y}_{t_i}^n$ is well defined. However, one needs an additional step, typically using Picard iteration, to solve the $\tilde{Y}_{t_i}^n$ in (5.3.4).

(iii) By induction one can easily see that $Y_{t_i}^n = u_i^n(X_{t_i}^n)$, where $u_n^n(x) := g(x)$ and $u_i^n(x) := \mathscr{T}_i^n(u_{i+1}^n)$, $i = n - 1, \cdots, 0$, and the operator $\mathscr{T}_i^n$ is defined as: for function $\varphi : \mathbb{R}^{d_1} \to \mathbb{R}^{d_2}$,

$$\mathscr{T}_i^n(\varphi)(x) := \mathbb{E}\Big[\varphi(x + \xi) + f\big(t_i, x, \varphi(x + \xi), \tfrac{1}{h}\mathbb{E}[\varphi(x + \xi) B_{t_i,t_{i+1}}^\top]\big) h\Big],$$

$$\text{where} \quad \xi := x + b(t_i, x) h + \sigma(t_i, x) B_{t_i,t_{i+1}}.$$

$$(5.3.5)$$

We shall note that $\mathscr{T}_i^n$ is not *monotone*, in the sense that

$$\varphi \geq \psi \quad \text{does not imply} \quad \mathscr{T}_i^n(\varphi) \geq \mathscr{T}_i^n(\psi). \quad (5.3.6)$$

We shall discuss more about monotone schemes in Section 11.5 below. The scheme $\mathscr{T}_i^n$ here will become monotone if we approximate B by random walks, namely replacing $B_{t_i,t_{i+1}}$ with $\sqrt{h}\eta$ where $\mathbb{P}(\eta = 1) = \mathbb{P}(\eta = -1) = \frac{1}{2}$. ∎

Theorem 5.3.3 *Let Assumption 5.0.1 hold and assume h is small enough. Then*

$$\max_{0\le i\le n} \mathbb{E}\Big[\sup_{t_i\le t\le t_{i+1}} |Y_t - Y_{t_i}^n|^2\Big] + \sum_{i=0}^{n-1}\mathbb{E}\Big[\int_{t_i}^{t_{i+1}} |Z_t - Z_{t_i}^n|^2 dt\Big] \le C[1 + |x|^2]h.$$

(5.3.7)

Proof Again, assume for simplicity that $d = d_1 = d_2 = 1$. Denote

$$\Delta Y_{t_i}^n := Y_{t_i}^n - Y_{t_i}, \quad \Delta Z_{t_i}^n := Z_{t_i}^n - \hat{Z}_{t_i}^n.$$

Similar to (5.3.2), we claim that

$$\mathbb{E}\Big[|\Delta Y_{t_i}^n|^2 + \frac{h}{2}|\Delta Z_{t_i}^n|^2\Big] \le (1 + Ch)\mathbb{E}[|\Delta Y_{t_{i+1}}^n|^2] + C\mathbb{E}\Big[\int_{t_i}^{t_{i+1}} |Z_t - \hat{Z}_{t_i}^n|^2 dt\Big] + C(1 + |x|^2)h^2.$$

(5.3.8)

Note that $|\Delta Y_n| = |g(X_T^n) - g(X_T)| \le C|\Delta X_T^n|$. Then, by the (backward) discrete Gronwall Inequality we have

$$\max_{0\le i\le n} \mathbb{E}[|\Delta Y_{t_i}^n|^2] \le C\mathbb{E}[|\Delta X_{t_n}^n|^2] + C\sum_{i=0}^{n-1}\mathbb{E}\Big[\int_{t_i}^{t_{i+1}} |Z_t - \hat{Z}_{t_i}^n|^2 dt\Big] + C(1 + |x|^2)h.$$

Now it follows from Theorems 5.3.1 and 5.2.4 that

$$\max_{0\le i\le n} \mathbb{E}[|\Delta Y_{t_i}^n|^2] \le C(1 + |x|^2)h.$$

(5.3.9)

Moreover, sum over $i = 0, \cdots, n-1$ in (5.3.8), we obtain

$$\sum_{i=0}^{n-1}\mathbb{E}[|\Delta Y_{t_i}^n|^2] + \frac{h}{2}\sum_{i=0}^{n-1}\mathbb{E}[|\Delta Z_{t_i}^n|^2]$$

$$\le (1 + Ch)\sum_{i=1}^{n}\mathbb{E}[|\Delta Y_{t_i}^n|^2] + C\sum_{i=0}^{n-1}\mathbb{E}\Big[\int_{t_i}^{t_{i+1}} |Z_t - \hat{Z}_{t_i}^n|^2 dt\Big] + C(1 + |x|^2)h.$$

Thus

$$\frac{h}{2}\sum_{i=0}^{n-1}\mathbb{E}[|\Delta Z_{t_i}^n|^2]$$

$$\le (1 + Ch)\mathbb{E}[|\Delta Y_{t_n}^n|^2] + Ch\sum_{i=1}^{n-1}\mathbb{E}[|\Delta Y_{t_i}^n|^2] + C\sum_{i=0}^{n-1}\mathbb{E}\Big[\int_{t_i}^{t_{i+1}} |Z_t - \hat{Z}_{t_i}^n|^2 dt\Big] + C(1 + |x|^2)h$$

By (5.3.9) and Theorem 5.2.4 we get

$$\frac{h}{2} \sum_{i=0}^{n-1} \mathbb{E}[|\Delta Z_{t_i}^n|^2] \le C(1 + |x|^2)h. \tag{5.3.10}$$

Now combine (5.3.9), (5.3.10), and Theorems 5.2.2, 5.2.4, we prove the theorem.

We now prove (5.3.8). First, by Martingale Representation Theorem, there exists $\bar{Z}^n$ such that

$$Y_{t_{i+1}}^n = \mathbb{E}_{t_i}[Y_{t_{i+1}}^n] + \int_{t_i}^{t_{i+1}} \bar{Z}_t^n dB_t. \tag{5.3.11}$$

Then

$$Z_{t_i}^n = \frac{1}{h} \mathbb{E}_{t_i}\left[Y_{t_{i+1}}^n B_{t_i,t_{i+1}}\right] = \frac{1}{h} \mathbb{E}_{t_i}\left[\int_{t_i}^{t_{i+1}} \bar{Z}_t^n dB_t \, B_{t_i,t_{i+1}}\right] = \frac{1}{h} \mathbb{E}_{t_i}\left[\int_{t_i}^{t_{i+1}} \bar{Z}_t^n dt\right],$$
$$\tag{5.3.12}$$

and thus, by the definition of $\hat{Z}^n$ in (5.2.8),

$$\mathbb{E}[|\Delta Z_{t_i}^n|^2] \le \frac{1}{h} \mathbb{E}\left[\int_{t_i}^{t_{i+1}} |\Delta \bar{Z}_t^n|^2 dt\right], \quad \text{where} \quad \Delta \bar{Z}_t^n := \bar{Z}_t^n - Z_t. \tag{5.3.13}$$

Note that

$$\Delta Y_{t_i}^n = \Delta Y_{t_{i+1}}^n + \int_{t_i}^{t_{i+1}} I_t dt - \int_{t_i}^{t_{i+1}} \Delta \bar{Z}_t^n dB_t,$$

where

$$I_t := \mathbb{E}_{t_i}[f(t_i, X_{t_i}^n, Y_{t_{i+1}}^n, Z_{t_i}^n)] - f(t, X_t, Y_t, Z_t).$$

Then, for any $\varepsilon > 0$, by (5.3.13) we have

$$\mathbb{E}\left[|\Delta Y_{t_i}^n|^2 + h|\Delta Z_{t_i}^n|^2 dt\right] \le \mathbb{E}\left[|\Delta Y_{t_i}^n|^2 + \int_{t_i}^{t_{i+1}} |\Delta \bar{Z}_t^n|^2 dt\right]$$

$$= \mathbb{E}\left[\left(\Delta Y_{t_{i+1}}^n + \int_{t_i}^{t_{i+1}} I_t dt\right)^2\right] \le (1 + \frac{h}{\varepsilon})\mathbb{E}[|\Delta Y_{t_{i+1}}^n|^2] + (1 + \frac{\varepsilon}{h})\mathbb{E}\left[\left(\int_{t_i}^{t_{i+1}} I_t dt\right)^2\right]$$

$$\le (1 + \frac{h}{\varepsilon})\mathbb{E}[|\Delta Y_{t_{i+1}}^n|^2] + (h + \varepsilon)\mathbb{E}\left[\int_{t_i}^{t_{i+1}} |I_t|^2 dt\right]. \tag{5.3.14}$$

Note that, by the regularity conditions of f in Assumption 5.0.1,

$$|I_t| \leq \left| \mathbb{E}_{t_i}[f(t_i, X_{t_i}^n, Y_{t_{i+1}}^n, Z_{t_i}^n)] - \mathbb{E}_{t_i}[f(t_i, X_{t_i}, Y_{t_{i+1}}, \hat{Z}_{t_i}^n)] \right|$$

$$+ \left| \mathbb{E}_{t_i}[f(t_i, X_{t_i}, Y_{t_{i+1}}, \hat{Z}_{t_i}^n)] - f(t_i, X_{t_i}, Y_{t_i}, \hat{Z}_{t_i}^n) \right| + \left| f(t_i, X_{t_i}, Y_{t_i}, \hat{Z}_{t_i}^n) - f(t, X_t, Y_t, Z_t) \right|$$

$$\leq C \left[|\Delta X_{t_i}^n| + \mathbb{E}_{t_i}[|\Delta Y_{t_{i+1}}^n|] + |\Delta Z_{t_i}^n| + \mathbb{E}_{t_i}[|Y_{t_i, t_{i+1}}|] + \sqrt{h} + |X_{t_i, t}| + |Y_{t_i, t}| + |Z_t - \hat{Z}_{t_i}^n| \right].$$

Then, by Theorems 5.2.2 and 5.3.1,

$$\mathbb{E}[|I_t|^2] \leq C_0 \mathbb{E} \left[|\Delta Y_{t_{i+1}}^n|^2 + |\Delta Z_{t_i}^n|^2 + |Z_t - \hat{Z}_{t_i}^n|^2 \right] + C(1 + |x|^2)h.$$

for some constant $C_0 > 0$. Plug this into (5.3.14):

$$\mathbb{E} \left[|\Delta Y_{t_i}^n|^2 + h|\Delta Z_{t_i}^n|^2 \right] \leq (1 + \frac{h}{\varepsilon}) \mathbb{E}[|\Delta Y_{t_{i+1}}^n|^2] + C_0(h + \varepsilon)h \mathbb{E} \left[|\Delta Y_{t_{i+1}}^n|^2 + |\Delta Z_{t_i}^n|^2 \right]$$

$$+ C(h + \varepsilon) \mathbb{E} \left[\int_{t_i}^{t_{i+1}} |Z_t - \hat{Z}_{t_i}^n|^2 dt \right] + C(1 + |x|^2)h^2.$$

Set $\varepsilon := \frac{1}{4C_0}$ for the above C_0, and assume $h \leq \varepsilon$ so that $C_0(\varepsilon + h) \leq \frac{1}{2}$. Then the above inequality leads to (5.3.8) immediately. ∎

5.4 Implementation of Backward Euler Scheme

In this section we discuss how to implement the Backward Euler scheme (5.3.3). Recalling Remark 5.3.2 (iii), this is equivalent to computing the functions u_i^n, which in the case $d_2 = 1$ is an approximation of the (viscosity) solution u to PDE (5.1.9). We first remark that there is a huge literature on numerical methods for PDEs. However, due to the well-known curse of dimensionality, the standard methods in PDE literature, e.g., finite difference method and finite elements method, typically work only for $d_1 \leq 3$. We shall instead use Monte Carlo methods, which is less sensitive to the dimension. The key is to compute the conditional expectations in (5.3.3). However, a naive application of Monte Carlo simulation on each conditional expectation will require simulating a huge number of paths which grows exponentially in n and thus is not feasible. At below we shall combine the least square regression and Monte Carlo simulation. Numerical examples show that the algorithm works reasonably well for problems with $d_1 = 10$ or even higher, see, e.g., Bender & Zhang [14] and Guo, Zhang, & Zhuo [99] (for somewhat more general equations).

5.4.1   Least Square Regression

Note that the space of functions on $\mathbb{R}^{d_1}$, which our object u_i^n lies in, is infinite
dimensional. The main idea here is to reduce the infinite dimensional problem to
a finite dimensional one. For each $i = 0, \cdots, n-1$, fix an appropriate set of basis
functions $e_j^i : \mathbb{R}^{d_1} \to \mathbb{R}$, $j = 1, \cdots, J_i$. Typically we set J_i and e_j^i independent
of i, but in general they may vary for different i. We intend to find the least
square regression of u_i^n on Span $(\{e_j^i\}_{1 \le j \le J_i})$, the space of the linear combinations
of $\{e_j^i\}_{1 \le j \le J_i}$.

To be precise, we fix $x_0 := X_{t_0}^n$ and denote $J := \{J_i\}_{1 \le i \le n}$, $Y_{t_n}^{n,J} := g(X_{t_n}^n)$. For
$i = n-1, \cdots, 1$, in light of (5.3.3), we define

$$Z_{t_i}^{n,J} := \sum_{j=1}^{J_i} \alpha_j^i e_j^i(X_{t_i}^n), \quad Y_{t_i}^{n,J} := \sum_{j=1}^{J_i} \beta_j^i e_j^i(X_{t_i}^n), \quad \text{where}$$

$$\{\alpha_j^i\}_{1 \le j \le J_i} := \operatorname*{argmin}_{\{\alpha_j\}_{1 \le j \le J_i} \subset \mathbb{R}^{d_2 \times d}} \mathbb{E}\Big[\big| \sum_{j=1}^{J_i} \alpha_j e_j^i(X_{t_i}^n) - \frac{1}{h} Y_{t_{i+1}}^{n,J} B_{t_i,t_{i+1}}^{\top} \big|^2\Big]; \quad (5.4.1)$$

$$\{\beta_j^i\}_{1 \le j \le J_i} := \operatorname*{argmin}_{\{\beta_j\}_{1 \le j \le J_i} \subset \mathbb{R}^{d_2}} \mathbb{E}\Big[\big| \sum_{j=1}^{J_i} \beta_j e_j^i(X_{t_i}^n) - Y_{t_{i+1}}^{n,J} - hf(t_i, X_{t_i}^n, Y_{t_{i+1}}^{n,J}, Z_{t_i}^{n,J}) \big|^2\Big].$$

By induction, one can easily see that

$$Y_{t_i}^{n,J} = u_i^{n,J}(X_{t_i}^n), \quad \text{where } u_i^{n,J}(x) := \sum_{j=1}^{J_i} \beta_j e_j^i(x) \text{ for some } \{\beta_j\}_{1 \le j \le J_i} \subset \mathbb{R}^{d_2}.$$

$$(5.4.2)$$

At $t_0 = 0$, since $X_{t_0}^n = x_0$, there is no function involved, we shall simply define

$$Z_{t_0}^{n,J} := \mathbb{E}\Big[\frac{1}{h} Y_{t_1}^{n,J} B_{t_1}^{\top}\Big]; \quad Y_{t_0}^{n,J} := \mathbb{E}\Big[Y_{t_1}^{n,J} + hf(t_0, x_0, Y_{t_1}^{n,J}, Z_{t_0}^{n,J})\Big]. \quad (5.4.3)$$

To obtain convergence, assume we actually have a sequence of basis functions
$\{e_j^i\}_{j \ge 1}$ for each i such that

Assumption 5.4.1 *For each i, $\{e_j^i(X_{t_i}^n)\}_{j \ge 1}$ is dense in $\mathbb{L}^2(\mathscr{F}_{t_i}^{X_{t_i}^n}, \mathbb{R})$. That is, for any
function $\varphi \in \mathscr{B}(\mathbb{R}^{d_1})$ satisfying $\mathbb{E}[|\varphi(X_{t_i}^n)|^2] < \infty$, there exist constants $\{\alpha_j\}_{j \ge 1} \subset \mathbb{R}$
such that*

$$\lim_{k \to \infty} \mathbb{E}\Big[\big| \sum_{j=1}^{k} \alpha_j e_j^i(X_{t_i}^n) - \varphi(X_{t_i}^n) \big|^2\Big] = 0. \quad (5.4.4)$$

Then, by backward induction, one can easily show that

Proposition 5.4.2 *Let Assumptions 5.0.1 and (5.4.1) hold. Then, for $0 \le i \le n$,*

$$\lim_{\min_{i \le k \le n} J_k \to \infty} \mathbb{E}\left[|Y_{t_i}^{n,J} - Y_{t_i}^n|^2\right] = 0, \text{ or equivalently, } \lim_{\min_{i \le k \le n} J_k \to \infty} \mathbb{E}\left[|u_i^{n,J}(X_{t_i}^n) - u_i^n(X_{t_i}^n)|^2\right] = 0.$$

$$(5.4.5)$$

We note that the convergence of $u_i^{n,J}$ is valid only at $X_{t_i}^n$, in particular, at t_0 it is valid only at x_0. The rate of convergence, however, is difficult to analyze. Clearly the efficiency of the algorithm relies on our choice of basis functions.

Remark 5.4.3

(i) To help solve the optimal arguments in (5.4.1) and thus improve the efficiency of the algorithm, in practice, people typically choose orthogonal basis functions in the sense that $\mathbb{E}[e_{j_1}^i(X_{t_i}^n)e_{j_2}^i(X_{t_i}^n)] = 0, j_1 \ne j_2$. This, however, does not increase the rate of convergence in (5.4.5).

(ii) The larger the $\{J_i\}_{1 \le i \le n}$ are, the more paths we will need to simulate in the next step. So it is crucial to have reasonably small $\{J_i\}_{1 \le i \le n}$.

(iii) Notice that the basis functions are used to approximate the u_i^n and v_i^n defined through $Z_{t_i}^n = v_i^n(X_{t_i}^n)$. Ideally one wants to have basis functions whose span include (u_i^n, v_i^n), or in light of (5.1.11), include $u(t_i, \cdot)$ and $\partial_x u(t_i, \cdot)$. Of course this is not feasible in practice because (u_i^n, v_i^n) are unknown and are exactly what we want to approximate. Nevertheless, we may always want to include g and $\partial_x g$ in the basis functions.

(iv) For given basis functions, if we are unlucky that the true solution is orthogonal to all of them in the sense of (i), then the numerical results won't be a good approximation of u_i^n. Overall speaking, how to choose good basis functions for a given problem is still an open problem. ∎

5.4.2 Monte Carlo Simulation

We now use Monte Carlo simulation to solve the optimal arguments in (5.4.1). For this purpose, we fix a parameter M and generate $M \times n$ independent d-dimensional standard normals N_i^m, $1 \le m \le M$, $1 \le i \le n$. These generate M paths of $X_{t_i}^n$ as follows. For $m = 1, \cdots, M$, $X_{m,t_0}^{M,n} := x_0$, and for $i = 0, \cdots, n-1$,

$$X_{m,t_{i+1}}^{M,n} := X_{m,t_i}^{M,n} + b(t_i, X_{m,t_i}^{M,n})h + \sqrt{h}\sigma(t_i, X_{m,t_i}^{M,n})N_{i+1}^m. \qquad (5.4.6)$$

Finally, we denote $Y_{m,t_n}^{M,n,J} := g(X_{m,t_n}^{M,n})$ and revise (5.4.1) and (5.4.3) as

$$Z_{m,t_i}^{M,n,J} := \sum_{j=1}^{J_i} \alpha_j^i e_j(X_{m,t_i}^{M,n}), \quad Y_{m,t_i}^{M,n,J} := \sum_{j=1}^{J_i} \beta_j^i e_j(X_{m,t_i}^{M,n}), \quad \text{where}$$

$$\{\alpha_j^i\}_{1 \leq j \leq J_i} := \operatorname*{argmin}_{\{\alpha_j\}_{1 \leq j \leq J_i} \subset \mathbb{R}^{d_2 \times d}} \frac{1}{M} \sum_{m=1}^{M} \Big[|\sum_{j=1}^{J_i} \alpha_j e_j(X_{m,t_i}^{M,n}) - \frac{1}{\sqrt{h}} Y_{m,t_{i+1}}^{M,n,J} (N_{i+1}^m)^\top|^2 \Big];$$

$$\{\beta_j^i\}_{1 \leq j \leq J_i} := \operatorname*{argmin}_{\{\beta_j\}_{1 \leq j \leq J_i} \subset \mathbb{R}^{d_2}} \frac{1}{M} \sum_{m=1}^{M} \Big[|\sum_{j=1}^{J_i} \beta_j e_j(X_{m,t_i}^{M,n}) - Y_{m,t_{i+1}}^{M,n,J} - hf(t_i, X_{m,t_i}^{M,n}, Y_{m,t_{i+1}}^{M,n,J}, Z_{m,t_i}^{M,n,J})|^2 \Big];$$

$$(5.4.7)$$

and

$$Z_{t_0}^{M,n,J} := \frac{1}{M} \sum_{m=1}^{M} \Big[\frac{1}{\sqrt{h}} Y_{m,t_1}^{M,n,J} (N_1^m)^\top \Big], \quad Y_{t_0}^{M,n,J} := \frac{1}{M} \sum_{m=1}^{M} \Big[Y_{m,t_1}^{M,n,J} + hf(t_0, x_0, Y_{m,t_1}^{M,n,J}, Z_{t_0}^{M,n,J}) \Big].$$

$$(5.4.8)$$

We emphasize that the optimal arguments $\{\alpha_j^i\}$ and $\{\beta_j^i\}$ in (5.4.7) depend on the random variables N_i^m and thus are random. Consequently, the $(Y_{t_0}^{M,n,J}, Z_{t_0}^{M,n,J})$ are also random. However, by applying the standard Law of Large Numbers, one can easily show that

Proposition 5.4.4 *Let Assumption 5.0.1 hold. Then, for any fixed basis functions,*

$$\lim_{M \to \infty} Y_{t_0}^{M,n,J} = Y_{t_0}^{n,J}, \quad a.s. \tag{5.4.9}$$

Finally, we remark that the optimization problem (5.4.7) is very standard in linear algebra and can be easily solved. See Problem 5.7.7 in the case $d_2 = 1$, which is the case in most applications.

5.5 Viscosity Property of BSDEs

In this section we assume

$$d_2 = 1, \tag{5.5.1}$$

and recall the PDE in the terminal value problem (5.1.9):

$$\mathbb{L}u(t,x) := \partial_t u + \frac{1}{2} \partial_{xx} u : (\sigma \sigma^\top) + \partial_x u b + f(t, x, u, \partial_x u \sigma) = 0, \quad (t,x) \in [0,T) \times \mathbb{R}^{d_1}.$$

$$(5.5.2)$$

We emphasize that the PDEs we consider in this book always have terminal conditions, rather than initial conditions. When needed, we shall call them backward PDEs so as to distinguish them from the forward ones.

Definition 5.5.1 *Let* $u \in C^{1,2}([0,T] \times \mathbb{R}^{d_1})$. *We say* u *is a classical solution (resp. subsolution, supersolution) of (backward) PDE (5.5.2) if*

$$\mathbb{L}u(t,x) = (resp. \geq, \leq) 0, \quad for\ all\ (t,x) \in [0,T) \times \mathbb{R}^{d_1}.$$

If the function u defined by (5.1.7) is in $C^{1,2}([0,T] \times \mathbb{R}^{d_1})$, then one can easily prove the opposite direction of Theorem 5.1.4, namely u is a classical solution of PDE (5.5.2). See Problem 5.7.8.

However, this function u is in general not in $C^{1,2}([0,T] \times \mathbb{R}^{d_1})$. In this section we show that u is a weak type of solution, called viscosity solution, of PDE (5.5.2). We remark that the general viscosity theory holds for semicontinuous functions, but for simplicity we shall only focus on continuous solutions.

For any $(t,x) \in [0,T) \times \mathbb{R}^{d_1}$ and $\delta > 0$, denote

$$O_\delta(t,x) := \{(t',x') \in [t,T] \times \mathbb{R}^{d_1} : t' - t + |x' - x|^2 \leq \delta\}. \tag{5.5.3}$$

Given $u : [0,T] \times \mathbb{R}^{d_1} \to \mathbb{R}$, introduce two classes of test functions: for any $(t,x) \in [0,T) \times \mathbb{R}^{d_1}$,

$$\begin{aligned}
\underline{\mathscr{A}}u(t,x) &:= \Big\{\varphi \in C^{1,2}([t,T] \times \mathbb{R}^{d_1}) : \exists \delta > 0 \text{ such that} \\
&\quad [\varphi - u](t,x) = 0 = \min_{(t',x') \in O_\delta(t,x)} [\varphi - u](t',x')\Big\}; \\
\overline{\mathscr{A}}u(t,x) &:= \Big\{\varphi \in C^{1,2}([t,T] \times \mathbb{R}^{d_1}) : \exists \delta > 0 \text{ such that} \\
&\quad [\varphi - u](t,x) = 0 = \max_{(t',x') \in O_\delta(t,x)} [\varphi - u](t',x')\Big\}.
\end{aligned} \tag{5.5.4}$$

We now define

Definition 5.5.2 *Let* $u \in C^0([0,T] \times \mathbb{R}^{d_1})$.

(i) *We say* u *is a viscosity subsolution of (backward) PDE (5.5.2) if, for any* $(t,x) \in [0,T) \times \mathbb{R}^{d_1}$ *and any* $\varphi \in \underline{\mathscr{A}}u(t,x)$, *it holds that* $\mathbb{L}\varphi(t,x) \geq 0$.

(ii) *We say* u *is a viscosity supersolution of (backward) PDE (5.5.2) if, for any* $(t,x) \in [0,T) \times \mathbb{R}^{d_1}$ *and any* $\varphi \in \overline{\mathscr{A}}u(t,x)$, *it holds that* $\mathbb{L}\varphi(t,x) \leq 0$.

(iii) *We say* u *is a viscosity solution to (5.5.2) if it is both a viscosity subsolution and a viscosity supersolution.*

We may also call u a *viscosity semi-solution* if it is either viscosity subsolution or supersolution.

Remark 5.5.3

(i) In the PDE literature, typically people study forward PDEs with initial conditions. That is, denoting $\tilde{\psi}(t, \cdot) := \psi(T - t, \cdot)$ for $\psi = u, b, \sigma, f$, then by straightforward calculation we see that $\tilde{u}$ satisfies the following (forward) PDE with initial condition $\tilde{u}(0, \cdot)$:

$$\tilde{\mathbb{L}}\tilde{u} := \partial_t \tilde{u} - \frac{1}{2}\partial_{xx}\tilde{u} : (\tilde{\sigma}\tilde{\sigma}^\top) - \partial_x\tilde{u}\tilde{b} - \tilde{f}(t, x, \tilde{u}, \partial_x\tilde{u}\tilde{\sigma}) = 0, \quad (t, x) \in (0, T] \times \mathbb{R}^{d_1}.$$
$$(5.5.5)$$

(ii) In the literature of viscosity solutions, quite often people write the PDE (5.5.2) as $-\mathbb{L}u = 0$ rather than $\mathbb{L}u = 0$. Indeed, if we start from the forward PDE (5.5.5), noticing that $\partial_t\tilde{u}(T - t, x) = -\partial_t u(t, x)$, so the direct correspondence of (5.5.5) is $-\mathbb{L}u = 0$. In fact by using this form it is more convenient to interpret the nature of viscosity solutions in terms of comparison principle, which we shall not get into in this book. For notational simplicity, in this book we change the sign by rewriting it as $\mathbb{L}u = 0$. Consequently, in our definitions of classical/viscosity semi-solutions, there is a difference of sign compared to the standard literature. Readers should be careful about it when they read the references in PDE literature. ∎

Remark 5.5.4

(i) For a viscosity subsolution u and $\varphi \in \underline{\mathscr{A}}u(t, x)$, we require $\mathbb{L}\varphi(t, x) \geq 0$ only at (t, x), not in $O_\delta(t, x)$.

(ii) It is possible that $\underline{\mathscr{A}}u(t, x) = \phi$. In this case u is automatically a viscosity subsolution at (t, x).

(iii) For viscosity solution, one has to check subsolution property and supersolution property separately. One cannot define certain $\mathscr{A}u(t, x)$ and verify the viscosity property by $\mathbb{L}\varphi(t, x) = 0$ for all $\varphi \in \mathscr{A}u(t, x)$. ∎

Remark 5.5.5 Notice that the $O_\delta(t, x)$ in (5.5.3) considers only the right neighborhood of t, which is due to the fact that we are considering PDEs with terminal conditions. In the standard literature, one uses the following alternative $\tilde{O}_\delta(t, x)$ in the definition of viscosity solution:

$$\tilde{O}_\delta(t, x) := \{(t', x') \in [0, T] \times \mathbb{R}^{d_1} : |t' - t| + |x' - x|^2 \leq \delta\}, \quad (5.5.6)$$

and correspondingly require test functions $\varphi \in C^{1,2}([0, T] \times \mathbb{R}^{d_1})$.

(i) At $(0, x)$, the two definitions are equivalent. For general (t, x), a viscosity semi-solution under O_δ is a viscosity semi-solution under $\tilde{O}_\delta$, but not vice versa. This means, under our definition, the uniqueness becomes easier and in particular is implied by the uniqueness in standard literature. The existence under our definition is slightly stronger, but nevertheless all results in the literature (for terminal value problems) should still hold true.

(ii) Since we require the test function φ to be defined only in the right time neighborhood, rigorously $\partial_t \varphi(t,x)$ should be understood as the right time derivative $\partial_t^+ \varphi(t,x) := \lim_{t' \downarrow t} \frac{\varphi(t',x) - \varphi(t,x)}{t'-t}$.

(iii) Since we are considering PDE with terminal condition, intuitively one solves the PDE backwardly. Then at t we may assume one has known the value of u at $t' > t$, but not for $t' < t$. In this sense, it is more natural to use O_δ than to use $\tilde{O}_\delta$. In fact, this is indeed true from the dynamic programming point of view, which is the key for verifying the viscosity property of u defined through certain probability representation. See Theorem 5.5.8 below.

(iv) For viscosity theory of the so-called path dependent PDEs in Chapter 11 below, it is crucial to use right time neighborhood. The alternative definition corresponding to $\tilde{O}_\delta$ does not work in path dependent case. ∎

Remark 5.5.6 The test function $\varphi \in C^{1,2}([t,T] \times \mathbb{R}^{d_1})$ can be replaced with quadratic test functions, which are called semi-jets in the PDE literature, see, e.g., Crandall, Ishii, & Lions [42]. To be rigorous, denote

$$\underline{J}u(t,x) := \Big\{\varphi \in \underline{\mathscr{A}}u(t,x) : \varphi(t',x') = a(t'-t) + b \cdot (x'-x) + \frac{1}{2}\gamma : (x'-x)(x'-x)^\top$$
$$\text{for some } a \in \mathbb{R}, b \in \mathbb{R}^{d_1}, \gamma \in \mathbb{S}^{d_1}\Big\}. \tag{5.5.7}$$

Then

u is a viscosity subsolution at (t,x) if and only if $\mathbb{L}\varphi(t,x) \geq 0$ for all $\varphi \in \underline{J}u(t,x)$.

$$\tag{5.5.8}$$

Similar statement for supersolution also holds true. The proof is deferred to Problem 5.7.9. ∎

We start with two basic properties of viscosity solutions.

Proposition 5.5.7 *Let Assumption 5.0.1 hold.*

(i) *Assume $u \in C^{1,2}([0,T] \times \mathbb{R}^{d_1})$. Then u is a viscosity solution (resp. subsolution, supersolution) to (5.5.2) if and only if it is a classical solution (resp. subsolution, supersolution).*

(ii) *Let $\lambda \in \mathbb{R}$ and $\tilde{u}(t,x) := e^{\lambda t} u(t,x)$. Then u is a viscosity solution (resp. subsolution, supersolution) to (5.5.2) if and only if $\tilde{u}$ is a viscosity solution (resp. subsolution, supersolution) to the following PDE:*

$$\tilde{\mathbb{L}}\tilde{u}(t,x) := \partial_t \tilde{u} + \frac{1}{2}\partial_{xx}\tilde{u} : (\sigma\sigma^\top) + \partial_x \tilde{u} b + \tilde{f}(t,x,\tilde{u},\partial_x \tilde{u}\sigma) = 0, \tag{5.5.9}$$
$$\text{where } \tilde{f}(t,x,y,z) := e^{\lambda t} f(t,x,e^{-\lambda t}y, e^{-\lambda t}z) - \lambda y.$$

Proof

(i) We prove only the subsolution property. The other statements can be proved similarly. First, if u is a viscosity subsolution at (t, x), then clearly $\varphi := u \in \mathscr{A}u(t, x)$, and thus by the viscosity subsolution property we have $\mathbb{L}u(t, x) \geq 0$. On the other hand, assume $\mathbb{L}u(t, x) \geq 0$. For any $\varphi \in \mathscr{A}u(t, x)$, by the minimum property we have

$$\partial_t(\varphi - u)(t, x) \geq 0, \quad \partial_x(\varphi - u)(t, x) = 0, \quad \partial_{xx}(\varphi - u)(t, x) \geq 0.$$

Thus

$$\mathbb{L}\varphi(t, x) \geq \mathbb{L}\varphi(t, x) - \mathbb{L}u(t, x) = \partial_t(\varphi - u)(t, x) + \frac{1}{2}\partial_{xx}(\varphi - u) : \sigma\sigma^{\top}(t, x) \geq 0.$$

That is, u is a viscosity subsolution at (t, x).

(ii) We prove only that the subsolution property of $\tilde{u}$ implies the subsolution property of u. The other statements can be proved similarly. Let $(t, x) \in [0, T) \times \mathbb{R}^{d_1}$ and $\varphi \in \mathscr{A}u(t, x)$. Denote $\tilde{\varphi}(s, y) := e^{\lambda s}\varphi(s, y)$, one may check straightforwardly that $\tilde{\varphi} \in \mathscr{A}\tilde{u}(t, x)$. Then $\tilde{\mathbb{L}}\tilde{\varphi}(t, x) \geq 0$. Note that

$$\partial_t\tilde{\varphi}(s, y) = \lambda\tilde{\varphi}(s, y) + e^{\lambda s}\partial_t\varphi(s, y), \quad \partial_x\tilde{\varphi}(s, y) = e^{\lambda s}\partial_x\varphi(s, y) \quad \partial_{xx}\tilde{\varphi}(s, y) = e^{\lambda s}\partial_{xx}\varphi(s, y).$$

Thus, at (t, x),

$$\mathbb{L}\varphi = \partial_t\varphi + \frac{1}{2}\sigma^2\partial_{xx}\varphi + b\partial_x\varphi + f(t, x, \varphi, \partial_x\varphi\sigma)$$

$$= e^{-\lambda t}[\partial_t\tilde{\varphi} - \lambda\tilde{\varphi}] + \frac{1}{2}e^{-\lambda t}\partial_{xx}\tilde{\varphi} : \sigma\sigma^{\top} + e^{-\lambda t}\partial_x\tilde{\varphi}b + f(t, x, e^{-\lambda t}\tilde{\varphi}, e^{-\lambda t}\partial_x\varphi\sigma)$$

$$= e^{-\lambda t}\tilde{\mathbb{L}}\tilde{\varphi} \geq 0.$$

That is, u is a viscosity subsolution.   ∎

Theorem 5.5.8 *Let Assumption 5.0.1 hold and u is defined by (5.1.7). Then u is a viscosity solution to (5.5.2) with terminal condition $u(T, x) = g(x)$.*

Proof First, it is obvious that $u(T, x) = Y_T^{T,x} = g(X_T^{T,x}) = g(x)$. By Theorems 5.2.1 and 5.2.2, we know u is continuous. We shall only prove that u is a viscosity subsolution, and the viscosity supersolution property can be proved similarly. For simplicity, assume $d = d_1 = 1$.

We prove by contradiction. Assume there exist $(t, x) \in [0, T) \times \mathbb{R}$ and $\varphi \in \mathscr{A}u(t, x)$ with corresponding δ such that

$$-c := \mathbb{L}\varphi(t, x) < 0. \tag{5.5.10}$$

Since $\mathbb{L}\varphi$ is continuous, we may assume δ is small enough so that $\mathbb{L}\varphi \leq \frac{-c}{2}$ on $O_\delta(t,x)$. Denote $(X,Y,Z) := (X^{t,x}, Y^{t,x}, Z^{t,x})$ and define

$$\tau := \inf\left\{s \geq t : (s,X_s) \notin O_\delta(t,x)\right\}. \tag{5.5.11}$$

Since X is continuous, we see that $\tau > t$ and $\mathbb{L}\varphi(s,X_s) \leq -\frac{c}{2}$ for $s \in [t,\tau]$. Note that

$$Y_s = Y_\tau + \int_s^\tau f(r,X_r,Y_r,Z_r)dr - \int_s^\tau Z_r dB_r. \tag{5.5.12}$$

Recall Remark 4.3.2 and let $(\bar{Y},\bar{Z})$ denote the unique solution of the following BSDE:

$$\bar{Y}_s = \varphi(\tau,X_\tau) + \int_s^\tau f(r,X_r,\bar{Y}_r,\bar{Y}_r)dr - \int_s^\tau \bar{Z}_r dB_r. \tag{5.5.13}$$

Since $Y_\tau = u(\tau,X_\tau) \leq \varphi(\tau,X_\tau)$, by the comparison principle of BSDEs we have

$$\bar{Y}_t \geq Y_t = u(t,x) = \varphi(t,x). \tag{5.5.14}$$

Now denote

$$\widehat{Y}_s := \varphi(s,X_s); \quad \widehat{Z}_s := \partial_x \varphi \sigma(s,X_s); \quad \Delta Y_s := \bar{Y}_s - \widehat{Y}_s; \quad \Delta Z_s := \bar{Z}_s - \widehat{Z}_s.$$

Then by the Itô formula we have

$$d\widehat{Y}_s = \left[\partial_t \varphi + \frac{1}{2}\partial_{xx}\varphi\sigma^2 + \partial_x\varphi b\right](s,X_s)ds + \partial_x\varphi\sigma(s,\bar{X}_s)dB_s$$

$$\leq \left[-\frac{c}{2} - f(s,X_s,\widehat{Y}_s,\widehat{Z}_s)\right]ds + \widehat{Z}_s dB_s,$$

thanks to (5.5.2) and (5.5.11). This implies

$$d\Delta Y_s \geq \left[-f(s,X_s,\bar{Y}_s,\bar{Z}_s) + \frac{c}{2} + f(s,X_s,\widehat{Y}_s,\widehat{Z}_s)\right]ds + \Delta Z_s dB_s$$

$$= \left[\alpha_s \Delta Y_s + \beta_s \Delta Z_s + \frac{c}{2}\right] + \Delta Z_s dB_s,$$

where α, β are defined in a standard way and are bounded. Define the adjoint process Γ as before:

$$d\Gamma_s = \Gamma_s[\alpha_s ds + \beta_s dB_s], \quad \Gamma_t = 1.$$

Then

$$d(\Gamma_s \Delta Y_s) \geq \frac{c}{2}\Gamma_s ds + \Gamma_s[\Delta Z_s + \beta_s \Delta Y_s]dB_s.$$

Note that $\Delta Y_\tau = 0$ and recall (5.5.14) that $\Delta Y_t \geq 0$. Then

$$0 \geq \mathbb{E}\Big[\Gamma_\tau \Delta Y_\tau - \Gamma_t \Delta Y_t\Big] \geq \mathbb{E}\Big[\int_t^\tau \frac{c}{2}\Gamma_s ds\Big] > 0, \qquad (5.5.15)$$

contradiction. Therefore, u is a viscosity subsolution of PDE (5.5.2).   ∎

Remark 5.5.9 The above proof used the following facts, which will be important for our definition of viscosity solution to the so-called path dependent PDEs in Chapter 11 below:

(i) The minimum property of $\varphi \in \mathscr{A}u(t,x)$ is used only at (τ, X_τ). Note that $\tau > t$ and is local, so we need the minimum property only in a small right neighborhood of t.
(ii) More precisely, the minimum property is used only to derive (5.5.14).
(iii) It is crucial that the solution (X, Y, Z) satisfies (5.5.12), which can be viewed as time consistency or dynamic programming principle.   ∎

The uniqueness of viscosity solution typically follows from the following comparison principle:

Let u be a viscosity subsolution of PDE (5.5.2) and v a viscosity supersolution. Assume both have linear growth in x and $u(T, \cdot) \leq v(T, \cdot)$. Then $u \leq v$.

$$(5.5.16)$$

The proof of the comparison principle is in general quite difficult, and we shall omit it in this book in order not to distract our main focus. We present below a partial comparison principle, just to give a flavor of the result.

Proposition 5.5.10 *Let Assumption 5.0.1 hold, and u and v be a viscosity subsolution and supersolution of PDE (5.5.2), respectively. Assume*

$$u(T, \cdot) \leq v(T, \cdot) \quad and \quad \limsup_{|x| \to \infty} \sup_{t \in [0,T]} [u - v](t, x) \leq 0. \qquad (5.5.17)$$

If either u or v is in $C^{1,2}([0, T] \times \mathbb{R}^{d_1})$, then $u \leq v$.
We remark that the second condition in *(5.5.17)* can be replaced with certain growth conditions of u, v in x, see Problem 5.7.10.

Proof Note that, by choosing λ appropriately, the $\tilde{f}$ in (5.5.9) is strictly decreasing in y. Then by Proposition 5.5.7 (ii) we may assume without loss of generality that

$$f \text{ is strictly decreasing in } y. \qquad (5.5.18)$$

Moreover, without loss of generality we assume $v \in C^{1,2}([0, T] \times \mathbb{R}^{d_1})$.

We prove by contradiction. Assume not, then $(u - v)(t_0, x_0) > 0$ for some (t_0, x_0) and denote

$$c_0 := \sup_{(t,x) \in [t_0, T] \times \mathbb{R}^{d_1}} (u - v)(t, x) \geq (u - v)(t_0, x_0) > 0. \qquad (5.5.19)$$

Notice that the state space $[t_0, T] \times \mathbb{R}^{d_1}$ is locally bounded. By the boundary condition (5.5.17) there exists $(t^*, x^*) \in [t_0, T) \times \mathbb{R}^{d_1}$ such that

$$(u - v)(t^*, x^*) = c_0 \geq (u - v)(t, x) \quad \text{for any } (t, x) \in [t_0, T] \times \mathbb{R}^{d_1}. \quad (5.5.20)$$

This implies that $\varphi := v + c_0 \in \mathscr{A} u(t^*, x^*)$. Then, it follows from the viscosity subsolution property of u that $\mathbb{L}\varphi(t^*, x^*) \geq 0$. On the other hand, by Proposition 5.5.7 (i) we have $\mathbb{L}v(t^*, x^*) \leq 0$. Thus, at (t^*, x^*),

$$0 \leq \mathbb{L}\varphi - \mathbb{L}v = f(t^*, x^*, \varphi, \partial_x \varphi \sigma) - f(t^*, x^*, v, \partial_x v \sigma) = f(t^*, x^*, v + c_0, \partial_x v \sigma)$$
$$-f(t^*, x^*, v, \partial_x v \sigma).$$

This contradicts with (5.5.18). ∎

5.6 Bibliographical Notes

The nonlinear Feynman-Kac formula, or more generally the connection between partial differential equations and stochastic differential equations, is deep and important in applications. The linear Feynman-Kac formula, which is not presented explicitly in this book, is standard in the literature, see, e.g., Karatzas & Shreve [117]. The nonlinear version in Section 5.1 was established by Peng [172] in classical solution case, and the viscosity solution results in Section 5.5 are from Pardoux & Peng [168]. See also Peng [173, 174] for more general results on the so-called stochastic HJB equations by using viscosity solution approach. There have been many works along this direction in more general cases, including the variational inequalities in Chapter 6, quasilinear PDEs in Chapter 8, and fully nonlinear PDEs as well as path dependent PDEs in Part III. We also mention the connection between elliptic PDEs and BSDEs with random terminal time established by Darling & Pardoux [54]. For general viscosity solution theory, we refer to the classical references Crandall, Ishii, & Lions [42], Fleming & Soner [89], and Yong & Zhou [242].

The regularity results in Section 5.2 are mainly from Ma & Zhang [149] and Zhang [244], and the representation formula in Lemma 5.2.3 is from El Karoui, Peng, & Quenez [81]. There are some further results along this direction in the PhD thesis of Zhang [243], including the pathwise regularity (in terms of time) in Ma & Zhang [150]. The Euler scheme for forward SDEs in Section 5.3.1 is standard, see, e.g., the classical reference Kloeden & Platen [125]. The backward Euler scheme for BSDEs in Section 5.3.2 was proposed independently by Zhang [244] and Bouchard & Touzi [18].

The least square regression approach in Section 5.4 was proposed by Gobet, Lemor, & Waxin [97], based on the work of Longstaff & Schwartz [138]. There have been numerous publications on efficient numerical schemes for BSDEs, see, e.g., Bally, Pages, & Printems [3] on the quantization method, Bender & Denk [13] on the forward scheme, Bouchard & Touzi [18] on the kernel method, Crisan & Manolarakis [44] on the curvature method, and Zhao, Zhang, & Ju [247] on the multi-step schemes, to mention a few. There are also many works on further analysis of efficiency of the schemes, see, e.g., Bouchard & Warin [20], Glasserman & Yu [96].

5.7 Exercises

Problem 5.7.1 Let $\mathscr{G} \subset \mathscr{F}$ be a sub-σ-algebra, $X \in \mathbb{L}^0(\mathscr{G}, \mathbb{R}^d)$, $\varphi : \mathbb{R}^d \times \Omega \to \mathbb{R}$ be bounded and $\mathscr{B}(\mathbb{R}^d) \times \mathscr{F}$-measurable. Assume, for each $x \in \mathbb{R}^d$, $\varphi(x, \cdot)$ is independent of $\mathscr{G}$. Show that

$$\mathbb{E}[\varphi(X, \omega)|\mathscr{G}] = \psi(X) \quad \text{where} \quad \psi(x) := \mathbb{E}[\varphi(x, \cdot)].$$

∎

Problem 5.7.2 Prove Theorem 5.1.3. ∎

Problem 5.7.3 In the multidimensional case, show that the ∇X_t in (5.2.3) is positive definite. ∎

Problem 5.7.4 Let Assumption 5.0.1 hold, and assume the coefficients b, σ, f, g are twice continuously differentiable in x. Following the arguments in Lemma 5.2.3 show that the function u defined by (5.1.7) is in $C^{1,2}([0, T] \times \mathbb{R}^{d_1}, \mathbb{R}^{d_2})$. ∎

Problem 5.7.5 In the setting of Theorems 5.3.1 and 5.3.3, show that

$$\mathbb{E}\left[\max_{0 \le i \le n-1} \sup_{t_i \le t \le t_{i+1}} [|X_t - X_{t_i}^n|^2 + |Y_t - Y_{t_i}^n|^2] \right] \le C[1 + |x|^2] \frac{\ln n}{n}.$$

(Hint: first prove the following fact:

$$\mathbb{E}\left[\max_{1 \le i \le n} |N_i|^2 \right] \le C \ln n,$$

where $N_1, \cdots, N_n$ are independent standard normals.) ∎

Problem 5.7.6 Prove Theorem 5.3.3 for the implicit scheme specified in Remark 5.3.2 (ii). ∎

Problem 5.7.7 Consider the following (deterministic) optimization problem:

$$\inf_{\{\alpha_j\}_{1 \le j \le J} \subset \mathbb{R}^d} \frac{1}{M} \sum_{m=1}^{M} \left| \sum_{j=1}^{J} \alpha_j x_j^m - c_m \right|^2,$$

where $x_j^m \in \mathbb{R}$ and $c_m \in \mathbb{R}^d$ are given. Find the optimal control $\{\alpha_j^*\}_{1 \le j \le J} \subset \mathbb{R}^d$. ∎

Problem 5.7.8 Let Assumption 5.0.1 hold and $d_2 = 1$. Assume further that the function u defined by (5.1.7) is in $C^{1,2}([0,T] \times \mathbb{R}^{d_1})$. Show that u is a classical solution of PDE (5.5.2). ∎

Problem 5.7.9 Prove (5.5.8). ∎

Problem 5.7.10 Prove Proposition 5.5.10 by replacing the second condition in (5.5.17) with the condition that both u and v have polynomial growth in x. (Hint: assume $|u(t,x)| + |v(t,x)| \le C[1 + |x|^k]$ and $(u - v)(t_0, x_0) > 0$. Consider the following optimization problem which has maximum arguments:

$$c_0 := \sup_{(t,x) \in [t_0, T] \times \mathbb{R}^{d_1}} \left[e^{\lambda(t-t_0)} [u - v](t,x) - |x - x_0|^m \right] \ge [u - v](t_0, x_0) > 0,$$

where $m > k$ is even. Derive the contradiction when λ is large enough. ∎

Problem 5.7.11 This problem proves the comparison principle (5.5.16) of viscosity solutions by using the special structure of semilinear PDEs. Assume $d_2 = 1$, Assumption 5.0.1 holds, and b, σ are twice differentiable in x. Let u be defined by (5.1.7), and $\underline{u}, \overline{u}$ be viscosity subsolution and supersolution, respectively, of PDE (5.5.2). For simplicity, assume

$$\underline{u}(T, \cdot) \le g \le \overline{u}(T, \cdot), \quad \limsup_{|x| \to \infty} \sup_{0 \le t \le T} [\underline{u}(t,x) - u(t,x)] \le 0 \le \liminf_{|x| \to \infty} \inf_{0 \le t \le T} [\overline{u}(t,x) - u(t,x)].$$

(i) Construct $\overline{f}_n, \overline{g}_n$ such that

   - for each n, $\overline{f}_n, \overline{g}_n$ are twice differentiable in x;
   - $\overline{f}_n, \overline{g}_n$ satisfy Assumption 5.0.1 with uniform bounds independent of n;
   - $\overline{f}_n \downarrow f, \overline{g}_n \downarrow g$, as $n \to \infty$.

(ii) Let $\overline{u}_n$ be defined by (5.1.7) corresponding to $(b, \sigma, \overline{f}_n, \overline{g}_n)$. Show that $\overline{u}_n \ge \underline{u}$.

(iii) Construct $\underline{f}_n, \underline{g}_n$ similarly such that $\underline{f}_n \uparrow f$, $\underline{g}_n \uparrow g$ and let $\underline{u}_n$ be the corresponding function. Show that $\underline{u}_n \le \overline{u}$.

(iv) Show that $\lim_{n \to \infty} [\overline{u}_n - \underline{u}_n] = 0$, and conclude that $u \le v$. ∎

Problem 5.7.12 This problem concerns the connection between elliptic PDEs and BSDEs. Let $d_2 = 1$, b, σ, f, g satisfy Assumption 5.0.1, and b, σ, f do not depend on t. Denote

$$D := O_1 := \{x \in \mathbb{R}^{d_1} : |x| < 1\}, \quad \partial D := \{x \in \mathbb{R}^{d_1} : |x| = 1\}, \quad \overline{D} := D \cup \partial D.$$

(i) For any $(t, x) \in [0, \infty) \times \overline{D}$, recall the $X^{t,x}$ defined in (5.1.1) and denote

$$\tau^{t,x} := \{s \geq t : X_s^{t,x} \in \partial D\}, \quad \text{and} \quad X^x := X^{0,x}, \ \tau^x := \tau^{0,x}. \quad (5.7.1)$$

Show that

$$\tau^x = \tau^{t,X_t^x}, \ 0 \leq t \leq \tau^x, \quad \text{and} \quad \tau^{t,x} = t \text{ for } x \in \partial D. \quad (5.7.2)$$

(ii) Assume $u \in C^2(D) \cap C^0(\overline{D})$ satisfy the following elliptic PDE:

$$\frac{1}{2}\partial_{xx}^2 : \sigma\sigma^\top(x) + \partial_x u b(x) + f(x, u(x), \partial_x u \sigma(x)) = 0, \ x \in D; \quad u(x) = g(x), \ x \in \partial D.$$

$$(5.7.3)$$

Denote

$$Y_t^x := u(X_t^x), \quad Z_t^x := \partial_x u \sigma(X_t^x), \quad 0 \leq t \leq \tau^x. \quad (5.7.4)$$

Show that (Y^x, Z^x) satisfy the following BSDE with random terminal time:

$$Y_t^x = g(X_{\tau^x}^x) + \int_t^{\tau^x} f(X_s^x, Y_s^x, Z_s^x)ds - \int_t^{\tau^x} Z_s^x dB_s, \quad 0 \leq t \leq \tau^x. \quad (5.7.5)$$

(Note: one needs to show that $\mathbb{E}[\int_0^{\tau^x} |Z_s^x|^2 ds] < \infty$, which is not a priori given.)

(iii) Consider a special case: $d_1 = d$, $b = 0$, $\sigma = I_d$, and $f = 0$. Show that the function

$$u(x) := \mathbb{E}[g(X_{\tau^x}^x)], \quad x \in \overline{D}, \quad (5.7.6)$$

is in $C^2(D) \cap C^0(\overline{D})$. (Note: this result is well known in PDE literature, but its probabilistic proof is far from easy because of the bad regularity of τ^x in x. The probabilistic argument is due to Krylov [130]. Moreover, we will study the regularity of such stopping times in Subsection 10.1.3 below.) ∎

Part II
Further Theory of BSDEs

Chapter 6
Reflected Backward SDEs

In this chapter again $(\Omega, \mathscr{F}, \mathbb{F}, \mathbb{P})$ is a filtered probability space, B is a d-dimensional Brownian motion. Since we are going to use the martingale representation theorem, we shall assume

$$\mathbb{F} = \mathbb{F}^B.$$

Moreover, as we will see, the theory of reflected BSDE relies heavily on comparison principle of BSDEs, thus we shall assume

$$d_2 = 1.$$

In fact, the multidimensional counterpart of RBSDEs remains a challenging open problem.

6.1 American Options and Reflected BSDEs

We first recall the financial model in Sections 2.8 and 4.5.1. For simplicity we assume $r = 0$ and $\mu = 0$. Then by Section 2.8.1, the unique fair price of the European option ξ at $t = 0$ is $Y_0^{Euro} = \mathbb{E}[\xi]$.

We next consider an American option with payoff process L in the above market, for example $L_t = (S_t - K)^+$ for an American call option. That is, the holder of the American option has the right to choose the exercise time (no later than the maturity time T), and if he chooses to exercise it at time t, then the payoff will be L_t. We shall note that, the exercise times are stopping times in $\mathscr{T} := \mathscr{T}(\mathbb{F})$. This is natural

© Springer Science+Business Media LLC 2017
J. Zhang, *Backward Stochastic Differential Equations*, Probability Theory and
Stochastic Modelling 86, DOI 10.1007/978-1-4939-7256-2_6

and crucial: the holder can decide the exercise time based on the information he observes, but not on future information. Then the price of the American option is an optimal stopping problem:

$$Y_0 := \sup_{\tau \in \mathscr{T}} \mathbb{E}[L_\tau]. \tag{6.1.1}$$

We now heuristically discuss how the above optimal stopping problem leads to a Reflected BSDE. The rigorous argument will become clear after we establish the well-posedness of RBSDEs. For any $0 \le t \le T$, denote

$$Y_t := \operatorname*{ess\,sup}_{\tau \in \mathscr{T}^t} \mathbb{E}_t[L_\tau], \quad \text{where } \mathscr{T}^t := \{\tau \in \mathscr{T} : t \le \tau \le T\}. \tag{6.1.2}$$

Then Y_t is the price of the option at time t, given that the option has not been exercised before t. This process Y is called the Snell envelope of L, which has the following properties. First, by choosing $\tau = t$, we have $Y_t \ge L_t$. Second, as we will prove later, Y is a supermartingale. Intuitively, for a European option, its price is a martingale. While for American option, it has time value which decreases as time evolves, and thus the price becomes a supermartingale. Now by the Doob-Meyer decomposition there exists a martingale M and an increasing process K such that $M_0 = K_0 = 0$ and

$$Y_t = Y_0 + M_t - K_t.$$

Here $K_T - K_t$ can roughly be viewed as the remaining time value. By Martingale Representation Theorem, $M_t = \int_0^t Z_s dB_s$ for some process Z. Also, note that $Y_T = L_T$, thus

$$Y_t = L_T - \int_t^T Z_s dB_s + K_T - K_t.$$

Finally, when $Y_t > L_t$, then the holder does not want to exercise at time t. Intuitively, this means there is no time value at t and consequently, $dK_t = 0$ when $Y_t > L_t$. This, together with the fact $Y \ge L$, implies the following minimum condition, also called Skorohod condition:

$$\int_0^T [Y_t - L_t] dK_t = 0. \tag{6.1.3}$$

Putting the above facts together, we obtain the following equation:

$$\begin{cases} Y_t = L_T - \displaystyle\int_t^T Z_s dW_s + K_T - K_t; \\ Y_t \ge L_t; \quad \displaystyle\int_0^T [Y_t - L_t] dK_t = 0. \end{cases} \tag{6.1.4}$$

This is a new type of BSDE which we call *Reflected BSDE* (RBSDE, for short), because the Y-component of the solution always stays above L, which is called the *lower barrier process*. By choosing $\tau = T$ in (6.1.2), we see that the American option price Y is greater than or equal to the corresponding European option price $Y_t^{Euro} = \mathbb{E}_t[L_T]$. In special cases, for example for call option $L_t = (S_t - K)^+$, we have $Y_t^{Euro} \geq L_t$, and $Y_t = Y_t^{Euro}$ and thus $K = 0$. In general, Y_t^{Euro} may go below L_t, and the time value dK is the external force which keeps Y staying above L. The Skorohod condition (6.1.3) indicates that K is the smallest external force which accomplishes this.

In this chapter, we study the following general RBSDE

$$\begin{cases} Y_t = \xi + \int_t^T f_s(Y_s, Z_s)ds - \int_t^T Z_s dB_s + K_T - K_t; \\ Y_t \geq L_t; \quad \int_0^T [Y_t - L_t]dK_t = 0; \end{cases} \tag{6.1.5}$$

where the solution triplet $(Y, Z, K) \in \mathbb{S}^2(\mathbb{F}, \mathbb{R}) \times \mathbb{L}^2(\mathbb{F}, \mathbb{R}^{1 \times d}) \times \mathbb{I}^2(\mathbb{F})$, and Y and K are required to be continuous with $K_0 = 0$. We shall always assume

Assumption 6.1.1

(i) $\mathbb{F} = \mathbb{F}^B$, $d_2 = 1$, and ξ, f satisfy Assumption 4.0.1.
(ii) $L \in \mathbb{L}^0(\mathbb{F}, \mathbb{R})$ is continuous with $L_T \leq \xi$, and $\mathbb{E}\big[|(L^+)_T^*|^2\big] < \infty$.

We remark that in this chapter we abuse the notation L to indicate both the Lipschitz constant of f and the lower barrier process. But its meaning should be clear in the contexts.

We first have the following simple result.

Theorem 6.1.2 *Let Assumption 6.1.1 hold and (Y, Z, K) be a solution to RBSDE* (6.1.5). *Denote*

$$\tau_t := \inf\{s \geq t : Y_s = L_s\} \wedge T \in \mathscr{T}^t. \tag{6.1.6}$$

Then

(i) $K_{\tau_t} = K_t$ *and* $Y_{\tau_t} = L_{\tau_t} 1_{\{\tau_t < T\}} + \xi 1_{\{\tau_t = T\}};$

(ii) $K_T - K_t = \displaystyle\sup_{t \leq s \leq T} \Big(\xi + \int_s^T f_r(Y_r, Z_r)dr - \int_s^T Z_r dB_r - L_s\Big)^-$ *and* $s^* := \tau_t(\omega)$ *is an optimal time;*

(iii) $Y_t = \text{ess}\sup_{\tau \in \mathscr{T}^t} \mathbb{E}_t\Big[\int_t^\tau f_s(Y_s, Z_s)ds + L_\tau 1_{\{\tau < T\}} + \xi 1_{\{\tau = T\}}\Big]$ *and* τ_t *is an optimal stopping time.*

Proof

(i) First, for $s \in [t, \tau_t)$ we have $Y_s > L_s$, then $dK_s = 0$ and thus $K_t = K_{\tau_t}$. Next, if $\tau_t < T$, by the continuity of Y and L, we have $Y_{\tau_t} = L_{\tau_t}$. Moreover, on $\{\tau_t = T\}$, it is clear that $Y_{\tau_t} = Y_T = \xi$.

(ii) On one hand, for any $s \in [t, T]$,

$$\left(\xi + \int_s^T f_r(Y_r, Z_r)dr - \int_s^T Z_r dB_r - L_s\right)^- = \left(Y_s - K_T + K_s - L_s\right)^-$$

$$= \left[(K_T - K_s) - (Y_s - L_s)\right] \vee 0 \leq (K_T - K_s) \vee 0 \leq K_T - K_t.$$

On the other hand, when $\tau_t < T$, by (i) we have

$$\left(Y_{\tau_t} - K_T + K_{\tau_t} - L_{\tau_t}\right)^- = \left(-K_T + K_{\tau_t}\right)^- = K_T - K_{\tau_t} = K_T - K_t.$$

Moreover, when $\tau_t = T$, by (i) we have $K_T = K_t$ and thus

$$\left(Y_{\tau_t} - K_T + K_{\tau_t} - L_{\tau_t}\right)^- = (\xi - L_T)^- = 0 = K_T - K_t.$$

This proves (ii).

(iii) Note that, for any $\tau \in \mathscr{T}^t$,

$$Y_t = Y_\tau + \int_t^\tau f_s(Y_s, Z_s)ds - \int_t^\tau Z_s dB_s + K_\tau - K_t. \tag{6.1.7}$$

On one hand,

$$Y_t \geq L_\tau 1_{\{\tau < T\}} + \xi 1_{\{\tau = T\}} + \int_t^\tau f_s(Y_s, Z_s)ds - \int_t^\tau Z_s dB_s.$$

Taking the conditional expectation on both sides we get

$$Y_t \geq \mathbb{E}_t\left[\int_t^\tau f_s(Y_s, Z_s)ds + L_\tau 1_{\{\tau < T\}} + \xi 1_{\{\tau = T\}}\right].$$

On the other hand, by (i) and (6.1.7) one can easily see that

$$Y_t = \mathbb{E}_t\left[\int_t^{\tau_t} f_s(Y_s, Z_s)ds + L_{\tau_t} 1_{\{\tau_t < T\}} + \xi 1_{\{\tau_t = T\}}\right].$$

This proves (iii). ∎

Remark 6.1.3 We remark that, in the case $f = 0$ and $\xi = L_T$, the above theorem shows that the solution to RBSDE (6.1.4), if it exists, indeed solves the American option pricing problem. In particular, in this setting Y stands for the option price, Z is the hedge portfolio, and τ_t is the optimal exercise time, provided the option has not been exercised at t. ∎

6.2 A Priori Estimates

We first show a connection between RBSDEs and BSDEs. The proof follows from comparison principle for (forward) SDEs and BSDEs, and is left to the readers.

Proposition 6.2.1 *Assume Assumption 6.1.1 holds and (Y, Z, K) is a solution to RBSDE (6.1.5). Let $\bar{Y}$ and $(\underline{Y}, \underline{Z})$ be the solutions to the following (forward) SDE and BSDE, respectively:*

$$\bar{Y}_t = Y_0 - \int_0^t f_s(\bar{Y}_s, Z_s) ds + \int_0^t Z_s dB_s;$$

$$\underline{Y}_t = \xi + \int_t^T f_s(\underline{Y}_s, \underline{Z}_s) ds - \int_t^T \underline{Z}_s dB_s.$$

Then

$$\underline{Y}_t \le Y_t \le \bar{Y}_t.$$

The a priori estimates in the next two theorems will be crucial for the well-posedness of RBSDEs.

Theorem 6.2.2 *Assume Assumption 6.1.1 holds and (Y, Z, K) is a solution to RBSDE (6.1.5). Then*

$$\|(Y, Z, K)\|^2 := \mathbb{E}\Big[|Y_T^*|^2 + |K_T^*|^2 + \int_0^T |Z_t|^2 dt \Big] \le C I_0^2 \tag{6.2.1}$$

$$\text{where} \quad I_0^2 := \mathbb{E}\Big[|\xi|^2 + \Big(\int_0^T |f_t^0| dt \Big)^2 + |(L^+)_T^*|^2 \Big].$$

We note that here $K_T^* = K_T$. However, the norm $\| \cdot \|$ is defined for non-monotone K as well, which will be needed for the next theorem.

Proof Similar to the proof of Theorem 4.2.1, we proceed in several steps.

Step 1. We first show that

$$\mathbb{E}[|Y_T^*|^2] \le C \mathbb{E}\Big[\int_0^T [|Y_t|^2 + |Z_t|^2] dt + K_T^2 \Big] + C I_0^2 < \infty. \tag{6.2.2}$$

Indeed, note that

$$|Y_t| \le |\xi| + C \int_t^T [|f_s^0| + |Y_s| + |Z_s|] ds + \Big| \int_t^T Z_s dB_s \Big| + K_T.$$

Then

$$Y_T^* \le |\xi| + C \int_0^T [|f_s^0| + |Y_s| + |Z_s|] ds + 2 \sup_{0 \le t \le T} \Big| \int_0^t Z_s dB_s \Big| + K_T.$$

Applying the Burkholder-Davis-Gundy inequality, we obtain (6.2.2) immediately.

Step 2. We next show that, for any $\varepsilon > 0$,

$$\sup_{0 \le t \le T} \mathbb{E}[|Y_t|^2] + \mathbb{E}\left[\int_0^T |Z_t|^2 dt\right] \le \varepsilon \mathbb{E}[|Y_T^*|^2 + K_T^2] + C\varepsilon^{-1} I_0^2. \quad (6.2.3)$$

Indeed, since K is a continuous increasing process, applying the extended Itô formula Theorem 2.3.4 on Y_t^2 we have

$$d(|Y_t|^2) = 2Y_t dY_t + |Z_t|^2 dt = [-2Y_t f_t(Y_t, Z_t) + |Z_t|^2]dt + 2Y_t Z_t dB_t - 2Y_t dK_t$$

$$= [-2Y_t f_t(Y_t, Z_t) + |Z_t|^2]dt + 2Y_t Z_t dB_t - 2L_t dK_t, \quad (6.2.4)$$

where the last equality thanks to the Skorohod condition (6.1.3). Thus

$$\mathbb{E}\left[|Y_t|^2 + \int_t^T |Z_s|^2 ds\right] = \mathbb{E}\left[|\xi|^2 + 2\int_t^T Y_s f_s(Y_s, Z_s)ds + 2\int_t^T L_s dK_s\right]$$

$$\le \mathbb{E}\left[|\xi|^2 + \int_t^T [C|Y_s|^2 + \frac{1}{2}|Z_s|^2]ds + 2Y_T^* \int_0^T |f_s^0|ds + 2(L^+)_T^* K_T\right].$$

Moving the Z part to the left side and applying the Gronwall Inequality, we get

$$\mathbb{E}\left[|Y_t|^2 + \int_t^T |Z_s|^2 ds\right] \le C\mathbb{E}\left[|\xi|^2 + Y_T^* \int_0^T |f_s^0|ds + (L^+)_T^* K_T\right],$$

which leads to (6.2.3) immediately.

Step 3. We now prove (6.2.1). Note that $K_0 = 0$, then

$$K_T = Y_0 - \xi - \int_0^T f_t(Y_t, Z_t)dt + \int_0^T Z_t dB_t.$$

This implies

$$\mathbb{E}[|K_T|^2] \le C \sup_{0 \le t \le T} \mathbb{E}[|Y_t|^2] + C\mathbb{E}\left[\int_0^T |Z_t|^2 dt\right] + CI_0^2.$$

Combine this and (6.2.3), and set ε small enough, we have

$$\sup_{0 \le t \le T} \mathbb{E}[|Y_t|^2] + \mathbb{E}\left[\int_0^T |Z_t|^2 dt\right] + \mathbb{E}[K_T^2] \le C\varepsilon \mathbb{E}[|Y_T^*|^2] + C\varepsilon^{-1} I_0^2.$$

This, together with (6.2.2), implies (6.2.1) by setting $\varepsilon > 0$ small enough.   ∎

We next establish the estimate for the difference of solutions to two RBSDEs, which leads to the uniqueness and stability of RBSDEs immediately.

Theorem 6.2.3 *For $i = 1, 2$, assume (ξ_i, f^i, L^i) satisfies Assumption 6.1.1, and (Y^i, Z^i, K^i) is a solution to RBSDE (6.1.5) with corresponding coefficients. Denote $\Delta \xi := \xi^1 - \xi^2$ and other terms in an obvious way. Then we have*

$$\|(\Delta Y, \Delta Z, \Delta K)\|^2 \leq C|\Delta I|^2 + C[I_1 + I_2]\left(\mathbb{E}\left[|(\Delta L)^*_T|^2\right]\right)^{\frac{1}{2}}, \quad where$$

$$|\Delta I|^2 := \mathbb{E}\left[|\Delta \xi|^2 + \left(\int_0^T |(\Delta f)_t(Y^1_t, Z^1_t)|dt\right)^2\right], \tag{6.2.5}$$

$$|I_i|^2 := \mathbb{E}\left[|\xi_i|^2 + \left(\int_0^T |f^i_t(0, 0)|dt\right)^2 + |((L^i)^+)^*_T|^2\right], \quad i = 1, 2.$$

Proof First we have

$$d\Delta Y_t = -\left[\alpha_t \Delta Y_t + \Delta Z_t \beta_t + (\Delta f)_t(Y^1_t, Z^1_t)\right]dt + \Delta Z_t dB_t - d\Delta K_t, \tag{6.2.6}$$

where α, β are bounded. Note that ΔK is not an increasing process, so one cannot apply Theorem 6.2.2 directly to (6.2.6). However, we shall apply similar arguments.

Applying the extended Itô formula Theorem 2.3.4 we have

$$d(\Delta Y_t)^2 = -2\Delta Y_t\left[\alpha_t \Delta Y_t + \Delta Z_t \beta_t + (\Delta f)_t(Y^1_t, Z^1_t)\right]dt \tag{6.2.7}$$

$$+ |\Delta Z_t|^2 dt + 2\Delta Y_t \Delta Z_t dB_t - 2\Delta Y_t d\Delta K_t.$$

Note that

$$\int_t^T \Delta Y_s d\Delta K_s = \int_t^T Y^1_s dK^1_s - \int_t^T Y^2_s dK^1_s - \int_t^T Y^1_s dK^2_s + \int_2^T Y^2_s dK^2_s$$

$$= \int_t^T L^1_s dK^1_s - \int_t^T Y^2_s dK^1_s - \int_t^T Y^1_s dK^2_s + \int_2^T L^2_s dK^2_s$$

$$\leq \int_t^T L^1_s dK^1_s - \int_t^T L^2_s dK^1_s - \int_t^T L^1_s dK^2_s + \int_2^T L^2_s dK^2_s \tag{6.2.8}$$

$$= \int_t^T \Delta L_s dK^1_s - \int_t^T \Delta L_s dK^2_s \leq |(\Delta L)^*_T|[K^1_T + K^2_T].$$

Combine this and (6.2.7), applying the Gronwall inequality one can easily have

$$\sup_{0 \leq t \leq T} \mathbb{E}[|\Delta Y_t|^2] + \mathbb{E}\left[\int_0^T |\Delta Z_t|^2 dt\right]$$

$$\leq C\mathbb{E}\left[|\Delta \xi|^2 + |(\Delta Y)^*_T|\int_0^T |\Delta f_t(Y^1_t, Z^1_t)|dt + |(\Delta L)^*_T|[K^1_T + K^2_T]\right].$$

Note that (6.2.1) implies that $\mathbb{E}[|K_T^i|^2] \leq |I_i|^2$. Then, for any $\varepsilon > 0$,

$$\sup_{0 \leq t \leq T} \mathbb{E}[|\Delta Y_t|^2] + \mathbb{E}\Big[\int_0^T |\Delta Z_t|^2 dt\Big] \tag{6.2.9}$$

$$\leq \varepsilon \mathbb{E}\Big[|(\Delta Y)_T^*|^2\Big] + C\varepsilon^{-1}|\Delta I|^2 + C[I_1 + I_2]\Big(\mathbb{E}\Big[|(\Delta L)_T^*|^2\Big]\Big)^{\frac{1}{2}}.$$

Moreover, by (6.2.7) we have

$$|(\Delta Y)_T^*|^2 \leq C\Big[|\Delta \xi|^2 + \int_0^T [|\Delta Y_t|^2 + |\Delta Z_t|^2]dt + (\Delta Y)_T^* \int_0^T |\Delta f_t(Y_t^1, Z_t^1)|dt$$

$$+ \sup_{0 \leq t \leq T} |\int_0^t \Delta Y_s \Delta Z_s dB_s| + \sup_{0 \leq t \leq T} \int_t^T \Delta Y_s d\Delta K_s\Big].$$

Plug (6.2.8) into the above estimate and apply the Burkholder-Davis-Gundy inequality, we get

$$\mathbb{E}\Big[|(\Delta Y)_T^*|^2\Big] \leq C\mathbb{E}\Big[|\Delta \xi|^2 + \int_0^T [|\Delta Y_t|^2 + |\Delta Z_t|^2]dt + (\Delta Y)_T^* \int_0^T |\Delta f_t(Y_t^1, Z_t^1)|dt$$

$$+ (\Delta Y)_T^*\Big(\int_0^T |\Delta Z_t|^2 dt\Big)^{\frac{1}{2}} + (\Delta L)_T^*[K_T^1 + K_T^2]\Big]$$

$$\leq \frac{1}{2}\mathbb{E}\Big[|(\Delta Y)_T^*|^2\Big] + C\mathbb{E}\Big[\int_0^T [|\Delta Y_t|^2 + |\Delta Z_t|^2]dt\Big]$$

$$+ C|\Delta I|^2 + C[I_1 + I_2]\Big(\mathbb{E}\Big[|(\Delta L)_T^*|^2\Big]\Big)^{\frac{1}{2}}.$$

Plug (6.2.9) into the right side of the second inequality and set ε small enough, we obtain the desired estimate for ΔY. This, together with (6.2.9), leads to the desired estimate for ΔZ. Finally, the desired estimate for ΔK follows from (6.2.6). ∎

Remark 6.2.4 The estimate for ΔK in Theorem 6.2.3 is under the norm $\mathbb{E}[|(\Delta K)_T^*|^2]$. Note that ΔK is a process with finite total variation, so a more natural norm should be $\mathbb{E}[|\bigvee_0^T(\Delta K)|^2]$. However, we are not able to provide a desired estimate for this norm. ∎

We conclude this section with the comparison principle for RBSDEs.

Theorem 6.2.5 *For $i = 1, 2$, assume (ξ_i, f^i, L^i) satisfies Assumption 6.1.1, and (Y^i, Z^i, K^i) is a solution to RBSDE (6.1.5) with corresponding coefficients. Assume further that $\xi_1 \leq \xi_2, f^1 \leq f^2$, and $L^1 \leq L^2$. Then $Y^1 \leq Y^2$.*

Proof We prove by contradiction, and assume without loss of generality that $Y_0^1 > Y_0^2$. Let

$$\tau := \inf\{t : Y_t^1 \leq Y_t^2\}.$$

Then $\tau > 0$. Note that $Y^2 \geq L^2 \geq L^1$, then $Y^1 > Y^2 \geq L^1$ in $[0, \tau)$ and thus $K^1 = 0$ in $[0, \tau)$, thanks to the Skorohod condition (6.1.3). Denote $\Delta Y := Y^1 - Y^2$ etc. as in Theorem 6.2.3. Now (6.2.6) becomes: for $t \in [0, \tau)$,

$$d\Delta Y_t = -\Big[\alpha_t \Delta Y_t + \Delta Z_t \beta_t + (\Delta f)_t(Y_t^1, Z_t^1)\Big]dt + \Delta Z_t dB_t + dK_t^2$$

$$\geq -\Big[\alpha_t \Delta Y_t + \beta_t \Delta Z_t\Big]dt + \Delta Z_t dB_t.$$

Since $\Delta Y_T = \Delta \xi \leq 0$, we have $\Delta Y_\tau = 0$. Then, similar to the proof of Proposition 6.2.1 in Problem 6.7.1, one can easily show that $\Delta Y_0 \leq 0$, contradiction. ∎

6.3 Well-Posedness of RBSDEs

We now establish the main well-posedness result for RBSDEs.

Theorem 6.3.1 *Assume Assumption 6.1.1. Then RBSDE (6.1.5) admits a unique solution.*

Clearly the uniqueness follows from Theorem 6.2.3, so the rest of this section is devoted to the existence of solutions. We shall introduce two methods, and in this section Assumption 6.1.1 is always in force.

6.3.1 The Snell Envelope Theory

As we can see from Theorem 6.1.2, the following Snell envelope for the barrier process L provides the candidate solution for RBSDE (6.1.5) with $f = 0$:

$$Y_t := \operatorname*{ess\,sup}_{\tau \in \mathscr{T}^t} \mathbb{E}_t[\widehat{L}_\tau], \quad \widehat{L}_t := L_t \mathbf{1}_{[0,T)} + \xi \mathbf{1}_{\{T\}}, \quad \tau^* := \inf\{t \geq 0 : Y_t = \widehat{L}_t\}.$$

$$(6.3.1)$$

Here, due to Assumption 6.1.1, we take the convention that $\mathbb{E}_t[\widehat{L}_\tau] = \mathbb{E}_t[\widehat{L}_\tau^+] - \mathbb{E}_t[\widehat{L}_\tau^-]$ could possibly take value $-\infty$. We have the following result.

Proposition 6.3.2 *With a possible modification, the Snell envelope Y satisfies:*

 (i) *Y is continuous in t with $Y_T = \xi$, a.s.*
 (ii) *$Y \geq \widehat{L}$ and $Y_{\tau^*} = \widehat{L}_{\tau^*}$, a.s.*
 (iii) *Y is a supermartingale on $[0, T]$ and a martingale on $[0, \tau^*]$. Consequently, τ^* is an optimal stopping time for Y_0.*

Remark 6.3.3 Proposition 6.3.2 holds true under weaker conditions, see, e.g., Karatzas & Shreve [118]. However, its proof relies on the dominated convergence theorem (under fixed $\mathbb{P}$). In the fully nonlinear case, we shall require some stronger regularities, see Section 10.3 below. ∎

Remark 6.3.4 There are some subtle measurability issues here.

(i) In the standard literature, see, e.g., Karatzas & Shreve [118], the Snell envelope Y is assumed to be $\overline{\mathbb{F}}^{\mathbb{P}}$-measurable and continuous in t for all ω. In this book we shall assume Y is $\mathbb{F}$-measurable, due to Proposition 1.2.1, but then it is continuous in t only for $\mathbb{P}$-a.e. ω. In general we cannot assume Y is $\mathbb{F}$-measurable and continuous in t for all ω. See Problem 6.7.3.

(ii) Due to the tradeoff between measurability and regularity of Y as discussed in (i), in both cases the τ^* defined in (6.3.1) is an $\overline{\mathbb{F}}^{\mathbb{P}}$-stopping time, not necessarily an $\mathbb{F}$-stopping time. However, note that τ^* is previsible, then by Proposition 1.2.5 there exists $\tilde{\tau}^* \in \mathscr{T}$ such that $\tilde{\tau}^* = \tau^*$, a.s. So rigorously the $\tilde{\tau}^*$ is our optimal stopping time. Nevertheless, the difference is not important when one fixes $\mathbb{P}$, and the readers may interpret the result in either way.

(iii) However, in the fully nonlinear case which we will study in Part III, the above difference becomes crucial. To overcome it, we shall assume pathwise regularity on ξ and L, then Y will have pathwise regularity and thus it is $\mathbb{F}$-measurable and continuous in t for all ω. See more details in Section 10.3 below. ∎

Proof of Proposition 6.3.2 As explained in Remark 6.3.4 (ii), in this proof we shall consider τ^* as a stopping time adapted to the augmented filtration $\overline{\mathbb{F}} := \overline{\mathbb{F}}^{\mathbb{P}}$, which satisfies the usual hypotheses by Corollary 2.1.10. We prove the proposition in several steps.

Step 1. Denote $Y_t^0 := \mathbb{E}_t[\xi]$. Then clearly $Y^0 \in \mathbb{S}^2(\overline{\mathbb{F}})$ with $Y_T^0 = \xi$. We claim that

$$Y_t = \operatorname*{ess\,sup}_{\tau \in \mathscr{T}^t} \mathbb{E}_t[\tilde{L}_\tau] \text{ and } \tau^* = \inf\{t : Y_t = \tilde{L}_t\}, \quad \text{where} \quad \tilde{L}_t = L_t \vee [Y_t^0 - (T-t)].$$

$$(6.3.2)$$

Indeed, since $\widehat{L} \le \tilde{L}$, it is clear that $Y_t \le \operatorname{ess\,sup}_{\tau \in \mathscr{T}^t} \mathbb{E}_t[\tilde{L}_\tau]$. On the other hand, for any $\tau \in \mathscr{T}$, denote

$$\tilde{\tau} := \tau \mathbf{1}_{\{L_\tau = \tilde{L}_\tau\}} + T\mathbf{1}_{\{L_\tau < \tilde{L}_\tau\}}.$$

By Problem 6.7.2 $\tilde{\tau} \in \mathscr{T}$. Then

$$\mathbb{E}_t[\tilde{L}_\tau] = \mathbb{E}_t\Big[L_\tau \mathbf{1}_{\{L_\tau = \tilde{L}_\tau\}} + [Y_\tau^0 - (T-\tau)]\mathbf{1}_{\{L_\tau < \tilde{L}_\tau\}}\Big]$$

$$\le \mathbb{E}_t\Big[L_\tau \mathbf{1}_{\{L_\tau = \tilde{L}_\tau\}} + Y_\tau^0 \mathbf{1}_{\{L_\tau < \tilde{L}_\tau\}}\Big] = \mathbb{E}_t\Big[L_\tau \mathbf{1}_{\{L_\tau = \tilde{L}_\tau\}} + \mathbb{E}_\tau[\xi]\mathbf{1}_{\{L_\tau < \tilde{L}_\tau\}}\Big]$$

$$= \mathbb{E}_t\Big[L_\tau \mathbf{1}_{\{L_\tau = \tilde{L}_\tau\}} + \xi \mathbf{1}_{\{L_\tau < \tilde{L}_\tau\}}\Big] = \mathbb{E}_t[\widehat{L}_{\tilde{\tau}}] \le Y_t.$$

This proves the first equality in (6.3.2).

The second equality is obvious because $Y_t \ge Y_t^0 > Y_t^0 - (T-t)$ for all $t < T$.

Now by (6.3.2) and Assumption 6.1.1 (ii), by otherwise replacing L with $\tilde{L}$, in the rest of this proof we assume without loss of generality that

$$L \in \mathbb{S}^2(\mathbb{F}) \text{ with } L_T = \xi, \quad \text{which implies further that} \quad \mathbb{E}[|Y_T^*|^2] < \infty. \text{ (6.3.3)}$$

Step 2. We next prove the dynamic programming principle in deterministic time:

$$Y_t = \operatorname*{ess\,sup}_{\tau \in \mathscr{T}^t} \mathbb{E}_t\Big[L_\tau \mathbf{1}_{\{\tau < t'\}} + Y_{t'} \mathbf{1}_{\{\tau \geq t'\}}\Big], \quad 0 \leq t < t' \leq T. \tag{6.3.4}$$

By choosing $\tau = t'$, this clearly implies that Y is a supermartingale.

To see this, for any $\tau \in \mathscr{T}^t$, note that $\tau \vee t' \in \mathscr{T}^{t'}$, we have

$$\mathbb{E}_t[L_\tau] = \mathbb{E}_t\Big[L_\tau \mathbf{1}_{\{\tau < t'\}} + L_{\tau \wedge t'} \mathbf{1}_{\{\tau \geq t'\}}\Big]$$

$$= \mathbb{E}_t\Big[L_\tau \mathbf{1}_{\{\tau < t'\}} + \mathbb{E}_{t'}[L_{\tau \wedge t'}] \mathbf{1}_{\{\tau \geq t'\}}\Big] \leq \mathbb{E}_t\Big[L_\tau \mathbf{1}_{\{\tau < t'\}} + Y_{t'} \mathbf{1}_{\{\tau \geq t'\}}\Big]$$

By the arbitrariness of τ this implies the "$\leq$" inequality of (6.3.4).

On the other hand, recall (6.3.3) and the definition (6.3.1). By Theorem 1.1.4 (i) there exist $\tau_n \in \mathscr{T}^{t'}$, $n \geq 1$, such that

$$Y_{t'} = \sup_{n \geq 1} \mathbb{E}_{t'}[L_{\tau_n}], \quad \text{a.s.}$$

Denote

$$Y_{t'}^N := \max_{1 \leq n \leq N} \mathbb{E}_{t'}[L_{\tau_n}], \quad E_n^N := \{\mathbb{E}_{t'}[L_{\tau_n}] = Y_{t'}^N\}, \quad \widehat{\tau}_N := \sum_{n=1}^N \tau_n \mathbf{1}_{E_n^N \setminus \cup_{i=1}^{n-1} E_i^N}.$$

Then $Y_{t'}^N \uparrow Y_{t'}$, a.s., $\widehat{\tau}_N \in \mathscr{T}^{t'}(\overline{\mathbb{F}})$ by Problem 6.7.2 again, and $Y_{t'}^N = \mathbb{E}_{t'}[L_{\widehat{\tau}_N}]$. For any $\tau \in \mathscr{T}^t$, we have

$$\mathbb{E}_t\Big[L_\tau \mathbf{1}_{\{\tau < t'\}} + Y_{t'}^N \mathbf{1}_{\{\tau \geq t'\}}\Big] = \mathbb{E}_t\Big[L_\tau \mathbf{1}_{\{\tau < t'\}} + \mathbb{E}_{t'}[L_{\widehat{\tau}_N}] \mathbf{1}_{\{\tau \geq t'\}}\Big]$$

$$= \mathbb{E}_t\Big[L_\tau \mathbf{1}_{\{\tau < t'\}} + L_{\widehat{\tau}_N} \mathbf{1}_{\{\tau \geq t'\}}\Big]$$

$$= \mathbb{E}_t\Big[L_{\tau \mathbf{1}_{\{\tau < t'\}} + \widehat{\tau}_N \mathbf{1}_{\{\tau \geq t'\}}}\Big] \leq Y_t,$$

where the last inequality thanks to the fact that $\tau \mathbf{1}_{\{\tau < t'\}} + \widehat{\tau}_N \mathbf{1}_{\{\tau \geq t'\}} \in \mathscr{T}^t(\overline{\mathbb{F}})$, again due to Problem 6.7.2. Now send $N \to \infty$, by (6.3.3) and applying the dominated convergence theorem, we obtain

$$\mathbb{E}_t\Big[L_\tau \mathbf{1}_{\{\tau < t'\}} + Y_{t'}^N \mathbf{1}_{\{\tau \geq t'\}}\Big] \leq Y_t.$$

This implies the "$\geq$" inequality of (6.3.4) and thus proves (6.3.4).

Step 3. In this step we show that Y is continuous in t, which immediately leads to $Y_{\tau^*} = L_{\tau^*}$, a.s. Indeed, for each $n \geq 1$, denote

$$D_n := \{t_i^n := \frac{i}{2^n}T, i = 0, \cdots, 2^n\}, \quad \mathscr{T}_t^n := \{\tau \in \mathscr{T}^t : \tau \text{ takes values in } D_n\},$$

$$Y_t^n := \operatorname*{ess\,sup}_{\tau \in \mathscr{T}_t^n} \mathbb{E}_t[L_\tau].$$

It is clear that

$$Y_t^n \leq Y_t, \ t \in [0, T], \quad \text{and} \quad Y_{t_i^n}^n \geq L_{t_i^n}, \ i = 0, \cdots, 2^n. \tag{6.3.5}$$

Note that, for $t_{i-1}^n < t \leq t_i^n$ and $\tau \in \mathscr{T}_t^n$, we have $\tau \geq t_i^n$. Thus, following similar arguments as in Step 2, one can easily show that,

$$Y_t^n = \mathbb{E}_t[Y_{t_i^n}^n], \ t_{i-1}^n < t \leq t_i^n; \quad Y_{t_{i-1}^n}^n = \sup_{\tau \in \mathscr{T}_{t_{i-1}^n}^n} \mathbb{E}_{t_{i-1}^n}\left[L_{t_{i-1}^n}\mathbf{1}_{\{\tau = t_{i-1}^n\}} + Y_{t_i^n}^n \mathbf{1}_{\{\tau > t_{i-1}^n\}}\right]$$

$$= L_{t_{i-1}^n} \vee \mathbb{E}_{t_{i-1}^n}[Y_{t_i^n}^n].$$

Denote

$$\tilde{Y}_t^n := \mathbb{E}_t[Y_{t_i^n}^n] + \frac{t_i^n - t}{t_i^n - t_{i-1}^n}\left(L_{t_{i-1}^n} - \mathbb{E}_{t_{i-1}^n}[Y_{t_i^n}^n]\right)^+, \ t_{i-1}^n \leq t \leq t_i^n.$$

Then $\tilde{Y}_{t_i^n}^n = Y_{t_i^n}^n$, $\tilde{Y}_{t_{i-1}^n}^n = Y_{t_{i-1}^n}^n$, and $\tilde{Y}^n$ is continuous on $[0, T]$, a.s. It suffices to show that

$$\lim_{n \to \infty} \mathbb{E}\left[|(\tilde{Y}^n - Y)_T^*|^2\right] = 0. \tag{6.3.6}$$

To see this, denote $h_n := \frac{T}{2^n}$ and, for any $\delta > 0$,

$$\xi_\delta := \sup_{0 \leq s, t \leq T, |t-s| \leq \delta} |L_s - L_t|.$$

Note that, for any t and $\tau \in \mathscr{T}^t$, there exists $\tilde{\tau} \in \mathscr{T}_t^n$ such that $0 \leq \tilde{\tau} - \tau \leq h_n$. Then

$$\mathbb{E}_t[L_\tau] - Y_t^n \leq \mathbb{E}_t[L_\tau] - \mathbb{E}_t[L_{\tilde{\tau}}] \leq \mathbb{E}_t[\xi_{h_n}],$$

and thus

$$0 \leq Y_t - Y_t^n \leq \mathbb{E}_t[\xi_{h_n}].$$

Now for $t = t_{i-1}^n$,

$$|\tilde{Y}_t^n - Y_t| = Y_t - Y_t^n \leq \mathbb{E}_t[\xi_{h_n}],$$

and for $t_{i-1}^n < t \le t_i^n$,

$$|\tilde{Y}_t^n - Y_t| \le |Y_t^n - Y_t| + \left(L_{t_{i-1}^n} - \mathbb{E}_{t_{i-1}^n}[Y_{t_i^n}^n]\right)^+ \le \mathbb{E}_t[\xi_{h_n}] + \left(L_{t_{i-1}^n} - \mathbb{E}_{t_{i-1}^n}[L_{t_i^n}^n]\right)^+$$

$$\le \mathbb{E}_t[\xi_{h_n}] + \mathbb{E}_{t_{i-1}^n}[\xi_{h_n}],$$

where the second inequality thanks to (6.3.5). Then, applying the Burkholder-Davis-Gundy inequality we have

$$\mathbb{E}\Big[|(\tilde{Y}^n - Y)_T^*|^2\Big] \le C\mathbb{E}\Big[\sup_{0 \le t \le T} |\mathbb{E}_t[\xi_{h_n}]|^2\Big] \le C\mathbb{E}[|\xi_{h_n}|^2].$$

This, together with (6.3.3) and the dominated convergence theorem, implies (6.3.6) and hence Y is continuous, a.s.

Step 4. We now extend the dynamic programming principle (6.3.4) to stopping times:

$$Y_t = \text{ess}\sup_{\tau \in \mathscr{T}^t} \mathbb{E}_t\Big[L_\tau \mathbf{1}_{\{\tau < \tilde{\tau}\}} + Y_{\tilde{\tau}} \mathbf{1}_{\{\tau \ge \tilde{\tau}\}}\Big], \quad \forall t \in [0, T], \ \tilde{\tau} \in \mathscr{T}^t. \quad (6.3.7)$$

We prove only the "$\le$" inequality here, which will be used in the next step, and the opposite inequality is left to Problem 6.7.5. First, assume $\tilde{\tau}$ takes only finitely many values $t_1, \cdots, t_n$. For any $\tau \in \mathscr{T}^t$,

$$\mathbb{E}_t[L_\tau] = \mathbb{E}_t\Big[\sum_{i=1}^n L_\tau \mathbf{1}_{\{\tilde{\tau}=t_i\}}\Big] = \sum_{i=1}^n \mathbb{E}_t\Big[[L_\tau \mathbf{1}_{\{\tau < t_i\}} + L_\tau \mathbf{1}_{\{\tau \ge t_i\}}]\mathbf{1}_{\{\tilde{\tau}=t_i\}}\Big]$$

$$= \sum_{i=1}^n \mathbb{E}_t\Big[[L_\tau \mathbf{1}_{\{\tau < t_i\}} + \mathbb{E}_{t_i}[L_\tau \vee t_i]\mathbf{1}_{\{\tau \ge t_i\}}]\mathbf{1}_{\{\tilde{\tau}=t_i\}}\Big]$$

$$\le \sum_{i=1}^n \mathbb{E}_t\Big[[L_\tau \mathbf{1}_{\{\tau < t_i\}} + Y_{t_i}\mathbf{1}_{\{\tau \ge t_i\}}]\mathbf{1}_{\{\tilde{\tau}=t_i\}}\Big]$$

$$= \mathbb{E}_t\Big[L_\tau \mathbf{1}_{\{\tau < \tilde{\tau}\}} + Y_{\tilde{\tau}}\mathbf{1}_{\{\tau \ge \tilde{\tau}\}}\Big].$$

For general $\tilde{\tau}$, by Problem 1.4.8 (i), there exist $\tilde{\tau}_n \in \mathscr{T}^t$ such that $\tilde{\tau}_n \downarrow \tilde{\tau}$ and each $\tilde{\tau}_n$ takes only finitely values. Then

$$\mathbb{E}_t[L_\tau] \le \mathbb{E}_t\Big[L_\tau \mathbf{1}_{\{\tau < \tilde{\tau}_n\}} + Y_{\tilde{\tau}_n}\mathbf{1}_{\{\tau \ge \tilde{\tau}_n\}}\Big].$$

Send $n \to \infty$, by the regularity of L, Y and applying the denominated convergence theorem, we have

$$\mathbb{E}_t[L_\tau] \le \mathbb{E}_t\Big[L_\tau \mathbf{1}_{\{\tau \le \tilde{\tau}\}} + Y_{\tilde{\tau}}\mathbf{1}_{\{\tau > \tilde{\tau}\}}\Big] \le \mathbb{E}_t\Big[L_\tau \mathbf{1}_{\{\tau < \tilde{\tau}\}} + Y_{\tilde{\tau}}\mathbf{1}_{\{\tau \ge \tilde{\tau}\}}\Big].$$

Since τ is arbitrary, we prove the $\le$ part of (6.3.7).

Step 5. It is obvious that $Y \geq L$ and $Y_T = \xi$. So it remains to prove Y is a martingale on $[0, \tau^*]$. Since we already know Y is a supermartingale on $[0, T]$, this is equivalent to

$$Y_0 \leq \mathbb{E}[Y_{\tau^*}]. \tag{6.3.8}$$

If $Y_0 = L_0$, then $\tau^* = 0$ and (6.3.8) is obvious. We now assume $Y_0 > L_0$. For each $n > \frac{1}{Y_0 - L_0}$, denote

$$\tau_n := \inf \left\{ t \geq 0 : Y_t - L_t \leq \frac{1}{n} \right\}.$$

Then $0 < \tau_n \uparrow \tau^*$ and $Y_t - L_t \geq \frac{1}{n}$ for $t \in [0, \tau_n]$. Applying Step 4 with $t = 0, \tilde{\tau} := \tau_n$, we have

$$Y_0 \leq \sup_{\tau \in \mathscr{T}} \mathbb{E} \left[L_\tau \mathbf{1}_{\{\tau < \tau_n\}} + Y_{\tau_n} \mathbf{1}_{\{\tau_n \leq \tau\}} \right]$$

For any $\varepsilon > 0$, there exists $\tau_\varepsilon \in \mathscr{T}$ such that

$$Y_0 \leq \mathbb{E} \left[L_{\tau_\varepsilon} \mathbf{1}_{\{\tau_\varepsilon < \tau_n\}} + Y_{\tau_n} \mathbf{1}_{\{\tau_n \leq \tau_\varepsilon\}} \right] + \varepsilon \leq \mathbb{E} \left[[Y_{\tau_\varepsilon} - \frac{1}{n}] \mathbf{1}_{\{\tau_\varepsilon < \tau_n\}} + Y_{\tau_n} \mathbf{1}_{\{\tau_n \leq \tau_\varepsilon\}} \right] + \varepsilon. \tag{6.3.9}$$

Since Y is a supermartingale, this implies

$$Y_0 \leq \mathbb{E} \left[Y_{\tau_\varepsilon \wedge \tau_n} - \frac{1}{n} \mathbf{1}_{\{\tau_\varepsilon < \tau_n\}} \right] + \varepsilon \leq Y_0 - \frac{1}{n} \mathbb{P}(\tau_\varepsilon < \tau_n) + \varepsilon.$$

Then

$$\mathbb{P}(\tau_\varepsilon < \tau_n) \leq n\varepsilon.$$

Moreover, by (6.3.9) we also have

$$Y_0 \leq \mathbb{E}[Y_{\tau_n}] + \mathbb{E} \left[[Y_{\tau_\varepsilon} - Y_{\tau_n} - \frac{1}{n}] \mathbf{1}_{\{\tau_\varepsilon < \tau_n\}} \right]$$

$$+ \varepsilon \leq \mathbb{E}[Y_{\tau_n}] + C \left(\mathbb{P}(\tau_\varepsilon < \tau_n) \right)^{\frac{1}{2}} + \varepsilon \leq \mathbb{E}[Y_{\tau_n}] + C\sqrt{n\varepsilon} + \varepsilon.$$

Send $\varepsilon \to 0$, we obtain

$$Y_0 \leq \mathbb{E}[Y_{\tau_n}].$$

Now send $n \to \infty$, since $\tau_n \uparrow \tau^*$ and Y is continuous, (6.3.8) follows from the dominated convergence theorem.   ∎

6.3.2 Existence via Picard Iteration

This method follows the approach in Section 4.3 for well-posedness of BSDEs. Namely we first solve the linear equation, in this reflected case we use the Snell envelope theory, and then construct the solution to general RBSDE via Picard iteration. We proceed in several steps.

Step 1. We first assume $f = 0$ and verify that the Y defined in (6.3.1) is a solution. The proof relies heavily on Proposition 6.3.2. First, we have $Y_T = \xi$. Next, note that

$$\mathbb{E}_t[\xi] \leq Y_t \leq \mathbb{E}_t\Big[(L^+)_T^* + |\xi|\Big],$$

then clearly $Y \in \mathbb{S}^2(\mathbb{F})$. Moreover, since Y is a continuous supermartingale, it follows from the Doob-Meyer decomposition Theorem 2.7.1 that there exist $Z \in \mathbb{L}^2(\mathbb{F}, \mathbb{R}^{1\times d})$ and $K \in \mathbb{I}^2(\mathbb{F})$ such that

$$dY_t = Z_t dB_t - dK_t.$$

Finally, for any t, following the same arguments of Proposition 6.3.2, Y is a martingale on $[t, \tau_t^*]$, where $\tau_t^* := \inf\{s \geq t : Y_s = \widehat{L}_s\}$ and it holds $Y_{\tau_t^*} = \widehat{L}_{\tau_t^*}$. Then $K_{\tau_t^*} = K_t$ and thus $\mathbf{1}_{\{\tau_t^* > t\}} dK_t = 0$. Therefore,

$$\int_0^T [Y_t - L_t] dK_t = \int_0^T [Y_t - \widehat{L}_t] dK_t = \int_0^T [Y_t - \widehat{L}_t] \mathbf{1}_{\{\tau_t^* > t\}} dK_t = 0, \quad \text{a.s.}$$

This verifies the Skorohod condition and thus (Y, Z, K) satisfies all the requirements.

Step 2. We next assume $f(y, z) = f^0 \in \mathbb{L}^{1,2}(\mathbb{F}, \mathbb{R})$ is independent of (y, z). Denote

$$\tilde{L}_t := L_t + \int_0^t f_s^0 ds, \quad \tilde{\xi} := \xi + \int_0^T f_s^0 ds.$$

Clearly $\tilde{L}, \tilde{\xi}$ satisfy Assumption 6.1.1. By Step 1, the following RBSDE has a solution $(\tilde{Y}, \tilde{Z}, \tilde{K})$:

$$\tilde{Y}_t = \tilde{\xi} - \int_t^T \tilde{Z}_s dB_s + \tilde{K}_T - \tilde{K}_t, \quad \tilde{Y}_t \geq \tilde{L}_t, \quad \int_0^T [\tilde{Y}_t - \tilde{L}_t] d\tilde{K}_t = 0.$$

Then it is straightforward to check that the following (Y, Z, K) is a solution to the original RBSDE (6.1.5) with generator $f = f^0$:

$$Y_t := \tilde{Y}_t - \int_0^t f_s^0 ds, \quad Z_t := \tilde{Z}_t, \quad K_t := \tilde{K}_t.$$

Step 3. We now prove the case that T is small by using Picard iteration. We first recall that the constant C in the right side of a priori estimate (6.2.5) does not depend on the terminal condition ξ, and is increasing in T. Let C_1 be the constant corresponding to $T = 1$. Now assume $T \leq \delta$ for some small constant $\delta \in (0, 1)$ which will be specified later. Set $Y^0 := 0$, $Z^0 := 0$, and $K^0 := 0$. For $n = 1, \cdots$, consider RBSDE with solution (Y^n, Z^n, K^n):

$$Y_t^n = \xi + \int_t^T f_s(Y_s^{n-1}, Z_s^{n-1})ds - \int_t^T Z_s^{n-1}dB_s + K_T^n - K_t^n, \ Y_t^n \geq L_t, \ [Y_t^n - L_t]dK_t^n = 0.$$

$$(6.3.10)$$

By induction and by applying Step 2 repeatedly we see that $f(Y^n, Z^n) \in \mathbb{L}^{1,2}(\mathbb{F})$ and the above RBSDE admits a unique solution for all n. Denote $\Delta Y^n := Y^n - Y^{n-1}$, $\Delta Z^n := Z^n - Z^{n-1}$, $\Delta K^n := K^n - K^{n-1}$, $n \geq 1$. Then by (6.2.5) we have, for some constant C_0 depending only on the Lipschitz constant of f in (y, z),

$$\|(\Delta Y^{n+1}, \Delta Z^{n+1}, \Delta K^{n+1})\|^2 \leq C_1 \mathbb{E}\left[\left(\int_0^T |f_t(Y_t^n, Z_t^n) - f_t(Y_t^{n-1}, Z_t^{n-1})|dt \right)^2 \right]$$

$$\leq C_0 C_1 \mathbb{E}\left[\left(\int_0^T [|\Delta Y_t^n| + |\Delta Z_t^n|]dt \right)^2 \right] \leq C_0 C_1 T \mathbb{E}\left[\int_0^T [|\Delta Y_t^n|^2 + |\Delta Z_t^n|^2]dt \right]$$

$$\leq C_0 C_1 \delta \|(\Delta Y^n, \Delta Z^n, \Delta K^n)\|^2.$$

Set $\delta := \frac{1}{4C_0 C_1}$, we have

$$\|(\Delta Y^{n+1}, \Delta Z^{n+1}, \Delta K^{n+1})\| \leq \frac{1}{2}\|(\Delta Y^n, \Delta Z^n, \Delta K^n)\|.$$

By induction,

$$\|(\Delta Y^{n+1}, \Delta Z^{n+1}, \Delta K^{n+1})\| \leq \frac{1}{2^n}\|(Y^1, Z^1, K^1)\|$$

Now for any $m \geq n$, we have

$$\|(Y^m - Y^n, Z^m - Z^n, K^m - K^n)\| \leq \sum_{i=n}^{m-1} \|(\Delta Y^{i+1}, \Delta Z^{i+1}, \Delta K^{i+1})\|$$

$$\leq \sum_{i=n}^{m-1} \frac{1}{2^i}\|(Y^1, Z^1, K^1)\| \leq \frac{1}{2^{n-1}}\|(Y^1, Z^1, K^1)\| \to 0,$$

as $n \to \infty$. Then there exist (Y, Z, K) such that

$$\lim_{n \to \infty} \|(Y^n - Y, Z^n - Z, K^n - K)\| = 0. \qquad (6.3.11)$$

Now by sending $n \to \infty$ in (6.3.10) we see immediately that

$$Y_t = \xi + \int_t^T f_s(Y_s, Z_s) ds - \int_t^T Z_s dB_s + K_T - K_t, \quad \text{and} \quad Y_t \geq L_t,$$

It remains to verify the Skorohod condition (6.1.3). Indeed, for each n, applying Theorem 6.1.2 (ii) on RBSDE (6.3.10) we have

$$Y_t^n = \operatorname*{ess\,sup}_{\tau \in \mathscr{T}^t} \mathbb{E}_t \left[\int_t^\tau f_s(Y_s^{n-1}, Z_s^{n-1}) ds + \widehat{L}_\tau \right].$$

Send $n \to \infty$, (6.3.11) leads to

$$Y_t = \operatorname*{ess\,sup}_{\tau \in \mathscr{T}^t} \mathbb{E}_t \left[\int_t^\tau f_s(Y_s, Z_s) ds + \widehat{L}_\tau \right].$$

Now the Skorohod condition follows from the arguments in Step 1.

Step 4. We now prove the general case with arbitrary large T. Fix $\delta \in (0, 1)$ as in Step 3. Let $0 = t_0 < \cdots < t_n = T$ be a time partition such that $t_{i+1} - t_i \leq \delta$ for all i. Applying Step 3 on $[t_{n-1}, t_n]$ with terminal condition ξ, we obtain a solution (Y, Z, K) on $[t_{n-1}, t_n]$. Next applying Step 3 on $[t_{n-2}, t_{n-1}]$ with terminal condition $Y_{t_{n-1}}$, we obtain a solution (Y, Z, K) on $[t_{n-2}, t_{n-1}]$. Repeating the arguments backwardly n times we obtain the solution over the whole interval $[0, T]$.   ∎

6.3.3   Existence via Penalization

This method approximates the reflected BSDE by a sequence of BSDEs without reflection, via the so-called penalization. We note that this proof does not rely on the Snell envelope theory. Again we shall proceed in several steps.

Step 1. For $n = 0, 1, \cdots$, let $f_t^n(\omega, y, z) := f_t(\omega, y, z) + n[y - L_t(\omega)]^-$. Then f^n is uniformly Lipschitz continuous in (y, z), uniformly on (t, ω) (but not uniformly on n). Let (Y^n, Z^n) be the solution to the following BSDE without reflection:

$$Y_t^n = \xi + \int_t^T f_s^n(Y_s^n, Z_s^n) ds - \int_t^T Z_s^n dB_s,$$

which can be rewritten as

$$Y_t^n = \xi + \int_t^T f_s(Y_s^n, Z_s^n) ds - \int_t^T Z_s^n dB_s + K_T^n - K_t^n, \quad \text{where } K_t^n :=$$

$$n \int_0^t [Y_s^n - L_s]^- ds. \tag{6.3.12}$$

Then for each n, $K^n \in \mathbb{L}^2(\mathbb{F})$ with $K_0^n = 0$. Note that

$$[Y_t^n - L_t][Y_t^n - L_t]^- \le 0, \quad \text{and thus} \quad Y_t^n dK_t^n \le L_t dK_t^n.$$

Now following the same arguments as in Theorem 6.2.2, in particular by changing the last equality in (6.2.4) to the following inequality:

$$d(|Y_t^n|^2) \ge [-2Y_t^n f_t(Y_t^n, Z_t^n) + |Z_t^n|^2]dt + 2Y_t^n Z_t^n dB_t - 2L_t dK_t^n,$$

we get, for some constant C independent of n,

$$\|(Y^n, Z^n, K^n)\|^2 \le C I_0^2. \tag{6.3.13}$$

We remark that we may also view (Y^n, Z^n, K^n) as the unique solution to the following RBSDE:

$$Y_t^n = \xi + \int_t^T f_s(Y_s^n, Z_s^n)ds - \int_t^T Z_s^n dB_s + K_T^n - K_t^n, \; Y^n \ge L^n,$$

$$\int_0^T [Y_t^n - L_t^n]dK_t^n = 0; \text{where} \quad L^n := Y^n \wedge L. \tag{6.3.14}$$

Step 2. We next show that

$$\lim_{n\to\infty} \mathbb{E}\left[|([Y^n - L]^-)_T^*|^2 \right] = 0. \tag{6.3.15}$$

Indeed, since $f^n \le f^{n+1}$. By the comparison Theorem 4.4.1, we have $Y_t^n \le Y_t^{n+1}$, $0 \le t \le T$, a.s. Then there exists Y such that $Y_t^n \uparrow Y_t$, $0 \le t \le T$, a.s. By (6.3.13) and applying Fatou's Lemma we see that $Y \in \mathbb{L}^2(\mathbb{F})$. Note that $Y_t^0 \le Y_t^n \le Y_t$, then it follows from the Dominated Convergence Theorem that

$$\mathbb{E}\left[\int_0^T [Y_t - L_t]^- dt \right] = \lim_{n\to\infty} \mathbb{E}\left[\int_0^T [Y_t^n - L_t]^- dt \right] = \lim_{n\to\infty} \frac{1}{n}\mathbb{E}[K_T^n] = 0.$$
$$\tag{6.3.16}$$

Denote $h^n := f(Y^n, Z^n) - f^0 \in \mathbb{L}^2(\mathbb{F})$. By (6.3.13) and Theorem 1.3.7 there exist $(h, Z) \in \mathbb{L}^2(\mathbb{F}) \times \mathbb{L}^2(\mathbb{F}, \mathbb{R}^{1\times d})$ such that $(h^n, Z^n) \to (h, Z)$, weakly in $\mathbb{L}^2(\mathbb{F})$. Denote

$$K_t := Y_t - Y_0 - \int_0^t [f_s^0 + h_s]ds + \int_0^t Z_s dB_s.$$

Clearly $K^n \to K$ weakly in $\mathbb{L}^2(\mathbb{F})$, then K is also increasing and thus has right (and left) limit. Therefore, for a.e. $\omega \in \Omega$, $Y(\omega)$ has right limit Y_{t+} and it satisfies $Y_{t+} \le Y_t$. This, together with (6.3.16) and the continuity of L, implies that $Y_t \ge L_t$,

$0 \leq t < T$, a.s. Since $Y_T = \xi \geq L_T$, we obtain $[Y_t - L_t]^- = 0, 0 \leq t \leq T$, a.s. and thus $[Y_t^n - L_t]^- \downarrow 0, 0 \leq t \leq T$, a.s. Now apply Dini's lemma, see Problem 6.7.6 , we have $([Y^n - L]^-)_T^* \downarrow 0$. Then (6.3.15) follows from the Dominated Convergence Theorem.

Step 3. Recall (6.3.14) that $L^n := Y^n \wedge L$. For any $n, m \to \infty$, by (6.3.15) we have

$$\mathbb{E}\Big[|(L^n - L^m)_T^*|^2\Big] \leq \mathbb{E}\Big[|([Y^n - L]^-)_T^*|^2 + |([Y^m - L]^-)_T^*|^2\Big] \to 0.$$

Now applying Theorem 6.2.3 on (6.3.14) we obtain

$$\lim_{n,m\to\infty} \|(Y^n - Y^m, Z^n - Z^m, K^n - K^m)\| = 0,$$

and thus

$$\lim_{n\to\infty} \|(Y^n - Y, Z^n - Z, K^n - K)\| = 0. \qquad (6.3.17)$$

Since Y^n, K^n are continuous, then so are Y and K. It is clear that $Y \in \mathbb{S}^2(\mathbb{F}), K \in \mathbb{I}^2(\mathbb{F})$, and

$$Y_t = \xi + \int_t^T f_s(Y_s, Z_s)ds - \int_t^T Z_s dB_s + K_T - K_t, \quad Y_t \geq L_t. \qquad (6.3.18)$$

It remains to verify the Skorohod condition (6.1.3). Denote

$$\tilde{L} := (Y - 1) \vee L \in \mathbb{S}^2(\mathbb{F}), \qquad (6.3.19)$$

thanks to Assumption 6.1.1 and the fact that $Y \in \mathbb{S}^2(\mathbb{F})$. Since $Y \geq L$, (6.1.3) is equivalent to

$$\mathbb{E}\Big[\int_0^T [Y_t - \tilde{L}_t]dK_t\Big] = 0. \qquad (6.3.20)$$

For any $m \geq 1$, denote $t_i := t_i^m := \frac{i}{m}T$. By (6.3.17) and recalling (6.2.1) for the norm $\|\cdot\|$, we have

$$\lim_{n\to\infty} \mathbb{E}\Big[\sum_{i=0}^{m-1}[Y_{t_i} - \tilde{L}_{t_i}][K_{t_i,t_{i+1}} - K_{t_i,t_{i+1}}^n]\Big] = 0.$$

This implies that

$$\mathbb{E}\Big[\int_0^T [Y_t - \tilde{L}_t]dK_t\Big] = \lim_{n\to\infty} \mathbb{E}\Big[\sum_{i=0}^{m-1}\int_{t_i}^{t_{i+1}} [(Y_t - \tilde{L}_t) - (Y_{t_i} - \tilde{L}_{t_i})]d(K_t - K_t^n) + \int_0^T [Y_t - \tilde{L}_t]dK_t^n\Big].$$

Note that

$$\limsup_{n\to\infty} \mathbb{E}\Big[\int_0^T [Y_t - \tilde{L}_t]dK_t^n\Big] \leq \limsup_{n\to\infty} \mathbb{E}\Big[\int_0^T [Y_t - L_t^n]dK_t^n\Big]$$

$$= \limsup_{n\to\infty} \mathbb{E}\Big[\int_0^T [Y_t - Y_t^n]dK_t^n\Big]$$

$$\leq \limsup_{n\to\infty} \mathbb{E}\Big[(Y - Y^n)_T^* K_T^n\Big] \leq \limsup_{n\to\infty} \Big(\mathbb{E}\big[|(Y-Y^n)_T^*|^2\big]\Big)^{\frac{1}{2}}\Big(\mathbb{E}[|K_T^n|^2]\Big)^{\frac{1}{2}} = 0,$$

thanks to (6.3.13) and (6.3.17). Then, by (6.3.13) again,

$$\mathbb{E}\Big[\int_0^T [Y_t - \tilde{L}_t]dK_t\Big] \leq \liminf_{n\to\infty} \mathbb{E}\Big[\sum_{i=0}^{m-1}\int_{t_i}^{t_{i+1}} [(Y_t - \tilde{L}_t) - (Y_{t_i} - \tilde{L}_{t_i})]d(K_t - K_t^n)\Big]$$

$$\leq \liminf_{n\to\infty} \mathbb{E}\Big[\sup_{0\leq i\leq m-1}\sup_{t_i\leq t\leq t_{i+1}} |(Y_t - \tilde{L}_t) - (Y_{t_i} - \tilde{L}_{t_i})|[K_T + K_T^n]\Big]$$

$$\leq C\Big(\mathbb{E}\Big[\sup_{0\leq i\leq m-1}\sup_{t_i\leq t\leq t_{i+1}} |(Y_t - \tilde{L}_t) - (Y_{t_i} - \tilde{L}_{t_i})|^2\Big]\Big)^{\frac{1}{2}}.$$

Now send $m \to \infty$. Since $Y, L \in \mathbb{S}^2(\mathbb{F})$, it follows from the Dominated Convergence Theorem that $\mathbb{E}\Big[\int_0^T [Y_t - \tilde{L}_t]dK_t\Big] \leq 0$. This, together with the fact $Y \geq \tilde{L}$, implies (6.3.20) and hence proves the Skorohod condition (6.1.3). Now by (6.3.18) we see that (Y, Z, K) is a solution to RBSDE (6.1.5). ∎

6.4   Markov RBSDEs and Obstacle Problem of PDEs

In this section we study the following Markov RBSDE:

$$\begin{cases} X_t = x + \int_0^t b(s, X_s)ds + \int_0^t \sigma(s, X_s)dB_s; \\ Y_t = g(X_T) + \int_t^T f(s, X_s, Y_s, Z_s)ds - \int_t^T Z_s dB_s + K_T - K_t; \qquad (6.4.1) \\ Y_t \geq l(t, X_t); \quad \int_0^T [Y_t - l(t, X_t)]dK_t = 0; \end{cases}$$

where b, σ, f, g, h are deterministic functions. We shall assume

Assumption 6.4.1

(i) b, σ, f, g satisfy Assumption 5.0.1 with $d_2 = 1$;
(ii) $l \in C^{1,2}([0, T] \times \mathbb{R}^{d_1}, \mathbb{R})$ with bounded derivatives and $l(T, \cdot) \leq g$.

Under the above assumption, by Theorems 3.3.1 and 6.3.1 it is clear that RBSDE (6.4.1) admits a unique solution.

We first establish the relationship between the Markov RBSDE (6.4.1) and PDEs. Following similar arguments as in Section 5.1, we see that $Y_t = u(t, X_t)$ for some deterministic function u. Assume $u \in C^{1,2}([0, T] \times \mathbb{R}^{d_1}, \mathbb{R})$ and apply Itô formula, we have

$$du(t, X_t) = [\partial_t u + \partial_x u b + \frac{1}{2} \partial_{xx} u : \sigma \sigma^\top](t, X_t) dt + \partial_x u \sigma (t, X_t) dB_t.$$

Compare this with

$$dY_t = -f(t, X_t, Y_t, Z_t) dt + Z_t dB_t - dK_t.$$

Following the arguments for the uniqueness proof in Theorem 2.7.1, we see that $Z_t = \partial_x u \sigma(t, X_t)$ and $\mathbb{L}u(t, X_t) dt = -dK_t \leq 0$, where the differential operator $\mathbb{L}u$ is defined by (5.1.9). Moreover, clearly $u \geq l$, and when $u > l$, we should have $dK_t = 0$, which implies $\mathbb{L}u(t, X_t) = 0$. Put together, we see that u satisfies the following obstacle problem of parabolic PDE:

$$\max\left(l - u, \mathbb{L}u\right) = 0; \quad u(T, x) = g(x); \tag{6.4.2}$$

The following result is in the reverse direction and is obvious.

Theorem 6.4.2 *Assume* $u \in C^{1,2}([0, T] \times \mathbb{R}^{d_1}, \mathbb{R})$ *is a classical solution to (6.4.2). Then*

$$Y_t := u(t, X_t), \quad Z_t := \partial_x u \sigma (t, X_t), \quad dK_t := -\mathbb{L}u(t, X_t) dt$$

are the solution to RBSDE (6.4.1).

In general, the obstacle PDE does not have a classical solution, we thus turn to its viscosity solution. For any (t, x), recall $X^{t,x}$ in (5.1.1) and define

$$u(t, x) := Y_t^{t,x}, \quad (t, x) \in [0, T] \times \mathbb{R}^d, \tag{6.4.3}$$

where $(Y^{t,x}, Z^{t,x}, K^{t,x})$ is the solution to the following RBSDE on $[t, T]$:

$$Y_s^{t,x} = g(X_T^{t,x}) + \int_s^T f(r, X_r^{t,x}, Y_r^{t,x}, Z_r^{t,x}) dr - \int_s^T Z_r^{t,x} dB_r + K_T^{t,x} - K_s^{t,x};$$
$$Y_s^{t,x} \geq l(s, X_s^{t,x}); \quad \int_t^T [Y_s^{t,x} - l(s, X_s^{t,x})] dK_s^{t,x} = 0. \tag{6.4.4}$$

Theorem 6.4.3 *Let Assumption 6.4.1 hold and u be defined by (6.4.3). Then $Y_t = u(t, X_t)$ and*

$$|u(t_1, x_1) - u(t_2, x_2)| \leq C(1 + |x_1| + |x_2|)^{\frac{1}{2}} [|t_1 - t_2|^{\frac{1}{4}} + |x_1 - x_2|^{\frac{1}{2}}]. \tag{6.4.5}$$

Proof We proceed in several steps.

Step 1. First, applying Theorem 3.2.4 and then Theorem 6.2.2 we can easily have, for any (t,x),

$$\mathbb{E}\left[|(X^{t,x})^*_T|^2 + |(Y^{t,x})^*_T|^2 + \int_t^T |Z^{t,x}_s|^2 ds + |K^{t,x}_T|^2\right] \leq C[1 + |x|^2]. \quad (6.4.6)$$

Next, applying Theorem 3.2.4 and then Theorem 6.2.3 we have, for any (t,x_1,x_2),

$$\mathbb{E}\left[|(X^{t,x_1} - X^{t,x_2})^*_T|^2\right] \leq C|x_1 - x_2|^2;$$

$$\mathbb{E}\left[|(Y^{t,x_1} - Y^{t,x_2})^*_T|^2 + |(K^{t,x_1} - K^{t,x_2})^*_T|^2 + \int_t^T |Z^{t,x_1}_s - Z^{t,x_2}_s|^2 ds\right] \quad (6.4.7)$$
$$\leq C(1 + |x_1| + |x_2|)|x_1 - x_2|.$$

In particular, this implies that

$$|u(t,x_1) - u(t,x_2)| = |Y^{t,x_1}_t - Y^{t,x_2}_t| \leq C(1 + |x_1| + |x_2|)^{\frac{1}{2}}|x_1 - x_2|^{\frac{1}{2}}. \quad (6.4.8)$$

Now following the arguments in Section 5.1.1, one may derive from the above regularity that

$$Y_t = u(t,X_t). \quad (6.4.9)$$

Step 2. This step is straightforward if we apply the more advanced Itô-Tanaka formula. To avoid introducing that, we follow the idea of its proof. Let η be as in Problem 1.4.14 and be symmetric (namely $\eta(-y) = \eta(y)$). For $\varphi(y) := y^-$, the smooth mollifier in Problem 1.4.14. becomes

$$\varphi_\varepsilon(y) := \int_{\mathbb{R}} (y - \varepsilon\tilde{y})^- \eta(\tilde{y})d\tilde{y} = \frac{1}{\varepsilon}\int_{\mathbb{R}} \tilde{y}^- \eta(\frac{y - \tilde{y}}{\varepsilon})d\tilde{y}. \quad (6.4.10)$$

Since φ is convex, by the first equality above φ_ε is also convex. Apply Itô formula we have,

$$dL_t := dl(t,X_t) = [\partial_t l + \partial_x lb + \frac{1}{2}\partial_{xx} l : \sigma\sigma^\top](t,X_t)dt + \partial_x l\sigma(,X_t)dB_t;$$

$$d\varphi_\varepsilon(Y_t - L_t) = \varphi'_\varepsilon(Y_t - L_t)d(Y_t - L_t) + \frac{1}{2}\varphi''_\varepsilon(Y_t - L_t)d\langle Y - L\rangle_t$$

$$\geq \varphi'_\varepsilon(Y_t - L_t)d(Y_t - L_t) = \varphi'_\varepsilon(Y_t - L_t)\Big[k_t dt - dK_t + [Z_t - \partial_x l\sigma(,X_t)dB_t]\Big],$$

where

$$k_t := -\Big[[\partial_t l + \partial_x lb + \frac{1}{2}\partial_{xx} l : \sigma\sigma^\top](t,X_t) + f(t,X_t,Y_t,Z_t)\Big] \quad (6.4.11)$$

Then, for any $t < s$,

$$\varphi_\varepsilon(Y_s - L_s) - \varphi_\varepsilon(Y_t - L_t) \geq \int_t^s \varphi_\varepsilon'(Y_r - L_r)\Big[k_r dr - dK_r + [Z_r - \partial_x l\sigma(., X_r)dB_r\Big]$$

(6.4.12)

By Problem 1.4.14 and (6.4.10), we have:

- $\lim_{\varepsilon \to 0} \varphi_\varepsilon(y) = y^-$. In particular, $\lim_{\varepsilon \to 0} \varphi_\varepsilon(y) = 0$ for $y \geq 0$.
- $\varphi_\varepsilon(y) = 0$ for $y \geq \varepsilon > 0$. In particular, $\lim_{\varepsilon \to 0} \varphi_\varepsilon'(y) = 0$ for $y > 0$.

Moreover, note that

$$\varphi_\varepsilon'(y) = \frac{1}{\varepsilon^2} \int_\mathbb{R} \tilde{y}^- \eta'(\frac{y - \tilde{y}}{\varepsilon}) d\tilde{y} = \frac{1}{\varepsilon} \int_\mathbb{R} (y - \varepsilon \tilde{y})^- \eta'(\tilde{y}) d\tilde{y},$$

and since η is symmetric, we see that

$$\varphi_\varepsilon'(0) = \frac{1}{\varepsilon} \int_\mathbb{R} (-\varepsilon \tilde{y})^- \eta'(\tilde{y}) d\tilde{y} = \int_0^1 \tilde{y} \eta'(\tilde{y}) d\tilde{y} = \tilde{y} \eta(\tilde{y}) \Big|_{\tilde{y}=0}^{\tilde{y}=1} - \int_0^1 \eta(\tilde{y}) d\tilde{y} = -\frac{1}{2}.$$

Now recall that $Y \geq L$ and send $\varepsilon \to 0$ in (6.4.12), we obtain

$$0 \geq -\frac{1}{2} \int_t^s \mathbf{1}_{\{Y_r = L_r\}}\Big[k_r dr - dK_r + [Z_r - \partial_x h\sigma(., X_r)dB_r\Big]$$

This implies that

$$0 \leq dK_t \leq \mathbf{1}_{\{Y_t = L_t\}} k_t dt.$$

(6.4.13)

Step 3. Now for $t_1 < t_2$, by (6.4.9) we have

$$u(t_1, x) = Y_{t_1}^{t_1, x} = \mathbb{E}\Big[u(t_2, X_{t_2}^{t_1, x}) + \int_{t_1}^{t_2} f(t, X_t^{t_1, x}, Y_t^{t_1, x}, Z_t^{t_1, x})dt + \int_{t_1}^{t_2} dK_t^{t_1, x}\Big]$$

Then, by (6.4.8),

$$|u(t_1, x) - u(t_2, x)| \leq \mathbb{E}\Big[|u(t_2, X_{t_2}^{t_1, x}) - u(t_2, x)|$$

$$+ \int_{t_1}^{t_2} |f(t, X_t^{t_1, x}, Y_t^{t_1, x}, Z_t^{t_1, x})|dt + \int_{t_1}^{t_2} |k_t^{t_1, x}|dt\Big]$$

$$\leq C\mathbb{E}\Big[(1 + |X_{t_2}^{t_1, x}| + |x|)^{\frac{1}{2}} |X_{t_2}^{t_1, x} - x|^{\frac{1}{2}}$$

$$+ \int_{t_1}^{t_2} |f(t, X_t^{t_1, x}, Y_t^{t_1, x}, Z_t^{t_1, x})|dt + \int_{t_1}^{t_2} |k_t^{t_1, x}|dt\Big].$$

This, together with (6.4.6), Theorem 5.2.2 (ii), and (6.4.11), implies (6.4.5) immediately. ∎

Remark 6.4.4

(i) By some more sophisticate arguments, one can improve the above regularity:

$$|u(t_1,x_1) - u(t_2,x_2)| \le C\Big[(1 + |x_1|^2 + |x_2|^2)|t_1 - t_2|^{\frac{1}{2}} + |x_1 - x_2|\Big]. \quad (6.4.14)$$

Moreover, one may prove the regularity of Z in the spirit of Theorem 5.2.4: recalling (5.2.7),

$$\sum_{i=0}^{n-1} \mathbb{E}\Big[\int_{t_i}^{t_{i+1}} |Z_t - \hat{Z}_{t_i}^n|^2 dt\Big] \le C(1+|x|^4)\sqrt{h} \text{ where } \hat{Z}_{t_i}^n := \frac{1}{h}\mathbb{E}_{t_i}\Big[\int_{t_i}^{t_{i+1}} Z_t dt\Big]. \quad (6.4.15)$$

(ii) Let X^n be defined by (5.3.1), and modify the backward Euler scheme (5.3.3) as follows: $Y_{t_n}^n := g(X_{t_n}^n)$, and, for $i = n-1, \cdots, 0$,

$$Z_{t_i}^n := \frac{1}{h}E_{t_i}\Big[Y_{t_{i+1}}^n B_{t_i,t_{i+1}}^\top\Big]; \quad Y_{t_i}^n := E_{t_i}\Big[Y_{t_{i+1}}^n + f(t_i, X_{t_i}^n, Y_{t_{i+1}}^n, Z_{t_i}^n)h\Big] \vee l(t_i, X_{t_i}^n). \quad (6.4.16)$$

We can show that

$$\sup_{0\le i\le n} \mathbb{E}\Big[\sup_{t_i \le t \le t_{i+1}} |Y_t - Y_{t_i}^n|^2\Big] + \sum_{i=0}^{n-1} \mathbb{E}\Big[\int_{t_i}^{t_{i+1}} |Z_t - Z_{t_i}^n|^2 dt\Big] \le C(1+|x|^4)\sqrt{h}. \quad (6.4.17)$$

Moreover, (Y^n, Z^n) can be computed by combing the least square regression and Monte Carlo simulation as in Section 5.4. ∎

We now define viscosity solutions for obstacle PDEs. Recall the sets of test functions $\overline{\mathscr{A}}u$ and $\underline{\mathscr{A}}u$ in (5.5.4).

Definition 6.4.5 *Let $u \in C^0([0,T] \times \mathbb{R}^{d_1})$ satisfy $u \ge l$.*

(i) *We say u is a viscosity subsolution of the obstacle PDE (6.4.2) if, for any $(t,x) \in [0,T) \times \mathbb{R}^{d_1}$ such that $u(t,x) > l(t,x)$ and any $\varphi \in \underline{\mathscr{A}}u(t,x)$, it holds that $\mathbb{L}\varphi(t,x) \ge 0$.*

(ii) *We say u is a viscosity supersolution of the obstacle PDE (6.4.2) if, for any $(t,x) \in [0,T) \times \mathbb{R}^{d_1}$ and any $\varphi \in \overline{\mathscr{A}}u(t,x)$, it holds that $\mathbb{L}\varphi(t,x) \le 0$.*

(iii) *We say u is a viscosity solution of the obstacle PDE (6.4.2) if it is both a viscosity subsolution and a viscosity supersolution.*

We emphasize that we do not require the viscosity subsolution property when $u(t, x) = l(t, x)$. The comparison principle and uniqueness of such viscosity solution hold true, but will not be discussed here. Instead, we shall show that (6.4.3) provides a viscosity solution for the obstacle PDE (6.4.2).

Theorem 6.4.6 *Let Assumption 6.4.1 hold and u be defined by (6.4.3). Then u is a viscosity solution to the obstacle PDE (6.4.2) with terminal condition $u(T, x) = g(x)$.*

Proof It is clear that $u(T, x) = g(x)$. We first prove the viscosity subsolution property by contradiction. Assume there exist $(t, x) \in [0, T) \times \mathbb{R}^{d_1}$ satisfying $u(t, x) > l(t, x)$ and $\varphi \in \underline{\mathscr{A}} u(t, x)$ with corresponding δ such that $-c := \mathbb{L}\varphi(t, x) < 0$. We replace the stopping time τ in (5.5.11) with a possibly smaller one τ':

$$\tau' := \tau \wedge \inf \left\{ s \geq t : u(s, X_s) \leq l(s, X_s) \right\}.$$

Note that $u(s, X_s) > l(s, X_s)$ and thus $dK_s^{t,x} = 0$ for $s \in (t, \tau')$. Then following exactly the same arguments as in Theorem 5.5.8 we derive the contradiction.

Next we prove the viscosity supersolution property by contradiction. Assume there exist $(t, x) \in [0, T) \times \mathbb{R}^{d_1}$ and $\varphi \in \overline{\mathscr{A}} u(t, x)$ with corresponding δ such that $c := \mathbb{L}\varphi(t, x) > 0$. Again we may assume $\delta > 0$ is small enough so that $\mathbb{L}\varphi \geq \frac{c}{2}$ on $O_\delta(t, x)$. Denote $(X, Y, Z) := (X^{t,x}, Y^{t,x}, Z^{t,x})$. Similar to (5.5.11) we define

$$\tau := \inf \left\{ s \geq t : (s, X_s) \notin O_\delta(t, x) \right\}.$$

Then $\tau > t$ and $\mathbb{L}\varphi(s, X_s) \geq \frac{c}{2}$ for $s \in [t, \tau]$. Note that

$$Y_s = Y_\tau + \int_s^\tau f(r, X_r, Y_r, Z_r) dr - \int_s^\tau Z_r dB_r + K_\tau - K_s,$$

and let $(\bar{Y}, \bar{Z})$ denote the unique solution of the following BSDE:

$$\bar{Y}_s = \varphi(\tau, X_\tau) + \int_s^\tau f(r, X_r, \bar{Y}_r, \bar{Y}_r) dr - \int_s^\tau \bar{Z}_r dB_r.$$

Since $Y_\tau = u(\tau, X_\tau) \geq \varphi(\tau, X_\tau)$, by the comparison principle of BSDEs we have

$$\bar{Y}_t \leq Y_t = u(t, x) = \varphi(t, x).$$

Then following the arguments in Theorem 5.5.8 we may derive

$$0 \leq -\mathbb{E}\left[\int_t^\tau \frac{c}{2} \Gamma_s ds \right],$$

for some process $\Gamma > 0$. This is a contradiction and thus u is a viscosity supersolution of the obstacle PDE (6.4.2). ∎

6.5   Semilinear Doob-Meyer Decomposition

In many applications, for example the American option pricing in Section 6.1, the
process Y has certain representation as the value function/process of some problem.
It is relatively easy to obtain the dynamics of Y once we have certain semimartingale
property of Y. The result here will be crucial for our study of second order BSDEs
in Chapter 12 later.

Let f satisfy Assumption 6.1.1. For $\tau \in \mathscr{T}$ and $\xi \in \mathbb{L}^2(\mathscr{F}_\tau, \mathbb{R})$, denote by
$(\mathscr{Y}^{\tau,\xi}, \mathscr{Z}^{\tau,\xi})$ the solution to the following BSDE on $[0, \tau]$:

$$\mathscr{Y}_t = \xi + \int_t^\tau f_s(\mathscr{Y}_s, \mathscr{Z}_s)ds - \int_t^\tau \mathscr{Z}_s dB_s, \quad 0 \le t \le \tau. \tag{6.5.1}$$

Definition 6.5.1 *We say a process $Y \in \mathbb{S}^2(\mathbb{F}, \mathbb{R})$ is an f-supermartingale if, for any
t and any $\tau \in \mathscr{T}$ satisfying $\tau \ge t$, $Y_t \ge \mathscr{Y}_t^{\tau, Y_\tau}$, a.s.*

Theorem 6.5.2 *Assume f satisfies Assumption 6.1.1 and Y is a continuous f-
supermartingale. Then there exists $Z \in \mathbb{L}^2(\mathbb{F}, \mathbb{R}^{1 \times d})$ and $K \in \mathbb{I}^2(\mathbb{F})$ such that*

$$dY_t = -f_t(Y_t, Z_t)dt + Z_t dB_t - dK_t;$$
$$\mathbb{E}\Big[\int_0^T |Z_t|^2 dt + |K_T|^2\Big] \le C\mathbb{E}\Big[|Y_T^*|^2 + \Big(\int_0^T |f_t(0,0)|dt\Big)^2\Big]. \tag{6.5.2}$$

Proof Consider the following RBSDE with barrier $L := Y$ and solution
$(\mathscr{Y}, \mathscr{Z}, \mathscr{K})$:

$$\begin{cases} \mathscr{Y}_t = Y_T + \int_t^T f_s(\mathscr{Y}_s, \mathscr{Z}_s)ds - \int_t^T \mathscr{Z}_s dB_s + \mathscr{K}_T - \mathscr{K}_t; \\ \mathscr{Y}_t \ge Y_t, \quad [\mathscr{Y}_t - Y_t]d\mathscr{K}_t = 0. \end{cases} \tag{6.5.3}$$

We claim that $\mathscr{Y} = Y$, then clearly $Z := \mathscr{Z}$ and $K := \mathscr{K}$ satisfy the requirement.

First, by definition $\mathscr{Y} \ge Y$. On the other hand, for any t, denote $\tau_t := \inf\{s \ge t : \mathscr{Y}_s = Y_s\}$. Then $\mathscr{Y}_{\tau_t} = Y_{\tau_t}$ and $d\mathscr{K}_s = 0$ on $[t, \tau_t)$. This implies $\mathscr{Y}_t = \mathscr{Y}_t^{\tau_t, \mathscr{Y}_{\tau_t}} = \mathscr{Y}_t^{\tau, Y_{\tau_t}} \le Y_t$, a.s. Therefore, $\mathscr{Y} = Y$. Finally, applying Theorem 6.2.1 on (6.5.3) we
obtain the desired estimate for (Z, K). ∎

6.6   Bibliographical Notes

The results of this chapter are mainly from El Karoui, Kapoudjian, Pardoux, Peng,
& Quenez [78]. The regularity and discretization in Remark 6.4.4 are due to Ma &
Zhang [151], and the notion of f-supermartingale in Section 6.5 is due to El Karoui,
Peng, & Quenez [81] and Peng [177], which use the name f-supersolution instead.
The regularity requirement on the barrier process L can be weakened significantly,
see, e.g., Hamadene [100] for càdlàg L and Peng & Xu [188] for measurable L.

Due to its reliance on the comparison principle, RBSDE is typically one dimensional (in terms of its Y component). However, by using the comparison principle for certain multidimensional BSDEs with special structure, one may consider multidimensional RBSDEs with the same structure, see, e.g., Hamadene & Jeablanc [102], Hu & Tang [111], and Hamadene & Zhang [104]. These works are mainly motivated from applications on the switching problems.

Another extension of RBSDE is RBSDE with two barriers in Cvitanic & Karatzas [45], motivated from its applications in Dynkin games. See also Hamadene & Hassani [101] for some general well-posedness result by using the idea of local solution. Notice that RBSDEs with one or two barriers can be viewed as a BSDE with a constraint in the Y component. One may also consider BSDE with general constraints on (Y, Z), see, e.g., Cvitanic, Karatzas, & Soner [46] and Peng & Xu [189]. For these problems there is no counterpart of the Skorohod condition (6.1.3), instead one considers minimum solutions.

6.7 Exercises

Problem 6.7.1 Prove Proposition 6.2.1. ∎

Problem 6.7.2 Let $\tau_n \in \mathscr{T}^t$, $1 \leq n \leq N$, and $\{E_n\}_{1 \leq n \leq N} \subset \mathscr{F}_t$ be a partition of Ω, namely they are disjoint and $\cup_{1 \leq n \leq N} E_n = \Omega$. Show that $\hat{\tau} := \sum_{n=1}^{N} \tau_n 1_{E_m} \in \mathscr{T}^t$. ∎

Problem 6.7.3 Find a counterexample $Y \in \mathbb{L}^0(\mathbb{F})$ such that Y is continuous a.s., but there is no modification $\tilde{Y} \in \mathbb{L}^0(\mathbb{F})$ such that $\tilde{Y}$ is continuous for all ω. ∎

Problem 6.7.4 Let (ξ, L) satisfy the conditions in Proposition 6.3.2 and Y be the Snell envelope defined in (6.3.1). Assume $\tilde{L} \in \mathbb{L}^0(\mathbb{F})$ such that $L \leq \tilde{L} \leq Y$ and let $\tilde{Y}$ denote the Snell envelope corresponding to $(\xi, \tilde{L})$. Show that $\tilde{Y}_t = Y_t$, $0 \leq t \leq T$, $\mathbb{P}$-a.s. for all t. ∎

Problem 6.7.5 Prove the "$\geq$" inequality of (6.3.7). ∎

Problem 6.7.6 (Dini's Lemma) Assume $f^n : [0, T] \to \mathbb{R}$, $n \geq 1$, satisfy $f^n(t) \downarrow 0$ for all $t \in [0, T]$. If f_n, $n \geq 1$, are upper semicontinuous on $[0, T]$, namely $\limsup_{s \to t} f^n(s) \leq f^n(t)$, then $\sup_{0 \leq t \leq T} f^n(t) \downarrow 0$. ∎

Problem 6.7.7 Assume (ξ, f) satisfy Assumption 4.0.1, $|\xi| \leq C_0$, and $d_2 = 1$. Let (Y, Z) denote the solution to BSDE (4.0.3), and (Y^n, Z^n, K^n) the solution to RBSDE (6.1.5) with barrier process $L_t^n = -n$, for each $n \geq C_0$. Show that

$$\lim_{n \to \infty} \mathbb{E}\left[|(Y^n - Y)_T^*|^2 + \int_0^T |Z_t^n - Z_t|^2 dt + |K_T^n|^2\right] = 0.$$

∎

Problem 6.7.8 Prove the partial comparison principle for viscosity solutions of the obstacle PDE (6.4.2), under Assumption 6.4.1 and in the spirit of Proposition 5.5.10. To be precise, let u and v be a viscosity subsolution and supersolution of PDE (6.4.2), respectively. Assume

$$u(T, \cdot) \le v(T, \cdot) \quad \text{and} \quad \limsup_{|x| \to \infty} \sup_{t \in [0,T]} [u - v](t, x) \le 0. \qquad (6.7.1)$$

If either u or v is in $C^{1,2}([0, T] \times \mathbb{R}^{d_1})$, then $u \le v$. ∎

Chapter 7
BSDEs with Quadratic Growth in Z

7.1 Introduction

In this chapter we study BSDE (4.0.3) whose generator f has quadratic growth in Z. In particular, in this case f is not uniformly Lipschitz continuous in Z. As in Chapter 6, the theory will rely on the martingale representation theorem and the comparison principle of BSDEs, thus we assume throughout this chapter that:

$$\mathbb{F} = \mathbb{F}^B, \quad d_2 = 1. \tag{7.1.1}$$

The well-posedness of multiple dimensional BSDEs with quadratic generator remains an important and difficult subject.

To motivate this type of BSDEs, we present two examples for the stochastic optimization problem (4.5.12). Recall that g can be viewed as a utility and $-f$ a cost function.

Example 7.1.1 (Quadratic Cost) *Let $\mathbb{K} := \mathbb{R}$, $\sigma := 1$, $b(t, x, k) := k$, and $f(t, x, k) := \frac{-\lambda}{2}|k|^2$. Then (4.5.15) becomes $f^*(t, x, z) = \frac{1}{2\lambda}|z|^2$, which grows quadratically in z.*

Example 7.1.2 (Exponential Utility) *Let $f := 0$, $g(x) := -e^{-\lambda x}$, and assume $\frac{b}{\sigma}$ is bounded. By (4.5.13) it is clear that $Y^k < 0$. One can actually show that $Y^* < 0$. In this case it is standard in the literature (see, e.g., Cvitanic & Zhang [52]) to consider the transformation: $\tilde{Y} := -\ln(-Y^*)$, $\tilde{Z} := -\frac{Z^*}{Y^*}$. Apply Itô formula one may rewrite (4.5.16) as:*

$$\tilde{Y}_t = \lambda X_T - \int_t^T [\frac{f^*(s, X_s, Z_s^*)}{Y_s^*} + \frac{1}{2}|\tilde{Z}_s|^2]ds - \int_t^T \tilde{Z}_s dB_s.$$

J. Zhang, *Backward Stochastic Differential Equations*, Probability Theory and Stochastic Modelling 86, DOI 10.1007/978-1-4939-7256-2_7

By (4.5.15) *one may check straightforwardly that* $-\frac{f^*(s,X_s,Z_s^*)}{Y_s^*} = f^*(s,X_s,\tilde{Z}_s)$, *and thus the above BSDE becomes:*

$$\tilde{Y}_t = \lambda X_T + \int_t^T [f^*(s,X_s,\tilde{Z}_s) - \frac{1}{2}|\tilde{Z}_s|^2]ds - \int_t^T \tilde{Z}_s dB_s,$$

which also has quadratic growth in $\tilde{Z}$.

The main technique to establish the well-posedness of such BSDEs is a nonlinear transformation of Y, in the spirit of the reverse order in Example 7.1.2. We shall always assume

Assumption 7.1.3

(i) (7.1.1) *holds.*
(ii) *There exists a constant C such that, for any (t, ω, y, z),*

$$|f_t(y,z)| \le C[1 + |y| + |z|^2]. \tag{7.1.2}$$

(iii) *There exists a constant C such that, for any (t, ω, y_i, z_i), $i = 1, 2$,*

$$|f_t(y_1,z_1) - f_t(y_2,z_2)| \le C\Big[|y_1 - y_2| + (1 + |y_1| + |y_2|$$

$$+ |z_1| + |z_2|)|z_1 - z_2|\Big]. \tag{7.1.3}$$

We remark that (7.1.2) implies $f_t^0 := f_t(0,0)$ is bounded, and when f is differentiable in (y,z), (7.1.3) is equivalent to

$$|\partial_y f_t(y,z)| \le C, \quad |\partial_z f_t(y,z)| \le C[1 + |y| + |z|]. \tag{7.1.4}$$

To motivate our further study, we first provide a very simple result:

Proposition 7.1.4 *Assume Assumption 7.1.3. If both (Y^i, Z^i), $i = 1, 2$, are bounded and satisfy BSDE (4.0.3), then $(Y^1, Z^1) = (Y^2, Z^2)$.*

Proof Let C_0 denote a common bound of (Y^i, Z^i). Denote

$$\tilde{f}_t(y,z) := f_t\Big((-C_0) \vee y \wedge C_0, (-C_0) \vee z \wedge C_0\Big).$$

Then $\tilde{f}$ is uniformly Lipschitz continuous in (y,z). The boundedness implies $\tilde{f}_t(Y_t^i, Z_t^i) = f_t(Y_t^i, Z_t^i)$, and thus (Y^i, Z^i) satisfies the following BSDE:

$$Y_t = \xi + \int_t^T \tilde{f}_s(Y_s, Z_s)ds - \int_t^T Z_s dB_s.$$

Now the result follows from the uniqueness of the above BSDE. ∎

As we see, the space of solutions is crucial for the uniqueness. However, in general we cannot expect to have bounded solutions. We shall restrict the process Z to satisfy the so-called BMO martingale property, as we will explain in the next section. Moreover, to focus on the main idea, we will study only the case that Y is bounded. For that purpose, we shall assume further that

Assumption 7.1.5 $\xi \in \mathbb{L}^{\infty}(\mathscr{F}_T)$.

7.2 BMO Martingales and A Priori Estimates

We first establish some a priori estimates.

Theorem 7.2.1 *Let Assumptions 7.1.3 and 7.1.5 hold. If $(Y, Z) \in \mathbb{L}^{\infty}(\mathbb{F}, \mathbb{R}) \times \mathbb{L}^2(\mathbb{F}, \mathbb{R}^{1 \times d})$ is a solution to BSDE (4.0.3), then*

$$|Y_t| \leq C \quad and \quad \mathbb{E}_t\left[\int_t^T |Z_s|^2 ds\right] \leq C,$$

where the constant C depends only on T, d, the constant in (7.1.2), and $\|\xi\|_{\infty}$.

Proof

(i) We first prove the estimate for Y. For any $t_0 \in [0, T]$, define a stopping time

$$\tau := \inf\{t \geq t_0 : |Y_t| \leq 1\} \wedge T.$$

We note that $\tau = t_0$ on $\{|Y_{t_0}| \leq 1\}$. Clearly

$$|Y_t| > 1, \quad t_0 \leq t < \tau, \quad and \quad |Y_\tau| \leq 1 \vee \|\xi\|_{\infty}. \tag{7.2.1}$$

Let C_0 denote the constant in (7.1.2) and $\varphi : [0, T] \times \mathbb{R} \to \mathbb{R}$ be a smooth function which will be specified later. Introduce the transformation:

$$\tilde{Y}_t := e^{\varphi(t, Y_t)} > 0, \quad \tilde{Z}_t = \partial_y \varphi(t, Y_t) \tilde{Y}_t Z_t. \tag{7.2.2}$$

Applying Itô formula, we obtain:

$$d\varphi(t, Y_t) = \left[\partial_t \varphi - \partial_y \varphi f_t(Y_t, Z_t) + \frac{1}{2}\partial_{yy}\varphi |Z_t|^2\right]dt + \partial_y f Z_t dB_t;$$

$$d\tilde{Y}_t = \tilde{Y}_t\left[\partial_t \varphi - \partial_y \varphi f_t(Y_t, Z_t) + \frac{1}{2}\partial_{yy}\varphi |Z_t|^2 + \frac{1}{2}|\partial_y \varphi Z_t|^2\right]dt + \tilde{Z}_t dB_t$$

$$\geq \tilde{Y}_t\left[\partial_t \varphi - C_0|\partial_y \varphi|[1 + |Y_t| + |Z_t|^2] + \frac{1}{2}\partial_{yy}\varphi |Z_t|^2 + \frac{1}{2}|\partial_y \varphi Z_t|^2\right]dt + \tilde{Z}_t dB_t,$$

where the last inequality thanks to (7.1.2). We want to choose φ so that the drift term in the last line above is nonnegative in (t_0, τ), namely when $|Y_t| > 1$. For this purpose, set $\varphi(t, y) := \lambda e^{\lambda t}[\psi(y) + 1]$ where $\lambda = 2(1 + C_0)$, $\psi \geq 0$, and $\psi(y) = |y|$ when $|y| \geq 1$. Then, when $|y| \geq 1$,

$$\partial_t \varphi - C_0 |\partial_y \varphi| [1 + y + |z|^2] + \frac{1}{2} \partial_{yy} \varphi |z|^2 + \frac{1}{2} |\partial_y \varphi z|^2$$

$$= \lambda^2 e^{\lambda t}[|y| + 1] - C_0 \lambda e^{\lambda t}[1 + y + |z|^2] + \frac{1}{2} \lambda^2 e^{2\lambda t} |z|^2 \geq 0.$$

This implies:

$$d\tilde{Y}_t \geq \tilde{Z}_t dB_t, \quad t_0 < t < \tau.$$

Since Y is bounded, one can easily see that $\tilde{Z} \in \mathbb{L}^2(\mathbb{F}, \mathbb{R}^{1 \times d})$, and thus $\tilde{Y}$ is a submartingale on $[t_0, \tau]$. Note that $|y| \leq \psi(y) + 1$ in all the cases. Then by (7.2.1) we have

$$e^{\lambda e^{\lambda t}|Y_{t_0}|} \leq e^{\lambda e^{\lambda t}[\psi(Y_{t_0})+1]} = \tilde{Y}_{t_0} \leq \mathbb{E}_{t_0}[\tilde{Y}_\tau] = \mathbb{E}_{t_0}\left[e^{\lambda e^{\lambda \tau}[|Y_\tau|+1]}\right] \leq e^{\lambda e^{\lambda T}[\|\xi\|_\infty + 2]}.$$

This implies that, for arbitrary t_0,

$$|Y_{t_0}| \leq e^{\lambda T}[\|\xi\|_\infty + 2] = e^{2(1+C_0)T}[\|\xi\|_\infty + 2].$$

(ii) We next prove the estimate for Z. For the same λ as in (i), applying Itô rule on $e^{\lambda Y_t}$ we obtain

$$de^{\lambda Y_t} = -\lambda e^{\lambda Y_t} f_t(Y_t, Z_t) dt + \lambda e^{\lambda Y_t} Z_t dB_t + \frac{1}{2} \lambda^2 e^{\lambda Y_t} |Z_t|^2 dt$$

$$\geq -\lambda e^{\lambda Y_t} C_0[1 + |Y_t| + |Z_t|^2] dt + \lambda e^{\lambda Y_t} Z_t dB_t + \frac{1}{2} \lambda^2 e^{\lambda Y_t} |Z_t|^2 dt$$

$$\geq \lambda e^{\lambda Y_t}\left[|Z_t|^2 - \lambda(1 + |Y_t|)\right] dt + \lambda e^{\lambda Y_t} Z_t dB_t.$$

Then, by the boundedness of Y again,

$$e^{\lambda Y_t} + \mathbb{E}_t\left[\int_t^T \lambda e^{\lambda Y_s} |Z_s|^2 ds\right] \leq \mathbb{E}_t\left[e^{\lambda \xi} + \int_t^T \lambda^2 e^{\lambda Y_s}[1 + |Y_s|] ds\right].$$

Thus,

$$\lambda e^{-\lambda \|Y\|_\infty} \mathbb{E}_t\left[\int_t^T |Z_s|^2 dt\right] \leq e^{\lambda \|\xi\|_\infty} + T\lambda^2 e^{\lambda \|Y\|_\infty}[1 + \|Y\|_\infty].$$

This implies the estimate for Z immediately.   ∎

The above estimate for Z implies that the martingale $\int_0^t Z_s dB_s$ is a so-called BMO martingale:

Definition 7.2.2 *A martingale M is called a BMO martingale if there exists a constant $C > 0$ such that $\mathbb{E}_t[(M_T - M_t)^2] \leq C$, a.s. for all $t \in [0,T]$.*

A BMO martingale satisfies the following important property, which extends Lemma 2.6.1.

Theorem 7.2.3 *Let $\theta \in \mathbb{L}^2(\mathbb{F}, \mathbb{R}^d)$ and recall the process M^θ defined by (2.6.1). Assume*

$$\mathbb{E}_t\left[\int_t^T |\theta_s|^2 ds\right] \leq C_0, \ t \in [0,T]. \tag{7.2.3}$$

Then

$$\mathbb{E}_\tau\left[\exp\left(\varepsilon \int_\tau^T |\theta_s|^2 ds\right)\right] \leq \frac{1}{1 - C_0\varepsilon}, \quad \forall \tau \in \mathscr{T}, 0 < \varepsilon < \frac{1}{C_0}, \tag{7.2.4}$$

and there exist constants $\varepsilon > 0$ and $C_\varepsilon > 0$, depending only on C_0, such that

$$\mathbb{E}[|(M^\theta)_T^*|^{1+\varepsilon}] \leq C_\varepsilon < \infty. \tag{7.2.5}$$

In particular, M^θ is a uniformly integrable martingale.

Proof We proceed in three steps.

Step 1. For simplicity, we prove (7.2.4) only at $\tau = 0$. We first claim that,

$$\mathbb{E}\left[\left(\int_0^T |\theta_t|^2 dt\right)^n\right] \leq n! C_0^n. \tag{7.2.6}$$

This implies (7.2.4) immediately: for $0 < \varepsilon < \frac{1}{C_0}$,

$$\mathbb{E}\left[\exp\left(\varepsilon \int_0^T |\theta_t|^2 dt\right)\right] = \sum_{n=0}^\infty \frac{\varepsilon^n}{n!} \mathbb{E}\left[\left(\int_0^T |\theta_t|^2 dt\right)^n\right] \leq \sum_{n=0}^\infty (\varepsilon C_0)^n = \frac{1}{1 - C_0\varepsilon} < \infty.$$

To see (7.2.6), denote $\Gamma_t := \int_0^t |\theta_s|^2 ds$. By applying Itô formula repeatedly we have

$$\Gamma_T^n = n! \int_0^T \int_0^{t_1} \cdots \int_0^{t_{n-1}} |\theta_{t_1}|^2 \cdots |\theta_{t_n}|^2 dt_n \cdots dt_1 = n! \int_0^T \int_{t_n}^T \cdots$$

$$\int_{t_2}^T |\theta_{t_1}|^2 \cdots |\theta_{t_n}|^2 dt_1 \cdots dt_n.$$

Thus, applying (7.2.3) repeatedly we obtain

$$\mathbb{E}[\Gamma_T^n] = n!\mathbb{E}\Big[\int_0^T \int_{t_n}^T \cdots \int_{t_2}^T |\theta_{t_1}|^2 \cdots |\theta_{t_n}|^2 dt_1 \cdots dt_n\Big]$$

$$= n!\mathbb{E}\Big[\int_0^T \int_{t_n}^T \cdots \int_{t_3}^T \mathbb{E}_{t_2}\Big[\int_{t_2}^T |\theta_{t_1}|^2 dt_1\Big]|\theta_{t_2}|^2 \cdots |\theta_{t_n}|^2 dt_2 \cdots dt_n\Big]$$

$$\leq C_0 n!\mathbb{E}\Big[\int_0^T \int_{t_n}^T \cdots \int_{t_3}^T |\theta_{t_2}|^2 \cdots |\theta_{t_n}|^2 dt_2 \cdots dt_n\Big]$$

$$\leq \cdots \leq n!C_0^n.$$

Step 2. We next show that there exist constants $\varepsilon > 0$ and $\tilde{C}_\varepsilon > 0$, depending only on C_0, such that

$$\mathbb{E}_\tau\Big[(M_T/M_\tau)^{-\varepsilon}\Big] \leq \tilde{C}_\varepsilon < \infty, \quad \forall \tau \in \mathscr{T}. \tag{7.2.7}$$

For simplicity, again we shall only prove it at $\tau = 0$. Indeed, let $p > 1$ be a constant which will be specified later, then

$$\mathbb{E}\Big[(M_T^\theta)^{-\varepsilon}\Big] = \mathbb{E}\Big[\exp\big(-\varepsilon\int_0^T \theta_t dB_t + \frac{\varepsilon}{2}\int_0^T |\theta_t|^2 dt\big)\Big] = \mathbb{E}\Big[(M_T^{-p\varepsilon\theta})^{\frac{1}{p}} \exp\big(\frac{\varepsilon(p+\varepsilon)}{2}\int_0^T |\theta_t|^2 dt\big)\Big]$$

$$\leq \big(\mathbb{E}[M_T^{-p\varepsilon\theta}]\big)^{\frac{1}{p}}\Big(\mathbb{E}\Big[\exp\big(\frac{p}{p-1}\frac{\varepsilon(p+\varepsilon)}{2}\int_0^T |\theta_t|^2 dt\big)\Big]\Big)^{\frac{p-1}{p}}.$$

Since $M^{-p\varepsilon\theta}$ is a positive local martingale, then by Problem 1.4.10 (iii) it is a supermartingale and thus $\mathbb{E}[M_T^{-p\varepsilon\theta}] \leq 1$. Set $p := 1 + \sqrt{1+\varepsilon}$ which minimizes $\frac{p}{p-1}\frac{\varepsilon(p+\varepsilon)}{2}$. We obtain

$$\mathbb{E}\Big[(M_T^\theta)^{-\varepsilon}\Big] \leq \Big(\mathbb{E}\Big[\exp\big(\frac{\varepsilon(1+\sqrt{1+\varepsilon})^2}{2}\int_0^T |\theta_t|^2 dt\big)\Big]\Big)^{\frac{\sqrt{1+\varepsilon}}{1+\sqrt{1+\varepsilon}}}.$$

Now choose $\varepsilon > 0$ small enough such that $\frac{\varepsilon(1+\sqrt{1+\varepsilon})^2}{2} < \frac{1}{C_0}$. Then, combined with (7.2.4), we obtain (7.2.7) immediately.

Step 3. For notational simplicity, in this step we denote $M := M^\theta$. By standard truncation procedure, we may assume without loss of generality that M_T^* is bounded. Of course, we emphasize that the a priori estimate (7.2.5) does not depend on the bound of M_T^*. Fix an $\varepsilon_0 > 0$ satisfying the requirement in Step 2. Let $a > 0$ be a constant which will be specified later. For $\lambda \geq 1$, denote $\tau := \tau_\lambda := \inf\{t \geq 0 : M_t \geq \lambda\} \wedge T$. Then

$$\mathbb{E}_\tau\Big[1_{\{M_T/M_\tau \leq \frac{1}{a}\}}\Big] = \mathbb{E}_\tau\Big[(M_T/M_\tau)^{-\frac{\varepsilon_0}{1+\varepsilon_0}}(M_T/M_\tau)^{\frac{\varepsilon_0}{1+\varepsilon_0}}1_{\{M_T/M_\tau \leq \frac{1}{a}\}}\Big]$$

$$\leq \big(\mathbb{E}_\tau[(M_T/M_\tau)^{-\varepsilon_0}]\big)^{\frac{1}{1+\varepsilon_0}}\big(\mathbb{E}_\tau[M_T/M_\tau 1_{\{M_T/M_\tau \leq \frac{1}{a}\}}]\big)^{\frac{\varepsilon_0}{1+\varepsilon_0}} \leq \tilde{C}_{\varepsilon_0}^{\frac{1}{1+\varepsilon_0}} a^{-\frac{\varepsilon_0}{1+\varepsilon_0}},$$

where the last inequality thanks to (7.2.7). Set $a := 2^{\frac{1+\varepsilon_0}{\varepsilon_0}} \tilde{C}_{\varepsilon_0}^{\frac{1}{\varepsilon_0}}$. Then $\mathbb{E}_\tau[\mathbf{1}_{\{M_T/M_\tau \le \frac{1}{a}\}}] \le \frac{1}{2}$, and thus $\mathbb{E}_\tau[\mathbf{1}_{\{M_T/M_\tau > \frac{1}{a}\}}] > \frac{1}{2}$. Since $M_0 = 1 \le \lambda$, we see that $M_\tau = \lambda$ on $\{M_T \ge \lambda\}$. Note that M is a martingale. Then we have

$$\mathbb{E}[M_T \mathbf{1}_{\{M_T \ge \lambda\}}] \le \mathbb{E}[M_T \mathbf{1}_{\{M_\tau = \lambda\}}] = \mathbb{E}[M_\tau \mathbf{1}_{\{M_\tau = \lambda\}}] = \lambda \mathbb{E}[\mathbf{1}_{\{M_\tau = \lambda\}}]$$

$$\le 2\lambda \mathbb{E}[\mathbb{E}_\tau[\mathbf{1}_{\{M_T/M_\tau > \frac{1}{a}\}}]\mathbf{1}_{\{M_\tau = \lambda\}}] = 2\lambda \mathbb{E}[\mathbf{1}_{\{M_T/M_\tau > \frac{1}{a}\}}\mathbf{1}_{\{M_\tau = \lambda\}}] \le 2\lambda \mathbb{E}[\mathbf{1}_{\{aM_T > \lambda\}}].$$

For $\varepsilon > 0$ small, multiply both sides above by $\varepsilon\lambda^{\varepsilon-1}$ and integrate for $\lambda \in [1, \infty)$, we obtain

$$\mathbb{E}[M_T[|M_T|^\varepsilon - 1]\mathbf{1}_{\{M_T \ge 1\}}] \le \frac{2\varepsilon a^{1+\varepsilon}}{1+\varepsilon}\mathbb{E}[|M_T|^{1+\varepsilon}]$$

Then

$$\mathbb{E}[|M_T|^{1+\varepsilon}] = \mathbb{E}[M_T[|M_T|^\varepsilon - 1]\mathbf{1}_{\{M_T \ge 1\}}] + \mathbb{E}[M_T[|M_T|^\varepsilon - 1]\mathbf{1}_{\{M_T < 1\}}] + \mathbb{E}[M_T]$$

$$\le \frac{2\varepsilon a^{1+\varepsilon}}{1+\varepsilon}\mathbb{E}[|M_T|^{1+\varepsilon}] + 1.$$

Choose ε small enough such that $\frac{2\varepsilon a^{1+\varepsilon}}{1+\varepsilon} \le \frac{1}{2}$, then

$$\mathbb{E}[|M_T|^{1+\varepsilon}] \le 2.$$

This, together with (2.4.4), proves (7.2.5) with $C_\varepsilon := 2(1 + \varepsilon^{-1})^{1+\varepsilon}$. ∎

7.3 Well-Posedness

We start with the comparison principle.

Theorem 7.3.1 *For $i = 1, 2$, assume (ξ_i, f^i) satisfies Assumptions 7.1.3 and 7.1.5, and $(Y^i, Z^i) \in \mathbb{L}^\infty(\mathbb{F}, \mathbb{R}) \times \mathbb{L}^2(\mathbb{F}, \mathbb{R}^{1 \times d})$ satisfies BSDE (4.0.3) corresponding to (ξ_i, f^i). If $\xi_1 \le \xi_2$ a.s. and, for any (y, z), $f_1(\cdot, y, z) \le f_2(\cdot, y, z)$, $dt \times d\mathbb{P}$-a.s. then $Y^1_t \le Y^2_t$, $0 \le t \le T$, a.s. In particular, $Y^1_0 \le Y^2_0$.*

Proof First, by Theorem 7.2.1 there exists a constant C such that

$$|Y^i_t|^2 + \mathbb{E}_t\left[\int_t^T |Z^i_s|^2 ds\right] \le C, \quad \forall t, \quad i = 1, 2. \tag{7.3.1}$$

Denote $\Delta Y := Y^1 - Y^2$, $\Delta Z := Z^1 - Z^2$, $\Delta \xi = \xi_1 - \xi_2$, $\Delta f := f_1 - f_2$. Then,

$$d\Delta Y_t = -\Big[f_t^1(Y_t^1, Z_t^1) - f_t^2(Y_t^2, Z_t^2)\Big]dt + \Delta Z_t dB_t$$

$$= -\Big[\Delta f_t(Y_t^1, Z_t^1) + \alpha_t \Delta Y_t + \Delta Z_t \theta_t\Big]dt + \Delta Z_t dB_t,$$

where α is bounded, and

$$|\theta_t| \le C[1 + |Y_t^1| + |Y_t^2| + |Z_t^1| + |Z_t^2|]. \tag{7.3.2}$$

Denote $\Gamma_t := \exp(\int_0^t \alpha_s ds)$, which is bounded. Then,

$$d(\Gamma_t M_t^\theta \Delta Y_t) = -\Gamma_t M_t^\theta \Delta f_t(Y_t^1, Z_t^1)dt + \Gamma_t M_t^\theta [\Delta Z_t + \Delta Y_t \theta_t^\mathsf{T}]dB_t.$$

By (7.3.1) and (7.3.2), it follows from (7.2.4) that $\mathbb{E}[|\Delta Z_t + \Delta Y_t \theta_t^\mathsf{T}|^p] < \infty$ for any $p \ge 1$. Then by (7.2.5) we see that $\int_0^t \Gamma_s M_s^\theta [\Delta Z_s + \Delta Y_s \theta_s^\mathsf{T}]dB_s$ is a true martingale, and thus

$$\Delta Y_0 = \mathbb{E}\Big[\Gamma_T M_T^\theta \Delta \xi + \int_0^T \Gamma_s M_s^\theta \Delta f_s(Y_s^1, Z_s^1)ds\Big] \le 0.$$

Similarly, $\Delta Y_t \le 0$ for any t.   ∎

We next turn to stability results. We shall first establish it under an additional monotonicity condition. The general result will be obtained in Theorem 7.3.4 below.

Lemma 7.3.2 *Let (ξ_n, f^n), $n \ge 1$, satisfy Assumptions 7.1.3 and 7.1.5 uniformly, and $(Y^n, Z^n) \in \mathbb{L}^\infty(\mathbb{F}, \mathbb{R}) \times \mathbb{L}^2(\mathbb{F}, \mathbb{R}^{1 \times d})$ satisfies BSDE (4.0.3) corresponding to (ξ_n, f^n). As $n \to \infty$, assume $\xi_n \to \xi$ in $\mathbb{P}$, $f^n(\cdot, y, z) \to f(\cdot, y, z)$ in measure $dt \times d\mathbb{P}$ for any (y, z), and $Y^n \uparrow Y$ (or $Y^n \downarrow Y$). Then there exists a process $Z \in \mathbb{L}^2(\mathbb{F}, \mathbb{R}^{1 \times d})$ such that (Y, Z) is a solution to BSDE (4.0.3) with coefficient (ξ, f).*

Proof We shall assume $Y^n \uparrow Y$. First, by Theorem 7.2.1,

$$(Y^n)_T^* \le C, \text{ a.s.} \quad \text{and} \quad \mathbb{E}\Big[\int_0^T |Z_t^n|^2 dt\Big] \le C, \quad n \ge 1. \tag{7.3.3}$$

Then, Y is also bounded, and $\{Z^n, n \ge 1\}$ has a weak limit $Z \in \mathbb{L}^2(\mathbb{F}, \mathbb{R}^{1 \times d})$, thanks to Theorem 1.3.7. By otherwise choosing a subsequence, we may assume that the whole sequence Z^n converges to Z weakly in $\mathbb{L}^2(\mathbb{F}, \mathbb{R}^{1 \times d})$. Denote

$$\Delta Y^n := Y - Y^n, \quad \Delta Y^{m,n} := Y^m - Y^n, \quad \text{and similarly for other notations.}$$

We claim that

$$\lim_{n \to \infty} \mathbb{E}\Big[\int_0^T |\Delta Z_t^n|^2 dt\Big] = 0. \tag{7.3.4}$$

Then, by (7.1.3),

$$\mathbb{E}\Big[\int_0^T |f_t^n(Y_t^n, Z_t^n) - f_t(Y_t, Z_t)|dt\Big]$$

$$\leq \mathbb{E}\Big[\int_0^T \Big[|\Delta f_t^n(Y_t, Z_t)| + C|\Delta Y_t^n| + C[1 + |Z_t^n| + |Z_t|]|\Delta Z_t^n|\Big]dt\Big]$$

$$\leq \mathbb{E}\Big[\int_0^T \Big[|\Delta f_t^n(Y_t, Z_t)| + C|\Delta Y_t^n|\Big]dt\Big] + C\Big(\mathbb{E}\Big[\int_0^T [1 + |Z_t^n|^2 + |Z_t|^2]dt\Big]\Big)^{\frac{1}{2}}\Big(\mathbb{E}\Big[\int_0^T |\Delta Z_t^n|^2 dt\Big]\Big)^{\frac{1}{2}}$$

$$\leq \mathbb{E}\Big[\int_0^T \Big[|\Delta f_t^n(Y_t, Z_t)| + C|\Delta Y_t^n|\Big]dt\Big] + C\Big(\mathbb{E}\Big[\int_0^T |\Delta Z_t^n|^2 dt\Big]\Big)^{\frac{1}{2}}.$$

Sending $n \to \infty$, by Problem 1.4.6 (iii), the convergence of Y^n, and (7.3.4), we get

$$\lim_{n \to \infty} \mathbb{E}\Big[\int_0^T |f_t^n(Y_t^n, Z_t^n) - f_t(Y_t, Z_t)|dt\Big] = 0.$$

Then it is straightforward to verify that (Y, Z) satisfies BSDE (4.0.3) with coefficient (ξ, f).

We now prove Claim (7.3.4). Similar to the proof of Theorem 7.2.1, we need a smooth transformation function $\varphi : \mathbb{R} \to \mathbb{R}$, which will be specified later. For $n < m$, applying Itô rule we have

$$d\varphi(\Delta Y_t^{m,n}) = -\varphi'(\Delta Y_t^{m,n})[f_t^m(Y_t^m, Z_t^m) - f_t^n(Y_t^n, Z_t^n)]dt$$

$$+ \frac{1}{2}\varphi''(\Delta Y_t^{m,n})|\Delta Z_t^{m,n}|^2 dt + \varphi'(\Delta Y_t^{m,n})\Delta Z_t^{m,n} dB_t$$

$$\geq -C|\varphi'(\Delta Y_t^{m,n})|\big[1 + |Z_t^m|^2 + |Z_t^n|^2\big]dt + \frac{1}{2}\varphi''(\Delta Y_t^{m,n})|\Delta Z_t^{m,n}|^2 dt$$

$$+ \varphi'(\Delta Y_t^{m,n})\Delta Z_t^{m,n} dB_t$$

$$\geq -C_0|\varphi'(\Delta Y_t^{m,n})|\big[1 + |\Delta Z_t^{m,n}|^2 + |\Delta Z_t^n|^2 + |Z_t|^2\big]dt$$

$$+ \frac{1}{2}\varphi''(\Delta Y_t^{m,n})|\Delta Z_t^{m,n}|^2 dt + \varphi'(\Delta Y_t^{m,n})\Delta Z_t^{m,n} dB_t.$$

Notice that we expect $\Delta Z^{m,n}$ and ΔZ^n to be close, we shall set φ so that $\varphi'' = 4C_0|\varphi'| + 2$. For this purpose, we define

$$\varphi(y) := \frac{1}{8C_0^2}[e^{4C_0 y} - 4C_0 y - 1], \quad \text{and thus } \varphi'(y) = \frac{1}{2C_0}[e^{4C_0 y} - 1], \quad \varphi''(y) = 2e^{4C_0 y}.$$

Then,

$$\varphi(0) = 0, \quad \varphi'(0) = 0, \quad \varphi(y) > 0, \varphi'(y) > 0 \text{ for } y > 0, \quad \text{and } \varphi''(y) = 4C_0\varphi'(y) + 2.$$

Note that $Y^m \geq Y^n$ for $m > n$. Then,

$$d\varphi(\Delta Y_t^{m,n}) \geq -C_0\varphi'(\Delta Y_t^{m,n})[1 + |\Delta Z_t^n|^2 + |Z_t|^2]dt$$
$$+[C_0\varphi'(\Delta Y_t^{m,n}) + 1]|\Delta Z_t^{m,n}|^2 dt + \varphi'(\Delta Y_t^{m,n})\Delta Z_t^{m,n}dB_t,$$

and thus

$$\mathbb{E}\left[\int_0^T [C_0\varphi'(\Delta Y_t^{m,n}) + 1]|\Delta Z_t^{m,n}|^2 dt\right]$$
$$\leq \mathbb{E}\left[\varphi(\Delta Y_T^{m,n}) - \varphi(\Delta Y_0^{m,n}) + C_0 \int_0^T \varphi'(\Delta Y_t^{m,n})[1 + |\Delta Z_t^n|^2 + |Z_t|^2]dt\right].$$

Now fix n and send $m \to \infty$. Since $|\Delta Y^{m,n}| \leq C$ and $\Delta Y^{m,n} \to \Delta Y^n$, by Problem 1.4.11 (ii) we obtain

$$\mathbb{E}\left[\int_0^T [C_0\varphi'(\Delta Y_t^n) + 1]|\Delta Z_t^n|^2 dt\right]$$
$$\leq \mathbb{E}\left[\varphi(\Delta Y_T^n) - \varphi(\Delta Y_0^n) + C_0 \int_0^T \varphi'(\Delta Y_t^n)[1 + |\Delta Z_t^n|^2 + |Z_t|^2]dt\right].$$

Then

$$\mathbb{E}\left[\int_0^T |\Delta Z_t^n|^2 dt\right] \leq \mathbb{E}\left[\varphi(\Delta Y_T^n) - \varphi(\Delta Y_0^n) + C_0 \int_0^T \varphi'(\Delta Y_t^n)[1 + |Z_t|^2]dt\right].$$

Now send $n \to \infty$, Claim (7.3.4) follows from the Dominated Convergence Theorem.   ∎

We now present the main well-posedness result.

Theorem 7.3.3 *Let Assumptions 7.1.3 and 7.1.5 hold. Then BSDE* (4.0.3) *admits a unique solution* $(Y, Z) \in \mathbb{L}^\infty(\mathbb{F}, \mathbb{R}) \times \mathbb{L}^2(\mathbb{F} \times \mathbb{R}^{1 \times d})$.

Proof First, by the comparison Theorem 7.3.1 it is clear that Y is unique, which in turn implies the uniqueness of Z immediately.

To prove the existence, for any $n, m, k \geq 1$, define

$$f^n := f \wedge n, \quad f^{n,m} := f^n \vee (-m), \quad f_t^{n,m,k}(y,z) := \inf_{z' \in \mathbb{R}^{1 \times d}} \left[f_t^{n,m}(y,z') + k|z - z'|\right].$$

Then,

$f^n \uparrow f$, $f^{n,m} \downarrow f^n$, $f^{n,m,k} \uparrow f^{n,m}$,   as n, m, k increase;

$-m \leq f^{n,m}, f^{n,m,k} \leq n$ and $|f^n|, |f^{n,m}| \leq C[1+|y|+|z|^2]$ for all (n, m, k); (7.3.5)

$f^{n,m,k}$ is uniformly Lipschitz continuous in (y, z) for each (n, m, k).

However, we should note that $f^{n,m,k}$ does not satisfy (7.1.2) uniformly, and thus we cannot apply Lemma 7.3.2 directly on it.

Let $(Y^{n,m,k}, Z^{n,m,k})$ denote the unique solution of the BSDE

$$Y_t^{n,m,k} = \xi + \int_t^T f_s^{n,m,k}(Y_s^{n,m,k}, Z_s^{n,m,k})ds - \int_t^T Z_s^{n,m,k}dB_s.$$

By Theorem 4.4.1, $Y^{n,m,k}$ is increasing in k, decreasing in m, and increasing in n, and thus we may define

$$Y^{n,m} := \lim_{k \to \infty} Y^{n,m,k}, \quad Y^n := \lim_{m \to \infty} Y^{n,m}, \quad Y := \lim_{n \to \infty} Y^n.$$

Note that

$$Y_t^{n,m,k} = \mathbb{E}_t\Big[\xi + \int_t^T f_s^{n,m,k}(Y_s^{n,m,k}, Z_s^{n,m,k})ds\Big].$$

Clearly,

$$|Y_t^{n,m,k}| \le \|\xi\|_\infty + (n \vee m)T, \quad \text{for all } k.$$

For any k_1, k_2, applying Itô formula on $|Y_t^{n,m,k_1} - Y_t^{n,m,k_2}|^2$ we obtain

$$\mathbb{E}\Big[\int_0^T |Z_t^{n,m,k_1} - Z_t^{n,m,k_2}|^2 dt\Big]$$

$$\le \mathbb{E}\Big[2\int_0^T |Y_t^{n,m,k_1} - Y_t^{n,m,k_2}||f_t^{n,m,k_1}(Y_t^{n,m,k_1}, Z_t^{n,m,k_1}) - f_t^{n,m,k_2}(Y_t^{n,m,k_2}, Z_t^{n,m,k_2})|dt\Big]$$

$$\le 2(n+m)\mathbb{E}\Big[\int_0^T |Y_t^{n,m,k_1} - Y_t^{n,m,k_2}|dt\Big] \to 0, \quad \text{as } k_1, k_2 \to \infty,$$

thanks to the Dominated Convergence Theorem. Then, there exists $Z^{n,m} \in \mathbb{L}^2(\mathbb{F} \times \mathbb{R}^{1 \times d})$ such that

$$\lim_{k \to \infty} \mathbb{E}\Big[\int_0^T |Z_t^{n,m,k} - Z_t^{n,m}|^2 dt\Big] = 0.$$

It is straightforward to check that $(Y^{n,m}, Z^{n,m})$ satisfies

$$Y_t^{n,m} = \xi + \int_t^T f_s^{n,m}(Y_s^{n,m}, Z_s^{n,m})ds - \int_t^T Z_s^{n,m}dB_s.$$

Now, fix n and send $m \to \infty$. By Lemma 7.3.2 there exists $Z^n \in \mathbb{L}^2(\mathbb{F} \times \mathbb{R}^{1 \times d})$ such that

$$Y_t^n = \xi + \int_t^T f_s^n(Y_s^n, Z_s^n)ds - \int_t^T Z_s^n dB_s.$$

Finally, send $n \to \infty$ and apply Lemma 7.3.2 again, there exists $Z \in \mathbb{L}^2(\mathbb{F} \times \mathbb{R}^{1 \times d})$ such that (Y, Z) is a solution to BSDE (4.0.3). ∎

We finally remove the monotonicity condition in Lemma 7.3.2.

Theorem 7.3.4 *Let (ξ_n, f^n), $n \geq 0$, satisfy Assumptions 7.1.3 and 7.1.5 uniformly, and $(Y^n, Z^n) \in \mathbb{L}^\infty(\mathbb{F}, \mathbb{R}) \times \mathbb{L}^2(\mathbb{F}, \mathbb{R}^{1 \times d})$ satisfies BSDE (4.0.3) corresponding to (ξ_n, f^n). As $n \to \infty$, assume $\xi_n \to \xi^0$ in $\mathbb{P}$, $f^n(\cdot, y, z) \to f^0(\cdot, y, z)$ in measure $dt \times d\mathbb{P}$ for any (y, z). Then, for any $p \geq 1$,*

$$\lim_{n \to \infty} \mathbb{E}\left[|(Y^n - Y^0)_T^*|^p + \left(\int_0^T |Z_t^n - Z_t^0|^2 dt\right)^{\frac{p}{2}}\right] = 0. \tag{7.3.6}$$

Proof First, by Theorem 7.2.1 there exists a constant C_0, independent of n, such that

$$|Y_t^n|^2 + \mathbb{E}_t\left[\int_t^T |Z_s^n|^2 ds\right] \leq C_0, \quad \forall t \in [0, T], n \geq 0. \tag{7.3.7}$$

For any $n \geq 1$, denote $\Delta Y^n := Y^n - Y^0$, $\Delta Z^n := Z^n - Z^0$, and similarly denote $\Delta \xi_n$, Δf^n. Then

$$\Delta Y_t^n = \Delta \xi_n + \int_t^T [\Delta f_s^n(Y_s, Z_s) + \alpha_s^n \Delta Y_s^n + \Delta Z_s^n \theta_s^n]ds - \int_t^T \Delta Z_s^n dB_s, \tag{7.3.8}$$

where α^n is uniformly bounded and

$$|\theta_t^n| \leq C[1 + |Y_t^n| + |Y_t| + |Z_t^n| + |Z_t|] \leq C[1 + |Z_t^n| + |Z_t|]. \tag{7.3.9}$$

Denote $\Gamma_t^n := \exp(\int_0^t \alpha_s^n ds)$ and $M^n := M^{\theta^n}$. Following the same arguments as in Theorem 7.3.1 we see that

$$\Gamma_t^n M_t^n \Delta Y_t^n = \mathbb{E}_t\left[\Gamma_T^n M_T^n \Delta \xi_n + \int_t^T \Gamma_s^n M_s^n \Delta f_s^n(Y_s, Z_s)ds\right].$$

By (7.3.7) and (7.3.9), it follows from Theorem 7.2.3 that there exist $\varepsilon, C_\varepsilon$, independent of n, such that $\mathbb{E}[|(M^n)_T^*|^{1+\varepsilon}] \leq C_\varepsilon$. Then we have

$$|\Delta Y_t^n| \leq C\mathbb{E}_t\left[\frac{M_T^n}{M_t^n}[|\Delta \xi_n| + \int_t^T |\Delta f_s^n(Y_s, Z_s)|ds]\right] \leq C_\varepsilon\left(\mathbb{E}_t\left[(|\Delta \xi_n| + \int_t^T |\Delta f_s^n(Y_s, Z_s)|ds)^{\frac{1+\varepsilon}{\varepsilon}}\right]\right)^{\frac{\varepsilon}{1+\varepsilon}}.$$

Thus

$$\mathbb{E}[|\Delta Y_t^n|^{\frac{1+\varepsilon}{\varepsilon}}] \le C_\varepsilon \mathbb{E}\Big[\Big(|\Delta \xi_n| + \int_t^T |\Delta f_s^n(Y_s, Z_s)|ds\Big)^{\frac{1+\varepsilon}{\varepsilon}}\Big].$$

Note that Y is bounded and thus $|\Delta f_s^n(Y_s, Z_s)| \le C[1 + |Z_s|^2]$. It follows from Problem 1.4.6 (iii) and the dominated convergence theory that $\lim_{n\to\infty} \mathbb{E}[|\Delta Y_t^n|^{\frac{1+\varepsilon}{\varepsilon}}] = 0$. Now since ΔY^n is uniformly bounded, we see that

$$\lim_{n\to\infty} \mathbb{E}[|\Delta Y_t^n|^p] = 0, \quad \forall p \ge 1. \tag{7.3.10}$$

Next, denote $dB_t^n := dB_t - \theta_t^n dt$ and $d\mathbb{P}^n := M_T^n d\mathbb{P}$. By (7.3.8) we have

$$\int_0^T \Delta Z_t^n dB_t^n = \Delta \xi_n - \Delta Y_0^n + \int_0^T [\Delta f_t^n(Y_t, Z_t) + \alpha_t^n \Delta Y_t^n]dt.$$

Following similar arguments as above one can easily show that, for any $p \ge 1$,

$$\mathbb{E}\Big[M_T^n\Big(\int_0^T |\Delta Z_t^n|^2 dt\Big)^{\frac{p}{2}}\Big] = \mathbb{E}^{\mathbb{P}^n}\Big[\Big(\int_0^T |\Delta Z_t^n|^2 dt\Big)^{\frac{p}{2}}\Big] \le C_p \mathbb{E}^{\mathbb{P}^n}\Big[\Big(\int_0^T \Delta Z_t^n dB_t^n\Big)^p\Big]$$

$$= C_p \mathbb{E}^{\mathbb{P}^n}\Big[\Big(\Delta \xi_n - \Delta Y_0^n + \int_0^T [\Delta f_t^n(Y_t, Z_t) + \alpha_t^n \Delta Y_t^n]dt\Big)^p\Big]$$

$$= C_p \mathbb{E}\Big[M_T^n\Big(\Delta \xi_n - \Delta Y_0^n + \int_0^T [\Delta f_t^n(Y_t, Z_t) + \alpha_t^n \Delta Y_t^n]dt\Big)^p\Big]$$

$$\le C_{p,\varepsilon}\Big(\mathbb{E}\Big[\Big(\Delta \xi_n - \Delta Y_0^n + \int_0^T [\Delta f_t^n(Y_t, Z_t) + \alpha_t^n \Delta Y_t^n]dt\Big)^{\frac{p(1+\varepsilon)}{\varepsilon}}\Big]\Big)^{\frac{\varepsilon}{1+\varepsilon}} \to 0, \quad \text{as } n \to \infty,$$

where ε again satisfies the requirement in (7.2.5). Now assume $\varepsilon > 0$ is small enough so that $\frac{\varepsilon}{1-\varepsilon}$ satisfies (7.2.7). Then, as $n \to \infty$,

$$\mathbb{E}\Big[\Big(\int_0^T |\Delta Z_t^n|^2 dt\Big)^{\frac{p}{2}}\Big] = \mathbb{E}\Big[(M_T^n)^{-\varepsilon}(M_T^n)^\varepsilon\Big(\int_0^T |\Delta Z_t^n|^2 dt\Big)^{\frac{p}{2}}\Big]$$

$$\le \Big(\mathbb{E}[(M_T^n)^{-\frac{\varepsilon}{1-\varepsilon}}]\Big)^{1-\varepsilon}\Big(\mathbb{E}[M_T^n\Big(\int_0^T |\Delta Z_t^n|^2 dt\Big)^{\frac{p}{2\varepsilon}}]\Big)^\varepsilon \le C_\varepsilon \Big(\mathbb{E}[M_T^n\Big(\int_0^T |\Delta Z_t^n|^2 dt\Big)^{\frac{p}{2\varepsilon}}]\Big)^\varepsilon \to 0.$$

$$\tag{7.3.11}$$

Finally, by (7.3.8) again we have

$$|(\Delta Y^n)_T^*| \le |\Delta Y_0^n| + C\int_0^T \Big[|\Delta f_t^n(Y_t, Z_t)| + |\Delta Y_t^n| + (1 + |Z_t^n| + |Z_t|)|\Delta Z_t^n|\Big]dt$$

$$+ \sup_{0 \le t \le T} \Big|\int_0^t \Delta Z_s^n dB_s\Big|.$$

By (7.3.10) and (7.3.11) one can easily show that $\lim_{n\to\infty} \mathbb{E}[|(\Delta Y^n)_T^*|^p] = 0$ for any $p \ge 1$. ∎

As a quick application of the above stability result, we consider Markovian BSDEs with quadratic growth in Z and recover Theorem 5.5.8 in this situation. To be precise, for any $(t, x) \in [0, T] \times \mathbb{R}^{d_1}$, consider

$$X^{t,x} = x + \int_t^s b(r, X_r^{t,x})dr + \int_t^s \sigma(r, X_r^{t,x})dB_r,$$

$$Y^{t,x} = g(X_T^{t,x}) + \int_s^T f(r, X_r^{t,x}, Y_r^{t,x}, Z_r^{t,x})dr - \int_s^T Z_r^{t,x}dB_r, \qquad t \le s \le T. \quad (7.3.12)$$

We shall assume

Assumption 7.3.5

 (i) b, σ, f, g are deterministic, and $d_2 = 1$;
 (ii) b, σ are uniformly Lipschitz continuous in x, and continuous in t;
 (iii) g is continuous in x and bounded;
 (iv) f is continuous in (t, x), uniformly Lipschitz continuous in y, and satisfy the following growth and regularity condition in terms of z:

$$|f(t, x, y, z)| \le C[1 + |y| + |z|^2],$$
$$|f(t, x, y, z_1) - f(t, x, y, z_2)| \le C[1 + |y| + |z_1| + |z_2|]|z_1 - z_2|. \quad (7.3.13)$$

Under the above conditions, it follows from Theorem 7.3.3 that the FBSDE (7.3.12) is well posed. Define

$$u(t, x) := Y_t^{t,x}. \quad (7.3.14)$$

Theorem 7.3.6 *Under Assumption 7.3.5, the function u defined by (7.3.14) is in $C_b^0([0, T] \times \mathbb{R}^{d_1})$ and is a viscosity solution of the PDE (5.5.2).*

The proof is rather standard, in particular the continuity of u relies on the stability Theorem 7.3.4. We leave the details to readers, see Problem 7.5.7.

7.4   Bibliographical Notes

This chapter is mainly based on Kobylanski [126], which initiated the study on BSDEs with quadratic growth in Z. The results on BMO martingales are mainly from Kazamaki [121]. There have been many further studies on this subject, especially when the terminal condition ξ is unbounded, see, e.g., Briand & Hu [22, 23], Delbaen, Hu, & Richou [58, 59], Barrieu & El Karoui [7], and Dos Reis [199]. Note that in all the above works the comparison principle plays a crucial role and thus the process Y is required to be scalar. There have been a few recent studies on multidimensional quadratic BSDEs, each requiring some special structure, see, e.g., Tevzadze [226], Frei [92], Cheridito & Nam [31], Hu & Tang [112], Kramkov & Pulido [127], Jamneshan, Kupper, & Luo [114], and Kardaras, Xing, & Zitkovic [120]. See also Frei and dos Reis [93] for a counterexample in multidimensional case.

The subject is also closely related to BSDEs with stochastic Lipschitz conditions, see, e.g., El Karoui & Huang [79] and Briand & Confortola [21]. While using quite different techniques, another related subject is Riccati type of BSDEs, which has superlinear growth in Y, see, e.g., Tang [222] and Ma, Wu, Zhang & Zhang [144].

7.5 Exercises

Problem 7.5.1 Under the condition (7.2.3), prove directly that M^θ is uniformly integrable by using the following simple fact due to Jensen's inequality: for any stopping time $\tau \in \mathscr{T}$,

$$\mathbb{E}_\tau[M_T^\theta / M_\tau^\theta] \geq \exp\left(\mathbb{E}_\tau\left[\int_\tau^T \theta_s dB_s - \frac{1}{2}\int_\tau^T |\theta_s|^2 ds\right]\right).$$

∎

Problem 7.5.2 Use the two steps specified in (i) and (ii) below, prove that M^θ is uniformly integrable under the following Novikov condition:

$$\mathbb{E}\left[\exp\left(\frac{1}{2}\int_0^T |\theta_t|^2 dt\right)\right] < \infty. \tag{7.5.1}$$

(i) First prove the uniform integrability under a stronger condition: for some $c > 1$,

$$\mathbb{E}\left[\exp\left(\frac{c}{2}\int_0^T |\theta_t|^2 dt\right)\right] < \infty. \tag{7.5.2}$$

(ii) By applying (i) on $M^{(1-\varepsilon)\theta}$ and thus obtaining $\mathbb{E}[M_T^{(1-\varepsilon)\theta}] = 1$ to prove that $\mathbb{E}[M_T^\theta] \geq 1$, and then prove further the general result. ∎

Problem 7.5.3 Let (ξ_n, f^n), $n \geq 0$, satisfy Assumptions 7.1.3 and 7.1.5 uniformly, and $(Y^n, Z^n) \in \mathbb{L}^\infty(\mathbb{F}, \mathbb{R}) \times \mathbb{L}^2(\mathbb{F}, \mathbb{R}^{1 \times d})$ satisfies BSDE (4.0.3) corresponding to (ξ_n, f^n). As $n \to \infty$, assume $\xi_n \to \xi^0$ and $f^n(\cdot, y, z) \to f^0(\cdot, y, z)$ uniformly, uniformly on (t, ω, y, z). Then,

$$\lim_{n \to \infty} \text{ess sup}\left((Y^n - Y^0)_T^* + \sup_{0 \leq t \leq T} \mathbb{E}_t\left[\int_t^T |Z_s^n - Z_s^0|^2 ds\right]\right) = 0, \tag{7.5.3}$$

where, for a random variable η, ess sup$\eta := \inf\{c > 0 : \eta \leq c, \text{a.s.}\}$. ∎

Problem 7.5.4 This problem concerns Riccati type of BSDEs (4.0.3). Assume

(i) $\mathbb{F} = \mathbb{F}^B$ and $d_2 = 1$;
(ii) $\xi \in \mathbb{L}^\infty(\mathscr{F}_T, \mathbb{R})$;

(iii) f is $\mathbb{F}$-measurable in (t, ω, y, z), uniformly Lipschitz continuous in z, and
$f(0, 0)$ is bounded;
(iv) f is differentiable in y with $\partial_y f \leq C - c|y|^2$ for some constants $C, c \geq 0$.

Show that BSDE (4.0.3) has a unique solution $(Y, Z) \in \mathbb{L}^\infty(\mathbb{F}, \mathbb{R}) \times \mathbb{L}^2(\mathbb{F}, \mathbb{R}^{1 \times d})$.
(Hint: truncate the BSDE in the spirit of Proposition 7.1.4.) ∎

Problem 7.5.5 This problem concerns a priori estimates for BSDE (4.0.3) under
weak conditions. Assume $\mathbb{F} = \mathbb{F}^B$, $d_2 = 1$, ξ is $\mathscr{F}_T$-measurable, f is $\mathbb{F}$-measurable,
and

$$|f_t(y, z)| \leq |f_t^0| + L[|y| + |z|^2]. \tag{7.5.4}$$

Assume further that the BSDE has a solution (Y, Z) and all the involved processes
have good enough integrability (but not necessarily bounded). Denote $\tau := \inf\{t \geq
0 : |Y_t| \leq 1\}$ and define

$$\tilde{Y}_t := \exp\left(e^{C_1 t}|Y_t| + \int_0^t e^{C_1 s}|f_s^0|ds\right) + C_2 t.$$

(i) Show that $\tilde{Y}$ is a submartingale on $[0, \tau]$ for certain large constants C_1 and C_2.
(ii) Prove the following a priori estimate:

$$|Y_0| \leq C + \ln \mathbb{E}\left[\exp\left(e^{C_1 T}|\xi| + \int_0^T e^{C_1 s}|f_s^0|ds\right)\right],$$

where the constants C_1 and C depend only on T, the constant L in (7.5.4), and
the dimension d.

∎

Problem 7.5.6 Consider the setting in Problem 7.5.5, except that (7.5.4) is replaced
with the following growth condition:

$$|f(t, y, z)| \leq |f_t^0| + C\left[|y| \ln(|y| \vee 1) + |z| \ln^{\frac{1}{2}}(|z| \vee 1)\right].$$

Find a smooth nonlinear transformation $\varphi : [0, T] \times [0, \infty) \to [0, \infty)$ such that

- For any t, $y \mapsto \varphi(t, y)$ is strictly increasing;
- The process $\tilde{Y}_t := \varphi(t, |Y_t|) + \int_0^t \varphi(s, |f_s^0|)ds$ is a submartingale on $[0, \tau]$.

Consequently, we obtain an a priori estimate: $\varphi(0, |Y_0|) \leq C + \mathbb{E}\Big[\varphi(T, |\xi|) +$
$\int_0^T \varphi(t, |f_t^0|)dt\Big]$. ∎

Problem 7.5.7 Prove Theorem 7.3.6. ∎

Chapter 8
Forward-Backward SDEs

8.1 Introduction

In this chapter we study coupled Forward-Backward SDEs of the form:

$$\begin{cases} X_t = x + \int_0^t b_s(X_s, Y_s, Z_s)ds + \int_0^t \sigma_s(X_s, Y_s, Z_s)dB_s; \\ Y_t = g(X_T) + \int_t^T f_s(X_s, Y_s, Z_s)ds - \int_t^T Z_s dB_s, \end{cases} \quad (8.1.1)$$

where b, σ, f, g are random fields with appropriate dimensions. Such equation is called coupled because the forward SDE depends on the backward components (Y, Z) as well. The solution triple $\Theta := (X, Y, Z)$ is equipped with the norm:

$$\|\Theta\|^2 := E\Big[|X_T^*|^2 + |Y_T^*|^2 + \int_0^T |Z_t|^2 dt\Big]. \quad (8.1.2)$$

We shall always assume the following standing assumptions:

Assumption 8.1.1

(i) $\mathbb{F} = \mathbb{F}^B$.

(ii) $g(0) \in \mathbb{L}^2(\mathscr{F}_T, \mathbb{R}^{d_2})$, $b^0 \in \mathbb{L}^{1,2}(\mathbb{F}, \mathbb{R}^{d_1})$, $f^0 \in \mathbb{L}^{1,2}(\mathbb{F}, \mathbb{R}^{d_2})$, and $\sigma^0 \in \mathbb{L}^2(\mathbb{F}, \mathbb{R}^{d_1 \times d})$, where $\varphi^0 := \varphi(0, 0, 0)$ for $\varphi = b, \sigma, f$.

(iii) b, σ, f, g are uniformly Lipschitz continuous in (x, y, z).

Such equations arise naturally in many applications. For example, let $d = d_1 = d_2 = 1$ and consider the optimization problem (4.5.7)–(4.5.8) in strong formulation, we shall see heuristically how it leads to an FBSDE via the so-called stochastic maximum principle. By (4.5.8) we have

© Springer Science+Business Media LLC 2017
J. Zhang, *Backward Stochastic Differential Equations*, Probability Theory and Stochastic Modelling 86, DOI 10.1007/978-1-4939-7256-2_8

$$J_S(k) = Y_0^k, \quad \text{where} \quad Y_t^k = g(X_T^k) + \int_t^T f(s, X_s^k, k_s)ds - \int_t^T Z_s^k dB_s. \quad (8.1.3)$$

Denote by $\Theta^k := (X^k, Y^k, Z^k)$ the solution to the decoupled FBSDE (4.5.7) and (8.1.3). Fix k, Δk, and denote $k^\varepsilon := k + \varepsilon \Delta k$. We assume k, k^ε are in the admissible set for all ε small and the coefficients b, σ, f, g are differentiable in (x, k). Differentiate Θ^{k^ε} formally in ε we obtain $\nabla \Theta^{k,\Delta k} = (\nabla X^{k,\Delta k}, \nabla Y^{k,\Delta k}, \nabla Z^{k,\Delta k})$ as follows:

$$\nabla X_t^{k,\Delta k} = \int_0^t [\partial_x b \nabla X_s^{k,\Delta k} + \partial_k b \Delta k_s]ds + \int_0^t [\partial_x \sigma \nabla X_s^{k,\Delta k} + \partial_k \sigma \Delta k_s]dB_s;$$

$$\nabla Y_t^{k,\Delta k} = \partial_x g \nabla X_T^{k,\Delta k} + \int_t^T [\partial_x f \nabla X_s^{k,\Delta k} + \partial_k f \Delta k_s]ds - \int_t^T \nabla Z_s^{k,\Delta k} dB_s.$$

Notice that the above equations are linear. Recall Section 3.1, in particular Case 3 there. Introduce adjoint processes $(\Gamma^k, \bar{Y}^k, \bar{Z}^k)$ via a decoupled FBSDE:

$$\Gamma_t^k = 1 - \int_0^t \Gamma_s^k [\partial_x b - |\partial_x \sigma|^2](s, X_s^k, k_s)ds - \int_0^t \Gamma_s^k \partial_x \sigma(s, X_s^k, k_s)dB_s;$$
$$\bar{Y}_t^k = \partial_x g(X_T^k)(\Gamma_T^k)^{-1} + \int_t^T \partial_x f(s, X_s^k, k_s)(\Gamma_s^k)^{-1}ds - \int_t^T \bar{Z}_s^k dB_s. \quad (8.1.4)$$

Applying Itô formula, we have

$$d(\Gamma_t^k \nabla X_t^{k,\Delta k} \bar{Y}_t^k) = \left[-\partial_x f \nabla X_t^{k,\Delta k} + \Gamma_t^k [\partial_k b - \partial_k \sigma \partial_x \sigma] \bar{Y}_t^k \Delta k_t + \bar{Z}_t^k \Gamma_t^k \partial_x \sigma \Delta k_t \right]dt$$
$$+ [\cdots]dB_t.$$

and thus

$$d(\nabla Y_t^{k,\Delta k} - \Gamma_t^k \nabla X_t^{k,\Delta k} \bar{Y}_t^k) = -\left[\partial_k f + \Gamma_t^k [\partial_k b - \partial_k \sigma \partial_x \sigma] \bar{Y}_t^k + \bar{Z}_t^k \Gamma_t^k \partial_x \sigma \right]\Delta k_t dt$$
$$+ [\cdots]dB_t.$$

Notice that $\nabla X_0^{k,\Delta k} = 0$ and $\Gamma_T^k \nabla X_T^{k,\Delta k} \bar{Y}_T^k = \nabla Y_T^{k,\Delta k}$. This leads to

$$\nabla Y_0^{k,\Delta k} = \mathbb{E}\left[\int_0^T [\partial_k f + \Gamma_t^k [\partial_k b - \partial_k \sigma \partial_x \sigma] \bar{Y}_t^k + \bar{Z}_t^k \Gamma_t^k \partial_x \sigma](t, X_t^k, k_t) \Delta k_t dt \right].$$
$$(8.1.5)$$

Denote $\Delta \Theta^\varepsilon := \Theta^{k^\varepsilon} - \Theta$, $\nabla \Theta^\varepsilon := \frac{1}{\varepsilon}\Delta \Theta^\varepsilon$. Under certain mild conditions, applying the BSDE stability Theorem 4.4.3 one can easily see that:

$$\lim_{\varepsilon \to 0} \|\Delta \Theta^\varepsilon\| = 0, \quad \lim_{\varepsilon \to 0} \|\nabla \Theta^\varepsilon - \nabla \Theta^{k,\Delta k}\| = 0.$$

Now assume k is an optimal control. Then $\Delta Y_0^\varepsilon = J_S(k^\varepsilon) - J_S(k) \leq 0$ for all ε and Δk, then we obtain $\nabla Y_0^{k, \Delta k} \leq 0$ for all Δk. Note that $\Gamma^k, \bar{Y}^k, \bar{Z}^k$ do not depend on Δk. Then by the arbitrariness of Δk, (8.1.5) leads to the following first order condition of the stochastic maximum principle:

$$\left[\partial_k f + \Gamma_t^k [\partial_k b - \partial_k \sigma \partial_x \sigma] \bar{Y}_t^k + \bar{Z}_t^k \Gamma_t^k \partial_x \sigma \right](t, X_t^k, k_t) = 0. \qquad (8.1.6)$$

Assume further that b, σ, f have certain structure so that the above first order condition uniquely determines k:

$$k_t = I(t, X_t^k, \Gamma_t^k, \bar{Y}_t^k, \bar{Z}_t^k) \quad \text{for some function } I. \qquad (8.1.7)$$

Denote $\widehat{\varphi}(t, x, \gamma, y, z) := \varphi(t, x, I(t, x, \gamma, y, z))$ for any function $\varphi(t, x, k)$. Plug (8.1.7) into (4.5.7), (8.1.3), and (8.1.4), we obtain the following (multidimensional) coupled FBSDE:

$$\begin{cases} X_t = x + \int_0^t \widehat{b}(s, X_s, \Gamma_s, \bar{Y}_s, \bar{Z}_s) ds + \int_0^t \widehat{\sigma}(s, X_s, \Gamma_s, \bar{Y}_s, \bar{Z}_s) dB_s; \\ \Gamma_t = 1 - \int_0^t \Gamma_s [\widehat{\partial_x b} - |\widehat{\partial_x \sigma}|^2](s, X_s, \Gamma_s, \bar{Y}_s, \bar{Z}_s) ds - \int_0^t \Gamma_s \widehat{\partial_x \sigma}(s, X_s, \Gamma_s, \bar{Y}_s, \bar{Z}_s) dB_s; \\ Y_t = g(X_T) + \int_t^T \widehat{f}(s, X_s, \Gamma_s, \bar{Y}_s, \bar{Z}_s) ds - \int_t^T Z_s dB_s; \\ \bar{Y}_t = \partial_x g(X_T)(\Gamma_T)^{-1} + \int_t^T \widehat{\partial_x f}(s, X_s, \Gamma_s, \bar{Y}_s, \bar{Z}_s)(\Gamma_s)^{-1} ds - \int_t^T \bar{Z}_s dB_s. \end{cases} \qquad (8.1.8)$$

If we can solve the above FBSDE, then the candidate solution to the optimization problem (4.5.8) is:

$$V_0^S = Y_0 \quad \text{with optimal control} \quad k_t^* = I(t, X_t, \Gamma_t, \bar{Y}_t, \bar{Z}_t). \qquad (8.1.9)$$

However, as pointed out in Remark 4.5.4, even when there is only drift control, the optimization problem in strong formulation typically does not have an optimal control, and thus the FBSDE (8.1.8) may not have a solution. Indeed, Assumption 8.1.1 is not sufficient for the well-posedness of coupled FBSDEs, as we see in the following very simple example.

Example 8.1.2 *Let $d = d_1 = d_2 = 1$ and $\xi \in \mathbb{L}^2(\mathscr{F}_T)$. Consider the following linear FBSDE:*

$$\begin{cases} X_t = \int_0^t Z_s dB_s; \\ Y_t = X_T + \xi - \int_t^T Z_s dB_s. \end{cases} \qquad (8.1.10)$$

(i) *When ξ is deterministic, the FBSDE has infinitely many solutions. Indeed, for any $Z \in \mathbb{L}^2(\mathbb{F})$, $X_t := \int_0^t Z_s dB_s$, $Y_t := X_t + \xi$ is a solution.*

(ii) *When ξ is not deterministic, the FBSDE has no solution. Indeed, if (X, Y, Z) is a solution, we have $Y_0 = X_T + \xi - \int_0^T Z_s dB_s = \xi$ which is not $\mathscr{F}_0$-measurable.*

The main objective of this chapter is to provide some sufficient conditions (in addition to Assumption 8.1.1) under which FBSDE (8.1.1) is well posed. There are three approaches in the literature, which will be introduced one by one in the next three sections. However, we should point out that the theory for general coupled FBSDE is still far from complete, and as we mentioned earlier many FBSDEs arising from applications may not have a solution.

8.2   Well-Posedness in Small Time Duration

The fixed point approach, or say the Picard iteration approach, works well for SDEs and BSDEs, as we see in Chapters 3 and 4. However, for FBSDEs one needs additional assumptions. In this section, we focus on the case when T is small, which has certain advantage as we saw in the proof of Theorem 4.3.1. Our main result is:

Theorem 8.2.1 *Let Assumption 8.1.1 hold. Assume*

$$c_0 := \|\partial_z \sigma\|_\infty \|\partial_x g\|_\infty < 1, \tag{8.2.1}$$

where $\|\partial_z \sigma\|_\infty$ and $\|\partial_x g\|_\infty$ denote the Lipschitz constants of σ in z and that of g in x, respectively. Then there exist $\delta_0 > 0$ and C, which depend only on the Lipschitz constants in Assumption 8.1.1, the dimensions, and the above constant c_0, such that whenever $T \le \delta_0$, FBSDE (8.1.1) admits a unique solution Θ and it holds

$$\|\Theta\|^2 \le C[I_0^2 + |x|^2] \text{ where } I_0^2 := \quad \mathbb{E}\Big[\Big(\int_0^T [|b_t^0| + |f_t^0|]dt\Big)^2$$
$$+ \int_0^T |\sigma_t^0|^2 dt + |g(0)|^2\Big]. \tag{8.2.2}$$

Proof For simplicity we assume $d = d_1 = d_2 = 1$. We emphasize that in this proof the generic constant C does not depend on T.

(i) We first prove the well-posedness of the FBSDE for small T. Let $\delta_0 > 0$ be a constant which will be specified later, and $T \le \delta_0$. Define a mapping F on $\mathbb{L}^2(\mathbb{F}) \times \mathbb{L}^2(\mathbb{F})$ by $F(\mathbf{y}, \mathbf{z}) := (Y^{\mathbf{y}, \mathbf{z}}, Z^{\mathbf{y}, \mathbf{z}})$, where $\Theta^{\mathbf{y}, \mathbf{z}} := (X^{\mathbf{y}, \mathbf{z}}, Y^{\mathbf{y}, \mathbf{z}}, Z^{\mathbf{y}, \mathbf{z}})$ is the unique solution to the following decoupled FBSDE:

$$\begin{cases} X_t^{\mathbf{y},\mathbf{z}} = x + \int_0^t b_s(X_s^{\mathbf{y},\mathbf{z}}, \mathbf{y}_s, \mathbf{z}_s)ds + \int_0^t \sigma_s(X_s^{\mathbf{y},\mathbf{z}}, \mathbf{y}_s, \mathbf{z}_s)dB_s; \\ Y_t^{\mathbf{y},\mathbf{z}} = g(X_T^{\mathbf{y},\mathbf{z}}) + \int_t^T f_s(X_s^{\mathbf{y},\mathbf{z}}, \mathbf{y}_s, \mathbf{z}_s)ds - \int_t^T Z_s^{\mathbf{y},\mathbf{z}}dB_s. \end{cases} \tag{8.2.3}$$

We shall show that F is a contraction mapping under the following norm, which is slightly weaker than the one in (4.2.1):

$$\|(\mathbf{y}, \mathbf{z})\|_w^2 := \sup_{0 \le t \le T} \mathbb{E}\Big[|\mathbf{y}_t|^2 + \int_t^T |\mathbf{z}_s|^2 ds\Big]. \tag{8.2.4}$$

For this purpose, let $(\mathbf{y}^i, \mathbf{z}^i) \in \mathbb{L}^2(\mathbb{F}) \times \mathbb{L}^2(\mathbb{F})$, $i = 1, 2$, and Θ^i be the solution to the corresponding decoupled FBSDE (8.2.3). Denote $\Delta\mathbf{y} := \mathbf{y}^1 - \mathbf{y}^2$, $\Delta\mathbf{z} := \mathbf{z}^1 - \mathbf{z}^2$, and $\Delta\Theta := \Theta^1 - \Theta^2$. Then

$$\Delta X_t = \int_0^t [\alpha_s^1 \Delta X_s + \beta_s^1 \Delta\mathbf{y}_s + \gamma_s^1 \Delta\mathbf{z}_s]ds + \int_0^t [\alpha_s^2 \Delta X_s + \beta_s^2 \Delta\mathbf{y}_s + \gamma_s^2 \Delta\mathbf{z}_s]dB_s;$$

$$\Delta Y_t = \lambda \Delta X_T + \int_t^T [\alpha_s^3 \Delta X_s + \beta_s^3 \Delta Y_s + \gamma_s^3 \Delta Z_s]ds - \int_t^T \Delta Z_s dB_s,$$

where $\alpha^i, \beta^i, \gamma^i$, $i = 1, 2, 3$, are bounded, and $|\gamma^2| \le \|\partial_z\sigma\|_\infty$, $|\lambda| \le \|\partial_x g\|_\infty$. For any $0 < \varepsilon < 1$, applying Itô formula we have

$$\mathbb{E}[|\Delta X_t|^2] = \mathbb{E}\Big[\int_0^t \Big[2\Delta X_s[\alpha_s^1 \Delta X_s + \beta_s^1 \Delta\mathbf{y}_s + \gamma_s^1 \Delta\mathbf{z}_s]$$
$$+ [\alpha_s^2 \Delta X_s + \beta_s^2 \Delta\mathbf{y}_s + \gamma_s^2 \Delta\mathbf{z}_s]^2\Big]ds\Big]$$
$$\le \mathbb{E}\Big[\int_0^t \Big[C\varepsilon^{-1}[|\Delta X_s|^2 + |\Delta\mathbf{y}_s|^2] + [\|\partial_z\sigma\|_\infty^2 + \varepsilon]|\Delta\mathbf{z}_s|^2\Big]ds\Big]$$
$$\le C\varepsilon^{-1}\delta_0 \sup_{0 \le s \le T} \mathbb{E}[|\Delta X_s|^2 + |\Delta\mathbf{y}_s|^2] + [\|\partial_z\sigma\|_\infty^2 + \varepsilon]\mathbb{E}\Big[\int_0^T |\Delta\mathbf{z}_s|^2 ds\Big]$$
$$\le C\varepsilon^{-1}\delta_0 \sup_{0 \le s \le T} \mathbb{E}[|\Delta X_s|^2] + [C\varepsilon^{-1}\delta_0 + \|\partial_z\sigma\|_\infty^2 + \varepsilon]\|(\Delta\mathbf{y}, \Delta\mathbf{z})\|_w^2;$$

$$\mathbb{E}\Big[|\Delta Y_t|^2 + \int_t^T |\Delta Z_s|^2 ds\Big] = \mathbb{E}\Big[\lambda^2 |\Delta X_T|^2 + \int_t^T 2\Delta Y_s[\alpha_s^3 \Delta X_s + \beta_s^3 \Delta Y_s$$
$$+ \gamma_s^3 \Delta Z_s]ds\Big]$$
$$\le \mathbb{E}\Big[\|\partial_x g\|_\infty^2 |\Delta X_T|^2 + \int_t^T [C\varepsilon^{-1}|\Delta Y_s|^2 + C|\Delta X_s|^2 + \varepsilon|\Delta Z_s|^2]ds\Big]$$
$$\le [\|\partial_x g\|_\infty^2 + C\delta_0] \sup_{0 \le s \le T} \mathbb{E}[|\Delta X_s|^2]$$
$$+ C\varepsilon^{-1}\delta_0 \sup_{0 \le s \le T} \mathbb{E}[|\Delta Y_s|^2] + \varepsilon\mathbb{E}\Big[\int_0^T |\Delta Z_s|^2 ds\Big]$$
$$\le [\|\partial_x g\|_\infty^2 + C\delta_0] \sup_{0 \le s \le T} \mathbb{E}[|\Delta X_s|^2] + [C\varepsilon^{-1}\delta_0 + \varepsilon]\|(\Delta Y, \Delta Z)\|_w^2.$$

Set $\delta_0 := \frac{\varepsilon^2}{C}$, then the above inequalities imply:

$$\sup_{0 \leq t \leq T} \mathbb{E}[|\Delta X_t|^2] \leq \frac{2\varepsilon + \|\partial_z \sigma\|_{\infty}^2}{1 - \varepsilon} \|(\Delta \mathbf{y}, \Delta \mathbf{z})\|_w^2;$$

$$\|(\Delta Y, \Delta Z)\|_w^2 \leq \frac{\|\partial_x g\|_{\infty}^2 + \varepsilon^2}{1 - 2\varepsilon} \sup_{0 \leq s \leq T} \mathbb{E}[|\Delta X_s|^2]$$

Plus the first estimate above into the second one, we have

$$\|(\Delta Y, \Delta Z)\|_w^2 \leq c_{\varepsilon} \|(\Delta \mathbf{y}, \Delta \mathbf{z})\|_w^2 \text{ where } c_{\varepsilon} := \frac{[\|\partial_x g\|_{\infty}^2 + \varepsilon^2][\|\partial_z \sigma\|_{\infty}^2 + 2\varepsilon]}{(1 - \varepsilon)(1 - 2\varepsilon)}.$$

$$(8.2.5)$$

By (8.2.1), we have $c_{\varepsilon} < 1$ when ε is small enough. Thus F is a contraction mapping when δ_0, hence $\varepsilon = \sqrt{C\delta_0}$, is small enough and $T \leq \delta_0$. Let $(Y, Z) \in \mathbb{L}^2(\mathbb{F}) \times \mathbb{L}^2(\mathbb{F})$ be the unique fixed point of F, namely $F(Y, Z) = (Y, Z)$. Let $X := X^{Y,Z}$ be determined by the first equation in (8.2.3), then it is straightforward to show that (X, Y, Z) is a solution to FBSDE (8.1.1). On the other hand, for any solution (X, Y, Z), clearly (Y, Z) is a fixed point of F and thus is unique, which in turn implies X is unique.

(ii) We now prove the estimate (8.2.2). Fix $\varepsilon := \sqrt{C\delta_0}$ as in (i) and let (X, Y, Z) be the unique solution to FBSDE (8.1.1). Denote $(Y^0, Z^0) := F(0, 0)$, namely corresponding to $\mathbf{y} = 0, \mathbf{z} = 0$. By (8.2.5),

$$\|(Y, Z)\|_w - \|(Y^0, Z^0)\|_w \leq \|(Y - Y^0, Z - Z^0)\|_w$$
$$\leq \sqrt{c_{\varepsilon}} \|(Y - 0, Z - 0)\|_w = \sqrt{c_{\varepsilon}} \|(Y, Z)\|_w.$$

Then

$$\|(Y, Z)\|_w \leq \frac{1}{1 - \sqrt{c_{\varepsilon}}} \|(Y^0, Z^0)\|_w = C.$$

It is straightforward to check that $\|(Y^0, Z^0)\|_w \leq CI_0$. Then $\|(Y, Z)\|_w \leq CI_0$, where the constant C depends on δ_0 through c_{ε}. Now apply the a priori estimates for SDEs and BSDEs, Theorems 3.2.2 and (4.2.1), we obtain (8.2.2) immediately.   ∎

Following the same arguments as in Theorems 3.2.4 and 4.2.3, we obtain immediately

Corollary 8.2.2 *Let all the conditions in Theorem 8.2.1 hold true for* (x, b, σ, f, g) *and* $(\tilde{x}, \tilde{b}, \tilde{\sigma}, \tilde{f}, \tilde{g})$. *Let* $\delta_0 > 0$ *be the common small number,* $T \leq \delta_0$, *and* $\Theta, \tilde{\Theta}$ *be the solution to FBSDE (8.1.1) with corresponding data sets. Denote* $\Delta\Theta := \tilde{\Theta} - \Theta$, $\Delta x := \tilde{x} - x$, *and* $\Delta\varphi := \tilde{\varphi} - \varphi$ *for* $\varphi = b, \sigma, f, g$. *Then,*

$$\|\Delta\Theta\|^2 \le C\mathbb{E}\Big[|\Delta x|^2 + \Big(\int_0^T [|\Delta b_t| + |\Delta f_t|](\tilde{\Theta}_t)dt\Big)^2 \tag{8.2.6}$$

$$+ \int_0^T |\Delta\sigma_t(\tilde{\Theta}_t)|^2 dt + |\Delta g(\tilde{X}_T)|^2\Big].$$

Now fix the δ_0 and assume $T \le \delta_0$. For any (t,x), let $\Theta^{t,x}$ be the unique solution to the following FBSDE on $[t,T]$:

$$\begin{cases} X_s^{t,x} = x + \displaystyle\int_t^s b_r(\Theta_r^{t,x})dr + \int_t^s \sigma_r(\Theta_r^{t,x})dB_r; \\ Y_s^{t,x} = g(X_T^{t,x}) + \displaystyle\int_s^T f_r(\Theta_r)dr - \int_s^T Z_r^{t,x}dB_r, \end{cases} \tag{8.2.7}$$

and define a random field $u : [0,T] \times \Omega \times \mathbb{R}^{d_1} \to \mathbb{R}^{d_2}$ by:

$$u_t(x) := Y_t^{t,x}. \tag{8.2.8}$$

Theorem 8.2.3 *Let all the conditions in Theorem 8.2.1 hold, and u is defined by (8.2.8). Then, for any $t \in [0,T]$,*

(i) $u_t(x) \in \mathbb{L}^2(\mathscr{F}_t, \mathbb{R}^{d_2})$ for any $x \in \mathbb{R}^{d_1}$.
(ii) For a.e. ω, u is uniformly Lipschitz continuous in x, uniformly in (t,ω).
(iii) $Y_t = u_t(X_t)$, a.s.

Proof First, for any fixed x, by definition (8.2.8) it is clear that $u_t(x) \in \mathbb{L}^2(\mathscr{F}_t, \mathbb{R}^{d_2})$. Next, for any x_1, x_2, by using $\mathbb{E}_t$ instead of $\mathbb{E}$ in (8.2.6) we have

$$|u_t(x_1) - u_t(x_2)| = |Y_t^{t,x_1} - Y_t^{t,x_2}| \le C|x_1 - x_2|.$$

That is, u is uniformly Lipschitz continuous in x. Then, following the arguments in Theorem 5.1.2 one can easily show that $Y_t = u(t, X_t)$, a.s. for any t. ∎

Remark 8.2.4 By Example 8.1.2, we see that the condition (8.2.1) cannot be removed for free, even when T is small. However, we shall note that (8.2.1) is by no means necessary. Indeed, consider the following linear FBSDE with constant coefficients:

$$\begin{cases} X_t = x + \displaystyle\int_0^t \gamma Z_s dB_s; \\ Y_t = \lambda X_T - \displaystyle\int_t^T Z_s dB_s. \end{cases} \tag{8.2.9}$$

One can show that it is well posed if and only if $\gamma\lambda \ne 1$. See Problem 8.6.3. ∎

8.3   The Decoupling Approach

In this section we shall extend the local well-posedness in the previous section to global well-posedness for arbitrary large T. We will follow the idea of Theorem 4.3.1 for BSDEs. We remark that the δ in Theorem 4.3.1 Step 1 does not depend on the terminal condition ξ, which is crucial for constructing the global solution in Step 2 there. However, the δ_0 in Theorem 8.2.1 depends on the terminal condition g, which prevents us from mimicking the arguments in Theorem 4.3.1 Step 2 directly. Indeed, the main difficulty in this section is to find sufficient conditions so that the δ_0 will be uniform in some sense. Our main tool will be the so-called decoupling field.

8.3.1   *The Four Step Scheme*

We start with the following Markov FBSDEs and establish its connection with PDEs:

$$\begin{cases} X_t = x + \int_0^t b(s, X_s, Y_s, Z_s)ds + \int_0^t \sigma(s, X_s, Y_s)dB_s; \\ Y_t = g(X_T) + \int_t^T f(s, X_s, Y_s, Z_s)ds - \int_t^T Z_s dB_s, \end{cases} \tag{8.3.1}$$

where the coefficients b, σ, f, g are deterministic measurable functions. We note that the forward diffusion coefficient σ does not depend on Z, which in particular implies $c_0 = 0$ in (8.2.1). In light of the nonlinear Feyman-Kac formula Theorem 5.1.4, we expect that $Y_t = u(t, X_t)$ for some deterministic function $u : [0, T] \times \mathbb{R}^{d_1} \to \mathbb{R}^{d_2}$. Assume u is smooth, applying Itô formula we have

$$du^i(t, X_t) = \left[\partial_t u^i + \partial_x u^i b + \frac{1}{2}\partial_{xx} u^i : (\sigma \sigma^t op) \right]dt + \partial_x u^i \sigma dB_t, \quad i = 1, \cdots, d_2.$$

Comparing this with the BSDE in (8.3.1), we have

$$\partial_t u^i + \partial_x u^i b + \frac{1}{2}\partial_{xx} u^i : (\sigma \sigma^\top) = -f^i, \quad \partial_x u^i \sigma = Z^i.$$

Thus the FBSDE is associated with the following system of quasi-linear parabolic PDE:

$$\partial_t u^i + \frac{1}{2}\partial_{xx} u^i : \sigma \sigma^\top (t, x, u) + \partial_x u^i b(t, x, u, \partial_x u \sigma(t, x, u))$$

$$+ f^i(t, x, u, \partial_x u \sigma(t, x, u)) = 0, i = 1, \cdots, d_2; u(T, x) = g(x). \tag{8.3.2}$$

We have the following result:

Theorem 8.3.1 *Assume*

(i) b, σ, f, g are deterministic and satisfy Assumption 8.1.1, and σ is bounded.
(ii) The PDE (8.3.2) has a classical solution u with bounded $\partial_x u$ and $\partial_{xx}^2 u$.

Then FBSDE (8.3.1) has a unique solution (X, Y, Z) and it holds that

$$Y_t = u(t, X_t), \quad Z_t = \partial_x u(t, X_t)\sigma(t, X_t, u(t, X_t)). \tag{8.3.3}$$

Proof For simplicity assume $d = d_1 = d_2 = 1$. We first prove existence. Consider SDE:

$$X_t = x + \int_0^t \tilde{b}(s, X_s)ds + \int_0^t \tilde{\sigma}(s, X_s)dB_s, \quad \text{where} \tag{8.3.4}$$
$$\tilde{\sigma}(t, x) := \sigma(t, x, u(t, x)), \quad \tilde{b}(t, x) := b(t, x, u(t, x), \partial_x u\tilde{\sigma}(t, x)).$$

By our conditions, one can easily check that $\tilde{b}$ and $\tilde{\sigma}$ are uniformly Lipschitz continuous in x, and $\int_0^T [|\tilde{b}(t, 0)| + |\tilde{\sigma}(t, 0)|^2]dt < \infty$. Then the above SDE has a unique solution X. Now define (Y, Z) by (8.3.3). Applying Itô formula one may check straightforwardly that (X, Y, Z) solves FSBDE (8.3.1).

To prove the uniqueness, let $\Theta := (X, Y, Z)$ be an arbitrary solution to FBSDE (8.3.1). Denote

$$\tilde{Y}_t := u(t, X_t), \quad \tilde{Z}_t := \partial_x u(t, X_t)\sigma(t, X_t, Y_t), \quad \hat{Z}_t := \partial_x u(t, X_t)\sigma(t, X_t, \tilde{Y}_t)$$
$$\Delta Y_t := \tilde{Y}_t - Y_t, \quad \Delta Z_t := \tilde{Z}_t - Z_t.$$

Applying Itô formula we get

$$d\tilde{Y}_t = \left[\partial_t u(t, X_t) + \partial_x u(t, X_t)b(t, \Theta_t) + \frac{1}{2}\partial_{xx}^2 u(t, X_t)\sigma^2(t, X_t, Y_t)\right]dt + \tilde{Z}_t dB_t.$$

Then, since u satisfies PDE (8.3.2),

$$d(\Delta Y_t) = \left[\partial_t u(t, X_t) + \partial_x u(t, X_t)b(t, \Theta_t) + \frac{1}{2}\partial_{xx}^2(t, X_t)\sigma^2(t, X_t, Y_t) + f(t, \Theta_t)\right]dt$$
$$+ \Delta Z_t dB_t$$
$$= -\left[\partial_x u(t, X_t)b(t, X_t, \tilde{Y}_t, \hat{Z}_t) + \frac{1}{2}\partial_{xx}^2 u(t, X_t)\sigma^2(t, X_t, \tilde{Y}_t) + f(t, X_t, \tilde{Y}_t, \hat{Z}_t)\right]dt$$
$$+ \left[\partial_x u(t, X_t)b(t, X_t, Y_t, Z_t) + \frac{1}{2}\partial_{xx}^2 u(t, X_t)\sigma^2(t, X_t, Y_t) + f(t, X_t, Y_t, Z_t)\right]dt$$
$$+ \Delta Z_t dB_t$$
$$= \left[\alpha_t \Delta Y_t + \beta_t[\hat{Z}_t - Z_t]\right]dt + \Delta Z_t dB_t,$$

where α, β are bounded. Note that

$$\hat{Z}_t - Z_t = \hat{Z}_t - \tilde{Z}_t + \Delta Z_t = \gamma_t \Delta Y_t + \Delta Z_t,$$

for some bounded γ. Then

$$d(\Delta Y_t) = \Big[(\alpha_t + \beta_t\gamma_t)\Delta Y_t + \beta_t\Delta Z_t\Big]dt + \Delta Z_t dB_t.$$

That is, $(\Delta Y, \Delta Z)$ satisfies the above linear BSDE. Note that $\Delta Y_T = u(T, X_T)$ $-g(X_T) = 0$. Then,

$$\Delta Y = 0, \quad \Delta Z = 0, \quad \text{and thus} \quad \hat{Z} = Z. \tag{8.3.5}$$

Therefore,

$$b(t, \Theta_t) = b(t, X_t, u(t, X_t), \partial_x u(t, X_t)\sigma(t, X_t, u(t, X_t))) = \tilde{b}(t, X_t),$$

$$\sigma(t, X_t, Y_t) = \sigma(t, X_t, u(t, X_t)) = \tilde{\sigma}(t, X_t).$$

That is, X satisfies SDE (8.3.4), and thus is unique. Moreover, it follows from (8.3.5) that (X, Y, Z) satisfies (8.3.3), and therefore (Y, Z) are also unique. ∎

8.3.2 The Decoupling Field

The main idea of Theorem 8.3.1 is to use the decoupling function u which decouples the originally coupled FBSDE (8.1.1) into a decoupled one (8.3.4). We now extend it to general non-Markov FBSDE (8.1.1).

Let $0 \le t_1 < t_2 \le T$, $\eta \in \mathbb{L}^2(\mathscr{F}_{t_1}, \mathbb{R}^{d_1})$, and $\varphi : \mathbb{R}^{d_1} \times \Omega \to \mathbb{R}^{d_2}$ such that $\varphi(x) \in \mathbb{L}^2(\mathscr{F}_{t_2}, \mathbb{R}^{d_2})$ for any $x \in \mathbb{R}^{d_1}$ and φ is uniformly Lipschitz continuous in x. Consider the following FBSDE:

$$\begin{cases} X_t = \eta + \displaystyle\int_{t_1}^t b_s(\Theta_s)ds + \int_{t_1}^t \sigma_s(\Theta_s)dB_s; \\ Y_t = \varphi(X_{t_2}) + \displaystyle\int_t^{t_2} f_s(\Theta_s)ds - \int_t^{t_2} Z_s dB_s, \end{cases} \quad t_1 \le t \le t_2. \tag{8.3.6}$$

Definition 8.3.2 *An $\mathbb{F}$-measurable random field $u : [0, T] \times \mathbb{R}^{d_1} \times \Omega \mapsto \mathbb{R}^{d_2}$ with $u_T(x) = g(x)$ is called a decoupling field of FBSDE (8.1.1) if there exists a constant $\delta > 0$ such that, for any $0 \le t_1 < t_2 \le T$ with $t_2 - t_1 \le \delta$ and any $\eta \in \mathbb{L}^2(\mathscr{F}_{t_1}, \mathbb{R}^{d_1})$, the FBSDE (8.3.6) with initial value η and terminal condition $\varphi(X_{t_2}) = u_{t_2}(X_{t_2})$ has a unique solution Θ, and it satisfies*

$$Y_t = u_t(X_t), \quad t \in [t_1, t_2], \quad \text{-a.s.} \tag{8.3.7}$$

Moreover, we say a decoupling field u is regular if it is uniformly Lipschitz continuous in x.

Remark 8.3.3 A decoupling field u, if it exists, is unique. Indeed, let δ be as in the definition and $0 = t_0 < \cdots < t_n = T$ be a time partition such that $t_i - t_{i-1} \leq \delta$, $i = 1, \cdots, n$. By (8.3.7), it is clear that u is determined by (8.2.8) on $[t_{n-1}, t_n]$ and thus is unique. In particular, $u_{t_{n-1}}$ is unique. Now applying the same arguments on FBSDE (8.3.6) on $[t_{n-2}, t_{n-1}]$ with terminal condition $u_{t_{n-1}}$, we see that u is unique on $[t_{n-2}, t_{n-1}]$. Repeating the arguments backwardly, we prove the uniqueness on $[0, T]$. ∎

Theorem 8.3.4 *Assume Assumption 8.1.1 holds, and FBSDE 8.1.1 has a decoupling field u. Then FBSDE (8.1.1) has a unique solution (X, Y, Z) and (8.3.7) holds on $[0, T]$.*

Proof We first prove existence. Let δ be a constant as in Definition 8.3.2 corresponding to the decoupling field u. Let $0 = t_0 < \cdots < t_n = T$ be partition of $[0, T]$ such that $t_i - t_{i-1} \leq \delta$ for $i = 1, \cdots, n$. Denote $X_0^0 := x$. For $i = 1, \cdots, n$, let Θ^i be the unique solution of the following FBSDE over $[t_{i-1}, t_i]$:

$$
\begin{cases}
X_t^i = X_{t_{i-1}}^{i-1} + \displaystyle\int_{t_{i-1}}^t b_s(\Theta_s^i) ds + \int_{t_{i-1}}^t \sigma_s(\Theta_s^i) dB_s; \\
Y_t^i = u_{t_i}(X_{t_i}^i) + \displaystyle\int_t^{t_i} f_s(\Theta_s^i) dr - \int_t^{t_i} Z_s^i dB_s.
\end{cases}
$$

Define

$$
\Theta_t := \sum_{i=1}^n \Theta_t^i 1_{[t_{i-1}, t_i)}(t) + \Theta_T^n 1_{\{T\}}(t).
$$

Note that $X_{t_{i-1}}^i = X_{t_{i-1}}^{i-1}$ and $Y_{t_{i-1}}^i = u_{t_{i-1}}(X_{t_{i-1}}^i) = u_{t_{i-1}}(X_{t_{i-1}}^{i-1}) = Y_{t_{i-1}}^{i-1}$. Then, X and Y are continuous, and one can check straightforwardly that (X, Y, Z) solves (8.1.1).

It remains to prove uniqueness. Let Θ be an arbitrary solution to FBSDE (8.1.1), and let $0 = t_0 < \cdots < t_n = T$ be as above. Since $u_T = g$, then on $[t_{n-1}, t_n]$ we have

$$
\begin{cases}
X_s = X_{t_{n-1}} + \displaystyle\int_{t_{n-1}}^s b_r(\Theta_r) dr + \int_{t_{n-1}}^s \sigma_r(\Theta_r) dB_r; \\
Y_s = u_{t_n}(X_{t_n}) + \displaystyle\int_s^{t_n} f_r(\Theta_r) dr - \int_s^{t_n} Z_r dB_r.
\end{cases}
$$

By our assumption, $Y_{t_{n-1}} = u_{t_{n-1}}(X_{t_{n-1}})$. By induction, one sees that, for $i = n, \cdots, 1$,

$$
\begin{cases}
X_t = X_{t_{i-1}} + \displaystyle\int_{t_{i-1}}^t b_s(\Theta_s) ds + \int_{t_{i-1}}^t \sigma_s(\Theta_s) dB_s; \\
Y_t = u_{t_i}(X_{t_i}) + \displaystyle\int_t^{t_i} f_s(\Theta_s) ds - \int_t^{t_i} Z_s dB_s;
\end{cases}
\quad t \in [t_{i-1}, t_i].
$$

Now, since $X_{t_0} = x$, for $i = 1, \cdots, n$, by forward induction one sees that Θ is unique on $[0, T]$.

Finally, for any $t \in [t_{i-1}, t_i]$, considering the FBSDE on $[t, t_i]$ we see that $Y_t = u_t(X_t)$. ∎

8.3.3 A Sufficient Condition for the Existence of Decoupling Field

Theorem 8.3.5 *Assume*

(i) $d = d_1 = d_2 = 1$; and σ does not depend on z;
(ii) Assumption 8.1.1 holds; b, σ, f, g are continuously differentiable in (x, y, z);
(iii) there exists a constant $c > 0$ such that

$$\partial_y \sigma \partial_z b \leq -c|\partial_y b + \partial_x \sigma \partial_z b + \partial_y \sigma \partial_z f|. \tag{8.3.8}$$

Then FBSDE (8.1.1) has a regular decoupling random field u, and consequently it admits a unique solution.

Proof We shall proceed in several steps.

Step 1. Let L denote the Lipschitz constant of b, σ, f with respect to (x, y, z), L_0 the Lipschitz constant of g with respect to x, and $L^* > L_0$ be a large constant which will be specified later. Note that in this case $\partial_z \sigma = 0$ and thus $c_0 = 0$ in (8.2.1). Let $\delta > 0$ be the constant in Theorem 8.2.1 corresponding to L and L^*. Set $0 = t_0 < \cdots < t_n = T$ such that $t_i - t_{i-1} \leq \delta$, $i = 1, \cdots, n$.

Consider the FBSDE (8.3.6) on $[t_{n-1}, t_n]$ with initial condition $\eta = x$ and terminal condition $\varphi = g$. Since $L^* > L^0$, by Theorem 8.2.1 this FBSDE is well posed on $[t_{n-1}, t_n]$ and thus one may define a random field u on $[t_{n-1}, t_n]$ via (8.2.8). By Theorem 8.2.3 we see that (8.3.3) holds on $[t_{n-1}, t_n]$ and u is Lipschitz continuous in x.

Our key step is the following more precise estimate on the Lipschitz constant of u under condition (8.3.8): for all $t \in [t_{n-1}, t_n]$,

$$\|\partial_x u_t\|_\infty \leq L_1 := e^{\bar{L}(t_n - t_{n-1})}(1 + L_0) - 1, \quad \text{where} \quad \bar{L} := 2L + L^2 + \frac{L + 2L^2}{4c}. \tag{8.3.9}$$

In particular, this implies that $u_{t_{n-1}}$ has a Lipschitz constant $L_1 > L_0$. We shall assume $L^* \geq L_1$. Then consider the FBSDE (8.3.6) on $[t_{n-2}, t_{n-1}]$ with initial condition $\eta = x$ and terminal condition $\varphi = u_{t_{n-1}}$, by the same arguments we will obtain u on $[t_{n-2}, t_{n-1}]$ and

$$\|\partial_x u_t\|_\infty \leq L_2 := e^{\bar{L}(t_n - t_{n-1})}(1 + L_1) - 1 = e^{\bar{L}(t_n - t_{n-2})}(1 + L_0) - 1,$$

$$t \in [t_{n-2}, t_{n-1}].$$

Repeating the arguments backwardly we obtain u on $[0, T]$ and

$$\|\partial_x u_t\|_\infty \le L_n := e^{\bar{L}T}(1 + L_0) - 1, \quad t \in [0, T]. \tag{8.3.10}$$

Now set $L^* := e^{\bar{L}T}(1 + L_0) - 1$, we see that u is a desired regular decoupling field.

Step 2. We prove (8.3.9) in this and the next steps. Since b, σ, f, g are continuously differentiable, by standard arguments one can easily see that u is differentiable in x with $\partial_x u_t(x) = \nabla Y_t^{t,x}$, for $(t, x) \in [t_{n-1}, t_n] \times \mathbb{R}$, where $\nabla\Theta := \nabla\Theta^{t,x}$ satisfies

$$\begin{cases} \nabla X_s = 1 + \displaystyle\int_t^s \left[\partial_x b_r(\Theta_r)\nabla X_r + \partial_y b_r(\Theta_r)\nabla Y_r + \partial_z b_r(\Theta_r)\nabla Z_r\right]ds \\ \qquad + \displaystyle\int_t^s \left[\partial_x \sigma_r(X_r, Y_r)\nabla X_r + \partial_y \sigma_r(X_r, Y_r)\nabla Y_r\right]dB_r; \\ \hfill s \in [t, t_n] \\ \nabla Y_s = \partial_x g(X_{t_n})\nabla X_{t_n} - \displaystyle\int_s^{t_n} \nabla Z_r dB_r \\ \qquad + \displaystyle\int_s^{t_n} \left[\partial_x f_r(\Theta_r)\nabla X_r + \partial_y f_r(\Theta_r)\nabla Y_r + \partial_z f_r(\Theta_r)\nabla Z_r\right]dr, \end{cases}$$

$$\tag{8.3.11}$$

In this step, we prove $\nabla X > 0$ on $[t, t_n]$. For this purpose, denote

$$\tau := \inf\{s > t : \nabla X_s = 0\} \wedge t_n, \quad \tau_n := \inf\{s > t : \nabla X_s = \frac{1}{n}\} \wedge t_n. \tag{8.3.12}$$

Then $\nabla X_s > 0$ in $[t, \tau)$. Define

$$\widehat{Y}_s := \nabla Y_s(\nabla X_s)^{-1},$$
$$\widehat{Z}_s := \nabla Z_s(\nabla X_s)^{-1} - \widehat{Y}_s[\partial_x \sigma_s(X_s, Y_s) + \partial_y \sigma_s(X_s, Y_s)\widehat{Y}_s], \quad s \in [t, \tau). \tag{8.3.13}$$

Applying Itô formula, it follows from lengthy but straightforward calculation that

$$d\widehat{Y}_s = -[\partial_x f(\Theta_s) + a_s(\widehat{Y}_s)\widehat{Y}_s + \theta_s(\widehat{Y}_s)\widehat{Z}_s]ds + \widehat{Z}_s dB_s, \tag{8.3.14}$$

where, by omitting Θ in the coefficients,

$$a(y) := [\partial_y f + \partial_x b + \partial_z f \partial_x \sigma] + [\partial_y b + \partial_z f \partial_y \sigma + \partial_z b \partial_x \sigma]y + \partial_z b \partial_y \sigma y^2,$$
$$\theta(y) := [\partial_z f + \partial_x \sigma] + [\partial_z b + \partial_y \sigma]y. \tag{8.3.15}$$

Then, for each n,

$$\widehat{Y}_s = \widehat{Y}_{\tau_n} + \int_s^{\tau_n} [\partial_x f_r(\Theta_r) + a_r(\widehat{Y}_r)\widehat{Y}_r + \theta_r(\widehat{Y}_r)\widehat{Z}_r]dr - \int_s^{\tau_n} \widehat{Z}_r dB_r, \quad s \in [t, \tau_n].$$

Note that $\widehat{Y}_s = \partial_x u_s(X_s)$ is bounded in $[t, \tau)$. Then one can easily check that

$$\mathbb{E}_t\Big[\int_t^{\tau_n} |\widehat{Z}_r|^2 dr\Big] \leq C < \infty,$$

where C is independent of n. Thus

$$\mathbb{E}_t\Big[\int_t^{\tau} |\widehat{Z}_r|^2 dr\Big] \leq C < \infty. \tag{8.3.16}$$

On the other hand, by (8.3.11) and (8.3.13) we have

$$\nabla X_s = 1 + \int_t^s \alpha_r \nabla X_r dr + \int_t^s \beta_r \nabla X_r dB_r, \quad s \in [t, \tau),$$

where, omitting the Θ in the coefficients again,

$$\alpha := \partial_x b + \partial_y b \widehat{Y}_r + \partial_z b\Big[\widehat{Z} + \widehat{Y}[\partial_x \sigma + \partial_y \sigma \widehat{Y}]\Big], \quad \beta := \partial_x \sigma + \partial_y \sigma \widehat{Y}$$

Then

$$\nabla X_\tau = \exp\Big(\int_t^\tau \beta_r dB_r + \int_t^\tau [\alpha_r - \frac{1}{2}|\beta_r|^2]dr\Big).$$

By the boundedness of $\widehat{Y}$ and (8.3.16), it is clear that

$$\int_t^\tau |\beta_r|^2 dr < \infty, \quad \int_t^\tau |\alpha_r - \frac{1}{2}|\beta_r|^2|dr < \infty.$$

Then $\nabla X_\tau > 0$, a.s. which, together with (8.3.12), implies that $\tau = t_n$ and $\nabla X_s > 0$, $s \in [t, t_n]$.

  Step 3. We now complete the proof for (8.3.9). By Step 2 and (8.3.14) we have

$$\widehat{Y}_s = \partial_x g(X_{t_n}) + \int_s^{t_n} [\partial_x f_r(\Theta_r) + a_r(\widehat{Y}_r)\widehat{Y}_r + \theta_r(\widehat{Y}_r)\widehat{Z}_r]dr$$

$$- \int_s^{t_n} \widehat{Z}_r dB_r, \quad s \in [t, t_n]. \tag{8.3.17}$$

Denote

$$\Gamma_s := \exp\Big(\int_t^s a_r(\tilde{Y}_r)ds\Big). \tag{8.3.18}$$

Recall the M^θ in (2.6.1), and by our assumptions $M^{\theta(Y)}$ is a true martingale. Applying Proposition 4.1.2 on (8.3.17), we have

$$\widehat{Y}_t = \mathbb{E}_t\Big[\Gamma_T M_T^{\theta(\widehat{Y})}\partial_x g(X_{t_n}) + \int_t^{t_n} \Gamma_s M_s^{\theta(\widehat{Y})}\partial_x f_s(\Theta_s)ds\Big].$$

By (8.3.8) we have

$$a(y) \leq [\partial_y f + \partial_x b + \partial_z f \partial_x \sigma] + [\partial_y b + \partial_z f \partial_y \sigma + \partial_z b \partial_x \sigma]y$$
$$- c|\partial_y b + \partial_z f \partial_y \sigma + \partial_z b \partial_x \sigma|y^2$$

$$\leq [\partial_y f + \partial_x b + \partial_z f \partial_x \sigma] + \frac{|\partial_y b + \partial_z f \partial_y \sigma + \partial_z b \partial_x \sigma|}{4c}$$

$$\leq 2L + L^2 + \frac{L + 2L^2}{4c} = \bar{L}.$$

Note further that $|\partial_x g| \leq L_0, |\partial_x f| \leq L$, then

$$|\partial_x u_t(x)| = |\widehat{Y}_t| \leq \mathbb{E}_t\Big[L_0 e^{\bar{L}(t_n-t)}M_T^{\theta(\widehat{Y})} + \int_t^{t_n} e^{\bar{L}(t_n-s)}L M_s^{\theta(\widehat{Y})}ds\Big]$$

$$= L_0 e^{\bar{L}(t_n-t)} + \int_t^{t_n} e^{\bar{L}(t_n-s)}L ds = L_0 e^{\bar{L}(t_n-t)} + \frac{L}{\bar{L}}[e^{\bar{L}(t_n-t)} - 1].$$

This implies (8.3.9) immediately. ∎

Remark 8.3.6

(i) If there exist some constants $c_1, c_2 > 0$ such that either $\partial_y \sigma \leq -c_1$ and $\partial_z b \geq c_2$, or $\partial_y \sigma \geq c_1$ and $\partial_z b \leq -c_2$, then (8.3.8) holds.

(ii) The following three classes of FBSDEs satisfy condition (8.3.8) with both sides equal to 0:

$$\begin{cases} X_t = x + \int_0^t b_s(X_s)ds + \int_0^t \sigma_s(X_s)dB_s; \\ Y_t = g(X_T) + \int_t^T f_s(\Theta_s)ds - \int_t^T Z_s dB_s. \end{cases}$$

$$\begin{cases} X_t = x + \int_0^t b_s(X_s, Z_s)ds + \int_0^t \sigma_s dB_s; \\ Y_t = g(X_T) + \int_t^T f_s(\Theta_s)ds - \int_t^T Z_s dB_s. \end{cases}$$

$$\begin{cases} X_t = x + \int_0^t b_s(X_s)ds + \int_0^t \sigma_s(X_s, Y_s)dB_s; \\ Y_t = g(X_T) + \int_t^T f_s(X_s, Y_s)ds - \int_t^T Z_s dB_s. \end{cases}$$

Also, instead of differentiability, it suffices to assume uniform Lipschitz continuity in these cases. ∎

We conclude this section with a comparison principle for FBSDEs.

Theorem 8.3.7 *Let* (b, σ, f^i, g^i), $i = 1, 2$, *satisfy all the conditions in Theorem 8.3.5 and* u^i *be the corresponding regular decoupling field. Assume* $f^1 \le f^2$ *and* $g^1 \le g^2$. *Then* $u^1 \le u^2$.

Proof Let δ be common for u^1 and u^2, and $0 = t_0 < \cdots < t_n = T$ be a time partition such that $t_i - t_{i-1} \le \delta$. We shall only prove $u^1 \le u^2$ on $[t_{n-1}, t_n]$, then by backward induction one can easily prove the inequality on the whole interval $[0, T]$. Thus, without loss of generality, we may assume $T \le \delta$. Moreover, it suffices to prove the comparison for $t = 0$. That is, fix x and let Θ^i be the solution to FBSDE with coefficients (b, σ, f^i, g^i) and initial value X_0^i, it suffices to prove $Y_0^1 \le Y_0^2$.

For this purpose, denote $\Delta\Theta := \Theta^1 - \Theta^2$ and $\Delta f := f^1 - f^2$, $\Delta g := g^1 - g^2$. Then we have

$$\begin{cases} \Delta X_t = \int_0^t [\alpha_s^1 \Delta X_s + \beta_s^1 \Delta Y_s + \gamma_s^1 \Delta Z_s]ds + \int_0^t [\alpha_s^2 \Delta X_s + \beta_s^2 \Delta Y_s]dB_s; \\ \Delta Y_t = \lambda \Delta X_T + \Delta g(X_T^1) + \int_t^T [\alpha_s^3 \Delta X_s + \beta_s^3 \Delta Y_s + \gamma_s^3 \Delta Z_s \\ \qquad\qquad + \Delta f_s(\Theta_s^1)]ds - \int_t^T \Delta Z_s dB_s, \end{cases} \qquad (8.3.19)$$

where $\alpha^i, \beta^i, \gamma^i$, and λ and bounded. Fix $\alpha^i, \beta^i, \gamma^i$, and λ, and consider the following linear FBSDE:

$$\begin{cases} X_t = 1 + \int_0^t [\alpha_s^1 X_s + \beta_s^1 Y_s + \gamma_s^1 Z_s]ds + \int_0^t [\alpha_s^2 X_s + \beta_s^2 Y_s]dB_s; \\ Y_t = \lambda \Delta X_T + \int_t^T [\alpha_s^3 X_s + \beta_s^3 Y_s + \gamma_s^3 Z_s]ds - \int_t^T Z_s dB_s. \end{cases} \qquad (8.3.20)$$

Since $T \le \delta$ and by our choice of δ, it follows from Theorem 8.2.1 that the above linear FBSDE is well posed. Moreover, by the arguments of Theorem 8.3.5 Step 2, we see that $X_t > 0$, $t \in [0, T]$. As in (8.3.13) we define

$$\cdot \ \widehat{Y}_s := Y_s(X_s)^{-1}, \ \widehat{Z}_s := Z_s(X_s)^{-1} - \widehat{Y}_s[\alpha_s^2 + \beta_s^2\widehat{Y}_s], \ s \in [0, T]. \qquad (8.3.21)$$

Then $\widehat{Y}$ is bounded and similar to (8.3.17) we have

$$\widehat{Y}_s = \lambda + \int_s^T [\alpha_r^3 + a_r(\widehat{Y}_r)\widehat{Y}_r + \theta_r(\widehat{Y}_r)\widehat{Z}_r]dr - \int_s^T \widehat{Z}_r dB_r, \quad s \in [0, T], \quad (8.3.22)$$

where

$$a(y) := [\beta^3 + \alpha^1 + \gamma^3\alpha^2] + [\beta^1 + \gamma^3\beta^2 + \gamma^1\alpha^2]y + \gamma^1\beta^2 y^2,$$
$$\theta(y) := [\gamma^3 + \alpha^2] + [\gamma^1 + \beta^2]y.$$

Now define

$$\delta Y := \Delta Y - \widehat{Y}\Delta X, \quad \delta Z := \Delta Z - \left[\widehat{Y}[\alpha^2\Delta X + \beta^2\Delta Y] + \widehat{Z}\Delta X\right]. \quad (8.3.23)$$

Then, by Itô formula,

$$d(\delta Y_s) = -[\alpha_s^3\Delta X_s + \beta_s^3\Delta Y_s + \gamma_s^3\Delta Z_s + \Delta f_s(\Theta_s^1)]ds$$
$$-\widehat{Y}_s[\alpha_s^1\Delta X_s + \beta_s^1\Delta Y_s + \gamma_s^1\Delta Z_s]ds$$
$$+[\alpha_s^3 + a_s(\widehat{Y}_s)\widehat{Y}_s + \theta_s(\widehat{Y}_s)\widehat{Z}_s]\Delta X_s ds - [\alpha_s^2\Delta X_s + \beta_s^2\Delta Y_s]\widehat{Z}_s ds + \delta Z_s dB_s.$$

$$(8.3.24)$$

By (8.3.23) we have

$$\Delta Y = \widehat{Y}\Delta X + \delta Y, \quad \Delta Z = \left[\widehat{Y}[\alpha^2 + \beta^2\widehat{Y}] + \widehat{Z}\right]\Delta X + \widehat{Y}\beta^2\delta Y + \delta Z.$$

Plug these into (8.3.24), and note that our choice of $\widehat{Y}, \widehat{Z}$ and the corresponding a, θ exactly imply that the coefficient of ΔX vanishes, and thus

$$d(\delta Y_s) = -\left[\beta_s^3\delta Y_s + \gamma_s^3[\widehat{Y}_s\beta_s^2\delta Y_s + \delta Z_s] + \Delta f_s(\Theta_s^1)\right]ds$$
$$-\widehat{Y}_s\left[\beta_s^1\delta Y_s + \gamma_s^1[\widehat{Y}_s\beta_s^2\delta Y_s + \delta Z_s]\right]ds - [\alpha_s^2\Delta X_s + \beta_s^2\Delta Y_s]\widehat{Z}_s ds + \delta Z_s dB_s$$
$$= -\left[\Delta f_s(\Theta_s^1) + [\beta_s^3 + \gamma_s^3\beta_s^2\widehat{Y}_s + \beta_s^1\widehat{Y}_s + \gamma_s^1\beta_s^2|\widehat{Y}_s|^2 + \beta_s^2\widehat{Z}_s]\delta Y_s\right.$$
$$\left. + [\gamma_s^3 + \gamma_s^1\widehat{Y}_s]\delta Z_s\right]ds + \delta Z_s dB_s.$$

Denote

$$\Gamma_t = \exp\left(\int_0^t [\beta_s^3 + \gamma_s^3\beta_s^2\widehat{Y}_s + \beta_s^1\widehat{Y}_s + \gamma_s^1\beta_s^2|\widehat{Y}_s|^2 + \beta_s^2\widehat{Z}_s]ds\right), \quad M_t := M_t^{\gamma^3 + \gamma^1\widehat{Y}}.$$

Then

$$d(\Gamma_s M_s \delta Y_s) = -\Gamma_s M_s \Delta f_s(\Theta_s^1) ds + \Gamma_s M_s \delta Z_s dB_s. \tag{8.3.25}$$

Since $\widehat{Y}$ is bounded, by (8.3.22) one can easily see that

$$\mathbb{E}_t\left[\int_t^T |\widehat{Z}_s|^2 ds \right] \le C < \infty.$$

Then applying Theorem 7.2.3 we have, for some $\varepsilon > 0$ small enough,

$$\mathbb{E}\left[e^{\varepsilon \int_0^T |\widehat{Z}_s|^2 ds} \right] < \infty.$$

Together with the boundedness of $\widehat{Y}$, this implies $\mathbb{E}[\sqrt{\int_0^T |\Gamma_s M_s \delta Z_s|^2 ds}] < \infty$ and thus $\Gamma_s M_s \delta Z_s dB_s$ is a true martingale. Then, by (8.3.25) and noting that $\delta Y_T = \Delta g(X_T^1)$, we have

$$\delta Y_0 = \mathbb{E}\left[\Gamma_T M_T \Delta g(X_T^1) + \int_0^T \Gamma_s M_s \Delta f_s(\Theta_s^1) ds \right] \le 0.$$

Since $\Delta X_0 = 0$, then $\Delta Y_0 = \delta Y_0 \le 0$.   ∎

Remark 8.3.8 We emphasize that the above comparison principle is for the decoupling field u, not for the solution Y. Clearly, under the notations in the theorem, we have $Y_0^1 = u_0^1(x) \le u_0^2(x) = Y_0^2$. However, since in general $X_t^1 \ne X_t^2$ for $t > 0$, we do not expect a general comparison between $Y_t^i = u_t^i(X_t^i)$, $i = 1, 2$.   ∎

8.4   The Method of Continuation

In this section we consider again general FBSDE (8.1.1) with random coefficients. The results can be extended to multidimensional cases. However, for simplicity we shall focus on 1-dimensional case again. Denote $\theta := (x, y, z)$ and $\Delta\theta := \theta_1 - \theta_2$. We adopt the following monotonicity assumptions.

Assumption 8.4.1 *There exists a constant $c > 0$ such that, for any θ_1, θ_2,*

$$[b_t(\theta_1) - b_t(\theta_2)]\Delta y + [\sigma_t(\theta_1) - \sigma_t(\theta_2)]\Delta z - [f_t(\theta_1) - f_t(\theta_2)]\Delta x \le -c[|\Delta x|^2$$
$$+ |\Delta y|^2 + |\Delta z|^2]; \ [g(x_1) - g(x_2)]\Delta x \ge 0. \tag{8.4.1}$$

Theorem 8.4.2 *Assume $d = d_1 = d_2 = 1$ and let Assumptions 8.1.1 and 8.4.1 hold. Then, FBSDE (8.1.1) admits a unique solution.*

Proof of uniqueness. Assume $\Theta^i, i = 1, 2$ are two solutions. Denote, for $\varphi = b, \sigma, f$,

$$\Delta\Theta := \Theta^1 - \Theta^2; \quad \Delta\varphi_t := \varphi_t(\Theta_t^1) - \varphi_t(\Theta_t^2); \quad \Delta g := g(X_T^1) - g(X_T^2).$$

Then,

$$\begin{cases} \Delta X_t = \int_0^t \Delta b_s ds + \int_0^t \Delta\sigma_s dB_s; \\ \Delta Y_t = \Delta g + \int_t^T \Delta f_s ds - \int_t^T \Delta Z_s dB_s. \end{cases}$$

Applying Itô formula on $\Delta X_t \Delta Y_t$ we have

$$d(\Delta X_t \Delta Y_t) = \Delta X_t d\Delta Y_t + \Delta Y_t d\Delta X_t + \Delta\sigma_t \Delta Z_t dt$$
$$= [-\Delta f_t \Delta X_t + \Delta b_t \Delta Y_t + \Delta\sigma_t \Delta Z_t]dt + [\Delta X_t \Delta Z_t + \Delta\sigma_t \Delta Y_t]dB_t.$$

Note that $\Delta X_0 = 0$ and $\Delta Y_T = \Delta g$. Thus,

$$E[\Delta g \Delta X_T] = E[\Delta Y_T \Delta X_T - \Delta Y_0 \Delta X_0]$$
$$= E\left[\int_0^T [-\Delta f_t \Delta X_t + \Delta b_t \Delta Y_t + \Delta\sigma_t \Delta Z_t]dt\right].$$

By Assumption 8.4.1 we get

$$0 \le -cE\left[\int_0^T [|\Delta X_t|^2 + |\Delta Y_t|^2 + |\Delta Z_t|^2]dt\right].$$

Then, obviously we have $\Delta X_t = \Delta Y_t = \Delta Z_t = 0$. ∎

The existence is first proved for a linear FSBDE.

Lemma 8.4.3 *Assume* $b^0, f^0 \in \mathbb{L}^{1,2}(\mathbb{F}, \mathbb{R})$, $\sigma^0 \in \mathbb{L}^2(\mathbb{F}, \mathbb{R})$, *and* $g^0 \in \mathbb{L}^2(\mathscr{F}_T, \mathbb{R})$. *Then, the following linear FBSDE admits a (unique) solution:*

$$\begin{cases} X_t = x + \int_0^t [-Y_s + b_s^0]ds + \int_0^t [-Z_s + \sigma_s^0]dB_s; \\ Y_t = X_T + g^0 + \int_t^T [X_s + f_s^0]ds - \int_t^T Z_s dB_s. \end{cases} \tag{8.4.2}$$

We note that, although we will not use it in the following proof, FBSDE (8.4.2) satisfies the monotonicity Assumption 8.4.1 with $c = 1$.

Proof We first notice that, if (X, Y, Z) is a solution,

$$X_t = X_T + \int_t^T [Y_s - b_s^0]ds + \int_t^T [Z_s - \sigma_s^0]dB_s.$$

Then, $\bar{Y} := Y - X$ satisfies

$$\bar{Y}_t = g^0 + \int_t^T [-\bar{Y}_s + f_s^0 + b_s^0]ds - \int_t^T [2Z_s + \sigma_s^0]dB_s.$$

We now solve (8.4.2) as follows. First, solve the following linear BSDE:

$$\bar{Y}_t = g^0 + \int_t^T [-\bar{Y}_s + f_s^0 + b_s^0]ds - \int_t^T \bar{Z}_s dB_s.$$

Next, set $Z := 2[\bar{Z} - \sigma^0]$ and solve the following linear (forward) SDE:

$$X_t = x + \int_0^t [-X_s - \bar{Y}_s + b_s^0]ds + \int_0^t [-\frac{1}{2}Z_s + \frac{3}{2}\sigma_s^0]dB_s.$$

Finally, let $Y := \bar{Y} + X$. Then, one can easily check that (X, Y, Z) is a solution to FBSDE (8.4.2). ∎

Now, we fix (b, σ, f, g) satisfying Assumptions 8.1.1 and 8.4.1. The method of continuation consists in building a bridge between FBSDEs (8.1.1) and (8.4.2). Namely, for $\alpha \in [0, 1]$, let

$$b_t^\alpha(\theta) := \alpha b_t(\theta) - (1-\alpha)y; \quad \sigma_t^\alpha(\theta) := \alpha\sigma_t(\theta) - (1-\alpha)z;$$

$$f_t^\alpha(\theta) := \alpha f_t(\theta) + (1-\alpha)x; \quad g^\alpha(x) := \alpha g(x) + (1-\alpha)x.$$

We note that $(b^\alpha, \sigma^\alpha, f^\alpha, g^\alpha)$ satisfies Assumptions 8.1.1 and 8.4.1 with constant

$$c_\alpha := \alpha c + 1 - \alpha \geq \min(c, 1). \tag{8.4.3}$$

Let FBSDE(α) denote the class of FBSDEs taking the following form with some $(b^0, \sigma^0, f^0, g^0)$:

$$\begin{cases} X_t = x + \int_0^t [b_s^\alpha(\Theta_s) + b_s^0]ds + \int_0^t [\sigma_s^\alpha(\Theta_s) + \sigma_s^0]dB_s; \\ Y_t = g^\alpha(X_T) + g^0 + \int_t^T [f_s^\alpha(\Theta_s) + f_s^0]ds - \int_t^T Z_s dB_s. \end{cases}$$

Then, FBSDE (8.4.2) is in class FBSDE(0), and FBSDE (8.1.1) is in class FBSDE(1) (with $b^0 = \sigma^0 = f^0 = g^0 = 0$). We say FBSDE($\alpha$) is solvable if the FBSDE has a solution for any $b^0, f^0 \in \mathbb{L}^{1,2}(\mathbb{F}, \mathbb{R})$, $\sigma^0 \in \mathbb{L}^2(\mathbb{F}, \mathbb{R})$, and $g^0 \in \mathbb{L}^2(\mathscr{F}_T, \mathbb{R})$. The following lemma plays a crucial role.

Lemma 8.4.4 *Assume $d = d_1 = d_2 = 1$ and let Assumptions 8.1.1 and 8.4.1 hold. If FBSDE(α_0) is solvable, then there exists $\delta_0 > 0$, depending only on the Lipschitz constants of (b, σ, f, g) and the constant c in Assumption 8.4.1, such that FBSDE(α) is solvable for any $\alpha \in [\alpha_0, \alpha_0 + \delta_0]$.*

Before we prove this lemma, we use it to prove the existence part of Theorem 8.4.2.

Proof of Existence in Theorem 8.4.2. By Lemma 8.4.3 FBSDE(0) is solvable. Assume $(n-1)\delta_0 < T \leq n\delta_0$. Applying Lemma 8.4.4 n times we know FBSDE(1) is also solvable. Therefore, FBSDE (8.1.1) admits a solution. ∎

Proof of Lemma 8.4.4. For any $\alpha \in [\alpha_0, \alpha_0 + \delta_0]$ where $\delta_0 > 0$ will be determined later, denote $\delta := \alpha - \alpha_0 \leq \delta_0$. For any $b^0, f^0 \in \mathbb{L}^{1,2}(\mathbb{F}, \mathbb{R})$, $\sigma^0 \in \mathbb{L}^2(\mathbb{F}, \mathbb{R})$, and $g^0 \in \mathbb{L}^2(\mathscr{F}_T, \mathbb{R})$, denote $\Theta^0 := (0,0,0)$ and for $n = 0, 1, \cdots$,

$$b_t^{n,0} := \delta[Y_t^n + b_t(\Theta_t^n)] + b_t^0; \quad \sigma_t^{n,0} := \delta[Z_t^n + \sigma_t(\Theta_t^n)] + \sigma_t^0;$$

$$f_t^{n,0} := \delta[X_t^n + f_t(\Theta_t^n)] + f_t^0; \quad g^{n,0} := \delta[-X_T^n + g(X_T^n)] + g^0,$$

and let Θ^{n+1} be the solution to the following FBSDE:

$$\begin{cases} X_t^{n+1} = x + \int_0^t [b_s^{\alpha_0}(\Theta_s^{n+1}) + b_s^{n,0}]ds + \int_0^t [\sigma_s^{\alpha_0}(\Theta_s^{n+1}) + \sigma_s^{n,0}]dB_s; \\ Y_t^{n+1} = g^{\alpha_0}(X_T^{n+1}) + g^{n,0} + \int_t^T [f_s^{\alpha_0}(\Theta_s^{n+1}) + f_s^{n,0}]ds - \int_t^T Z_s^{n+1}dB_s. \end{cases}$$

By our assumption FBSDE(α_0) is solvable and thus Θ^n are well defined for all $n \geq 1$. Denote $\Delta\Theta^n := \Theta^{n+1} - \Theta^n$. Then,

$$d\Delta X_t^n = \left[[b_t^{\alpha_0}(\Theta_t^{n+1}) - b_t^{\alpha_0}(\Theta_t^n)] + \delta[\Delta Y_t^{n-1} + b_t(\Theta_t^n) - b_t(\Theta_t^{n-1})] \right]dt$$

$$+ \left[[\sigma_t^{\alpha_0}(\Theta_t^{n+1}) - \sigma_t^{\alpha_0}(\Theta_t^n)] + \delta[\Delta Z_t^{n-1} + \sigma_t(\Theta_t^n) - \sigma_t(\Theta_t^{n-1})] \right]dB_t;$$

$$d\Delta Y_t^n = -\left[[f_t^{\alpha_0}(\Theta_t^{n+1}) - f_t^{\alpha_0}(\Theta_t^n)] + \delta[\Delta X_t^{n-1} + f_t(\Theta_t^n) - f_t(\Theta_t^{n-1})] \right]dt + \Delta Z_t^n dB_t.$$

Applying Itô formula we have

$$d(\Delta X_t^n \Delta Y_t^n) = [\cdots]dB_t + \left[-[f_t^{\alpha_0}(\Theta_t^{n+1}) - f_t^{\alpha_0}(\Theta_t^n)]\Delta X_t^n \right.$$

$$+ [b_t^{\alpha_0}(\Theta_t^{n+1}) - b_t^{\alpha_0}(\Theta_t^n)]\Delta Y_t^n + [\sigma_t^{\alpha_0}(\Theta_t^{n+1}) - \sigma_t^{\alpha_0}(\Theta_t^n)]\Delta Z_t^n \right]dt$$

$$+ \delta\left[-[\Delta X_t^{n-1} + f_t(\Theta_t^n) - f_t(\Theta_t^{n-1})]\Delta X_t^n \right.$$

$$+ [\Delta Y_t^{n-1} + b_t(\Theta_t^n) - b_t(\Theta_t^{n-1})]\Delta Y_t^n$$

$$+ [\Delta Z_t^{n-1} + \sigma_t(\Theta_t^n) - \sigma_t(\Theta_t^{n-1})]\Delta Z^n \right]dt.$$

Recall (8.4.3), we have

$$
\begin{aligned}
d(\Delta X_t^n \Delta Y_t^n) &\leq [\cdots]dB_t - c_{\alpha_0} |\Delta\Theta_t^n|^2 dt \\
&\quad + C\delta\Big[|\Delta X_t^{n-1}||\Delta X_t^n| \\
&\quad + |\Delta Y_t^{n-1}||\Delta Y_t^n| + |\Delta Z_t^{n-1}||\Delta Z_t^n| dt \\
&\leq [\cdots]dB_t + \Big[(\delta - c_{\alpha_0})|\Delta\Theta_t^n|^2 + C\delta|\Delta\Theta_t^{n-1}|^2\Big]dt.
\end{aligned}
$$

Note that $\Delta X_0^n = 0$ and

$$
\Delta X_T^n \Delta Y_T^n = \Delta X_T^n [g^{\alpha_0}(X_T^{n+1}) - g^{\alpha_0}(X_T^n)] \geq 0.
$$

Then,

$$
\mathbb{E}\Big[(c_{\alpha_0} - \delta)\int_0^T |\Delta\Theta_t^n|^2 dt\Big] \leq C\delta\mathbb{E}\Big[\int_0^T |\Delta\Theta_t^{n-1}|^2 dt\Big].
$$

Without loss of generality we assume $c \leq 1$. Then, $c_{\alpha_0} \geq c$ and thus

$$
(c - \delta)\mathbb{E}\Big[\int_0^T |\Delta\Theta_t^n|^2 dt\Big] \leq C_1\delta\mathbb{E}\Big[\int_0^T |\Delta\Theta_t^{n-1}|^2 dt\Big].
$$

Choose $\delta_0 := \frac{c}{1+4C_1} > 0$. Then, for any $\delta \leq \delta_0$, we have $C_1\delta \leq \frac{1}{4}(c - \delta)$. Therefore,

$$
\mathbb{E}\Big[\int_0^T |\Delta\Theta_t^n|^2 dt\Big] \leq \frac{1}{4}\mathbb{E}\Big[\int_0^T |\Delta\Theta_t^{n-1}|^2 dt\Big].
$$

By induction we get

$$
\mathbb{E}\Big[\int_0^T |\Delta\Theta_t^n|^2 dt\Big] \leq \frac{C}{4^n} \quad\text{and thus}\quad \Big(\mathbb{E}\Big[\int_0^T |\Delta\Theta_t^n|^2 dt\Big]\Big)^{\frac{1}{2}} \leq \frac{C}{2^n}.
$$

Then, for any $n > m$,

$$
\Big(\mathbb{E}\Big[\int_0^T |\Theta_t^n - \Theta_t^m|^2 dt\Big]\Big)^{\frac{1}{2}} \leq \sum_{i=m}^{n-1}\Big(\mathbb{E}\Big[\int_0^T |\Delta\Theta_t^i|^2 dt\Big]\Big)^{\frac{1}{2}} \leq C\sum_{i=m}^{n-1}\frac{1}{2^i} \leq \frac{C}{2^m} \to 0,
$$

as $m \to \infty$. So, there exists Θ such that

$$
\lim_{n\to\infty} \mathbb{E}\Big[\int_0^T |\Theta_t^n - \Theta_t|^2 dt\Big] = 0.
$$

Note that, as $n \to \infty$,

$$
\begin{aligned}
b_t^{\alpha_0}(\Theta_t^{n+1}) + b_t^{n,0} &= b_t^{\alpha_0}(\Theta_t^{n+1}) + \delta[Y_t^n + b_t(\Theta_t^n)] + b_t^0 \\
&\to b_t^{\alpha_0}(\Theta_t) + \delta[Y_t + b_t(\Theta_t)] + b_t^0 = b_t^{\alpha_0+\delta}(\Theta_t) + b_t^0 = b_t^\alpha(\Theta_t) + b_t^0.
\end{aligned}
$$

Similar results hold for the other terms. Thus, Θ satisfies FBSDE(α) for any $\alpha \in [\alpha_0, \alpha_0 + \delta_0]$. Finally, it is straightforward to check that $\|\Theta\| < \infty$. ∎

Remark 8.4.5

(i) The monotonicity conditions in Assumption 8.4.1 can be replaced by

$$[b_t(\theta_1) - b_t(\theta_2)]\Delta y + [\sigma_t(\theta_1) - \sigma_t(\theta_2)]\Delta z - [f_t(\theta_1) - f_t(\theta_2)]\Delta x \geq c[|\Delta x|^2$$
$$+|\Delta y|^2 + |\Delta z|^2]; [g(x_1) - g(x_2)]\Delta x \leq 0. \qquad (8.4.4)$$

and we can still obtain existence and uniqueness results in a similar way.

(ii) In the case that $d = d_1 = d_2 = 1$, it follows from Ma, Wu, Zhang, & Zhang [144] that the results hold true even if $c = 0$ in (8.4.1) and (8.4.4). This is the case considered in Jeanblanc & Yu [116]. ∎

8.5 Bibliographical Notes

The coupled FBSDE was first studied by Antonelli [1], and a classical reference is the book Ma & Yong [148]. For the fixed point approach, the materials in Section 8.2 with small time duration is based on [1]; the result is extended to arbitrary T but with certain monotonicity condition or weakly coupling property by Pardoux & Tang [171]. The decoupling approach was first introduced by Ma, Protter, & Yong [141], see the four step scheme in Section 8.3.1. For the existence of classical solutions of PDE (8.3.2) required in Theorem 8.3.1, we refer to the book Ladyzenskaja, Solonnikov, & Uralceva [133]. The idea of Section 8.3.2 is due to Delarue [55], which provided the uniform estimates for the Lipschitz constant of the decoupling function u in the nondegenerate Markovian case by using PDE arguments. Section 8.3.3 is based on Zhang [245], which extends the idea to degenerate non-Markovian case, but requires all processes to be scalar. For all three works above, σ is required to be independent of z. The work Ma, Wu, Zhang, & Zhang [144] extended [245] further and allows σ to depend on z, still requiring scalar processes. The works Yong [240, 241] consider multiple dimensions, but restricted to linear case. For the method of continuation, it works in multiple dimensional situation as well and the key assumption is the monotonicity condition. We refer to Hu & Peng [109], Peng & Wu [187], and Yong [239] on this approach. We also refer to Ma & Yong [146] for another method based on stochastic control.

Besides the strong well-posedness, there have been serious efforts on weak solutions of FBSDEs, see, e.g., Antonelli & Ma [2], Delarue & Guatteri [56], Ma, Zhang, & Zheng [153], and Ma & Zhang [152]. In the spirit of weak solutions, the recent work Wang & Zhang [231] studied FBSDEs in weak formulation. The numerical method for coupled FBSDE is also much more challenging than that for BSDEs, especially in high dimensional case. We refer to Douglas, Ma, & Protter [66], Milstein & Tretyakov [157], and Ma, Shen, & Zhao [143] on the finite difference method, Cvitanic & Zhang [51] on the steepest decent method, Delarue & Menozzi [57] on the quantization method, Bender & Zhang [14] on the least square regression method, and Guo, Zhang, & Zhuo [99] on monotone schemes.

8.6 Exercises

Problem 8.6.1 Consider a special case of the stochastic control problem (4.5.7)–(4.5.8) in strong formulation:

$$X_t^k = x + \int_0^t k_s ds + B_t; \ V_0^S := \sup_{k \in \mathbb{L}^2(\mathbb{F}^B)} J_S(k) \tag{8.6.1}$$

$$\text{where } J_S(k) := \mathbb{E}\Big[g(X_T^k) - \tfrac{1}{2}\int_0^T [g''(X_t^k) + |g'(X_t^k)|^2 + |k_t|^2]dt\Big].$$

Assume $d = d_1 = d_2 = 1$ and g is smooth enough with all involved derivatives bounded.

(i) Derive the coupled FBSDE (8.1.8) corresponding to this problem;
(ii) Show that the FBSDE (8.1.8) derived in (i) is well posed, in particular it holds that $\overline{Y}_t = g'(X_t)$.
(iii) Find some sufficient condition on g such that $k_t := \overline{Y}_t$ is indeed an optimal control, and find the corresponding V_0^S.
(iv) Formulate the problem (8.6.1) in weak formulation, in the spirit of (4.5.12). Is the V_0 defined in (4.5.12) equal to V_0^S? ∎

Problem 8.6.2 Let Assumption 8.1.1 hold. Assume further that FBSDE (8.1.1) is weekly coupling, in the sense that either $\partial_y b, \partial_z b, \partial_y \sigma, \partial_z \sigma$ are small enough, or $\partial_x g, \partial_x f$ are small enough (which may depend on T). Show that FBSDE (8.1.1) admits unique solution (X, Y, Z). ∎

Problem 8.6.3 Prove directly that FBSDE (8.2.9) is well posed if and only if $\lambda \gamma \neq 1$. ∎

Problem 8.6.4 Consider the following Markov FBSDE with $d_2 = 1$:

$$\begin{cases} X_t = x + \int_0^t \sigma(s, X_s, Y_s) dB_s; \\ Y_t = g(X_T) + \int_t^T f(s, X_s, Y_s) ds - \int_t^T Z_s dB_s. \end{cases} \tag{8.6.2}$$

Assume all the coefficients are smooth enough with bounded derivatives, and the corresponding PDE (8.3.2) has a classical solution u which is also smooth enough with all the related derivatives bounded. Then it follows Theorem 8.3.1 that FBSDE (8.6.2). Given $y_0 \in \mathbb{R}$ and $Z^0 \in \mathbb{L}^2(\mathbb{F}^B)$, denote

$$V(y_0, Z^0) := \frac{1}{2}\mathbb{E}\Big[|Y_T^0 - g(X_T^0)|^2\Big], \tag{8.6.3}$$

where (X^0, Y^0) is the solution to the following SDE:

$$X_t^0 = x + \int_0^t \sigma(s, X_s^0, Y_s^0)dB_s, \quad Y_t^0 = y_0 - \int_0^t f(s, X_s^0, Y_s^0)ds + \int_0^t Z_s^0 dB_s. \quad (8.6.4)$$

Show that

$$\mathbb{E}\left[|(X - X^0)_T^*|^2 + |(Y - Y^0)_T^*|^2 + \int_0^T |Z_t - Z_t^0|^2 dt\right] \le CV(y_0, Z^0). \quad (8.6.5)$$

■

Part III
The Fully Nonlinear Theory of BSDEs

Chapter 9
Stochastic Calculus Under Weak Formulation

The fully nonlinear theory will be built on the canonical space under weak formulation, which has many advantages both in theory and in applications. In this chapter we present some basic materials crucial for the theory. While we will try our best to make the presentation self-contained, due to the limit of the pages we will have to borrow some results from the standard literature of stochastic calculus, e.g., Karatzas & Shreve [117] and Revuz & Yor [206].

9.1 Some Motivations for Weak Formulation

In this section we provide some heuristic arguments to justify why we prefer weak formulation than strong formulation when there are controls (or uncertainty) involved. The rigorous treatment will be carried out in the rest of the book. We remark that when there is only drift control, as we see in Section 4.5.2 the stochastic optimization problem under weak formulation leads to BSDEs. The fully nonlinear theory in Part III is mainly for problems with diffusion control. In the control literature, see, e.g., Yong & Zhou [242], people consider open loop controls and close loop controls. For open controls the strong formulation is natural, while for close loop controls, the weak formulation is more convenient. We shall remark though, as we will carry out rigorously later, the control set in weak formulation could be larger than the set of close loop controls. While the choice of the controls should in general depend on the specific applications, in this section we will explain that the weak formulation and/or close loop controls are more appropriate than strong formulation and/or open loop controls, from several considerations we have.

For simplicity, in this section we mainly restrict to Markov case, but all the arguments remain true in path dependent case after appropriate modifications. Moreover, unless otherwise stated, in this section we shall always use one-dimensional notations.

© Springer Science+Business Media LLC 2017
J. Zhang, *Backward Stochastic Differential Equations*, Probability Theory and Stochastic Modelling 86, DOI 10.1007/978-1-4939-7256-2_9

9.1.1 Practical Considerations on Information

Consider a diffusion:

$$X_t = x + \int_0^t b(s, X_s)ds + \int_0^t \sigma(s, X_s)dB_s, \qquad (9.1.1)$$

where B is a Brownian motion and X is the state process. When there is a control k, the coefficients b and σ may depend on k. In strong formulation, the probability measure, denoted as $\mathbb{P}_0$, and the Brownian motion B are given, and one needs to find a state process X adapted to the filtration $\mathbb{F}^B$ (or to a given larger filtration $\mathbb{F}$). In weak formulation, the state process X is given (typically on canonical space as we will see in details), and one needs to find a probability measure $\mathbb{P}$ such that $\frac{1}{\sigma(t,X_t)}[dX_t - b(t, X_t)dt]$ is a $\mathbb{P}$-Brownian motion (assuming $\sigma > 0$), or equivalently, one wants to find the distribution of X.

In applications, one crucial issue is the information we observe. In particular, the controls can depend only on the observed information. Consider the financial model in Sections 2.8 and 4.5.1, where $X = S$ is the stock price. In the Black-Scholes model, $\mathbb{F}^S = \mathbb{F}^B$ and thus B and S contain the same information. However, in many general models, they may contain different information. As explained in Section 2.8.3, in practice people typically observe S, but not B which is the random noise. Indeed, one possible justification for using Brownian motion B to model stock price is the central limit theorem, where the convergence is in distribution, rather than pointwise. In that case, the noise B does not exist physically but is only used to model the distribution of S, which is exactly the spirit of weak formulation. So in the situation where people observe the state process X, but not the noise B directly (especially when the noise B does not exist physically), it is more natural and more convenient to use weak formulation.

We remark that weak formulation is largely used in economics literature. At below we provide an example where weak formulation is required: the principal agent problem. In this problem, the principal hires an agent to work for her. The agent controls the state process:

$$X_t^k = x + \int_0^t b(s, X_s^k, k_s)ds + \int_0^t \sigma(s, X_s^k, k_s)dB_s, \qquad (9.1.2)$$

in the form of strong formulation at the moment. The principal needs to pay the agent at time T through certain contract ξ. Then the agent's problem is: given a contract ξ,

$$V_0^A(\xi) := \sup_k \mathbb{E}^{\mathbb{P}_0}\left[U_A(\xi) - \int_0^T f(t, X_t^k, k_t)dt\right]. \qquad (9.1.3)$$

Here U_A is the agent's utility function, and f is his running cost. Assuming the agent has optimal control k^ξ, which depends on ξ, then the principal's problem is:

$$V_0^P := \sup_\xi \mathbb{E}^{\mathbb{P}_0}\left[U_P(X_T^{k^\xi} - \xi)\right] \text{ subject to } V_0^A(\xi) \ge R, \qquad (9.1.4)$$

where R is the individual reservation which stands for the market value of the agent.

One crucial issue of the problem is: what information do the principal and agent have and consequently what information can ξ and k depend on? The standard literature assumes moral hazard: the principal observes only the agent's outcome X^k, not the agent's action k. We note that, even if $\mathbb{F}^{X^k} = \mathbb{F}^B$ for some admissible control k, the principal does not observe B because she does not know k. So ξ can only depend on X^k, not on B, and thus it is natural to use weak formulation for the principal's problem. We remark that, the fact that ξ depends on X^k is also crucial for providing incentives to the agent, which is one of the key issues in contract theory. If $\xi = \xi(B)$ (assuming the principal could observe B), then the agent will be paid randomly, which has nothing to do with his effort k but is only by luck. Then in (9.1.3) the agent will just try to minimize his cost f. This is obviously not desirable in practice. So to provide the incentive the contract should depend on X, and due to moral hazard the contract can depend only on X.

The agent's control k depends on the information the agent observes. The possible choices are $k(X)$, $k(B)$, and $k(X, B)$, and it depends on the real situation in applications. Especially, as explained in Section 2.8.3, when the noise B does not exist physically but is used only to model the distribution of X, the agent has to use weak formulation. Moreover, mathematically the weak formulation is much easier for the agent's problem. Recall again that the contract is in the form $\xi(X)$. In the strong formulation, the agent's problem (9.1.3) is:

$$V_0^A(\xi) := \sup_k \mathbb{E}^{\mathbb{P}_0}\left[U_A(\xi(X_\cdot^k)) - \int_0^T f(t, X_t^k, k_t)dt \right].$$

Note that ξ depends on the whole path of X^k and as a control typically ξ does not have good regularity. It is very difficult to solve the above optimization problem. The weak formulation, however, is much easier. For simplicity let us assume $\sigma = 1$ and b is bounded. Fix $X_t = B_t$ (we will add the initial value x later), and for each k, denote

$$\frac{d\mathbb{P}^k}{d\mathbb{P}_0} := M_T^k := \exp\left(\int_0^T b(s, x + X_s, k_s)dB_s - \frac{1}{2}\int_0^T |b(s, x + X_s, k_s)|^2 ds \right).$$

That is, $B_t^k := B_t - \int_0^t b(s, x + X_s, k_s)ds$ is a $\mathbb{P}^k$-Brownian motion. Then in weak formulation the agent's problem becomes:

$$\tilde{V}_0^A(\xi) := \sup_k \mathbb{E}^{\mathbb{P}^k}\left[U_A(\xi(x + X_\cdot)) - \int_0^T f(t, x + X_t, k_t)dt \right]$$

$$= \sup_k \mathbb{E}^{\mathbb{P}_0}\left[M_T^k \left[U_A(\xi(x + X_\cdot)) - \int_0^T f(t, x + X_t, k_t)dt \right] \right]. \quad (9.1.5)$$

Note that both ξ and k depend on X which is a Brownian motion under $\mathbb{P}$, so by using BSDE the above problem is rather easy to solve, see Section 4.5.2.

9.1.2   Stochastic Controls

We have explained in Section 4.5.2 that for stochastic control problem weak formulation is easier than strong formulation. The main message is that, under mild conditions, the values of the optimization problem under strong formulation and weak formulation are typically equal; however, it is much more likely to have optimal control under weak formulation than under strong formulation. In this subsection we shall explore these ideas further.

Let B be a $\mathbb{P}_0$-Brownian motion, and $\mathscr{H}$ an appropriate set of admissible controls k taking values in $\mathbb{K}$. We consider the following control problem in strong formulation:

$$V_0^S := \sup_{k \in \mathscr{H}} J_S(k), \tag{9.1.6}$$

where $X_t^k := \displaystyle\int_0^t b(s, X_s^k, k_s)ds + \int_0^t \sigma(s, X_s^k, k_s)dB_s,$

$$J_S(k) := \mathbb{E}^{\mathbb{P}_0}\Big[g(X_T^k) + \int_0^T f(t, X_t^k, k_t)dt\Big].$$

Under mild conditions, by standard literature (see, e.g., Fleming & Soner [90] and Yong & Zhou [242]) we have $V_0^S = v(0,0)$, where v is the unique viscosity solution to the following HJB equation:

$$\partial_t v(t, x) + H(t, x, \partial_x v, \partial_{xx}^2 v) = 0, \quad v(T, x) = g(x), \tag{9.1.7}$$

where $H(t, x, z, \gamma) := \displaystyle\sup_{k \in \mathbb{K}}\Big[\frac{1}{2}\sigma^2(t, x, k)\gamma + b(t, x, k)z + f(t, x, k)\Big].$

We now introduce the weak formulation corresponding to (9.1.6). Let X be a given process, and denote by $\mathbb{P}^k$ a probability measure such that, under $\mathbb{P}^k$, X is a semimartingale and $\frac{1}{\sigma(t, X_t, k_t)}[dX_t - b(t, X_t, k_t)dt]$ is a Brownian motion. The optimization problem under weak formulation is:

$$V_0^W := \sup_{k \in \mathscr{H}} J_W(k), \quad \text{where} \quad J_W(k) := \mathbb{E}^{\mathbb{P}^k}\Big[g(X_T) + \int_0^T f(t, X_t, k_t)dt\Big]. \tag{9.1.8}$$

Under mild conditions, we still have $V_0^W = v(0,0)$, where v is the unique viscosity solution to the same PDE (9.1.7). In particular, we have

$$V_0^S = V_0^W. \tag{9.1.9}$$

The above arguments rely on the PDE. We now try to understand the equality (9.1.9) directly. Let $\mathscr{H}_0$ denote the subset of $k \in \mathscr{H}$ such that k is piecewise constant, and define

$$\tilde{V}_0^S := \sup_{k \in \mathscr{H}_0} J_S(k), \quad \tilde{V}_0^W := \sup_{k \in \mathscr{H}_0} J_W(k). \tag{9.1.10}$$

Then clearly $\tilde{V}_0^S \leq V_0^S$ and $\tilde{V}_0^W \leq V_0^W$. One can easily show that $\tilde{V}_0^S = \tilde{V}_0^W$. Moreover, when f and g are continuous, then by density arguments we see that $\tilde{V}_0^S = V_0^S$ and $\tilde{V}_0^W = V_0^W$, which implies (9.1.9) immediately. However, we shall note that the density argument in weak formulation is actually quite tricky and may require some additional technical conditions. See Problem 9.6.10 for some results along this direction.

The above discussion is about the value of the optimization problem, for which the difference between the two formulations is not important since anyway the values are equal. We now turn to the optimal control and provide two arguments to explain why the optimal control is more likely to exist in weak formulation.

First, let us assume the PDE (9.1.7) has a classical solution v, and the Hamiltonian H in (9.1.7) has an optimal argument $k^* = I(t, x, z, \gamma)$ for some measurable function I. Then, intuitively, the optimal control should take the form $k_t^* = I(t, X_t^*, \partial_x v(t, X_t^*), \partial_{xx}^2 v(t, X_t^*))$. That is, the optimal control depends on X, rather than on B directly, or say, the optimal control should be close loop, not open loop. Moreover, to find the optimal control we need to solve the following SDE:

$$X_t^* = \int_0^t \hat{b}(s, X_s^*)ds + \int_0^t \hat{\sigma}(s, X_s^*)dB_s, \qquad (9.1.11)$$

where $\hat{\varphi}(s, x) := \varphi(s, x, I(s, x, \partial_x v(s, x), \partial_{xx}^2 v(s, x))$ for $\varphi = b, \sigma$.

Roughly speaking, the existence of optimal control in strong formulation amounts to saying the above SDE admits a strong solution, while the weak formulation requires only weak solution. So it is more likely to have optimal control in weak formulation.

Next, denote $\mathscr{P}_S := \{\mathbb{P}_0 \circ (X^k)^{-1} : k \in \mathscr{K}\}$, $\mathscr{P}_W := \{\mathbb{P}^k : k \in \mathscr{K}\}$. Then, assuming $f = 0$ for simplicity, we have

$$V_0^S = \sup_{\mathbb{P} \in \mathscr{P}_S} J(\mathbb{P}), \quad , \quad V_0^W = \sup_{\mathbb{P} \in \mathscr{P}_W} J(\mathbb{P}), \quad \text{where} \quad J(\mathbb{P}) := \mathbb{E}^{\mathbb{P}}[g(X_T)].$$

When g is continuous, J is continuous in $\mathbb{P}$ under the topology induced by the weak convergence in the sense of Definition 9.2.13 below. As we will see later, the set $\mathscr{P}_W$ is compact under this topology, and thus V_0^W has an optimal argument $\mathbb{P}^* \in \mathscr{P}_W$. However, the set $\mathscr{P}_S$ is typically not compact, which explains why optimal control does not exist in general in strong formulation. We shall remark though, to obtain the weak compactness of $\mathscr{P}_W$ we may need to extend the set of controls and thus the control k may not be $\mathbb{F}^X$-measurable, see Definition 9.2.9 and Remark 9.2.10 (ii) below.

9.1.3 Two Person Zero-Sum Stochastic Differential Games

For control problem, if we are interested only in the value, then as mentioned we may use either strong formulation or weak formulation. However, for stochastic differential games, the close loop control is necessary even for the value itself and thus we shall use weak formulation. Let $\mathbb{K}_1, \mathbb{K}_2$ be two sets and $\mathcal{K}_1, \mathcal{K}_2$ be the corresponding sets of admissible controls. In strong formulation, the upper and lower values of a zero sum game are defined as: for $k := (k^1, k^2)$, and recalling the X^k and $J_S(k)$ in (9.1.6),

$$\overline{V}_0^S := \inf_{k^1 \in \mathcal{K}_1} \sup_{k^2 \in \mathcal{K}_2} J_S(k^1, k^2), \quad \underline{V}_0^S := \sup_{k^2 \in \mathcal{K}_2} \inf_{k^1 \in \mathcal{K}_1} J_S(k^1, k^2). \quad (9.1.12)$$

It is clear that $\underline{V}_0^S \le \overline{V}_0^S$. We say the game value exists if $\underline{V}_0^S = \overline{V}_0^S$. However, as the following simple example shows, under strong formulation the game value typically does not exist.

Example 9.1.1 *Set* $\mathbb{K}_1 := \mathbb{K}_2 := [-1, 1]$, *and* $d = 1, d_1 = 2$ *(namely X is 2-dimensional) with*

$$X_t^{i,k^1,k^2} := \int_0^t k_s^i ds + cB_t, \ i = 1, 2, \quad J_S(k) := \mathbb{E}\Big[|X_T^{1,k} - X_T^{2,k}|\Big], \quad (9.1.13)$$

where $c \ge 0$ *is a constant. Then* $\underline{V}_0^S = 0 < T \le \overline{V}_0^S$.

Proof First, for any $k^1 \in \mathcal{K}_1$, set $k^2 := k^1$, we see that $X^{1,k} = X^{2,k}$ and thus $J_S(k) = 0$. Then $\inf_{k^2 \in \mathcal{K}_2} J_S(k^1, k^2) \le 0$ for any $k^1 \in \mathcal{K}_1$. It is obvious that $\underline{V}_0^S \ge 0$. Then we must have $\underline{V}_0^S = 0$.

On the other hand, for any $k^2 \in \mathcal{K}_2$, note that X^{2,k^1,k^2} depends only on k^2 and thus can be denoted as X^{2,k^2}. Set $k_t^1 := \text{sign}(\mathbb{E}[\int_0^T k_t^2 dt])$, where $\text{sign}(x) := \mathbf{1}_{\{x \ge 0\}} - \mathbf{1}_{\{x < 0\}}$ for any $x \in \mathbb{R}$. Then

$$J_S(k) = \mathbb{E}\Big[|\int_0^T k_t^1 dt - \int_0^T k_t^2 dt|\Big] = \mathbb{E}\Big[|T\text{sign}(\mathbb{E}[\int_0^T k_t^2 dt]) - \int_0^T k_t^2 dt|\Big]$$

$$\ge \Big|T\text{sign}(\mathbb{E}[\int_0^T k_t^2 dt]) - \mathbb{E}[\int_0^T k_t^2 dt]\Big| = T + \Big|\mathbb{E}[\int_0^T k_t^2 dt]\Big| \ge T.$$

Then $\sup_{k^2 \in \mathcal{K}_2} J_S(k^1, k^2) \ge T$ for any $k^1 \in \mathcal{K}_1$, and thus $\overline{V}_0^S \ge T$. ∎

Remark 9.1.2

(i) The above example is valid even when $c = 0$ and thus the game is deterministic. So the issue is indeed about the structure of the problem, not due to the information.

(ii) The underlying reason is that, under strong formulation, the dynamic values $\overline{V}_t^S$ and $\underline{V}_t^S$ are time inconsistent, in the sense that the dynamic programming principle fails. ∎

Now consider the weak formulation of the game problem: for the $\mathbb{P}^k$ and $J_W(k)$ in (9.1.8),

$$\overline{V}_0^W := \inf_{k^1 \in \mathcal{K}_1} \sup_{k^2 \in \mathcal{K}_2} J_W(k^1, k^2), \quad \underline{V}_0^W := \sup_{k^2 \in \mathcal{K}_2} \inf_{k^1 \in \mathcal{K}_1} J_W(k^1, k^2). \quad (9.1.14)$$

Then, under certain technical conditions, we will have (see Subsection 11.3.4 below for general path dependent case):

(i) $\overline{V}_0^W = \overline{v}(0,0)$, $\underline{V}_0^W = \underline{v}(0,0)$, where $\overline{v}$, $\underline{v}$ are the unique viscosity solutions to the following Isaacs equations with terminal condition $\overline{v}(T,x) = \underline{v}(t,x) = g(x)$:

$$\partial_t \overline{v} + \overline{H}(t,x,\partial_x\overline{v},\partial_{xx}^2\overline{v}) = 0, \quad \partial_t\underline{v} + \underline{H}(t,x,\partial_x\underline{v},\partial_{xx}^2\underline{v}) = 0, \quad (9.1.15)$$

where $\overline{H}(t,x,z,\gamma) := \inf_{k_1 \in \mathbb{K}_1} \sup_{k_2 \in \mathbb{K}_2} H(t,x,z,\gamma,k_1,k_2), \underline{H}(t,x,z,\gamma) := \sup_{k_2 \in \mathbb{K}_2} \inf_{k_1 \in \mathbb{K}_1} H(t,x,z,\gamma,k_1,k_2), H(t,x,z,\gamma,k_1,k_2) := \frac{1}{2}\sigma^2(t,x,k_1,k_2)\gamma + b(t,x,k_1,k_2)z + f(t,x,k_1,k_2)$.

(ii) If the Isaacs condition holds:

$$H := \overline{H} = \underline{H}, \qquad (9.1.16)$$

then $\overline{v} = \underline{v}$ and consequently $\overline{V}_0^W = \underline{V}_0^W$, namely the game value exists.

In summary, for a zero sum game, if we use strong formulation, typically the game value does not exist, while if we use weak formulation, the results are very natural and are very much similar to the results for the stochastic control problem under weak formulation.

Remark 9.1.3 If we restrict the controls $k = (k^1, k^2)$ to piecewise constant processes, by Problem 9.6.10 there exists unique $\tilde{k} = (\tilde{k}^1, \tilde{k}^2)$ such that $J_W(\tilde{k}) = J_S(k)$. However, unlike stochastic control problems, even in this case we still do not have $\overline{V}_0^W = \overline{V}_0^S$. The reason is that in general $\tilde{k}^1$ depends on both k^1 and k^2. So, given k^1, typically we cannot find a $\tilde{k}^1$ such that $\sup_{k^2 \in \mathscr{K}_2} J_S(k^1, k^2) = \sup_{\tilde{k}^2 \in \mathscr{K}_2} J_W(\tilde{k}^1, \tilde{k}^2)$. See Problem 9.6.12. ∎

Remark 9.1.4 In this remark, we revisit Example 9.1.1, but using the close loop controls. Then the state process becomes: for $k = (k^1, k^2)$,

$$X_t^{W,i,k} := \int_0^t k_s^i(X_\cdot^{W,1,k}, X_\cdot^{W,2,k})ds + cB_t, \quad i = 1,2. \qquad (9.1.17)$$

We emphasize that, unlike in (9.1.13) where X^i depends only on k^i, here each X^i depends on both k^1 and k^2. To ensure the well-posedness of (9.1.17), we use piecewise constant k^i as in the previous Remark. That is, let $\mathscr{K}_1, \mathscr{K}_2$ be piecewise constant taking values in $[-1, 1]$. Define

$$J_W(k) := \mathbb{E}^{\mathbb{P}_0}\Big[|X_T^{W,1,k} - X^{W,2,k}|\Big], \quad \underline{V}_0^W := \sup_{k^1 \in \mathscr{K}_1} \inf_{k^2 \in \mathscr{K}_2} J_W(k), \quad \overline{V}_0^W := \inf_{k^2 \in \mathscr{K}_2} \sup_{k^1 \in \mathscr{K}_1} J_W(k).$$

Then one can show that $\underline{V}_0^W = \overline{V}_0^W = u(0,0,0)$, where $u(t, x_1, x_2)$ is the unique viscosity solution of the following Isaacs equation:

$$\partial_t u + \frac{c^2}{2}[\partial_{x_1 x_1}^2 u + 2\partial_{x_1 x_2}^2 u + \partial_{x_2 x_2}^2 u] + |\partial_{x_1} u| - |\partial_{x_2} u| = 0, \quad u(T, x_1, x_2) = |x_1 - x_2|.$$

$$(9.1.18)$$

The viscosity property follows similar arguments as in Proposition 11.3.13 below, and the uniqueness of viscosity solution is standard in PDE literature, see, e.g., Crandall, Ishii, & Lions [42].

We shall also note that, since $d_2 > d$, the PDE (9.1.18) is degenerate, even when $c > 0$. For the same reason, one cannot apply the Girsanov Theorem on (9.1.17). ∎

Remark 9.1.5 In the standard literature of zero sum game, see, e.g., Fleming & Souganidis [90], people use the so-called strategy versus controls. Let $\mathscr{A}_1$ denote the set of mappings $\theta^1 : \mathscr{K}_2 \to \mathscr{K}_1$ such that, for any $k^2, \tilde{k}^2 \in \mathscr{K}_2$ satisfying $k_s^2 = \tilde{k}_s^2, 0 \le s \le t$, it holds that $\theta_s^1(k^2) = \theta_s^1(\tilde{k}^2), 0 \le s \le t$. Denote $\mathscr{A}_2$ similarly, and define

$$\overline{V}_0^{SC} := \sup_{\theta^2 \in \mathscr{A}_2} \inf_{k^1 \in \mathscr{K}_1} J_S(k^1, \theta^2(k^1)), \quad \underline{V}_0^{SC} := \inf_{\theta^1 \in \mathscr{A}_1} \sup_{k^2 \in \mathscr{K}_2} J_S(\theta^1(k^2), k^2).$$

(9.1.19)

Here the superscript ^{SC} stands for strategy versus control. Then, under natural conditions, we have $\overline{V}_0^{SC} = \overline{v}(0,0) = \overline{V}_0^W, \underline{V}_0^{SC} = \underline{v}(0,0) = \underline{V}_0^W$, and in particular, if the Isaacs condition (9.1.16) holds, then $\overline{V}_0^{SC} = \underline{V}_0^{SC}$, and is equal to the game value under weak formulation.

We remark that, while the strategy versus control formulation induce the same value as the weak formulation, we prefer the weak formulation due to the following reasons.

(i) The strategy versus control formulation requires the strategy player observes the other player's control. This is not implementable in practice, especially since this is a zero sum game and the two players are noncooperative. In fact, in the principal agent problem, this is exactly excluded by the moral hazard. In weak formulation, both players determine their controls based on the state process X, which is public information observable to both players.

(ii) The upper value $\overline{V}_0^{SC}$ and lower value $\underline{V}_0^{SC}$ are defined in terms of different information, which is somewhat unnatural. In particular, it is not convenient even just to define the saddle point of the game.

(iii) The strategy versus control is still in strong formulation. As we discussed in the previous subsection, this makes it less likely to admit saddle point (even if we can define saddle point in some appropriate sense).

(iv) Under weak formulation, the PDE and the saddle point of the Hamiltonian H naturally lead to a saddle point of the game, provided all the technical conditions are satisfied, see, e.g., Hamadene & Lepeltier [103] for a path dependent game with drift controls. However, the saddle point of H provides no clue at all for potential saddle point for strategy versus control formulation. ∎

Remark 9.1.6 There is one pitfall in the weak formulation though. The stability of the game value in terms of the coefficients b, σ is quite subtle, and typically requires some strong conditions. In particular, if we define $\overline{v}(t, x)$ as the upper value of the game in weak formulation whose state process starts at t with value x, then the regularity of $\overline{v}$ requires some special structure of the coefficients. ∎

9.2 The Canonical Setting and Semimartingale Measures

9.2.1 The Canonical Setting

For weak formulation, we shall fix a canonical space Ω, a canonical process X, and consider various probability measures. For the rest of the book, unless otherwise stated, we shall always set:

- $\Omega := \{\omega \in C([0, T], \mathbb{R}^d) : \omega_0 = 0\}$ is the canonical space, $\Lambda := [0, T) \times \Omega$, and $\overline{\Lambda} := [0, T] \times \Omega$.
- X is the canonical process, namely $X_t(\omega) := \omega_t$.
- $\mathbb{F} = \{\mathscr{F}_t\}_{0 \le t \le T} := \mathbb{F}^X = \{\mathscr{F}_t^X\}_{0 \le t \le T}$ is the natural filtration generated by X.
- $\mathbb{P}_0$ is the Wiener measure, namely X is a $\mathbb{P}_0$-Brownian motion. Quite often, in order to emphasize it is a Brownian motion, we may also use the notation $B := X$ when $\mathbb{P}_0$ is considered.
- $\mathscr{T}$ is the set of all $\mathbb{F}$-stopping times.
- Ω and $\overline{\Lambda}$ are equipped with the following seminorm and pseudometric, respectively:

$$\|\omega\|_t := \sup_{0 \le s \le t} |\omega_s|, \quad \mathbf{d}((t, \omega), (\tilde{t}, \tilde{\omega})) := \sqrt{|t - \tilde{t}|} + \|\omega_{\cdot \wedge t} - \tilde{\omega}_{\cdot \wedge \tilde{t}}\|_T, \quad \forall (t, \omega), (\tilde{t}, \tilde{\omega}) \in \overline{\Lambda}.$$

$$(9.2.1)$$

We note that when $t = T$, $\mathbf{d}((T, \omega), (T, \tilde{\omega})) = \|\omega - \tilde{\omega}\|_T$ and $\|\cdot\|_T$ is a norm on Ω. In particular, $(\Omega, \|\cdot\|_T)$ is a normed vector space.

- $C^0(\Omega)$ is the set of random variables $\xi : \Omega \to \mathbb{R}$ continuous in ω under $\|\cdot\|_T$. That is,

$$\lim_{n \to \infty} \xi(\omega^n) = \xi(\omega) \quad \text{for any } \omega, \omega^n \in \Omega \text{ satisfying } \lim_{n \to \infty} \|\omega^n - \omega\|_T = 0.$$

- $UC(\Omega)$ is the subset of uniformly continuous $\xi \in C^0(\Omega)$. That is,

$$|\xi(\omega) - \xi(\tilde{\omega})| \le \rho(\|\omega - \tilde{\omega}\|_T), \quad \forall \omega, \tilde{\omega} \in \Omega,$$

where the modulus of continuity function $\rho : [0, \infty) \to [0, \infty)$ is a continuous increasing function satisfying $\rho(0) = 0$.
- $C_b^0(\Omega)$ is the subset of bounded $\xi \in C^0(\Omega)$; and $UC_b(\Omega) := C_b^0(\Omega) \cap UC(\Omega)$.
- $C^0(\Lambda)$ is the set of processes $u : \Lambda \to \mathbb{R}$ continuous in (t, ω) under $\mathbf{d}$, and similarly define $C_b^0(\Lambda)$, $C^0(\overline{\Lambda})$, $C_b^0(\overline{\Lambda})$, $UC(\overline{\Lambda})$, $UC_b(\overline{\Lambda})$.
- In multidimensional case, we use $C^0(\Omega, \mathbb{R}^n)$, etc. to denote the sets for $\mathbb{R}^n$-valued processes.

We remark that $C^0(\Omega) \subset \mathbb{L}^0(\mathscr{F}_T)$ and $C^0(\overline{\Lambda}) \subset \mathbb{L}^0(\mathbb{F})$, see Problem 9.6.1. In particular, $X \in UC(\overline{\Lambda})$. Note that a process $\eta \in \mathbb{L}^0(\mathbb{F})$ can also be written as $\eta = \eta(X.)$, since $X(\omega) = \omega$. Unless otherwise stated, we will take the convention that $\eta(\omega)$ indicates the value for a fixed path ω, while for $\eta(X)$ we will talk about its distribution, typically its expectation under certain probability measure. Moreover, let $\tilde{X}$ be another $\mathbb{R}^d$-valued continuous process with $\tilde{X}_0 = 0$. Since $\tilde{X}(\omega) \in \Omega$ for all $\omega \in \Omega$, then $\tilde{\eta}(\omega) := \eta(\tilde{X}(\omega))$ is well defined and it is clear that $\tilde{\eta} \in \mathbb{L}^0(\mathbb{F}^{\tilde{X}})$.

The following simple result will be crucial.

Lemma 9.2.1 $(\Omega, \|\cdot\|_T)$ *is a Polish space, namely it is a complete and separable metric space.*

Proof The completeness is obvious. We shall only show the separability, namely there exist a dense countable sequence $\{\omega^i, i \geq 1\} \subset \Omega$. For any $\varepsilon > 0$ and $n \geq 1$, set

$$\Omega_n^\varepsilon := \{\omega \in \Omega : \|\omega\|_T \leq n, \mathrm{OSC}_{\frac{1}{n}}(\omega) \leq \varepsilon\}, \tag{9.2.2}$$

where OSC_δ is defined in (1.2.4). Clearly $\cup_{n\geq 1}\Omega_n^\varepsilon = \Omega$ for any $\varepsilon > 0$. For each ε, n, set $t_i := t_i^n := \frac{i}{n}T, i = 0, \cdots, n$, and let $O_j := O_j^{\varepsilon,n}, j = 1, \cdots, m$, be a measurable ε-partition of $\{x \in \mathbb{R}^d : |x| \leq n\}$, namely $|x - \tilde{x}| \leq \varepsilon$ for all $x, \tilde{x} \in O_j$. Now let $E_{j_1,\cdots j_n} := \{\omega \in \Omega_n^\varepsilon : \omega_{t_i} \in O_{j_i}, i = 1, \cdots, n\}$, for any $j_i \in \{1, \cdots, m\}, i = 1, \cdots, n$, and pick an arbitrary $\omega^{j_1,\cdots j_n} \in E_{j_1,\cdots j_n}$. Then $\{E_{j_1,\cdots j_n}\}$ form a finite partition of Ω_n^ε, and for any $\omega \in E_{j_1,\cdots j_n}$, denoting $\tilde{\omega} := \omega^{j_1,\cdots j_n}$,

$$\|\omega - \tilde{\omega}\|_T = \max_{0\leq i\leq n-1} \sup_{t_i\leq t\leq t_{i+1}} |\omega_t - \tilde{\omega}_t| \leq \max_{0\leq i\leq n-1} \sup_{t_i\leq t\leq t_{i+1}} \left[|\omega_{t_i,t}| + |\omega_{t_i} - \tilde{\omega}_{t_i}| + |\tilde{\omega}_{t_i,t}|\right] \leq 3\varepsilon.$$
$$\tag{9.2.3}$$

Now denote by E_n^ε the collection of all $\omega^{j_1,\cdots j_n}$, which is finite. Then the set $\cup_{n\geq 1,k\geq 1}E_n^{\frac{1}{k}}$ satisfies the desired properties. ∎

9.2.2 Semimartingale Measures

We say a probability measure $\mathbb{P}$ on the canonical setting $(\Omega, \mathbb{F}, X)$, as introduced in the previous subsection, is a semimartingale measure if X is a $\mathbb{P}$-semimartingale, and is a martingale measure if X is a $\mathbb{P}$-martingale. In particular, $\mathbb{P}_0$ is a martingale measure. Given a semimartingale measure $\mathbb{P}$, following the arguments in Chapter 2 one can show that X has quadratic variation $\langle X \rangle$ under $\mathbb{P}$ and we may introduce stochastic integration $\eta \cdot dX_t$ under $\mathbb{P}$. We shall refer to the book Revuz & Yor [206] for the general continuous martingale theory and in the rest of the book we may use some results without explicitly citing the references.

For our future purpose, in this subsection we introduce three classes of semimartingale measures. We start with the martingale measures induced from strong formulation. Recall that $B := X$ is a $\mathbb{P}_0$-Brownian motion.

Definition 9.2.2 *Let* $0 \leq \underline{\sigma} \leq \overline{\sigma}$ *be two constant matrices in* $\mathbb{S}^d$. *Define*

$$\mathscr{P}^S_{[\underline{\sigma},\overline{\sigma}]} := \left\{ \mathbb{P}^\sigma : \sigma \in \mathbb{L}^0(\mathbb{F}, \mathbb{S}^d) \text{ such that } \underline{\sigma} \leq \sigma \leq \overline{\sigma}, \sigma > 0 \right\},$$

where $\mathbb{P}^\sigma := \mathbb{P}_0 \circ (X^\sigma)^{-1}$ *and* $X^\sigma_t := \int_0^t \sigma_s dB_s$, $\mathbb{P}_0$-*a.s.* (9.2.4)

The above definition means

the $\mathbb{P}^\sigma$-distribution of the canonical process X is equal to the $\mathbb{P}_0$-distribution of X^σ.

(9.2.5)

That is, for any $0 < t_1 < \cdots < t_n \leq T$ and any bounded Borel measurable function $g : \mathbb{R}^{nd} \to \mathbb{R}$,

$$\mathbb{E}^{\mathbb{P}^\sigma}\left[g(X_{t_1}, \cdots, X_{t_n}) \right] = \mathbb{E}^{\mathbb{P}_0}\left[g(X^\sigma_{t_1}, \cdots, X^\sigma_{t_n}) \right].$$

Remark 9.2.3 Typically we will require ξ to be continuous in ω, or equivalently ξ is a continuous function of the canonical process X. Note that we are in weak formulation. In strong formulation, this means that we are considering random variables of the form $\tilde{\xi} := \xi(X^\sigma)$, where ξ is continuous (in X^σ), but we are not requiring the continuity of the mapping $\omega \mapsto \tilde{\xi}(\omega) = \xi(X^\sigma(\omega))$, which is quite unlikely because typically X^σ is not continuous in ω. ∎

Remark 9.2.4 Here we assume $\sigma \in \mathbb{S}^d$ is symmetric. In the language of Parts I and II, this implies that $d_1 = d$. Note that in Itô formula (2.3.17) and in PDE (5.1.9), the coefficient of the hessian is $\sigma\sigma^\top$ which is always symmetric. That is, given an arbitrary σ (possibly $d_1 \neq d$), we may always set $\tilde{\sigma} := (\sigma\sigma^\top)^{\frac{1}{2}} \in \mathbb{S}^{d_1}$ so that $\tilde{\sigma}^2 = \sigma\sigma^\top$. Then, when f does not depend on σ in PDE (5.1.9), σ and $\tilde{\sigma}$ will lead to the same PDE. So our restriction here is not very serious and indeed we can extend our formulation to more general σ. However, under this restriction the presentation is much easier. ∎

Recall the augmented filtration $\overline{\mathbb{F}}^\mathbb{P}$ and Problem 3.7.8 for path dependent SDEs.

Lemma 9.2.5 *Each* $\mathbb{P} = \mathbb{P}^\sigma \in \mathscr{P}^S_{[\underline{\sigma},\overline{\sigma}]}$ *is a martingale measure. Moreover,*

(i) *There exists* $\tilde{\sigma}$ *such that* $\mathbb{P}^\sigma = \mathbb{P}_0 \circ (\tilde{X})^{-1}$, *where* $\tilde{X}$ *is a strong solution of the following SDE:*

$$\tilde{X}_t = \int_0^t \tilde{\sigma}(s, \tilde{X}.) dB_s, \quad \mathbb{P}_0\text{-a.s.} \quad (9.2.6)$$

(ii) $\mathbb{P}$ *has martingale representation property, namely for any* $\xi \in \mathbb{L}^2(\mathscr{F}_T, \mathbb{P}; \mathbb{R})$, *there exists unique* $\eta \in \mathbb{L}^0(\mathbb{F}, \mathbb{P}; \mathbb{R}^d)$ *such that* $\mathbb{E}^\mathbb{P}[\int_0^T \eta_t \eta_t^\top : d\langle X \rangle_t] < \infty$ *and* $\xi = \mathbb{E}^\mathbb{P}[\xi] + \int_0^T \eta_t \cdot dX_t$, $\mathbb{P}$-*a.s.*

Proof It is clear that X^σ is a $\mathbb{P}_0$-martingale, then (9.2.5) implies that X is a $\mathbb{P}^\sigma$-martingale.

(i) By Problem 2.10.6, we have $\overline{\mathbb{F}^{X^\sigma}}^{\mathbb{P}_0} = \overline{\mathbb{F}^B}^{\mathbb{P}_0}$. This implies that $\sigma \in \mathbb{L}^0(\mathbb{F}^B) \subset \mathbb{L}^0(\overline{\mathbb{F}^{X^\sigma}}^{\mathbb{P}_0})$. Then there exists $\tilde\sigma$ such that

$$\sigma(B) = \tilde\sigma(X^\sigma), \quad \mathbb{P}_0\text{-a.s.} \tag{9.2.7}$$

and thus $\tilde X := X^\sigma$ is a strong solution of the path dependent SDE (9.2.6).

(ii) Denote $\tilde\xi := \xi(X^\sigma)$. By (9.2.5) we have $\tilde\xi \in \mathbb{L}^2(\mathbb{F}^{X^\sigma}, \mathbb{P}_0) \subset \mathbb{L}^2(\overline{\mathbb{F}^B}^{\mathbb{P}_0}, \mathbb{P}_0)$. Applying the standard martingale representation Theorem 2.5.2, there exists unique $\tilde\eta \in \mathbb{L}^2(\mathbb{F}^B, \mathbb{P}_0; \mathbb{R}^d)$ such that

$$\tilde\xi = \mathbb{E}^{\mathbb{P}_0}[\tilde\xi] + \int_0^T \tilde\eta_t(B.) \cdot dB_t = \mathbb{E}^{\mathbb{P}_0}[\tilde\xi] + \int_0^T [\sigma^{-1}\tilde\eta]_t(B.) \cdot dX_t^\sigma.$$

Again since $\overline{\mathbb{F}^{X^\sigma}}^{\mathbb{P}_0} = \overline{\mathbb{F}^B}^{\mathbb{P}_0}$, there exists η such that $[\sigma^{-1}\tilde\eta](B.) = \eta(X^\sigma)$, $\mathbb{P}_0$-a.s. That is,

$$\tilde\xi = \mathbb{E}^{\mathbb{P}_0}[\tilde\xi] + \int_0^T \eta_t(X_.^\sigma) \cdot dX_t^\sigma.$$

By (9.2.5) again, this implies the result immediately. ∎

Remark 9.2.6

(i) The measures in $\mathscr{P}_{[\underline\sigma,\overline\sigma]}^S$ are in general not equivalent. Indeed, let $\mathbb{P}^1, \mathbb{P}^2$ be corresponding to $\sigma = 1$ and $\sigma = 2$ in (9.2.4), respectively. Denote

$$E_1 := \{\limsup_{t\downarrow 0} \frac{X_t}{\sqrt{2t \ln\ln\frac{1}{t}}} = 1\}, \quad E_2 := \{\limsup_{t\downarrow 0} \frac{\frac{1}{2}X_t}{\sqrt{2t \ln\ln\frac{1}{t}}} = 1\}.$$

Then, by the law of iterated logarithm (see, e.g., Karatzas & Shreve [117]),

$$E_1 \cap E_2 = \emptyset, \quad \mathbb{P}^1(E_1) = \mathbb{P}^2(E_2) = 1, \quad \mathbb{P}^1(E_2) = \mathbb{P}^2(E_1) = 0.$$

That is, $\mathbb{P}^1$ and $\mathbb{P}^2$ have disjoint supports and thus are mutually singular.

(ii) Unless $\underline\sigma = \overline\sigma$, the class $\mathscr{P}_{[\underline\sigma,\overline\sigma]}^S$ has no dominating measure. That is, there is no probability measure $\mathbb{P}^*$ on $\mathscr{F}_T$ such that all $\mathbb{P} \in \mathscr{P}_{[\underline\sigma,\overline\sigma]}^S$ is absolutely continuous with respect to $\mathbb{P}^*$. This is a consequence of Example 10.1.4 below. This feature makes the fully nonlinear theory much more involved technically. ∎

The major drawback of the class $\mathscr{P}^S_{[\underline{\sigma},\overline{\sigma}]}$ is that it is not weakly compact, see Problem 9.6.11. The notion of weak compactness will be introduced in Subsection 9.2.3 below and will be crucial for the rest of the book. For this purpose, we extend $\mathscr{P}^S_{[\underline{\sigma},\overline{\sigma}]}$ to $\mathscr{P}^W_{[\underline{\sigma},\overline{\sigma}]}$, consisting of martingale measures induced from weak formulation. Recall Definition 3.5.1 (ii) for weak solution of SDEs.

Definition 9.2.7 *Let* $0 \leq \underline{\sigma} \leq \overline{\sigma}$ *be two constant matrices in* $\mathbb{S}^d$. *Let* $\mathscr{P}^W_{[\underline{\sigma},\overline{\sigma}]}$ *denote the class of martingale measures* $\mathbb{P}$ *such that* $(\Omega, \mathscr{F}_T, \mathbb{P}, X)$ *is a weak solution to SDE (9.2.6) for some* $\tilde{\sigma} \in \mathbb{L}^0(\mathbb{F}, \mathbb{S}^d)$ *satisfying* $\underline{\sigma} \leq \tilde{\sigma} \leq \overline{\sigma}$.

Remark 9.2.8

(i) It is clear that $\mathscr{P}^S_{[\underline{\sigma},\overline{\sigma}]} \subset \mathscr{P}^W_{[\underline{\sigma},\overline{\sigma}]}$, then $\mathscr{P}^W_{[\underline{\sigma},\overline{\sigma}]}$ is also a non-dominated class of (possibly mutually singular) measures, in the sense of Remark 9.2.6. Moreover, the above inclusion is strict, see Barlow [6] for a counterexample. In particular, the measures $\mathbb{P} \in \mathscr{P}^W_{[\underline{\sigma},\overline{\sigma}]}$ may not satisfy the martingale representation property.

(ii) For any semimartingale measure $\mathbb{P}$, denote

$$\sigma^{\mathbb{P}}_t := \sqrt{\frac{d\langle X\rangle_t}{dt}}, \quad \mathbb{P}\text{-a.s.} \tag{9.2.8}$$

provided that $d\langle X\rangle$ is absolutely continuous with respect to dt under $\mathbb{P}$. Then $\mathbb{P} \in \mathscr{P}^W_{[\underline{\sigma},\overline{\sigma}]}$ if and only if $\mathbb{P}$ is a martingale measure with $\underline{\sigma} \leq \sigma^{\mathbb{P}} \leq \overline{\sigma}$, $\mathbb{P}$-a.s. and in this case $\sigma^{\mathbb{P}} = \tilde{\sigma}$, $\mathbb{P}$-a.s., for the $\tilde{\sigma}$ in Definition 9.2.7. Moreover, when $\sigma^{\mathbb{P}} > 0$, $\mathbb{P}$-a.s., it follows from Levy's martingale characterization of Brownian motion in Remark 2.6.5 that the following process $B^{\mathbb{P}}$ is a $\mathbb{P}$-Brownian motion:

$$B^{\mathbb{P}}_t := \int_0^t (\sigma^{\mathbb{P}}_t)^{-1} dX_t, \quad \mathbb{P}\text{-a.s.} \tag{9.2.9}$$

(iii) When $\sigma^{\mathbb{P}} > 0$, $\mathbb{P}$-a.s., it is clear that $\overline{\mathbb{F}^{B^{\mathbb{P}}}}^{\mathbb{P}} \subset \overline{\mathbb{F}}^{\mathbb{P}}$, or equivalently $\overline{\mathbb{F}^B}^{\mathbb{P}_0} \subset \overline{\mathbb{F}^{\tilde{X}}}^{\mathbb{P}_0}$ for the weak solution $\tilde{X}$ of (9.2.6). Equality holds if and only if $\mathbb{P} \in \mathscr{P}^S_{[\underline{\sigma},\overline{\sigma}]}$, and in this case the $\mathbb{P}$-distribution of $(X, B^{\mathbb{P}})$ is equal to the $\mathbb{P}_0$-distribution of $(\tilde{X}, B)$. However, when $\sigma^{\mathbb{P}}$ is degenerate, $\overline{\mathbb{F}^B}^{\mathbb{P}_0}$ and $\overline{\mathbb{F}^{\tilde{X}}}^{\mathbb{P}_0}$ in (9.2.6) may not include each other. ∎

The class $\mathscr{P}^W_{[\underline{\sigma},\overline{\sigma}]}$ contains only martingale measures, we next extend it further by including drifts.

Definition 9.2.9

(i) *For any constant* $L \geq 0$, *let* $\mathscr{P}_L$ *denote the class of semimartingale measures* $\mathbb{P}$ *such that its drift is bounded by* L *and diffusion is bounded by* $\sqrt{2L}$. *To be precise,* $\mathbb{P}$ *takes the form* $\mathbb{P} = \mathbb{P}' \circ (X')^{-1}$, *where* $(\Omega', \mathbb{F}', \mathbb{P}')$ *is a filtered* *probability space, B' is a d-dimensional $\mathbb{P}'$-Brownian motion, $b' \in \mathbb{L}^0(\mathbb{F}', \mathbb{R}^d)$, $\sigma' \in \mathbb{L}^0(\mathbb{F}', \mathbb{S}^d)$, and*

$$|b'| \le L, \quad \sigma' \ge 0, \quad |\sigma'| \le \sqrt{2L}, \quad \text{and } X_t' = \int_0^t b_s' ds + \int_0^t \sigma_s' dB_s', \quad \mathbb{P}'\text{-}a.s.$$

$$(9.2.10)$$

(ii) $\mathscr{P}_\infty := \bigcup_{L \ge 0} \mathscr{P}_L.$

Remark 9.2.10

(i) It is clear that $\mathscr{P}^W_{[\underline{\sigma},\overline{\sigma}]} \subset \mathscr{P}_L$ for $L \ge \frac{1}{2}|\overline{\sigma}|^2$. Moreover, for $L \ge \frac{d}{2}$ and $\sigma' = I_d$, clearly $\mathbb{P} = \mathbb{P}^{b'}$ in the sense of (2.6.3). We shall denote

$$\mathscr{P}_L^{drift} := \{\mathbb{P} \in \mathscr{P}_L : \sigma' = I_d\} \subset \mathscr{P}_L. \qquad (9.2.11)$$

(ii) Note that σ' is always $\mathbb{F}^{X'}$-measurable, where X' is defined by (9.2.10). In fact, since $d\langle X'\rangle_t = (\sigma_t')^2 dt$, $\mathbb{P}'$-a.s., is $\mathbb{F}^{X'}$-measurable, we have $\sigma' = \sigma^{\mathbb{P}}(X')$, $\mathbb{P}'$-a.s. for the $\sigma^{\mathbb{P}}$ defined by (9.2.8). If b' is also $\mathbb{F}^{X'}$-measurable, namely there exists a mapping $b^{\mathbb{P}}$ such that $b' = b^{\mathbb{P}}(X')$, $\mathbb{P}'$-a.s., then $(\Omega', \mathscr{F}', \mathbb{P}', X')$ is a weak solution to the following SDE:

$$X_t' = \int_0^t b_s^{\mathbb{P}}(X')ds + \int_0^t \sigma_s^{\mathbb{P}}(X')dB_s', \quad \mathbb{P}'\text{-a.s.} \qquad (9.2.12)$$

However, we emphasize that in general the drift b' in (9.2.10) may not be $\mathbb{F}^{X'}$-measurable, and in this sense the class $\mathscr{P}_L$ is larger than the set induced from close loop controls.

(iii) Alternatively, $\mathscr{P}_L$ can be characterized as follows. Let $\Omega' := \Omega \times \Omega$ be an enlarged canonical space with canonical processes (A, M), namely for $\omega' = (\omega_1, \omega_2) \in \Omega', A(\omega') = \omega_1, M(\omega') = \omega_2$. Let $\mathbb{P}'$ be a probability measure on it such that,

- A is absolutely continuous in t with $|b'| \le L$, where $b_t' := \frac{dA_t}{dt}$, $\mathbb{P}'$-a.s.;
- M is a $\mathbb{P}'$-martingale with $|\sigma'| \le \sqrt{2L}$, where $\sigma_t' := \sqrt{\frac{d\langle M\rangle_t}{dt}}$, $\mathbb{P}'$-a.s.

Then one may construct $\mathbb{P} := \mathbb{P}' \circ (X')^{-1} \in \mathscr{P}_L$ with $X' := A + M$.

(iv) One may extend the construction in (iii) to obtain $(\Omega', \mathbb{F}', \mathbb{P}', B', X')$ in Definition 9.2.9 as follows. Let $\Omega' := \Omega \times \Omega \times \Omega$ be an enlarged canonical space with canonical processes (A, M, W). Let $\mathbb{P}'$ be a probability measure on it such that A and M satisfy the requirement in (iii) and

- W is a $\mathbb{P}'$-Brownian motion independent of (A, M).

Define b', σ' as in (iii) and $\mathbb{P} := \mathbb{P}' \circ (X')^{-1}$ with $X' := A + M$. We shall construct B' to satisfy (9.2.10). Denote by $[a_1, \cdots, a_d]$ the diagonal matrix with (i,i)-th entry a_i. Since σ' is symmetric and $\overline{\mathbb{F}^M}^{\mathbb{P}'}$-measurable, there exists $\overline{\mathbb{F}^M}^{\mathbb{P}'}$-measurable and orthogonal matrix Q such that $\tilde{\sigma}' := Q^{\top}\sigma' Q = [\sigma_1, \cdots, \sigma_d]$ is diagonal. Now define

$$B'_t := \int_0^t Q_s [\sigma_1^{-1} \mathbf{1}_{\sigma_1 \neq 0}, \cdots, \sigma_d^{-1} \mathbf{1}_{\sigma_d \neq 0}] Q_s^{\top} dM_s$$
$$+ \int_0^t Q_s [\mathbf{1}_{\sigma_1=0}, \cdots, \mathbf{1}_{\sigma_d=0}] Q_s^{\top} dW_s, \quad \mathbb{P}'\text{-a.s.}$$

It follows from Levy's martingale characterization of Brownian motion in Remark 2.6.5 again that B' is a $\mathbb{P}'$-Brownian motion. Now it is straightforward to verify (9.2.10). ∎

Remark 9.2.11 In the case that $b' = b^{\mathbb{P}}(X')$ in Remark 9.2.10 (ii), SDE (9.2.12) admits a weak solution is equivalent to that $\mathbb{P}$ is a solution to the following so-called martingale problem: for any $\varphi \in C^{1,2}([0,T] \times \mathbb{R}^d)$,

$$Y_t := \varphi(t, X_t) - \int_0^t \Big[\partial_t \varphi(s, X_s) + \partial_x \varphi(s, X_s) b_s^{\mathbb{P}}(X.) + \frac{1}{2}\partial_{xx}^2 \varphi(s, X_s) : (\sigma_s^{\mathbb{P}})^2(X.)\Big] ds$$
(9.2.13)

is a local martingale under $\mathbb{P}$, see Stroock & Varadahn [218]. ∎

Another important subclass of $\mathscr{P}_L$ consists of those induced by piecewise constant processes b and σ: recalling the elementary processes $\mathbb{L}_0^2(\mathbb{F})$ in Definition 2.2.2,

$$\mathscr{P}_L^{PC} := \Big\{\mathbb{P}^{b,\sigma} : b \in \mathbb{L}_0^2(\mathbb{F}, \mathbb{R}^d), \sigma \in \mathbb{L}_0^2(\mathbb{F}, \mathbb{S}^d), |b| \le L, \sigma > 0, |\sigma| \le \sqrt{2L}\Big\},$$

where $\mathbb{P}^{b,\sigma} := \mathbb{P}_0 \circ (X^{b,\sigma})^{-1}$ and $X_t^{b,\sigma} := \int_0^t b_s ds + \int_0^t \sigma_s dB_s$, $\mathbb{P}_0$-a.s.

(9.2.14)

One can easily show that $\overline{\mathbb{F}^{X^{b,\sigma}}}^{\mathbb{P}_0} = \overline{\mathbb{F}^B}^{\mathbb{P}_0}$, see Problem 9.6.10. For control problems, as analyzed in Section 9.1.2, typically $\mathscr{P}_L^{PC}$ and $\mathscr{P}_L$ induce the same value function. However, $\mathscr{P}_L^{PC}$ is not compact, and in general there is no optimal control in $\mathscr{P}_L^{PC}$.

We conclude this subsection with a few estimates, whose proofs are left to readers in Problem 9.6.5. First, we may define quadratic variation pathwise as follows: denoting $t_i^n := \frac{i}{2^n}T, i = 0, \cdots, 2^n$,

$$\langle X\rangle_t := \limsup_{n\to\infty} \sum_{i=0}^{2^n-1} |X_{t_i^n \wedge t, t_{i+1}^n \wedge t}|^2, \quad 0 \le t \le T, \quad \mathbb{P}\text{-a.s. for all } \mathbb{P} \in \mathscr{P}_\infty. \quad (9.2.15)$$

In the rest of the book, we shall always define $\langle X \rangle$ as the right side of (9.2.15). In particular, $\langle X \rangle$ is $\mathbb{F}$-measurable. Moreover, recalling (1.2.4), for any $L > 0, p \geq 1$, and $\delta > 0$,

$$\sup_{\mathbb{P} \in \mathscr{P}_L} \mathbb{E}^{\mathbb{P}}[\|X\|_T^p] \leq C_p L^p, \quad \sup_{\mathbb{P} \in \mathscr{P}_L} \mathbb{E}^{\mathbb{P}}[|\mathrm{OSC}_\delta(X)|^p] \leq C_p L^p \delta^{\frac{p}{2}}. \quad (9.2.16)$$

9.2.3 Weak Compactness

The weak compactness will be crucial for the later chapters. We start with the following simple result, which is from real analysis and the proof is left to readers in Problem 9.6.6.

Lemma 9.2.12 *Assume an operator* $\mathbb{E} : C_b^0(\Omega) \to \mathbb{R}$ *satisfies the following properties:*

(i) *(Constant preserving)* $\mathbb{E}[1] = 1$.
(ii) *(Monotonicity)* $\mathbb{E}[\xi_1] \leq \mathbb{E}[\xi_2]$ *for any* $\xi_1, \xi_2 \in C_b^0(\Omega)$ *such that* $\xi_1 \leq \xi_2$.
(iii) *(Linearity)* $\mathbb{E}[\xi_1 + \xi_2] = \mathbb{E}[\xi_1] + \mathbb{E}[\xi_2]$ *for any* $\xi_1, \xi_2 \in C_b^0(\Omega)$.
(iv) *(Tightness) For any* $\varepsilon > 0$, *there exists a compact set* $K_\varepsilon \subset\subset \Omega$ *such that* $\mathbb{E}[\xi] \leq \varepsilon$ *for any* $\xi \in C_b^0(\Omega)$ *with* $0 \leq \xi \leq 1$ *and* $\xi \mathbf{1}_{K_\varepsilon} = 0$.

Then there exists a unique probability measure $\mathbb{P}$ *on* $(\Omega, \mathscr{F}_T)$ *such that* $\mathbb{E}^{\mathbb{P}} = \mathbb{E}$.

The weak compactness involves the following concepts.

Definition 9.2.13 *Let* $\mathscr{P}$ *be a set of probability measures on* Ω.

(i) *Let* $\mathbb{P}, \mathbb{P}_n, n \geq 1$ *be a sequence of probability measures on* Ω. *We say* $\mathbb{P}_n \to \mathbb{P}$ *weakly if* $\mathbb{E}^{\mathbb{P}_n}[\xi] \to \mathbb{E}^{\mathbb{P}}[\xi]$ *for all* $\xi \in C_b^0(\Omega)$.
(ii) *We say* $\mathscr{P}$ *is relatively weak compact if, for any* $\{\mathbb{P}_n, n \geq 1\} \subset \mathscr{P}$, *there exists a subsequence* $\{\mathbb{P}_{n_k}, k \geq 1\}$ *and a probability measure* $\mathbb{P}$ *(not necessarily in* $\mathscr{P}$) *such that* $\mathbb{P}_{n_k} \to \mathbb{P}$ *weakly as* $k \to \infty$.
(iii) *We say* $\mathscr{P}$ *is weakly compact if for any sequence* $\{\mathbb{P}_n, n \geq 1\} \subset \mathscr{P}$, *there exists a subsequence* $\{\mathbb{P}_{n_k}, k \geq 1\}$ *and a measure* $\mathbb{P} \in \mathscr{P}$ *such that* $\mathbb{P}_{n_k} \to \mathbb{P}$ *weakly as* $k \to \infty$.

The following simple results are very helpful.

Lemma 9.2.14

(i) *There exists* $\{K_n\}_{n \geq 1} \subset \Omega$ *such that*

$$K_n \text{ is increasing, each } K_n \text{ is compact, and } \sup_{\mathbb{P} \in \mathscr{P}_L} \mathbb{P}[K_n^c] \leq C_L 2^{-n}.$$

$$(9.2.17)$$

(ii) *For any* $\xi \in C_b^0(\Omega)$, *there exists* $\{\xi_n\}_{n \geq 1} \subset UC_b(\Omega)$ *such that, for any* $L > 0$ *and* $p \geq 1$,

$$\|\xi_n\|_\infty \leq \|\xi\|_\infty, \quad \lim_{n \to \infty} \sup_{\mathbb{P} \in \mathscr{P}_L} \mathbb{E}^{\mathbb{P}}[|\xi_n - \xi|^p] = 0.$$

(iii) Let $\{\mathbb{P}_n\}_{n\geq 1} \subset \mathscr{P}_L$. If $\lim_{n\to\infty} \mathbb{E}^{\mathbb{P}_n}[\xi]$ exists for all $\xi \in UC_b(\Omega)$, then there exists a probability measure $\mathbb{P}$ such that $\mathbb{P}_n \to \mathbb{P}$ weakly.

Proof

(i) Recall (1.2.4) and define

$$K_n := \Big\{\omega \in \Omega : \|\omega\|_T \leq 2^{\frac{n}{2}}, \ \mathrm{OSC}_{3^{-m}}(\omega) \leq m^{-1}, \ \forall m \geq n\Big\}.$$

$$(9.2.18)$$

Clearly K_n is increasing, and the elements in K_n are bounded and equicontinuous and thus K_n is compact, thanks to the Arzela-Ascoli Theorem. Moreover, for any $\mathbb{P} \in \mathscr{P}_L$, by (9.2.16) we have

$$\mathbb{P}(K_n^c) \leq \mathbb{P}\Big(\{\|X\|_T \geq 2^{\frac{n}{2}}\} \bigcup \big(\bigcup_{m\geq n}\{\mathrm{OSC}_{3^{-m}}(X) > m^{-1}\}\big)\Big)$$

$$\leq \mathbb{E}^{\mathbb{P}}\Big[2^{-n}\|X\|_T^2 + \sum_{m\geq n} m^2 |\mathrm{OSC}_{3^{-m}}(X)|^2\Big] \leq C2^{-n}$$

$$+ \sum_{m\geq n} Cm^2 3^{-m} \leq C2^{-n}.$$

(ii) Let $\xi \in C_b^0(\Omega)$. Since $\xi = \xi^+ - \xi^-$, clearly we may assume without loss of generality that $\xi \geq 0$. For any m, n, define

$$\xi_{m,n}(\omega) := \sup_{\tilde{\omega}\in K_n}\Big[(1 - m\|\omega - \tilde{\omega}\|_T)^+\xi(\tilde{\omega})\Big], \quad \omega \in \Omega. \qquad (9.2.19)$$

It is clear that $\|\xi_{m,n}\|_\infty \leq \|\xi\|_\infty$, $\xi_{m,n} \geq \xi$ on K_n, and $\xi_{m,n}$ is uniformly continuous on Ω. Since ξ is continuous and K_n is compact, ξ is uniformly continuous on K_n. Then there exists m_n such that $|\xi(\omega^1) - \xi(\omega^2)| \leq \frac{1}{n}$ for all $\omega^1, \omega^2 \in K_n$ such that $\|\omega^1 - \omega^2\|_T \leq m_n^{-1}$. Define $\xi_n := \xi_{m_n,n}$. For any $\omega \in \Omega$, again due to the compactness of K_n, the optimization (9.2.19) of ξ_n has an optimal argument $\omega^* \in K_n$. If $\xi_n(\omega) = 0$, then $\xi_n(\omega) = \xi(\omega) = 0$. If $\xi_n(\omega) > 0$, one must have $\|\omega^* - \omega\|_T < \frac{1}{m_n}$. Thus

$$0 \leq \xi_n(\omega) - \xi(\omega) \leq \xi(\omega^*) - \xi(\omega) \leq \frac{1}{n}.$$

So in both cases, we have $0 \leq \xi_n(\omega) - \xi(\omega) \leq n^{-1}$ for all $\omega \in K_n$. Therefore, for any $p \geq 1$,

$$\sup_{\mathbb{P}\in\mathscr{P}_L} \mathbb{E}^{\mathbb{P}}\big[|\xi_n - \xi|^p\big] = \sup_{\mathbb{P}\in\mathscr{P}_L} \mathbb{E}^{\mathbb{P}}\Big[|\xi_n - \xi|^p[\mathbf{1}_{K_n} + \mathbf{1}_{K_n^c}]\Big] \leq n^{-p}$$

$$+ C\|\xi\|_\infty^p \sup_{\mathbb{P}\in\mathscr{P}_L} \mathbb{P}(K_n^c) \to 0,$$

as $n \to \infty$.

(iii) For any $\xi \in C_b^0(\Omega)$, any $\varepsilon > 0$, by (ii) there exists $\xi_\varepsilon \in UC_b(\Omega)$ such that $\sup_{n \geq 1} \mathbb{E}^{\mathbb{P}_n}[|\xi_\varepsilon - \xi|] \leq \varepsilon$. By our assumption, there exists N_ε such that $|\mathbb{E}^{\mathbb{P}_n}[\xi_\varepsilon] - \mathbb{E}^{\mathbb{P}_m}[\xi_\varepsilon]| \leq \varepsilon$ for all $m, n \geq N_\varepsilon$. Then, for all $m, n \geq N_\varepsilon$,

$$|\mathbb{E}^{\mathbb{P}_n}[\xi] - \mathbb{E}^{\mathbb{P}_m}[\xi]| \leq \mathbb{E}^{\mathbb{P}_n}[|\xi_\varepsilon - \xi|] + \mathbb{E}^{\mathbb{P}_m}[|\xi_\varepsilon - \xi|] + |\mathbb{E}^{\mathbb{P}_n}[\xi_\varepsilon] - \mathbb{E}^{\mathbb{P}_m}[\xi_\varepsilon]| \leq 3\varepsilon.$$

Therefore, $\mathbb{E}[\xi] := \lim_{n \to \infty} \mathbb{E}^{\mathbb{P}_n}[\xi]$ exists for all $\xi \in C_b^0(\Omega)$. It is clear that the operator $\mathbb{E}$ satisfies Lemma 9.2.12 (i)-(iii). Moreover, the tightness condition (iv) also follows directly from (9.2.17). Then applying Lemma 9.2.12 we obtain the desired probability measure $\mathbb{P}$. ∎

Our main result of this subsection is:

Theorem 9.2.15 *For any $L > 0$ and $0 \leq \underline{\sigma} \leq \overline{\sigma}$, the classes $\mathscr{P}_L$ and $\mathscr{P}_{[\underline{\sigma}, \overline{\sigma}]}^W$ are weakly compact.*

Proof We will only prove the result for $\mathscr{P}_L$ in several steps. The proof for $\mathscr{P}_{[\underline{\sigma}, \overline{\sigma}]}^W$ is similar. For notational simplicity, we assume $d = 1$ in the proof.

Step 1. Let $\mathscr{P}_L'$ denote the set of all $\mathbb{P}'$ corresponding to $\mathbb{P} \in \mathscr{P}_L$ as in Remark 9.2.10 (iii). We first claim that

$$\mathscr{P}_L' \text{ is weakly compact.} \tag{9.2.20}$$

Then, for any $\{\mathbb{P}_n\}_{n \geq 1} \subset \mathscr{P}_L$, let $\{\mathbb{P}_n'\}_{n \geq 1} \subset \mathscr{P}_L'$ be the corresponding measures. By (9.2.20) there exists n_k and $\mathbb{P}' \in \mathscr{P}_L'$ such that $\mathbb{P}_{n_k}'$ converges to $\mathbb{P}'$ weakly as $k \to \infty$. Now for any $\xi \in C_b^0(\Omega)$, clearly the mapping $(A, M) \in \Omega \times \Omega \mapsto \xi(A + M) \in \mathbb{R}$ is continuous, then $\lim_{k \to \infty} \mathbb{E}^{\mathbb{P}_n'}[\xi(X')] = \mathbb{E}^{\mathbb{P}'}[\xi(X')]$, where $X' := A + M$. Denote $\mathbb{P} := \mathbb{P}' \circ (X')^{-1} \in \mathscr{P}_L$, we see that $\lim_{k \to \infty} \mathbb{E}^{\mathbb{P}_n}[\xi(X)] = \mathbb{E}^{\mathbb{P}}[\xi(X)]$, namely $\mathbb{P}_n$ converges to $\mathbb{P}$ weakly. Therefore, $\mathscr{P}_L$ is weakly compact.

Step 2. We next show that $\mathscr{P}_L'$ is relatively weak compact. Fix $\{\mathbb{P}_n', n \geq 1\} \subset \mathscr{P}_L'$. By Lemma 9.2.14, it suffices to find a subsequence $\{n_k\}$ such that $\lim_{k \to \infty} \mathbb{E}^{\mathbb{P}_{n_k}'}[\xi]$ exists for all $\xi \in UC_b(\Omega')$.

For this purpose, we let $\mathscr{A}$ denote the set of ξ taking the following form:

$$\xi(\omega') = \psi\Big(\varphi_m(\omega_{t_1}'), \cdots, \varphi_m(\omega_{t_m}')\Big), \tag{9.2.21}$$

where $m \geq 1$, $t_i := \frac{i}{m}T$, $i = 0, \cdots, m$, $\psi : \mathbb{R}^m \to \mathbb{R}$ is a polynomial with rational coefficients, and $\varphi_m(x) := \frac{x}{|x|}(|x| \wedge m)$ is a truncation function. Clearly $\mathscr{A} \subset UC_b(\Omega')$ is countable, and we numerate its elements as $\{\xi_n, n \geq 1\}$.

We next show that $\mathscr{A}$ is dense in $UC_b(\Omega')$. Indeed, let $\xi \in UC_b(\Omega')$ with an integer bound C_0 and modulus of continuity function ρ. For any $m \geq 1$, let t_i be as above. Given $x = (x_1, \cdots, x_m) \in \mathbb{R}^m$, denote by $\omega'(x)$ the linear interpolation of $(0,0)$, $(t_i, x_i)_{1 \leq i \leq m}$ and define $g_m(x) := \xi(\omega'(x))$. Then clearly g is uniformly continuous and bounded, and thus there exists a polynomial ψ_m on $\mathbb{R}^m$ with rational coefficients such that $\sup\{|\psi_m(x) - g_m(x)| : |x_i| \leq m, i = 1, \cdots, m\} \leq \frac{1}{m}$. Denote

$\xi_m(\omega') := \psi_m\big(\varphi_m(\omega'_{t_1}), \cdots, \varphi_m(\omega'_{t_m})\big)$. Then $\xi_m \in \mathscr{A}$ and $\|\xi_m\|_\infty \le \|\xi\|_\infty + \frac{1}{m}$.
Note that

$$\begin{aligned}
|\xi_m - \xi| &\le \frac{1}{m} + \Big|g_m\big(\varphi_m(\omega'_{t_1}), \cdots, \varphi_m(\omega'_{t_m})\big) - \xi(\omega)\Big| \\
&\le \frac{1}{m} + 2\|\xi\|_\infty \mathbf{1}_{\{\|\omega\|_T > m\}} + \Big|g_m\big(\omega'_{t_1}, \cdots, \omega'_{t_m}\big) - \xi(\omega)\Big| \\
&\le \frac{1}{m} + 2\|\xi\|_\infty \mathbf{1}_{\{\|\omega\|_T > m\}} + \rho\big(\mathrm{OSC}_{\frac{1}{m}}(\omega')\big).
\end{aligned} \tag{9.2.22}$$

Then by (9.2.16) we have

$$\lim_{m\to\infty} \sup_{n\ge 1} \mathbb{E}^{\mathbb{P}_{n'}}\Big[|\xi_m - \xi|\Big] = 0. \tag{9.2.23}$$

Now for each $\xi_m \in \mathscr{A}$, since $\{\mathbb{E}^{\mathbb{P}_{n'}}[\xi_m], n \ge 1\}$ is bounded, there exists a subsequence $\{n_k, k \ge 1\}$ such that $\{\mathbb{E}^{\mathbb{P}_{n_k'}}[\xi_m]$ converges when $k \to \infty$. By using the diagonal arguments, one can assume without loss of generality that $\mathbb{E}[\xi_m] := \lim_{n\to\infty} \mathbb{E}^{\mathbb{P}_{n'}}[\xi_m]$ exists for all $m \ge 1$. By (9.2.23), we see that $\mathbb{E}[\xi] := \lim_{n\to\infty} \mathbb{E}^{\mathbb{P}_{n'}}[\xi]$ exists for all $\xi \in UC_b(\Omega')$. Then it follows from Lemma 9.2.14 that $\mathscr{P}'_L$ is relatively weak compact.

Step 3. We finally show that $\mathscr{P}'_L$ is closed under weak convergence. That is, if $\mathbb{P}_n \in \mathscr{P}'_L$ and $\mathbb{P}'_n \to \mathbb{P}'$ weakly, then $\mathbb{P}' \in \mathscr{P}'_L$. This, together with Step 3, implies (9.2.20) and completes the proof.

First, for any $0 \le s < t \le T$ and any $\varepsilon > 0$, there exists a uniformly continuous function $\varphi_\varepsilon : [0, \infty) \to [0, 1]$ such that $\varphi_\varepsilon(x) = 1$ for $x \le L(t-s)$ and $\varphi_\varepsilon(x) = 0$ for $x \ge (L + \varepsilon)(t - s)$. Note that $|A_t - A_s| \le L(t-s)$, $\mathbb{P}'_n$-a.s. Then $\varphi_\varepsilon(|A_t - A_s|) = 1$, $\mathbb{P}'_n$-a.s. for all n. This implies that

$$\mathbb{E}^{\mathbb{P}'}[\varphi_\varepsilon(|A_t - A_s|)] = \lim_{n\to\infty} \mathbb{E}^{\mathbb{P}'_n}[\varphi_\varepsilon(|A_t - A_s|)] = 1.$$

By our construction of φ_ε, we see that $|A_t - A_s| \le (L + \varepsilon)(t - s)$, $\mathbb{P}'$-a.s. Since $\varepsilon > 0$ is arbitrary, we have $|A_t - A_s| \le L(t - s)$, $\mathbb{P}'$-a.s. for any $s < t$. Clearly this implies

$$\mathbb{P}'\Big(|A_t - A_s| \le L|t - s|, \ s, t \in \mathbb{Q} \cap [0, T]\Big) = 1.$$

Moreover, by definition A is continuous, then

$$\mathbb{P}'\Big(|A_t - A_s| \le L|t - s|, \ s, t \in [0, T]\Big) = 1.$$

Next, note that

$$\sup_{n\ge 1} \mathbb{E}^{\mathbb{P}'_n}[|M_T^*|^2] \le C \sup_{n\ge 1} \mathbb{E}^{\mathbb{P}'_n}\Big[\int_0^T |\sigma'_t|^2 dt\Big] \le C;$$

$$\mathbb{E}^{\mathbb{P}'}[|M_T^*|^2] = \lim_{m\to\infty} \mathbb{E}^{\mathbb{P}'}[|M_T^* \wedge m|^2] = \lim_{m\to\infty}\lim_{n\to\infty} \mathbb{E}^{\mathbb{P}'_n}[|M_T^* \wedge m|^2] \le \sup_{n\ge 1} \mathbb{E}^{\mathbb{P}'_n}[|M_T^*|^2] \le C.$$

$$\tag{9.2.24}$$

For any $m \geq 1$, note that $\varphi_m(x) := \frac{x}{|x|}(|x| \wedge m)$ is bounded and uniformly continuous. Then, for any $0 \leq s < t \leq T$ and $\eta \in C_b^0(\Omega') \cap \mathbb{L}^0(\mathscr{F}_s')$, by (9.2.24) and the fact that M is a $\mathbb{P}_n'$ martingale, we have

$$\mathbb{E}^{\mathbb{P}'}\Big[\eta[M_t - M_s]\Big] = \lim_{m \to \infty} \mathbb{E}^{\mathbb{P}'}\Big[\eta \varphi_m(M_t - M_s)\Big] = \lim_{m \to \infty} \lim_{n \to \infty} \mathbb{E}^{\mathbb{P}_n'}\Big[\eta \varphi_m(M_t - M_s)\Big]$$

$$= \lim_{m \to \infty} \lim_{n \to \infty} \mathbb{E}^{\mathbb{P}_n'}\Big[\eta\big(\varphi_m(M_t - M_s) - (M_t - M_s)\big)\Big]$$

This leads to, with the constant C depending on η,

$$\Big|\mathbb{E}^{\mathbb{P}'}\Big[\eta[M_t - M_s]\Big]\Big| \leq \liminf_{m \to \infty} \sup_{n \geq 1} \mathbb{E}^{\mathbb{P}_n'}\Big[\Big|\eta\big(\varphi_m(M_t - M_s) - (M_t - M_s)\big)\Big|\Big]$$

$$\leq C \liminf_{m \to \infty} \sup_{n \geq 1} \mathbb{E}^{\mathbb{P}_n'}\Big[|M_t - M_s|\mathbf{1}_{\{|M_t - M_s| \geq m\}}\Big|\Big]$$

$$\leq C \liminf_{m \to \infty} \sup_{n \geq 1} \frac{1}{m} \mathbb{E}^{\mathbb{P}_n'}[|M_T^*|^2] \leq C \liminf_{m \to \infty} \frac{1}{m} = 0,$$

which clearly implies that M is a $\mathbb{P}'$-martingale.

It remains to check that $\langle M \rangle$ is Lipschitz continuous in t under $\mathbb{P}'$ with Lipschitz constant $2L$. However, since $\langle M \rangle$ is not uniformly continuous in (A, M), our arguments should be different from that for A. Note that $M^2 - \langle M \rangle$ is a martingale under $\mathbb{P}'$ and $\mathbb{P}_n'$, and $\langle M \rangle_t - \langle M \rangle_s \leq L(t - s)$, $\mathbb{P}_n'$-a.s. for all $0 \leq s < t \leq T$. Following the arguments in the previous paragraph one can easily show that, for any $\eta \in C_b^0(\Omega') \cap \mathbb{L}^0(\mathscr{F}_s')$ with $0 \leq \eta \leq 1$,

$$\mathbb{E}^{\mathbb{P}'}\Big[\eta[\langle M \rangle_t - \langle M \rangle_s]\Big] = \mathbb{E}^{\mathbb{P}'}\Big[\eta[M_t^2 - M_s^2]\Big] \leq \limsup_{n \geq 1} \mathbb{E}^{\mathbb{P}_n'}\Big[\eta[M_t^2 - M_s^2]\Big]$$

$$\leq \limsup_{n \geq 1} \mathbb{E}^{\mathbb{P}_n'}\Big[\eta[\langle M \rangle_t - \langle M \rangle_s]\Big] \leq 2L \limsup_{n \geq 1} \mathbb{E}^{\mathbb{P}_n'}\Big[\eta(t - s)\Big] = 2L\mathbb{E}^{\mathbb{P}'}\Big[\eta(t - s)\Big].$$

Then, by standard approximation, one can see that, for any $\eta \in \mathbb{L}^0(\mathbb{F}')$ with $0 \leq \eta \leq 1$,

$$\mathbb{E}^{\mathbb{P}'}\Big[\int_0^T \eta_t d\langle M \rangle_t\Big] \leq 2L\mathbb{E}^{\mathbb{P}'}\Big[\int_0^T \eta_t dt\Big].$$

This implies that $d\langle M \rangle_t \leq 2Ldt$, $\mathbb{P}'$-a.s., which is the desired property. ∎

9.2.4   The Localized Spaces

For future purpose, we introduce some localized spaces. Recall the notations in Subsection 9.2.1. Let $H \in \mathscr{T}$, we define:

$$\mathscr{T}_H := \{\tau \in \mathscr{T} : \tau \le H\}; \; \Lambda_H := \{(t,\omega) \in \Lambda : t < H(\omega)\}; \; \overline{\Lambda}_H := \{(t,\omega) \in \Lambda : t \le H(\omega)\}.$$

$$(9.2.25)$$

We emphasize that, when $H = t$, the $\mathscr{T}_t$ here is different from the $\mathscr{T}^t$ in (6.1.2).

We denote by $C^0(\Lambda_H)$ the set of mappings $u : \Lambda_H \to \mathbb{R}$ continuous under $\mathbf{d}$. The other spaces $C_b^0(\Lambda)$, $C^0(\overline{\Lambda}_H)$, $C_b^0(\overline{\Lambda}_H)$, $UC(\overline{\Lambda}_H)$, and $UC_b(\overline{\Lambda}_H)$ are defined in an obvious way. We note that, for $H_1 \le H_2$, obviously $\mathscr{T}_{H_1} \subset \mathscr{T}_{H_2}$ and $\Lambda_{H_1} \subset \Lambda_{H_2}$, $\overline{\Lambda}_{H_1} \subset \overline{\Lambda}_{H_2}$. For any $u \in C^0(\Lambda_{H_2})$, we may restrict it to Λ_{H_1} and in this sense, obviously $C^0(\Lambda_{H_2}) \subset C^0(\Lambda_{H_1})$. Similar inclusions for the other localized spaces also hold true.

9.3   Regular Conditional Probability Distributions

9.3.1   The Shifting Operators

We first introduce a few notations concerning the concatenation and shifting operators. First, given $\omega, \tilde{\omega} \in \Omega$ and $t \in [0, T]$, define the concatenation path $\omega \otimes_t \tilde{\omega} \in \Omega$ by:

$$(\omega \otimes_t \tilde{\omega})(s) := \omega_s \mathbf{1}_{[0,t)}(s) + (\omega_t + \tilde{\omega}_{s-t})\mathbf{1}_{[t,T]}(s), \quad s \in [0, T]. \qquad (9.3.1)$$

Next, for any $\xi \in \mathbb{L}^0(\mathscr{F}_T)$, $\eta \in \mathbb{L}^0(\mathbb{F})$, and any $(t, \omega) \in \Lambda$, denote

$$\xi^{t,\omega}(\tilde{\omega}) := \xi(\omega \otimes_t \tilde{\omega}), \quad \eta_s^{t,\omega}(\tilde{\omega}) := \eta_{t+s}(\omega \otimes_t \tilde{\omega}), \; s \in [0, T-t], \quad \forall \tilde{\omega} \in \Omega. $$

$$(9.3.2)$$

It is clear that $\xi^{t,\omega} \in \mathbb{L}^0(\mathscr{F}_{T-t})$ and $\eta_s^{t,\omega}$ is $\mathscr{F}_s$-measurable. In particular,

$$X_s^{t,\omega}(\tilde{\omega}) = X_{t+s}(\omega \otimes_t \tilde{\omega}) = (\omega \otimes_t \tilde{\omega})_{t+s} = \omega_t + \tilde{\omega}_s, \text{ namely } X_s^{t,\omega} = \omega_t + X_s, \; s \in [0, T-t].$$

$$(9.3.3)$$

We denote

$$\mathscr{F}_s^t := \mathscr{F}_{s-t}, \; s \in [t, T], \quad \Lambda^t := [0, T-t) \times \Omega, \quad \overline{\Lambda}^t := [0, T-t] \times \Omega.$$

$$(9.3.4)$$

In particular, $\Lambda^t = \Lambda_{T-t}$ and $\overline{\Lambda}^t = \overline{\Lambda}_{T-t}$. Moreover, one can check that

$$\tau^{t,\omega} - t \in \mathcal{T} \quad \text{for all } \tau \in \mathcal{T} \text{ and } (t, \omega) \in \Lambda \text{ such that } t < \tau(\omega). \quad (9.3.5)$$

For $0 \le t \le T$, $X_{t\wedge\cdot}$ denotes the path stopping at t, that is, $(X_{t\wedge\cdot})_s := X_{t\wedge s} :=$ $X_s \mathbf{1}_{[0,t]}(s) + X_t \mathbf{1}_{(t,T]}(s)$. For $\xi \in \mathbb{L}^0(\mathcal{F}_T)$ and $0 \le t \le T$, note that $\xi(X_{t\wedge\cdot})$ is $\mathcal{F}_t$-measurable. For a probability measure $\mathbb{P}$ on $\mathcal{F}_T^t$, we shall always extend it to $\mathcal{F}_T$ by taking the convention:

$$\mathbb{E}^{\mathbb{P}}[\xi] := \mathbb{E}^{\mathbb{P}}\Big[\xi(X_{(T-t)\wedge\cdot})\Big], \quad \forall \xi \in C_b^0(\Omega).k \quad (9.3.6)$$

9.3.2 Regular Conditional Probability Distribution

We first introduce the regular conditional probability distribution (r.c.p.d. for short). Fix arbitrary $\mathbb{P} \in \mathscr{P}_\infty$ and $\tau \in \mathcal{T}$.

Definition 9.3.1 *The r.c.p.d. of $\mathbb{P}$ given $\mathcal{F}_\tau$ is a family $\{\mathbb{P}^{\tau,\omega} : \omega \in \Omega\}$ satisfying:*

(i) *For each $\omega \in \Omega$, $\mathbb{P}^{\tau,\omega}$ is a probability measure on $\mathcal{F}_T^{\tau(\omega)}$, and extended to $\mathcal{F}_T$ in the sense of (9.3.6).*

(ii) *For each $\xi \in \mathbb{L}^1(\mathcal{F}_T, \mathbb{P})$, the mapping $\omega \to \mathbb{E}^{\mathbb{P}^{\tau,\omega}}[\xi^{\tau,\omega}]$ is $\mathcal{F}_\tau$-measurable;*

(iii) *For $\mathbb{P}$-a.e. $\omega \in \Omega$, $\mathbb{P}^{\tau,\omega}$ is the conditional probability measure of $\mathbb{P}$ on $\mathcal{F}_\tau$, that is,*

$$\mathbb{E}^{\mathbb{P}}[\xi | \mathcal{F}_\tau](\omega) = \mathbb{E}^{\mathbb{P}^{\tau,\omega}}[\xi^{\tau,\omega}], \text{ for } \mathbb{P}\text{-a.e. } \omega \in \Omega \text{ and for all } \xi \in \mathbb{L}^1(\mathcal{F}_T, \mathbb{P}).$$

$$(9.3.7)$$

Theorem 9.3.2 *There exists unique (in $\mathbb{P}$-a.s. sense) r.c.p.d. $\{\mathbb{P}^{\tau,\omega} : \omega \in \Omega\}$. Moreover, for fixed τ, one may choose an appropriate version of $\{\mathbb{P}^{t,\omega} : \omega \in \Omega\}_{t\in[0,T]}$ such that*

$$\mathbb{P}^{\tau,\omega} = \mathbb{P}^{\tau(\omega),\omega} \quad \text{for } \mathbb{P}\text{-a.e. } \omega \in \Omega. \quad (9.3.8)$$

Proof The uniqueness is obvious. We prove the existence in three steps.

Step 1. In this step we construct certain probability measures $\mathbb{P}_\tau^\omega$ by using Lemmas 9.2.12 and 9.2.14. Assume $\mathbb{P} \in \mathscr{P}_L$. Let $K_n \subset \Omega$ satisfy (9.2.20), and by abusing the notations we let $\mathscr{A} \subset UC_b(\Omega)$ be the set of ξ taking the form (9.2.21), namely $\xi(\omega) = \psi(\varphi_m(\omega_{t_1}), \cdots, \varphi_m(\omega_{t_m}))$. Then clearly $\mathscr{A}$ is countable, and we numerate it as $\mathscr{A} = \{\xi_n\}_{n\ge 0}$, with $\xi_0 = 0$, $\xi_1 = 1$. Now for each n, fix a version of the following conditional expectations:

$$\eta_n := \mathbb{E}^{\mathbb{P}}[\xi_n | \mathcal{F}_\tau], \quad I_n := \mathbb{E}^{\mathbb{P}}[\mathbf{1}_{K_n^c} | \mathcal{F}_\tau]. \quad (9.3.9)$$

Our idea is to construct a version of $\mathbb{E}^{\mathbb{P}}[\xi | \mathcal{F}_\tau]$ through $\{\eta_n\}_{n\ge 1}$ for all $\xi \in C_b^0(\Omega)$.

Note that $\mathscr{A}$ is closed under linear combination with rational coefficients. Motivated by the conditions in Lemma 9.2.12, let Ω_1 denote the set of $\omega \in \Omega$ satisfying:

$$\eta_0(\omega) = 0, \quad \eta_1(\omega) = 1, \qquad \lim_{n\to\infty} I_n(\omega) = 0;$$

$$\xi_N(\tilde{\omega}) = \sum_{i=1}^{m} r_i \xi_{n_i}(\tilde{\omega}) \text{ for all } \tilde{\omega} \in \Omega \implies \eta_N(\omega) = \sum_{i=1}^{m} r_i \eta_{n_i}(\omega); \quad (9.3.10)$$

$$\eta_n(\omega) - \eta_m(\omega) \le \sup_{\tilde{\omega} \in K_k} [\xi_n(\tilde{\omega}) - \xi_m(\tilde{\omega})]^+ + \|\xi_n - \xi_m\|_\infty I_k(\omega), \quad \forall n, m, k.$$

We remark that, by sending $k \to \infty$, the last line above implies that $\eta_n \le \eta_m$ on Ω_1 if $\xi_n \le \xi_m$ on Ω. Note that $\mathbb{E}^{\mathbb{P}}\big[\sum_{n\ge 1} I_n\big] < \infty$. One can easily see that $\Omega_1 \in \mathscr{F}_\tau$ and $\mathbb{P}(\Omega_1) = 1$.

Now let $\xi \in C_b^0(\Omega)$. By the proof of Lemma 9.2.14 (ii) there exists $\{\tilde{\xi}_k\}_{k\ge 1} \subset UC_b(\Omega)$ such that

$$\|\tilde{\xi}_k\|_\infty \le \|\xi\|_\infty, \quad |\tilde{\xi}_k - \xi| \le k^{-1} \text{ on } K_k,$$

and similar to (9.2.22) there exists n_k such that

$$\|\xi_{n_k}\|_\infty \le \|\tilde{\xi}_k\|_\infty + 1, \quad |\xi_{n_k} - \tilde{\xi}_k| \le k^{-1} \text{ on } K_k.$$

Combining the above estimates, we have

$$\|\xi_{n_k}\|_\infty \le \|\xi\|_\infty + 1, \quad |\xi_{n_k} - \xi| \le 2k^{-1} \text{ on } K_k. \qquad (9.3.11)$$

For any k, and $i, j \ge k$, noting that K_k is increasing, on K_k we have

$$|\xi_{n_i} - \xi_{n_j}| \le |\xi_{n_i} - \tilde{\xi}_i| + |\tilde{\xi}_i - \xi| + |\tilde{\xi}_j - \xi| + |\tilde{\xi}_j - \xi_{n_j}| \le 4k^{-1}.$$

Then, by the last line of (9.3.10),

$$|\eta_{n_i}(\omega) - \eta_{n_j}(\omega)| \le \|(\xi_{n_i} - \xi_{n_j})\mathbf{1}_{K_k}\|_\infty + \|\xi_{n_i} - \xi_{n_j}\|_\infty I_k(\omega) \le 4k^{-1}$$
$$+ 2[\|\xi\|_\infty + 1]I_k(\omega).$$

This implies

$$\lim_{i,j\to\infty} |\eta_{n_i}(\omega) - \eta_{n_j}(\omega)| \le 4k^{-1} + 2[\|\xi\|_\infty + 1]I_k(\omega) \to 0 \quad \text{as } k \to \infty.$$

Thus $\lim_{i\to\infty} \eta_{n_i}(\omega)$ exists, and we may define an operator E_τ^ω:

$$\mathbb{E}_\tau^\omega[\xi] := \lim_{i\to\infty} \eta_{n_i}(\omega), \quad \forall \omega \in \Omega_1, \xi \in C_b^0(\Omega). \qquad (9.3.12)$$

We shall remark though the sequence n_i depends on ξ, but does not depend on $\omega \in \Omega_1$. Moreover, as long as (9.3.11) hold true, the choice of $\{n_i\}_{i \geq 1}$ does not affect the value of $\mathbb{E}^\omega_\tau[\xi]$.

We now fix $\omega \in \Omega_1$, and check that $\mathbb{E}^\omega_\tau$ satisfies the conditions in Lemma 9.2.12. First, $\mathbb{E}^\omega_\tau[1] = \mathbb{E}^\omega_\tau[\xi_1] = \eta_1(\omega) = 1$. Second, let $\xi, \tilde{\xi} \in C^0_b(\Omega)$ be such that $\xi \leq \tilde{\xi}$, and $\{\xi_{n_i}\}_{i \geq 1}, \{\tilde{\xi}_{\tilde{n}_i}\}_{i \geq 1} \subset \mathscr{A}$ be corresponding approximations satisfying (9.3.11). For any k and $i \geq k$, by (9.3.11) we have $|\xi_{n_i} - \xi| \leq 2k^{-1}$, $|\tilde{\xi}_{\tilde{n}_i} - \tilde{\xi}| \leq 2k^{-1}$ on K_k. Since $\xi \leq \tilde{\xi}$, then $[\xi_{n_i} - \tilde{\xi}_{\tilde{n}_i}] \leq 4k^{-1}$ on K_k. Then it follows from the last line of (9.3.10) that

$$\mathbb{E}^\omega_\tau[\xi] - \mathbb{E}^\omega_\tau[\tilde{\xi}] = \lim_{i \to \infty} \Big[\eta_{n_i}(\omega) - \eta_{\tilde{n}_i}(\omega) \Big] \leq 4k^{-1} + [\|\xi\|_\infty + \|\tilde{\xi}\|_\infty + 2]I_k(\omega)$$

Send $k \to \infty$, we obtain $\mathbb{E}^\omega_\tau[\xi] \leq \mathbb{E}^\omega_\tau[\tilde{\xi}]$. Moreover, note that $\xi_{n_i} + \tilde{\xi}_{\tilde{n}_i} \in \mathscr{A}$ is a desired approximation of $\xi + \tilde{\xi}$, then it follows from the second line of (9.3.10) that $\mathbb{E}^\omega_\tau[\xi + \tilde{\xi}] = \mathbb{E}^\omega_\tau[\xi] + \mathbb{E}^\omega_\tau[\tilde{\xi}]$. Finally, for any $\varepsilon > 0$, let k be large enough such that $2k^{-1} + 6I_k(\omega) \leq \varepsilon$. Now for $\xi \in C^0_b(\Omega)$ with $0 \leq \xi \leq 1$ and $\xi \mathbf{1}_{K_k} = 0$, let $\{\xi_{n_i}\}_{i \geq 1} \subset \mathscr{A}$ be a desired approximation. Then for $i \geq k$, we have

$$0 \leq \mathbb{E}^\omega_\tau[\xi] = \lim_{i \to \infty} \eta_{n_i}(\omega) = \lim_{i \to \infty} [\eta_{n_i}(\omega) - \eta_0(\omega)]$$

$$\leq \liminf_{i \to \infty} \Big[\sup_{\tilde{\omega} \in K_k} [\xi_{n_i}(\tilde{\omega})]^+ + 2\|\xi_{n_i}\|_\infty I_k(\omega) \Big] \leq 2k^{-1} + 2[\|\xi\|_\infty + 1]I_k(\omega) \leq \varepsilon.$$

Now applying Lemma 9.2.12 there exists a probability measure $\mathbb{P}^\omega_\tau$ on Ω such that $\mathbb{E}^{\mathbb{P}^\omega_\tau} = \mathbb{E}^\omega_\tau$, for each $\omega \in \Omega_1$.

Step 2. In this step we derive the desired properties of $\mathbb{P}^\omega_\tau$. First, for any $\xi \in C^0_b(\Omega)$, note that the approximation ξ_{n_i} in (9.3.11) does not depend on ω. Since η_{n_i} is $\mathscr{F}_\tau$-measurable, thus the mapping $\omega \in \Omega_1 \mapsto \mathbb{E}^\omega_\tau[\xi] = \lim_{i \to \infty} \eta_{n_i}(\omega)$ is also $\mathscr{F}_\tau$-measurable. Moreover, note that $\mathbb{E}^\mathbb{P}[\xi_{n_i} | \mathscr{F}_\tau](\omega) = \eta_{n_i}(\omega)$ for $\mathbb{P}$-a.e. ω, and by (9.3.11) clearly $\lim_{i \to \infty} \mathbb{E}^\mathbb{P}[\xi_{n_i} | \mathscr{F}_\tau] = \mathbb{E}^\mathbb{P}[\xi | \mathscr{F}_\tau]$, $\mathbb{P}$-a.s. Thus

$$\mathbb{E}^\mathbb{P}[\xi | \mathscr{F}_\tau](\omega) = \mathbb{E}^\omega_\tau[\xi] \quad \text{for } \mathbb{P}\text{-a.e. } \omega, \quad \forall \xi \in C^0_b(\Omega). \tag{9.3.13}$$

Next, since $\mathscr{A}$ is dense in $C^0_b(\Omega)$ in the sense of (9.3.11), one can easily see that $\mathscr{F}_\tau$ is generated by $\{\xi^\tau_n\}_{n \geq 1}$, where $\xi^\tau_n(\omega) := \xi_n(X_{\cdot \wedge \tau}(\omega))$. By (9.3.13) we have, for each $n \geq 1$,

$$\mathbb{E}^\omega_\tau[\xi^\tau_n] = \mathbb{E}^\mathbb{P}[\xi^\tau_n | \mathscr{F}_\tau](\omega) = \xi^\tau_n(\omega), \quad \text{for } \mathbb{P}\text{-a.e. } \omega.$$

Then

$$\mathbb{P}(\Omega_2) = 1, \quad \text{where } \Omega_2 := \Big\{ \omega \in \Omega_1 : \mathbb{E}^\omega_\tau[\xi^\tau_n] = \xi^\tau_n(\omega) \text{ for all } n \geq 1 \Big\}. \tag{9.3.14}$$

For each $\omega \in \Omega_2$, applying the monotone class theorem we see that

$$\mathbb{P}^\omega_\tau(E) = \mathbb{E}^{\mathbb{P}^\omega_\tau}[\mathbf{1}_E] = \mathbb{E}^\omega_\tau[\mathbf{1}_E] = \mathbf{1}_E(\omega) \text{ for all } E \in \mathscr{F}_\tau. \tag{9.3.15}$$

Denote

$$E^\omega := \left\{ \tilde{\omega} \in \Omega : \tilde{\omega}_t = \omega_t, 0 \le t \le \tau(\omega) \right\} = \left\{ \omega \otimes_{\tau(\omega)} \tilde{\omega} : \tilde{\omega} \in \Omega \right\} \in \mathscr{F}_\tau.$$
(9.3.16)

Clearly $\omega \in E^\omega$ for all $\omega \in \Omega$. Then by (9.3.15) we have,

$$\mathbb{P}_\tau^\omega(E^\omega) = \mathbf{1}_{E^\omega}(\omega) = 1 \text{ for all } \omega \in \Omega_2.$$
(9.3.17)

Now for each $\omega \in \Omega_2$ we may transform $\mathbb{P}_\tau^\omega$ on $\mathscr{F}_T$ to a probability measure $\mathbb{P}^{\tau,\omega}$ on $\mathscr{F}_T^{\tau(\omega)}$:

$$\mathbb{P}^{\tau,\omega}(A) := \mathbb{P}_\tau^\omega\left(\left\{ \omega \otimes_{\tau(\omega)} \tilde{\omega} : \tilde{\omega} \in A \right\}\right), \quad \forall A \in \mathscr{F}_T^{\tau(\omega)}.$$
(9.3.18)

It is clear that $\mathbb{E}^{\mathbb{P}_\tau^\omega}[\xi] = \mathbb{E}^{\mathbb{P}^{\tau,\omega}}[\xi^{\tau(\omega),\omega}]$ for all $\xi \in C_b^0(\Omega)$. Moreover, for $\omega \in \Omega_2^c$, we may define $\mathbb{P}^{\tau,\omega} := \delta_0$, namely $\mathbb{E}^{\mathbb{P}^{\tau,\omega}}[\xi] = \xi(0)$ for any $\mathscr{F}_T^{\tau(\omega)}$-measurable $\xi \in C_b^0(\Omega)$. Then it is straightforward to verify that $\{\mathbb{P}^{\tau,\omega}, \omega \in \Omega\}$ is the r.c.p.d. of $\mathbb{P}$ given $\mathscr{F}_\tau$.

Step 3. It remains to prove (9.3.8). Clearly it suffices to show that, for each $n \ge 1$,

$$\mathbb{E}_\tau^\omega[\xi_n] = \mathbb{E}_{\tau(\omega)}^\omega[\xi_n], \quad \text{for } \mathbb{P}\text{-a.e. } \omega.$$
(9.3.19)

Fix n. First, if $\tau = t$ is a constant, clearly (9.3.19) holds true. Next, assume τ takes finally many values $t_1 < \cdots < t_m$. Note that, for each $i = 1, \cdots, m$,

$$\mathbb{E}_\tau^\omega[\xi_n]\mathbf{1}_{\{\tau=t_i\}}(\omega) = \mathbb{E}^\mathbb{P}[\xi_n|\mathscr{F}_\tau](\omega)\mathbf{1}_{\{\tau=t_i\}}(\omega)$$
$$= \mathbb{E}^\mathbb{P}[\xi_n|\mathscr{F}_{t_i}](\omega)\mathbf{1}_{\{\tau=t_i\}}(\omega) = \mathbb{E}_{t_i}^\omega[\xi_n]\mathbf{1}_{\{\tau=t_i\}}(\omega), \text{ for } \mathbb{P}\text{-a.e. } \omega.$$

This implies (9.3.19) immediately. Finally, for an arbitrary $\tau \in \mathscr{T}$, denote $\tau_m := \sum_{i=1}^{2^m} t_i^m \mathbf{1}_{\{t_{i-1}^m < \tau \le t_i^m\}} \in \mathscr{T}$, where $t_i^m := i 2^{-m} T$, $i = 0, \cdots, 2^m$. Then $\tau_m \downarrow \tau$ as $m \to \infty$, and

$$\mathbb{P}(\Omega_3) = 1, \text{ where } \Omega_3 := \left\{ \omega \in \Omega_2 : \mathbb{E}_{\tau_m}^\omega[\xi_n] = \mathbb{E}_{\tau_m(\omega)}^\omega[\xi_n] \text{ for all } m \ge 1 \right\}.$$

Moreover, recall (9.3.9), we define

$$\eta_n(t) := \mathbb{E}^\mathbb{P}[\xi_n|\mathscr{F}_t], \quad t \in [0, T].$$

By Problem 9.6.2 (ii), we may fix a version of $\eta_n(t)$ such that $t \mapsto \eta_n(t)$ is continuous, $\mathbb{P}$-a.s. By (9.3.13) and Problem 9.6.2 (i), we have $\mathbb{P}(\Omega_4) = 1$, where Ω_4 consists of all those $\omega \in \Omega_3$ satisfying:

$$\mathbb{E}^\mathbb{P}[\xi_n|\mathscr{F}_{\tau_m}](\omega) = \mathbb{E}_{\tau_m}^\omega[\xi_n], \text{ for all } m \ge 1; \quad \mathbb{E}^\mathbb{P}[\xi_n|\mathscr{F}_\tau](\omega) = \mathbb{E}_\tau^\omega[\xi_n];$$

$$\lim_{m \to \infty} \mathbb{E}^\mathbb{P}[\xi_n|\mathscr{F}_{\tau_m}](\omega) = \mathbb{E}^\mathbb{P}[\xi_n|\mathscr{F}_\tau](\omega); \quad t \mapsto \eta_n(t, \omega) \text{ is continuous.}$$

Note that by definition $\mathbb{E}_t^\omega[\xi_n] = \eta_n(t, \omega)$. Then, for $\omega \in \Omega_4 \subset \Omega_3$,

$$\mathbb{E}_\tau^\omega[\xi_n] = \mathbb{E}^{\mathbb{P}}[\xi_n | \mathscr{F}_\tau](\omega) = \lim_{m \to \infty} \mathbb{E}^{\mathbb{P}}[\xi_n | \mathscr{F}_{\tau_m}](\omega) = \lim_{m \to \infty} \mathbb{E}_{\tau_m}^\omega[\xi_n]$$

$$= \lim_{m \to \infty} \mathbb{E}_{\tau_m(\omega)}^\omega[\xi_n] = \mathbb{E}_{\tau(\omega)}^\omega[\xi_n].$$

This verifies (9.3.19) and thus completes the proof.   ∎

9.3.3 Dynamic Sets of Probability Measures

One of the major tools to study dynamic problems is the dynamic programming principle. For that purpose, we fix $\mathscr{P} \subset \mathscr{P}_\infty$ and define

$$\mathscr{P}(t, \omega) := \{\mathbb{P}^{t,\omega} : \mathbb{P} \in \mathscr{P}\}. \tag{9.3.20}$$

The above set, however, is not pathwise defined because $\mathbb{P}^{t,\omega}$ is only in $\mathbb{P}$-a.s. sense. To get around of this, we shall fix a particular version of r.c.p.d. by assuming:

Assumption 9.3.3 $\mathscr{P} \subset \mathscr{P}_L$ for some $L > 0$, and there exist a version of r.c.p.d. such that

(i) $\mathscr{P}_t := \mathscr{P}(t, \omega) \subset \mathscr{P}_L$ does not depend on ω.
(ii) For any $0 \le s < t \le T$, $\mathbb{P} \in \mathscr{P}_s$, $\{E_i, i \ge 1\} \subset \mathscr{F}_t^s$ disjoint, and $\mathbb{P}_i \in \mathscr{P}_t$, $i \ge 1$, the following concatenation measure $\widehat{\mathbb{P}}$ is also in $\mathscr{P}_s$:

$$\widehat{\mathbb{P}} := \left[\mathbb{P} \otimes_t \sum_{i=1}^{\infty} \mathbb{P}_i \mathbf{1}_{E_i}\right] + \mathbb{P}\mathbf{1}_{\cap_{i=1}^n E_i^c} \tag{9.3.21}$$

that is, $\widehat{\mathbb{P}}$ is determined by: for any $\xi \in UC_b(\Omega)$,

$$\mathbb{E}^{\widehat{\mathbb{P}}}[\xi] := \sum_{i=1}^{\infty} \mathbb{E}^{\mathbb{P}}[\xi_i \mathbf{1}_{E_i}] + \mathbb{E}^{\mathbb{P}}[\xi \mathbf{1}_{\cap_{i=1}^\infty E_i^c}], \quad \text{where} \quad \xi_i(\omega) := \mathbb{E}^{\mathbb{P}_i}[\xi^{t-s,\omega}]. \tag{9.3.22}$$

We note that, for the $\widehat{\mathbb{P}}$ defined in (9.3.21), one can easily check that: denoting $r := t - s$,

$$\widehat{\mathbb{P}} = \mathbb{P} \text{ on } \mathscr{F}_r; \widehat{\mathbb{P}}^{r,\omega} = \mathbb{P}_i \text{ for } \mathbb{P}\text{-a.e. } \omega \in E_i; \widehat{\mathbb{P}}^{r,\omega} = \mathbb{P}^{r,\omega} \text{ for } \mathbb{P}\text{-a.e. } \omega \in \cap_{i=1}^\infty E_i^c. \tag{9.3.23}$$

Remark 9.3.4 In this book we assume $\mathscr{P}(t,\omega)$ does not depend on ω. This is mainly because of the regularity issue: in general it is difficult to characterize the regularity of the mapping $\omega \mapsto \mathscr{P}(t,\omega)$. Consequently, it will be difficult to establish the regularity (in ω) of the processes considered in the next chapters when $\mathscr{P}(t,\omega)$ depends on ω. We shall note though, for many results we require $\mathscr{P}$ to be weakly compact. However, in the situations that the weak compactness is not crucial and the $\mathscr{P}$ is induced from controls in strong formulation, in the spirit of $\mathscr{P}^S_{[\underline{\sigma},\overline{\sigma}]}$, then the regularity is a lot easier and quite often we may allow $\mathscr{P}(t,\omega)$ to depend on ω, see for example Proposition 11.3.8 below. ∎

We have the following result.

Proposition 9.3.5 *For any* $L > 0$ *and* $0 \leq \underline{\sigma} \leq \overline{\sigma}$, *the classes* $\mathscr{P}_L$, $\mathscr{P}^W_{[\underline{\sigma},\overline{\sigma}]}$ *and* $\mathscr{P}^S_{[\underline{\sigma},\overline{\sigma}]}$ *all satisfy Assumption 9.3.3.*

Proof We shall only prove the result for $\mathscr{P} = \mathscr{P}_L$. The other two cases can be proved similarly.

Consider the setting $(\Omega', A, M, X', \mathbb{P}')$ in Remark 9.2.10 (iii), with ω' denoting a generic element of Ω', and recall $\mathscr{P}'_L$ in Theorem 9.2.15 Step 1. Denote by $\mathbf{x} := X'(\omega') \in \Omega$ a generic sample path of X'.

Step 1. Fix $\mathbb{P}' \in \mathscr{P}'_L$ and $0 < t < T$. Following the arguments in Theorem 9.3.2, there exists a family $\{(\mathbb{P}')^{\mathbf{x}}_t : \mathbf{x} \in \Omega\}$ satisfying:

- For each $\mathbf{x} \in \Omega$, $(\mathbb{P}')^{\mathbf{x}}_t$ is a probability measure on $\mathscr{F}'_T$;
- For each $\xi \in \mathbb{L}^1(\mathscr{F}'_T, \mathbb{P}')$, the mapping $\omega' \in \Omega' \mapsto \mathbb{E}^{(\mathbb{P}')^{X'(\omega')}_t}[\xi]$ is $\mathscr{F}'_t$-measurable;
- For $\mathbb{P}'$-a.e. $\omega' \in \Omega'$, $(\mathbb{P}')^{X'(\omega')}_t$ is the conditional probability measure of $\mathbb{P}'$ on $\mathscr{F}^{X'}_t$, that is,

$$\mathbb{E}^{\mathbb{P}'}[\xi | \mathscr{F}^{X'}_t](\omega') = \mathbb{E}^{(\mathbb{P}')^{X'(\omega')}_t}[\xi], \quad \mathbb{P}'\text{-a.s. for all } \xi \in \mathbb{L}^1(\mathscr{F}'_T, \mathbb{P}').$$

- the family $\{(\mathbb{P}')^{\mathbf{x}}_t : \mathbf{x} \in \Omega\}$ is regular in the following sense:

$$(\mathbb{P}')^{X'(\omega')}_t\left(E'_{t,X'(\omega')}\right) = 1, \quad \text{for } \mathbb{P}'\text{-a.e. } \omega' \in \Omega',$$
$$\text{where } E'_{t,\mathbf{x}} := \left\{\omega' \in \Omega' : X'_s(\omega') = \mathbf{x}_s, 0 \leq s \leq t\right\}. \tag{9.3.24}$$

Denote $\tilde{\mathbb{P}} := \mathbb{P}'\big|_{\mathscr{F}^{X'}_T}$, then the $\tilde{\mathbb{P}}$-distribution of X' is equal to the $\mathbb{P}$-distribution of X. Thanks to (9.3.24), as in (9.3.18) we may introduce probability measures $(\mathbb{P}')^{t,\mathbf{x}}$ on $\mathscr{F}'_{T-t}$ as follows:

$$(\mathbb{P}')^{t,\mathbf{x}}(A) := (\mathbb{P}')^{\mathbf{x}}_t\left(\left\{\omega' \otimes_t \tilde{\omega}' : \omega' \in E'_{t,\mathbf{x}}, \tilde{\omega}' \in A\right\}\right), \quad \forall A \in \mathscr{F}'_{T-t}. \tag{9.3.25}$$

One can easily see that $\tilde{\mathbb{P}}^{t,\mathbf{x}} = (\mathbb{P}')^{t,\mathbf{x}}\big|_{\mathscr{F}^{X'}_{T-t}}$, then $\mathbb{P}^{t,\omega} = (\mathbb{P}')^{t,\mathbf{x}} \circ (X')^{-1}$ for $\mathbf{x} = \omega$.

Step 2. In this step we show that $(\mathbb{P}')^{t,\mathbf{x}} \in \mathscr{P}'_L$, for $\tilde{\mathbb{P}}$-a.e. $\mathbf{x} \in \Omega$, in the sense of (9.3.6). Then $\mathbb{P}^{t,\omega} \in \mathscr{P}_L$ for $\mathbb{P}$-a.e. $\omega \in \Omega$, and thus $\mathscr{P}_t \subset \mathscr{P}_L$.

First, for any $r_1, r_2 \in \mathbb{Q} \cap [0, T-t]$ with $r_1 < r_2$, and any $\mathbf{x} \in \Omega$, recalling (9.3.3) we have

$$(\mathbb{P}')_t^{\mathbf{x}}\big(|A_{t+r_1} - A_{t+r_2}| \le L|r_2 - r_1|\big)$$
$$= (\mathbb{P}')_t^{\mathbf{x}}\big(\omega' \otimes_t \tilde{\omega}' : \omega' \in E'_{t,\mathbf{x}}, \tilde{\omega}' \in \Omega', |A_{t+r_1}(\omega' \otimes_t \tilde{\omega}') - A_{t+r_2}(\omega' \otimes_t \tilde{\omega}')| \le L|r_2 - r_1|\big)$$
$$= (\mathbb{P}')^{t,\mathbf{x}}\big(\tilde{\omega}' \in \Omega' : |A_{r_1}(\tilde{\omega}') - A_{r_2}(\tilde{\omega}')| \le L|r_2 - r_1|\big).$$
$$(9.3.26)$$

Then

$$1 = \mathbb{P}'\big(|A_{t+r_1} - A_{t+r_2}| \le L|r_2 - r_1|\big) = \mathbb{E}^{\mathbb{P}'}\Big[(\mathbb{P}')^{t,X'(\omega')}\big(|A_{r_1} - A_{r_2}| \le L|r_2 - r_1|\big)\Big]$$
$$= \mathbb{E}^{\tilde{\mathbb{P}}}\Big[(\mathbb{P}')^{t,\mathbf{x}}\big(|A_{r_1} - A_{r_2}| \le L|r_2 - r_1|\big)\Big],$$

where $\mathbb{E}^{\mathbb{P}'}$ and $\mathbb{E}^{\tilde{\mathbb{P}}}$ are with respect to ω' and $\mathbf{x}$, respectively. This implies that

$$(\mathbb{P}')^{t,\mathbf{x}}\big(|A_{r_1} - A_{r_2}| \le L|r_2 - r_1|\big) = 1, \quad \text{for } \tilde{\mathbb{P}}\text{-a.e. } \mathbf{x}.$$

Then

$$(\mathbb{P}')^{t,\mathbf{x}}\Big(|A_{r_1} - A_{r_2}| \le L|r_2 - r_1|, \forall r_1, r_2 \in \mathbb{Q} \cap [0, T-t]\Big) = 1, \quad \text{for } \tilde{\mathbb{P}}\text{-a.e. } \mathbf{x}.$$

Since A is continuous, we see that

$$(\mathbb{P}')^{t,\mathbf{x}}\Big(|A_{t_1} - A_{t_2}| \le L|t_2 - t_1|, \forall t_1, t_2 \in [0, T-t]\Big) = 1, \quad \text{for } \tilde{\mathbb{P}}\text{-a.e. } \mathbf{x}.$$
$$(9.3.27)$$

Next, applying the arguments in (9.3.26) on M and by the Burkholder-Davis-Gundy equality (for general martingales), we have

$$\mathbb{E}^{\tilde{\mathbb{P}}}\Big[\mathbb{E}^{(\mathbb{P}')^{t,\mathbf{x}}}\big[\sup_{0 \le s \le T-t} |M_s|^2\big]\Big] = \mathbb{E}^{\mathbb{P}'}\Big[\sup_{t \le s \le T} |M_s - M_t|^2\Big] \le C\mathbb{E}^{\mathbb{P}'}\Big[\langle M \rangle_T - \langle M \rangle_t\Big] < \infty,$$

where the $\mathbb{E}^{\tilde{\mathbb{P}}}$ is with respect to $\mathbf{x}$. Then

$$\mathbb{E}^{(\mathbb{P}')^{t,\mathbf{x}}}\big[\sup_{0 \le s \le T-t} |M_s|^2\big] < \infty, \quad \text{for } \tilde{\mathbb{P}}\text{-a.e. } \mathbf{x} \in \Omega. \qquad (9.3.28)$$

Moreover, for any $r_1, r_2 \in \mathbb{Q} \cap [0, T-t]$ with $r_1 < r_2$, any $\xi \in UC_b(\Omega') \cap \mathbb{L}^0(\mathscr{F}'_{r_1})$, and any bounded $\eta \in \mathbb{L}^0(\mathscr{F}_t^{X'})$, denoting $\tilde{\xi}(\omega') := \xi\big((\omega'_{t,s})_{t \le s \le T}\big)$, we have

$$0 = \mathbb{E}^{\mathbb{P}'}\Big[(M_{t+r_2} - M_{t+r_1})\tilde{\xi}\eta\Big] = \mathbb{E}^{\tilde{\mathbb{P}}}\Big[\mathbb{E}^{(\mathbb{P}')^{t,\mathbf{x}}}[(M_{r_2} - M_{r_1})\xi]\eta(\mathbf{x})\Big].$$

Set $\eta(\mathbf{x}) := \text{sign}\big(\mathbb{E}^{(\mathbb{P}')^{t,\mathbf{x}}}[(M_{r_2} - M_{r_1})\xi]\big)$. The above equality implies

$$\mathbb{E}^{(\mathbb{P}')^{t,\mathbf{x}}}[(M_{r_2} - M_{r_1})\xi] = 0, \quad \text{for } \tilde{\mathbb{P}}\text{-a.e. } \mathbf{x} \in \Omega.$$

Now let $\{\xi_n^{r_1}\}_{n \geq 1}$ denote the set $\mathscr{A}$ in Theorem 9.3.2 Step 1, but restricting to the space of continuous paths on $[0, r_1]$. Then we have

$$\mathbb{E}^{(\mathbb{P}')^{t,\mathbf{x}}}[(M_{r_2} - M_{r_1})\xi_n^{r_1}] = 0, \quad \forall r_1, r_2 \in \mathbb{Q} \cap [0, T - t] \text{ with } r_1 < r_2, \forall n \geq 1,$$
$$\text{for } \tilde{\mathbb{P}}\text{-a.e. } \mathbf{x} \in \Omega.$$

Since $\{\xi_n^{r_1}\}_{n \geq 1}$ is dense and M is continuous, by (9.3.28) we see that, for $\tilde{\mathbb{P}}$-a.e. $\mathbf{x} \in \Omega$,

$$\mathbb{E}^{(\mathbb{P}')^{t,\mathbf{x}}}[(M_{t_2} - M_{t_1})\xi^{t_1}] = 0, \quad \forall 0 \leq t_1 < t_2 \leq T - t \text{ and } \forall \xi^{t_1} \in \mathbb{L}^2(\mathscr{F}_{t_1}', (\mathbb{P}')^{t,\mathbf{x}}).$$

That is,

$$M \text{ is a } (\mathbb{P}')^{t,\mathbf{x}}\text{-martingale on } [0, T - t], \text{ for } \tilde{\mathbb{P}}\text{-a.e. } \mathbf{x} \in \Omega. \qquad (9.3.29)$$

Finally, denote $\tilde{\xi} := 0 \vee \xi \wedge 1$ for any ξ. For any $0 \leq t_1 < t_2 \leq T - t$, let $r_1, \cdots, r_m \in \mathbb{Q} \cap [0, T - t]$ with $t_1 = r_0 < \cdots < r_m = t_2$. Then for arbitrary $\eta \in \mathbb{L}^0(\mathscr{F}_t)$ and $\xi_{n_i}^{r_i}, i = 0, \cdots, m$, we have,

$$\mathbb{E}^{\tilde{\mathbb{P}}}\Big[\eta(\mathbf{x})\mathbb{E}^{(\mathbb{P}')^{t,\mathbf{x}}}\Big[\sum_{i=0}^{m-1} \tilde{\xi}_{n_i}^{r_i}(\langle M\rangle_{r_{i+1}} - \langle M\rangle_{r_i})\Big]\Big]$$

$$= \mathbb{E}^{\mathbb{P}'}\Big[\eta(X')\sum_{i=0}^{m-1} \tilde{\xi}_{n_i}^{r_i}((\omega_s' - \omega_t')_{t \leq s \leq T})[\langle M\rangle_{t+r_{i+1}} - \langle M\rangle_{t+r_i}]\Big]$$

$$\leq 2L \sum_{i=0}^{m-1}(r_{i+1} - r_i) = 2L(t_2 - t_1).$$

Since η is arbitrary, we see that

$$\mathbb{E}^{(\mathbb{P}')^{t,\mathbf{x}}}\Big[\sum_{i=0}^{m-1} \tilde{\xi}_{n_i}^{r_i}(\langle M\rangle_{r_{i+1}} - \langle M\rangle_{r_i})\Big] \leq 2L(t_2 - t_1), \quad \text{for } \tilde{\mathbb{P}}\text{-a.e. } \mathbf{x} \in \Omega.$$

By the arbitrariness of r_i and $\xi_{n_i}^{t_i}$, we see that, for all $\xi \in \mathbb{L}^0(\mathbb{F})$ such that $0 \leq \xi \leq 1$,

$$\mathbb{E}^{(\mathbb{P}')^{t,\mathbf{x}}}\Big[\int_{t_1}^{t_2} \xi_s d\langle M\rangle_s\Big] \leq 2L(t_2 - t_1), \quad \text{for } \tilde{\mathbb{P}}\text{-a.e. } \mathbf{x} \in \Omega.$$

Now by first restricting t_1, t_2 to rationals and then extend to general real numbers, we have

$$\mathbb{E}^{(\mathbb{P}')^{t,\mathbf{x}}}\left[\int_{t_1}^{t_2} \xi_s d\langle M\rangle_s\right] \leq 2L(t_2 - t_1), \quad 0 \leq t_1 < t_2 \leq T - t, \text{ for } \tilde{\mathbb{P}}\text{-a.e. } \mathbf{x} \in \Omega.$$

Therefore, it follows from the arbitrariness of ξ that:

$$d\langle M\rangle_s \leq 2L ds, \ 0 \leq s \leq T - t, \ (\mathbb{P}')^{t,\mathbf{x}}\text{-a.s.} \quad \text{for } \mathbb{P}'\text{-a.e. } \mathbf{x} \in \Omega. \quad (9.3.30)$$

Combining (9.3.27), (9.3.29), and (9.3.30), we see that $(\mathbb{P}')^{t,\mathbf{x}} \in \mathscr{P}'_L$, for $\tilde{\mathbb{P}}$-a.e. $\mathbf{x} \in \Omega$.

Step 3. We now verify Assumption 9.3.3 (ii) together with a slightly stronger statement than (i):

$$\mathscr{P}_t = \mathscr{P}^t_L := \{\mathbb{P} \in \mathscr{P}_L : b'_s = 0, \sigma'_s = 0, s \in [T - t, T]\}. \quad (9.3.31)$$

Here b', σ' are as in Definition 9.2.9. We remark that in Step 2 we actually have proved $\mathscr{P}_t \subset \mathscr{P}^t_L$.

We first assume $s = 0$. For any $\mathbb{P} \in \mathscr{P}_L$ and $\mathbb{P}_i \in \mathscr{P}^t_L$, $i \geq 1$, let $\mathbb{P}' \in \mathscr{P}'_L$ and $\mathbb{P}'_i \in \mathscr{P}'_L$ be the corresponding measures and recall the setting $(\Omega', A, M, X', \mathbb{P}')$ in Remark 9.2.10 (iii) again. Define $\hat{\mathbb{P}}$ by (9.3.21) and

$$\hat{\mathbb{P}}' := \left[\mathbb{P}' \otimes_t \sum_{i=1}^{\infty} \mathbb{P}'_i \mathbf{1}_{E_i}(X')\right] + \mathbb{P}' \mathbf{1}_{\cap_{i=1}^{n} E_i^c(X')}.$$

It is clear that $\hat{\mathbb{P}} = \hat{\mathbb{P}}' \circ (X')^{-1}$. Following the arguments in Step 2, one can easily show that $\hat{\mathbb{P}}' \in \mathscr{P}'_L$ and thus $\hat{\mathbb{P}} \in \mathscr{P}_L$. In particular, $\mathbb{P} \in \mathscr{P}_0 = \mathscr{P}_L$ and $\mathbb{P}_i \in \mathscr{P}_t \subset \mathscr{P}'_L$, we have $\hat{\mathbb{P}} \in \mathscr{P}_L = \mathscr{P}_0$, verifying Assumption 9.3.3 (ii) at $s = 0$.

Next, for any $\mathbb{P} \in \mathscr{P}$ and $\tilde{\mathbb{P}} \in \mathscr{P}^t_L$, denote $\hat{\mathbb{P}} := \mathbb{P} \otimes_t \tilde{\mathbb{P}}$. By above we have $\hat{\mathbb{P}} \in \mathscr{P}_L$. It follows from (9.3.23) that $(\hat{\mathbb{P}})^{t,\omega} = \tilde{\mathbb{P}}$ on $\mathscr{F}^t_T$, for $\mathbb{P}$-a.e. $\omega \in \Omega$. Then by the definition of $\mathscr{P}_t$ we see that $\tilde{\mathbb{P}} \in \mathscr{P}_t$. That is, $\mathscr{P}^t_L \subset \mathscr{P}_t$, and thus by Step 2 they are equal.

Finally, given (9.3.31), one can easily verify Assumption 9.3.3 (ii) for general s in the same way as $s = 0$. ∎

9.4 Functional Itô Formula

As we saw in the previous chapters of the book, the Itô formula is a corner stone for stochastic calculus. In this section we extend it to the so-called functional Itô formula, which is convenient and powerful for the pathwise analysis in weak formulation. We first note that $C^0(\overline{\Lambda}) \subset \bigcap_{\mathbb{P} \in \mathscr{P}_\infty} \mathbb{L}^2_{loc}(\mathbb{F}, \mathbb{P})$. Then, for any $u \in C^0(\overline{\Lambda}, \mathbb{R}^d)$, the stochastic integral $u_t \cdot dX_t$ exists $\mathbb{P}$-a.s. for all $\mathbb{P} \in \mathscr{P}_\infty$. In fact, as we will see in Theorem 12.1.2 below, one may find a common version of $u_t \cdot dX_t$ independent of $\mathbb{P}$.

Recall (9.2.25). The following definition is crucial.

Definition 9.4.1 *Let* $u \in C^0(\overline{\Lambda})$. *We say* $u \in C^{1,2}(\Lambda)$ *if there exist* $\partial_t u \in C^0(\Lambda)$, $\partial_\omega u \in C^0(\Lambda, \mathbb{R}^{1 \times d})$, *and* $\partial^2_{\omega\omega} u \in C^0(\Lambda, \mathbb{S}^d)$ *such that the following holds:*

$$du_t = \partial_t u dt + \partial_\omega u_t dX_t + \frac{1}{2} \partial^2_{\omega\omega} u_t : d\langle X \rangle_t, \quad 0 \le t < T, \mathbb{P}\text{-a.s. for all } \mathbb{P} \in \mathscr{P}_\infty.$$
$$(9.4.1)$$

In this case, we call $\partial_t u$, $\partial_\omega u$ *the path derivatives of* u *with respect to* t *and* ω, *respectively,* $\partial^2_{\omega\omega} u$ *the second order path derivative of* u *with respect to* ω, *and we call* (9.4.1) *the functional Itô formula.*

Moreover, we say $u \in C^{1,2}(\overline{\Lambda})$ *if the path derivatives* $\partial_t u$, $\partial_\omega u$, *and* $\partial^2_{\omega\omega} u$ *are continuous on* $\overline{\Lambda}$.

Similarly, for any $\mathrm{H} \in \mathscr{T}$, *we say* $u \in C^{1,2}(\Lambda_\mathrm{H})$ *if* $u \in C^0(\overline{\Lambda}_\mathrm{H})$ *and there exist* $\partial_t u \in C^0(\Lambda_\mathrm{H})$, $\partial_\omega u \in C^0(\Lambda_\mathrm{H}, \mathbb{R}^{1 \times d})$, *and* $\partial^2_{\omega\omega} u \in C^0(\Lambda_\mathrm{H}, \mathbb{S}^d)$ *such that* (9.4.1) *holds for* $0 \le t < \mathrm{H}$.

By definition $u \in C^{1,2}(\Lambda)$ is a semimartingale under each $\mathbb{P} \in \mathscr{P}_\infty$. We emphasize that at above we require u itself is continuous on $\overline{\Lambda}$ but the path derivatives are continuous only in Λ. We also note that while it is straightforward to extend the smoothness to multidimensional processes, in this book we consider only scalar u, and $\partial_\omega u$ is viewed as a row vector.

Lemma 9.4.2 *For any* $u \in C^{1,2}(\Lambda)$, *the path derivatives* $\partial_t u$, $\partial_\omega u$, *and* $\partial^2_{\omega\omega} u$ *are unique.*

Proof First, let $\mathbb{P}^0$ denote the measure corresponding to $b' = 0, \sigma' = 0$ in (9.2.10). For any t, apply the functional Itô formula (9.4.1) on $\mathbb{P}_0 \otimes_t \mathbb{P}^0$, where the concatenation is in the sense of (9.3.21)–(9.3.22) with $i = 1$ and $E_1 = \Omega$, for any $\varepsilon > 0$ we have

$$u(t + \varepsilon, \omega_{\cdot \wedge t}) - u(t, \omega) = \int_t^{t+\varepsilon} \partial_t u(s, \omega_{\cdot \wedge t}) ds, \quad \text{for } \mathbb{P}_0\text{-a.e. } \omega.$$

Since both u and $\partial_t u$ are continuous, by Problem 9.6.13 we see that

$$u(t + \varepsilon, \omega_{\cdot \wedge t}) - u(t, \omega) = \int_t^{t+\varepsilon} \partial_t u(s, \omega_{\cdot \wedge t}) ds, \quad \text{for all } \omega.$$

Then, for all $\omega \in \Omega$, we have the following representation of $\partial_t u(t, \omega)$ and thus $\partial_t u$ is unique:

$$\partial_t u(t, \omega) = \lim_{\varepsilon \to 0} \frac{u(t + \varepsilon, \omega_{\cdot \wedge t}) - u(t, \omega)}{\varepsilon}. \tag{9.4.2}$$

Next, applying functional Itô formula (9.4.1) on $\mathbb{P}_0$, we see that $\partial_\omega u$ is unique, $\mathbb{P}_0$-a.s. Then it follows from Problem 9.6.13 again that $\partial_\omega u$ is unique for all (t, ω).

Finally, for any $\sigma \in \mathbb{S}^d$ such that $\sigma > 0$, let $\mathbb{P}^\sigma$ denote the measure corresponding to constants $b' = 0, \sigma' = \sigma$ in (9.2.10). Applying (9.4.1) on $\mathbb{P}^\sigma$ and by the uniqueness of $\partial_t u$ and $\partial_\omega u$, we see that $\partial^2_{\omega\omega} u(t, \omega) : \sigma^2$ is unique, $\mathbb{P}^\sigma$-a.s. Now since $\sigma > 0$, similar to Problem 9.6.13 we can show that any support of $\mathbb{P}^\sigma$ is dense. Then the continuity of $\partial^2_{\omega\omega} u$ implies that $\partial^2_{\omega\omega} u(t, \omega) : \sigma^2$ is unique for all (t, ω). Now since σ is arbitrary and $\partial^2_{\omega\omega} u(t, \omega)$ is symmetric, we see that $\partial^2_{\omega\omega} u(t, \omega)$ is unique. ∎

Remark 9.4.3

(i) The path derivatives are initiated by Dupire [71], which defines the time derivative by (9.4.2) and spatial derivatives through perturbation and thus involves càdlàg paths. Our definition of $\partial_\omega u$ and $\partial^2_{\omega\omega} u$ is due to Ekren, Touzi, & Zhang [75], which is consistent with Dupire's definition, provided they are smooth enough, and is more convenient for some applications, as we will see in Chapter 11 below.

(ii) When u is smooth enough, $\partial_\omega u(t, \omega) = D_t u(t, \omega)$, where D_t is the Malliavin differential operator. We refer to Nualart [159] for Malliavin calculus.

(iii) From (9.4.2) we see that the time derivative $\partial_t u$ is actually the right time derivative. This is crucial and appropriate because we will use it to study backward problems. For forward problems, for example forward SDEs or SPDEs, one needs to use left time derivatives, defined in an appropriate way. See also Remark 9.4.7 below. ∎

At below we provide some examples.

Example 9.4.4

(i) Let $u(t, \omega) = v(t, \omega_t)$ be Markovian. If $v \in C^{1,2}([0, T] \times \mathbb{R}^d)$, by the standard Itô formula we see that $u \in C^{1,2}(\overline{\Lambda})$ with

$$\partial_\omega u(t, \omega) = \partial_x v(t, \omega_t), \quad \partial^2_{\omega\omega} u(t, \omega) = \partial^2_{xx} v(t, \omega_t), \quad \partial_t u(t, \omega) = \partial_t v(t, \omega_t)$$

$$(9.4.3)$$

(ii) Let $\eta \in C^0(\overline{\Lambda})$ and $u(t, \omega) := \int_0^t \eta_s(\omega)ds$. It is clear that $u \in C^{1,2}(\overline{\Lambda})$ with $\partial_\omega u = 0$, $\partial^2_{\omega\omega} u = 0$, and $\partial_t u = \eta$.

(iii) Let $\eta \in C^0(\overline{\Lambda}, \mathbb{R}^d)$, $A_t := \int_0^t \eta_s ds$, and $u(t, \omega) := A_t(\omega) \cdot \omega_t - \int_0^t \omega_s \cdot \eta_s(\omega)ds$. Then clearly $A, u \in C^0(\overline{\Lambda})$. Moreover, recall that $X(\omega) = \omega$ and thus $u_t = A_t \cdot X_t - \int_0^t X_s \cdot \eta_s ds$. For each $\mathbb{P} \in \mathscr{P}_\infty$, applying Itô formula we have $du = A_t \cdot dX_t$, $\mathbb{P}$-a.s. Then it follows from Definition 9.4.1 that $u \in C^{1,2}(\overline{\Lambda})$ with $\partial_\omega u = A^\top$, $\partial^2_{\omega\omega} u = 0$, $\partial_t u = 0$.

(iv) Let $d = 1$ and consider the running maximum process: $u(t, \omega) := \omega_t^* := \max_{0 \le s \le t} |\omega_s|$, $(t, \omega) \in \overline{\Lambda}$. Clearly $u \in C^0(\overline{\Lambda})$. We claim that $u \notin C^{1,2}(\Lambda)$. Indeed, assume $u \in C^{1,2}(\Lambda)$, then by (9.4.1) we have: $du(t, \omega) = [\partial_t u_t +$

$\frac{1}{2}\partial^2_{\omega\omega}u]dt + \partial_\omega u dB_t$, $\mathbb{P}_0$-a.s. Since u itself is nondecreasing, we must have $\partial_\omega u = 0$, and thus $du(t, \omega) = [\partial_t u_t + \frac{1}{2}\partial^2_{\omega\omega}u]dt$, $\mathbb{P}_0$-a.s. That is, ω^*_t is absolutely continuous with respect to dt, which is a contradiction. ∎

Remark 9.4.5 Let $d = 1$ for simplicity. As we will see in Section 12.1 below, given $\eta \in C^0(\overline{\Lambda}, \mathbb{R}^d)$, one may define $\eta_t dX_t$, $\eta_t d\langle X \rangle_t$ in a pathwise way. However, typically they are not continuous in ω. In other words, given $\eta^i \in C^0(\overline{\Lambda})$, $i = 1, 2, 3$, and define

$$u_t = u_0 + \int_0^t \eta^1_s ds + \int_0^t \eta^2_s dX_s + \frac{1}{2}\int_0^t \eta^3_s d\langle X \rangle_s.$$

Typically u is not in $C^0(\overline{\Lambda})$, hence we cannot claim that $u \in C^{1,2}(\Lambda)$ with $\partial_t u = \eta^1$, $\partial_\omega u = \eta^2$, $\partial^2_{\omega\omega}u = \eta^3$. ∎

Remark 9.4.6

(i) Typically ∂_t and ∂_ω do not commute. For example, consider $u(t, \omega) := \int_0^t \omega_s ds$, which is in $C^{1,2}(\overline{\Lambda})$ as we see in Example 9.4.4 (ii). Then we have

$$\partial_t u = \omega_t, \quad \partial_\omega u = 0, \quad \text{and thus} \quad \partial_\omega \partial_t u = 1 \neq 0 = \partial_t \partial_\omega u.$$

(ii) When u is smooth enough, one can show that ∂_{ω^i} and ∂_{ω^j} commute, and $\partial^2_{\omega\omega}u = \partial_\omega(\partial_\omega u)^\top$ is symmetric. See also Remark 9.4.7 below.

(iii) For $u \in C^{1,2}(\Lambda)$, from Definition 9.4.1 it is not clear that $\partial^2_{\omega\omega}u = \partial_\omega(\partial_\omega u)^\top$. In particular, this cannot be true when $\partial_\omega(\partial_\omega u)^\top$ is not symmetric. However, note that $\langle X \rangle_t$ is symmetric, then

$$\partial_\omega(\partial_\omega u)^\top_t : d\langle X \rangle_t = \frac{1}{2}\Big[\partial_\omega(\partial_\omega u)^\top_t + [\partial_\omega(\partial_\omega u)^\top_t]^\top\Big] : d\langle X \rangle_t.$$

Roughly speaking, in this case we may view $\partial^2_{\omega\omega}u = \frac{1}{2}\Big[\partial_\omega(\partial_\omega u)^\top_t + [\partial_\omega(\partial_\omega u)^\top_t]^\top\Big]$, which is by definition symmetric. ∎

Remark 9.4.7

(i) An alternative way to define the path derivatives is to use the pathwise Taylor expansion under $\mathbb{P}_0$, see Buckdahn, Ma, & Zhang [26]: for $\delta > 0$,

$$u(t + \delta, \omega) - u(t, \omega) = \partial_t u(t, \omega)\delta + \partial_\omega u(t, \omega)\omega_{t,t+\delta}$$

$$+ \frac{1}{2}\partial^2_{\omega\omega}u(t, \omega) : [\omega_{t,t+\delta}\omega^\top_{t,t+\delta}] + o(\delta), \quad \mathbb{P}_0\text{-a.e. } \omega \in \Omega.$$

$$(9.4.4)$$

In particular, the spatial derivative $\partial_\omega u$ is equal to the Gubinelli's derivative in rough path theory, for the latter we refer to Friz & Hairer [94]. Provided

enough regularity, the above definition is equivalent to the functional Itô
formula (9.4.1). Moreover, in this case we have $\partial^2_{\omega\omega} u = \partial_\omega[(\partial_\omega u)^\top]$ and is
symmetric. See also Remark 9.4.6 (ii).

(ii) The Taylor expansion (9.4.4) is from the right. For forward problems, for
example forward stochastic PDEs, it is more appropriate to use left Taylor
expansion: for $\delta > 0$,

$$u(t - \delta, \omega) - u(t, \omega) = -\partial_t u(t, \omega)\delta - \partial_\omega u(t, \omega)\omega_{t-\delta,t}$$

$$+ \frac{1}{2}\partial^2_{\omega\omega} u(t, \omega) : [\omega_{t-\delta,t}\omega^\top_{t-\delta,t}] + o(\delta), \quad \mathbb{P}_0\text{-a.e. } \omega \in \Omega.$$

$$(9.4.5)$$

When u is smooth enough, both (9.4.4) and (9.4.5) hold true. However, (9.4.5)
is more appropriate for forward problems because the $\partial_t u$ here is left time
derivative. See also Remark 9.4.3 (iii).

(iii) One may extend the above pathwise Taylor expansion to general rough paths.
In that case we shall replace the $\partial_t u(t, \omega)\delta$ in (9.4.4) with $\partial_t u(t, \omega)\langle\omega\rangle_{t,t+\delta}$ for
some appropriately defined $\langle\omega\rangle$, see Keller & Zhang [123]. ∎

As an application of the functional Itô formula, we note that the Itô-Ventzell
formula is equivalent to the chain rule of the path derivatives.

Remark 9.4.8 Let $\tilde{X} : \Lambda \to \mathbb{R}^n$ and $u : \Lambda \times \mathbb{R}^n \to \mathbb{R}$ be $\mathbb{F}$-measurable taking the
form:

$$d\tilde{X}_t = b_t dt + \sigma_t dX_t, \quad du(t, x) = \alpha(t, x)dt + \beta(t, x)dX_t, \quad \mathbb{P}_0\text{-a.s.} \quad (9.4.6)$$

Assume enough regularity and integrability, then the following Itô-Ventzell formula:

$$du(t, \tilde{X}_t) = \left[\alpha + \partial_x u b_t + \frac{1}{2}\partial^2_{xx} u : \sigma_t\sigma_t^\top + \partial_x \beta^\top : \sigma_t\right](t, \tilde{X}_t)dt$$

$$+ \left[\partial_x u \sigma_t + \beta\right](t, \tilde{X}_t)dX_t, \mathbb{P}_0\text{-a.s.}$$

$$(9.4.7)$$

is equivalent to the chain rule of path derivatives: denoting $Y_t(\omega) := u(t, \omega, \tilde{X}_t(\omega))$,

$$\partial_t Y_t = (\partial_t u)(t, \tilde{X}_t) + (\partial_x u)(t, \tilde{X}_t)\partial_t\tilde{X}_t,$$

$$\partial_\omega Y_t = (\partial_x u)(t, \tilde{X}_t)\partial_\omega\tilde{X}_t + (\partial_\omega u)(t, \tilde{X}_t).$$

$$(9.4.8)$$

∎

Proof A rigorous proof of the chain rule involves detailed estimates. We refer to
Keller & Zhang [123] for the analysis in the framework of rough path theory. Here
we shall only sketch a proof to illustrate the connection between (9.4.7) and (9.4.8).
In fact, by accepting the Ito-Ventzell formula (9.4.7), the arguments below can be
viewed as a proof for the chain rule (9.4.8). For simplicity, we assume $n = d = 1$.

Assume $\tilde{X}$, u, and Y are smooth in terms of (t, ω) and x. By (9.4.6) and (9.4.1), we have

$$\partial_t \tilde{X} + \frac{1}{2}\partial^2_{\omega\omega}\tilde{X} = b, \quad \partial_\omega \tilde{X} = \sigma; \qquad \partial_t u + \frac{1}{2}\partial^2_{\omega\omega}u = \alpha, \quad \partial_\omega u = \beta.$$

Applying (9.4.1) on $Y = u(t, \tilde{X}_t)$, then the martingale part of (9.4.7) amounts to say:

$$\partial_\omega Y_t = (\partial_x u)(t, \tilde{X}_t)\partial_\omega \tilde{X}_t + (\partial_\omega u)(t, \tilde{X}_t),$$

which is exactly the second equality in (9.4.8). Moreover, this implies further that (assuming $\partial^2_{\omega\omega} Y = \partial_\omega(\partial_\omega Y)$, see Remarks 9.4.6 (iv) and 9.4.7 (i)):

$$\partial^2_{\omega\omega} Y_t = (\partial^2_{xx}u)(t, \tilde{X}_t)|\partial_\omega \tilde{X}_t|^2 + 2(\partial^2_{x\omega}u)(t, \tilde{X}_t)\partial_\omega \tilde{X}_t + (\partial_x u)(t, \tilde{X}_t)\partial^2_{\omega\omega}\tilde{X}_t + (\partial^2_{\omega\omega}u)(t, \tilde{X}_t)$$

$$= (\partial^2_{xx}u)(t, \tilde{X}_t)|\sigma_t|^2 + 2(\partial_x\beta)(t, \tilde{X}_t)\sigma_t + (\partial_x u)(t, \tilde{X}_t)\partial^2_{\omega\omega}\tilde{X}_t + (\partial^2_{\omega\omega}u)(t, \tilde{X}_t).$$

Here we used the fact that $\partial_x\partial_\omega = \partial_\omega\partial_x$, which can be proved easily.

Now applying (9.4.1) on $Y_t = u(t, \tilde{X}_t)$ again, we see that the drift part of (9.4.7) amounts to saying:

$$\partial_t Y_t = \left[\partial_t Y_t + \frac{1}{2}\partial^2_{\omega\omega}Y_t\right] - \frac{1}{2}\partial^2_{\omega\omega}Y_t$$

$$= \left[\alpha + \partial_x u b_t + \frac{1}{2}\partial^2_{xx}u|\sigma_t|^2 + \partial_x\beta\sigma_t\right](t, \tilde{X}_t) - \frac{1}{2}\left[\partial^2_{xx}u|\sigma_t|^2\right.$$

$$\left. + 2\partial_x\beta\sigma_t + \partial_x u\partial^2_{\omega\omega}\tilde{X}_t + \partial^2_{\omega\omega}u\right](t, \tilde{X}_t)$$

$$= \left[\left[\alpha - \frac{1}{2}\partial^2_{\omega\omega}u\right] + \partial_x u\left[b_t - \frac{1}{2}\partial^2_{\omega\omega}\tilde{X}_t\right]\right](t, \tilde{X}_t)$$

$$= \left[\partial_t u + \partial_x u\partial_t\tilde{X}_t\right](t, \tilde{X}_t).$$

This is the chain rule for time derivative. ∎

To conclude this section, we extend the functional Itô calculus on the shifting operators.

Lemma 9.4.9 *Let* $\mathrm{H} \in \mathscr{T}$ *and* $u \in C^{1,2}(\Lambda)$. *For any* $(t, \omega) \in \Lambda$, *we have* $u^{t,\omega} \in C^{1,2}(\Lambda^t)$ *with*

$$\partial_t u^{t,\omega} = (\partial_t u)^{t,\omega}, \quad \partial_\omega u^{t,\omega} = (\partial_\omega u)^{t,\omega}, \quad \partial^2_{\omega\omega}u^{t,\omega} = (\partial^2_{\omega\omega}u)^{t,\omega}.$$

Proof It is clear that $u^{t,\omega} \in C^0(\overline{\Lambda}^t)$, and $(\partial_t u)^{t,\omega}, (\partial_\omega u)^{t,\omega}, (\partial^2_{\omega\omega}u)^{t,\omega} \in C^0(\Lambda^t)$. Then it suffices to show that, for any $\mathbb{P} \in \mathscr{P}_\infty$,

$$u^{t,\omega}_s - u(t, \omega) = \int_0^s \left[(\partial_t u)^{t,\omega}_r dr + (\partial_\omega u)^{t,\omega}_r dX_r + \frac{1}{2}(\partial^2_{\omega\omega}u)^{t,\omega}_r : d\langle X\rangle_r\right], 0 \le s < T - t, \mathbb{P}\text{-a.s.}$$

$$(9.4.9)$$

We proceed in two steps. For notational simplicity we assume $d = 1$.

Step 1. We first assume $u \in C_b^{1,2}(\Lambda)$. Consider the independent concatenation $\widehat{\mathbb{P}} := \mathbb{P}_0 \otimes_t \mathbb{P} \in \mathscr{P}_\infty$. Apply the functional Itô formula (9.4.1) with $\widehat{\mathbb{P}}$, we have

$$u_{t+s} - u_t = \int_t^{t+s} \left[\partial_t u_r dr + \partial_\omega u_r dX_r + \frac{1}{2} \partial_{\omega\omega}^2 u_r d\langle X \rangle_r \right], \quad \widehat{\mathbb{P}}\text{-a.s.}$$

By Problem 9.6.14 we see that (9.4.9) holds for $\mathbb{P}_0$-a.e. $\omega \in \Omega$. That is, $\mathbb{P}_0(\Omega_0) = 1$, where Ω_0 is the set of all $\omega \in \Omega$ such that (9.4.9) holds. Then it follows from Problem 9.6.13 that Ω_0 is dense in Ω. Now for any $\omega \in \Omega$, there exist $\omega^n \in \Omega_0$ such that $\lim_{n \to \infty} \|\omega^n - \omega\|_t = 0$. For each n, we have

$$u_s^{t,\omega^n} - u(t,\omega^n) = \int_0^s \left[(\partial_t u)_r^{t,\omega^n} dr + (\partial_\omega u)_r^{t,\omega^n} dX_r + \frac{1}{2}(\partial_{\omega\omega}^2 u)_r^{t,\omega^n} d\langle X \rangle_r \right], \quad \mathbb{P}\text{-a.s.}$$

Recall that u and its path derivatives are all continuous. Send $n \to \infty$, the left side above obviously converges to $u_s^{t,\omega} - u(t,\omega)$, and it follows from the dominated convergence theorem under $\mathbb{P}$ that the right side above also converges to the right side of (9.4.9) corresponding to ω. Therefore, (9.4.9) holds for all $\omega \in \Omega$.

Step 2. We now prove (9.4.9) for general $u \in C^{1,2}(\Lambda)$. For each n, let $\varphi_n : \mathbb{R} \to \mathbb{R}$ be smooth such that $\varphi_x(x) = x$ for $|x| \le n$ and $\varphi_n(x) = 0$ for $|x| \ge 2n$. Then $u^n := \varphi_n(u) \in C_b^{1,2}(\Lambda)$. Moreover, denote $H_n(\omega) := \inf\{t > 0 : |\omega_t| \ge n\}$, then we have

$$\partial_t u^n = \partial_t u, \quad \partial_\omega u^n = \partial_\omega u, \quad \partial_{\omega\omega}^2 u^n = \partial_{\omega\omega}^2 u, \quad 0 \le t \le H_n.$$

Now for each (t,ω), each $\mathbb{P} \in \mathscr{P}$, and each $n \ge \|\omega\|_t$, note that (9.4.9) holds for u^n. Then

$$u_s^{t,\omega} - u(t,\omega) = \int_0^s \left[(\partial_t u)_r^{t,\omega} dr + (\partial_\omega u)_r^{t,\omega} dX_r + \frac{1}{2}(\partial_{\omega\omega}^2 u)_r^{t,\omega} : d\langle X \rangle_r \right],$$

$$0 \le s < (H_n)^{t,\omega}, \quad \mathbb{P}\text{-a.s.}$$

Send $n \to \infty$, clearly $H_n \uparrow T$, or say $(H_n)^{t,\omega} \uparrow T - t$, then we obtain (9.4.9). ∎

9.5   Bibliographical Notes

It is standard to use weak formulation for principal agent problems in economics literature. We refer to Cvitanic, Wan, & Zhang [50] and Cvitanic & Zhang [52] for the rigorous formulation, and Cvitanic, Possamai, & Touzi [48, 49] for the problem with diffusion controls. For stochastic control, both strong and weak formulations are standard, see, e.g., Yong & Zhou [242]. For zero sum games, a large literature uses the strategy versus control approach, initiated by Elliott & Kalton [84] for deterministic case and Fleming & Souganidis [90] for stochastic case, see also Buckdahn & Li [25] for a more general model. Hamadene & Lepeltier [103] used BSDEs to study games with drift controls in weak formulation, and Pham & Zhang [192] used path dependent PDEs for games with diffusion controls in weak formulation, both are in path dependent case. The work of Cardaliaguet & Rainer [28] is also in this spirit, and Sirbu [211] uses the so-called stochastic Perron's method which allows to avoid the subtle regularity issue. The latter two works are for Markovian setting only.

It is natural and convenient to use the canonical space for weak formulation. Our presentation in Section 9.2 follow from Soner, Touzi, & Zhang [213, 215], and Ekren, Touzi, & Zhang [74]. For general theory on weak convergence and weak compactness on the space $C([0, T], \mathbb{R}^d)$, we refer to the books Billingsley [15] and Jacod & Shiryaev [113]. The particular weak compactness result Theorem 9.2.15 is due to Meyer & Zheng [156] and Zheng [248].

The regular conditional probability distribution in Section 9.3 is due to Stroock & Varadhan [218]. Our Definition 9.3.1 is slightly different (but equivalent), and is due to Soner, Touzi, & Zhang [215]. Moreover, the shifting operator here follows from Nutz [161], which is slightly different from the shifting spaces in [74, 213, 215].

The functional Itô calculus in Section 9.4 is first initiated by Dupire [71], and further developed by Cont & Fournie [36, 37] and Cont [35]. Our Definition 9.4.1 is due to Ekren, Touzi, & Zhang [75, 76], which is more appropriate for the study of PPDEs in Chapter 11 below. The connection with Itô-Ventzell formula, Remark 9.4.8, is due to Buckdahn, Ma, & Zhang [26]. The theory has received very strong attentions in recent years, besides the above-mentioned works, see, e.g., Oberhauser [166], Jazaerli & Saporito [115], Leao, Ohashi, & Simas [134], Saporito [208], and Keller & Zhang [123]. In particular, [123] provides a unified language for the functional Itô calculus and the rough path theory in terms of Gubinelli's derivatives [98]. The path derivatives are also related to the Malliavin calculus, for the latter we refer to Nualart [159].

9.6 Exercises

Problem 9.6.1

(i) Show that $C^0(\Omega) \subset \mathbb{L}^0(\mathscr{F}_T)$ and $C^0(\overline{\Lambda}) \subset \mathbb{L}^0(\mathbb{F})$.

(ii) Construct a counterexample to show that $\mathscr{F}_{0+} \neq \mathscr{F}_0$.

(iii) Show that $C^0(\Omega) \cap \mathbb{L}^0(\mathscr{F}_{t+}) \subset \mathbb{L}^0(\mathscr{F}_t)$. ∎

Problem 9.6.2 Let $\mathbb{P} \in \mathscr{P}_\infty$ and $\xi \in C_b^0(\Omega)$.

(i) For any $\tau, \tau_n \in \mathscr{T}$ such that $\tau_n \downarrow \tau$, we have $\mathbb{E}^{\mathbb{P}}[\xi | \mathscr{F}_{\tau_n}] \to \mathbb{E}^{\mathbb{P}}[\xi | \mathscr{F}_\tau]$ as $n \to \infty$, $\mathbb{P}$-a.s.

(ii) The martingale $\mathbb{E}^{\mathbb{P}}[\xi | \mathscr{F}_t]$ is continuous in t, $\mathbb{P}$-a.s. ∎

Problem 9.6.3

(i) Construct a counterexample such that the drift b' in (9.2.10) is not $\mathbb{F}^{X'}$-measurable.

(ii) Construct a counterexample such that $\overline{\mathbb{F}^B}^{\mathbb{P}_0}$ and $\overline{\mathbb{F}^{\tilde{X}}}^{\mathbb{P}_0}$ in (9.2.6) do not include each other. ∎

Problem 9.6.4 Prove Remark 9.2.10 (iii) and (iv). ∎

Problem 9.6.5 Prove (9.2.15) and (9.2.16). ∎

Problem 9.6.6 Prove Lemma 9.2.12 in the following two steps.
  Step 1. For open sets O and closed sets D, define

$$\mathbb{P}(O) := \sup\left\{\mathbb{E}[\xi] : \xi \in C_b^0(\Omega), 0 \le \xi \le 1, \xi \mathbf{1}_{O^c} = 0\right\},$$

$$\mathbb{P}(D) := 1 - \mathbb{P}(D^c) = \inf\left\{\mathbb{E}[\xi] : \xi \in C_b^0(\Omega), 0 \le \xi \le 1, \xi \mathbf{1}_D = 1\right\}.$$

Show that, for any $A \in \mathscr{B}(\mathbb{R}^d)$,

$$\mathbb{P}(A) := \inf\{\mathbb{P}(O) : O \text{ is open and } A \subset O\} = \sup\{\mathbb{P}(D) : D \text{ is closed and } D \subset A\}.$$

$$(9.6.1)$$

  Step 2. Prove that the $\mathbb{P}$ defined by (9.6.1) is indeed a probability measure and $\mathbb{E}^{\mathbb{P}} = \mathbb{E}$. ∎

Problem 9.6.7 Prove (9.3.23). ∎

Problem 9.6.8

(i) Show that the set $\mathscr{P}_L^{drift}$ defined in (9.2.11) is weakly compact and satisfies Assumption 9.3.3.

(ii) Show that the set $\mathscr{P}_L^{PC}$ in (9.2.14) also satisfies Assumption 9.3.3. ∎

Problem 9.6.9

(i) For any compact set $K \subset \Omega$, show that $\mathscr{C}^{\mathscr{P}_\infty}[K^c] = \infty$.

(ii) Show that $\mathscr{P}_\infty$ is not weakly compact. ∎

Problem 9.6.10

(i) Let $\mathbb{P}^{b,\sigma} \in \mathscr{P}_L^{PC}$ as in (9.2.14). Show that $\overline{\mathbb{F}^{X^{b,\sigma}}}^{\mathbb{P}_0} = \overline{\mathbb{F}^B}^{\mathbb{P}_0}$.

(ii) This part proves the opposite direction of (i) by starting with the weak formulation, and we extend it to partition with stopping times. Let $\tau_n \in \mathscr{T}$ be increasing satisfying $\tau_0 = 0$ and $\tau_n = T$ when n is large enough. Let α_n, β_n be $\mathscr{F}_{\tau_n}$-measurable and $|\alpha_n| \le L, 0 \le \beta_n \le \sqrt{2L}$. Denote

$$b_t(\omega) := \sum_{n \ge 0} \alpha_n(\omega) \mathbf{1}_{[\tau_n(\omega), \tau_{n+1}(\omega))}(t), \quad \sigma_t(\omega) := \sum_{n \ge 0} \beta_n(\omega) \mathbf{1}_{[\tau_n(\omega), \tau_{n+1}(\omega))}(t).$$

Then the following path dependent SDE has a unique strong solution and $\mathbb{P} := \mathbb{P}_0 \circ \tilde{X}^{-1} \in \mathscr{P}_L$:

$$\tilde{X}_t = \int_0^t b(s, \tilde{X}.)ds + \int_0^t \sigma(s, \tilde{X}.)dX_s, \quad \mathbb{P}_0\text{-a.s.}$$

(iii) Recall (9.1.6), (9.1.8), and (9.1.10). Assume b, σ are bounded and uniformly Lipschitz continuous in x, and f, g are bounded and continuous in x. Show that $\tilde{V}_0^S = \tilde{V}_0^W$.

(iv) In addition to the conditions in (iii), assume further that $\mathbb{K} \subset \mathbb{R}^m$ is a convex set for certain $m \geq 1$, and b, σ, f are continuous in k. Show that $\tilde{V}_0^S = V_0^S$.

(v) Construct a counterexample $\xi \in \mathbb{L}^\infty(\mathscr{F}_T)$ such that $\sup_{\mathbb{P} \in \mathscr{P}_L^{PC}} \mathbb{E}^\mathbb{P}[\xi] < \sup_{\mathbb{P} \in \mathscr{P}_L} \mathbb{E}^\mathbb{P}[\xi]$. ∎

Problem 9.6.11 Recall that Barlow [6] constructed a continuous function $\sigma_0 : \mathbb{R} \to [1, 2]$ such that the following SDE has a unique weak solution but no strong solution:

$$\tilde{X}_t = \int_0^t \sigma_0(\tilde{X}_s)dB_s, \quad \mathbb{P}_0\text{-a.s.}$$

(i) Use this fact to construct a counterexample such that the problem (9.1.8) has an optimal control but (9.1.6) has no optimal control. (Hint: construct the example so that the PDE (9.1.7) has a classical solution v and we may use the fact that $V_0^S = V_0^W = v(0,0)$.)

(ii) Show that $\mathscr{P}_{[1,2]}^S$ is not weakly compact. ∎

Problem 9.6.12 Consider the setting in Section 9.1.3. Assume b and σ are bounded and uniformly Lipschitz continuous in x, $\sigma > 0$, and $\mathscr{K}_i$ is set of piecewise constant $\mathbb{K}_i$-valued processes, $i = 1, 2$.

(i) For any $(k^1, k^2) \in \mathscr{K}_1 \times \mathscr{K}_2$, show that there exists $(\tilde{k}^1, \tilde{k}^2) \in \mathscr{K}_1 \times \mathscr{K}_2$ such that

$$J_S(k^1, k^2) = J_W(\tilde{k}^1, \tilde{k}^2). \tag{9.6.2}$$

(ii) Construct a counterexample such that the above $\tilde{k}^1$ depends on both k^1 and k^2. In other words, given k^1, there is no $\tilde{k}^1$ (independent of k^2) such that, for all k^2, there exists a $\tilde{k}^2$ satisfying (9.6.2). ∎

Problem 9.6.13 Show that any support of $\mathbb{P}_0$ is dense in Ω. Or equivalently, for any $\omega^0 \in \Omega$ and any $\varepsilon > 0$, $\mathbb{P}_0(\{\omega \in \Omega : \|\omega - \omega^0\|_T < \varepsilon\}) > 0$. ∎

Problem 9.6.14 Let $0 \le t \le T, 0 \le s \le T - t, \eta \in \mathbb{L}^0(\mathbb{F})$ bounded, and $\mathbb{P}_1, \mathbb{P}_2 \in \mathscr{P}_L$. Show that

$$\Big(\int_t^{t+s} \eta_r dr\Big)^{t,\omega} = \int_0^s \eta_r^{t,\omega} dr, \quad \forall \omega \in \Omega;$$

$$\Big(\int_t^{t+s} \eta_r dX_r\Big)^{t,\omega} = \int_0^s \eta_r^{t,\omega} dX_r, \quad \mathbb{P}_2\text{-a.s. for } \mathbb{P}_1\text{-a.e. } \omega;$$

$$\Big(\int_t^{t+s} \eta_r d\langle X \rangle_r\Big)^{t,\omega} = \int_0^s \eta_r^{t,\omega} d\langle X \rangle_r, \quad \mathbb{P}_2\text{-a.s. for } \mathbb{P}_1\text{-a.e. } \omega.$$

∎

Chapter 10
Nonlinear Expectation

In this chapter we adopt the canonical setting in Sections 9.2 and 9.3. In particular, X is the canonical process, and we may also denote it as B when it is under Wiener measure $\mathbb{P}_0$.

10.1 Nonlinear Expectation

Recall that X is the state process in applications. Let $\mathscr{P}$ denote the set of all possible distributions of X under consideration. We say

a property holds $\mathscr{P}$-quasi surely (abbreviated as $\mathscr{P}$-q.s.) if it holds $\mathbb{P}$-a.s. for all $\mathbb{P} \in \mathscr{P}$. $\qquad$ (10.1.1)

Define nonlinear expectation $\mathscr{E}^{\mathscr{P}}$ and capacity $\mathscr{C}^{\mathscr{P}}$ as follows:

$$\mathscr{E}^{\mathscr{P}}[\xi] := \sup_{\mathbb{P} \in \mathscr{P}} \mathbb{E}^{\mathbb{P}}[\xi], \ \forall \xi \in \bigcap_{\mathbb{P} \in \mathscr{P}} \mathbb{L}^1(\mathscr{F}_T, \mathbb{P}, \mathbb{R});$$

$$\mathscr{C}^{\mathscr{P}}[E] := \mathscr{E}^{\mathscr{P}}[\mathbf{1}_E], \ \forall E \in \mathscr{F}_T. \qquad (10.1.2)$$

We emphasize that since supremum is used,

$$\mathscr{E}^{\mathscr{P}} \text{ is defined only for scalar random variables } \xi.$$

We also note that $\mathscr{E}^{\mathbb{P}_L}$ and $\mathscr{C}^{\mathscr{P}_L}$ have already been used in Section 9.2, see e.g. (9.2.16) and (9.2.20). Here are a few more typical examples of such nonlinear expectation.

© Springer Science+Business Media LLC 2017
J. Zhang, *Backward Stochastic Differential Equations*, Probability Theory and Stochastic Modelling 86, DOI 10.1007/978-1-4939-7256-2_10

Example 10.1.1

(i) $\mathscr{P} := \{\mathbb{P}\}$ *for some* $\mathbb{P} \in \mathscr{P}_\infty$, *then* $\mathscr{E}^{\mathscr{P}} = \mathbb{E}^{\mathbb{P}}$.

(ii) $\mathscr{P} := \{\mathbb{P}^\theta : |\theta| \leq L\} \subset \mathscr{P}_{LV1}$, *where* $\mathbb{P}^\theta$ *is defined in Section 2.6. Recall the setting in Section 4.5.2 with* $\sigma = 1, f = 0$. *Then* $H^*(z) = \sup_{|\theta| \leq L}[\theta z] = L|z|$, *and thus* $\mathscr{E}^{\mathscr{P}}[\xi] = Y_0$, *where*

$$Y_t = \xi + \int_t^T L|Z_s|ds - \int_t^T Z_s dB_s, \quad \mathbb{P}_0\text{-}a.s.$$

This is a special case of the so-called Peng's g-expectation, see Peng [176].

(iii) $\mathscr{P} := \mathscr{P}_{[\underline{\sigma},\overline{\sigma}]}^W$. *In this case* $\mathscr{E}^{\mathscr{P}}$ *is the so-called Peng's G-expectation, see Peng [181].*

We next introduce the following spaces: for any $p, q \geq 1$,

$$
\begin{aligned}
\mathbb{L}^p(\mathscr{F}_T, \mathscr{P}) &:= \left\{ \xi \in \mathbb{L}^0(\mathscr{F}_T) : \mathscr{E}^{\mathscr{P}}[|\xi|^p] < \infty \right\}; \\
\mathbb{L}^{p,q}(\mathbb{F}, \mathscr{P}) &:= \left\{ X \in \mathbb{L}^0(\mathbb{F}) : \left(\int_0^T |X_t|^p dt \right)^{\frac{1}{p}} \in \mathbb{L}^q(\mathscr{F}_T, \mathscr{P}) \right\}; \\
\mathbb{L}^p(\mathbb{F}, \mathscr{P}) &:= \mathbb{L}^{p,p}(\mathbb{F}, \mathscr{P}); \\
\mathbb{S}^p(\mathbb{F}, \mathscr{P}) &:= \left\{ X \in \mathbb{L}^0(\mathbb{F}) : X \text{ is continuous}, \mathscr{P}\text{-q.s. and } \mathscr{E}^{\mathscr{P}}[|X_T^*|^p] < \infty \right\}.
\end{aligned}
\tag{10.1.3}
$$

For vector valued random variables or processes, say $\mathbb{R}^n$-valued, we use the notation $\mathbb{L}^p(\mathscr{F}_T, \mathscr{P}, \mathbb{R}^n)$, etc. to indicate the dimension.

It is straightforward to check that $\mathscr{E}^{\mathscr{P}}$ satisfies the properties of the so-called sublinear expectation, a special type of nonlinear expectation.

Proposition 10.1.2

(i) $\mathscr{E}^{\mathscr{P}}[c] = c$ *for any constant c.*

(ii) $\mathscr{E}^{\mathscr{P}}[c\xi] = c\mathscr{E}^{\mathscr{P}}[\xi]$ *for any constant* $c \geq 0$ *and any* $\xi \in \mathbb{L}^1(\mathscr{F}_T, \mathscr{P})$.

(iii) (Monotonicity) $\mathscr{E}^{\mathscr{P}}[\xi_1] \leq \mathscr{E}^{\mathscr{P}}[\xi_2]$ *for any* $\xi_1, \xi_2 \in \mathbb{L}^1(\mathscr{F}_T, \mathscr{P})$ *such that* $\xi_1 \leq \xi_2, \mathscr{P}$-*q.s.*

(iv) (Sublinearity) $\mathscr{E}^{\mathscr{P}}[\xi_1 + \xi_2] \leq \mathscr{E}^{\mathscr{P}}[\xi_1] + \mathscr{E}^{\mathscr{P}}[\xi_2]$ *for any* $\xi_1, \xi_2 \in \mathbb{L}^1(\mathscr{F}_T, \mathscr{P})$.

The following properties are also immediate.

Proposition 10.1.3 *Provided appropriate integrability at below:*

(i) If $\mathscr{P}_1 \subset \mathscr{P}_2$, then $\mathscr{E}^{\mathscr{P}_1}[\xi] \leq \mathscr{E}^{\mathscr{P}_2}[\xi]$ for all $\xi \in \mathbb{L}^1(\mathscr{F}_T, \mathscr{P}_2)$, and thus $\mathbb{L}^p(\mathscr{F}_T, \mathscr{P}_2) \subset \mathbb{L}^p(\mathscr{F}_T, \mathscr{P}_1)$ for any $p \geq 1$.

(ii) Let $p, q \geq 1$ be conjugates. Then $\mathscr{E}^{\mathscr{P}}[\xi_1 \xi_2] \leq \left(\mathscr{E}^{\mathscr{P}}[|\xi_1|^p] \right)^{\frac{1}{p}} \left(\mathscr{E}^{\mathscr{P}}[|\xi_2|^q] \right)^{\frac{1}{q}}$.

(iii) If $\varphi : \mathbb{R} \to \mathbb{R}$ is convex, then $\varphi(\mathscr{E}^{\mathscr{P}}[\xi]) \leq \mathscr{E}^{\mathscr{P}}[\varphi(\xi)]$.

Proof (i) is obvious. (ii) follows from the Höolder Inequality under each $\mathbb{P} \in \mathscr{P}$. To see (iii), let $\mathbb{P}_n \in \mathscr{P}$ be such that $\lim_{n\to\infty} \mathbb{E}^{\mathbb{P}_n}[\xi] = \mathscr{E}^{\mathscr{P}}[\xi]$. For each n, by Jensen's inequality under $\mathbb{P}_n$, we have $\varphi(\mathbb{E}^{\mathbb{P}_n}[\xi]) \leq \mathbb{E}^{\mathbb{P}_n}[\varphi(\xi)] \leq \mathscr{E}^{\mathscr{P}}[\varphi(\xi)]$. Sending $n \to \infty$, we obtain the desired estimate. ∎

10.1.1 Convergence Under Nonlinear Expectation

It is well understood that convergence theorems are crucial in real analysis and stochastic analysis. Convergence under nonlinear expectation, unfortunately, is more involved. Note that $\mathscr{P}^W_{[\underline{\sigma},\overline{\sigma}]}$ contains mutually singular measures and has no dominating measure, see Remarks 9.2.6 and 9.2.8 (i). As a consequence, the dominated convergence theorem under standard conditions fails under the G-expectation $\mathscr{E}^{\mathscr{P}^W_{[\underline{\sigma},\overline{\sigma}]}}$. Recall the pathwise quadratic variation $\langle X \rangle$ as in (9.2.15).

Example 10.1.4 *Let* $d = 1$, $\mathscr{P} := \mathscr{P}^W_{[0,1]}$, *and* $E_n := \{\langle X \rangle_T = \frac{T}{n^2}\}$, $\xi_n := \mathbf{1}_{\cup_{m \geq n} E_m}$, $n \geq 1$. *Then* $0 \leq \xi_n \leq 1$ *and* $\xi_n(\omega) \downarrow 0$ *for all* $\omega \in \Omega$. *However,* $\mathscr{E}^{\mathscr{P}}[\xi_n] = 1$ *for all* n, *and thus*

$$\lim_{n \to \infty} \mathscr{E}^{\mathscr{P}}[\xi_n] = 1 \neq 0 = \mathscr{E}^{\mathscr{P}}[\lim_{n \to \infty} \xi_n].$$

In this subsection we introduce several convergence results, under certain additional conditions. We start with the definition of several types of convergence.

Definition 10.1.5 *Let* $\xi_n, \xi \in \mathbb{L}^0(\mathscr{F}_T)$, $n \geq 1$.

(i) *We say* $\xi_n \to \xi$ *in* $\mathbb{L}^1(\mathscr{F}_T, \mathscr{P})$ *if* $\lim_{n \to \infty} \mathscr{E}^{\mathscr{P}}[|\xi_n - \xi|] = 0$;
(ii) *We say* $\xi_n \to \xi$ *in (capacity)* $\mathscr{C}^{\mathscr{P}}$ *if, for any* $\varepsilon > 0$, $\lim_{n \to \infty} \mathscr{C}^{\mathscr{P}}[|\xi_n - \xi| \geq \varepsilon] = 0$;
(iii) *We say* $\xi_n \to \xi \mathscr{P}$-*q.s if* $\mathscr{C}^{\mathscr{P}}\left[\limsup_{n \to \infty} |\xi_n - \xi| > 0\right] = 0$.

Proposition 10.1.6 *Let* $\xi_n, \xi \in \mathbb{L}^0(\mathscr{F}_T)$, $n \geq 1$.

(i) *If* $\xi_n \to \xi$ *in* $\mathbb{L}^1(\mathscr{F}_T, \mathscr{P})$, *then* $\xi_n \to \xi$ *in (capacity)* $\mathscr{C}^{\mathscr{P}}$.
(ii) *If* $\xi_n \to \xi$ *in* $\mathbb{L}^1(\mathscr{F}_T, \mathscr{P})$ *or in* $\mathscr{C}^{\mathscr{P}}$, *then there exists a subsequence* n_k *such that* $\xi_{n_k} \to \xi \mathscr{P}$-*q.s*

The proof is similar to the corresponding results in real analysis and is left to readers. However, unlike in real analysis, the $\mathscr{P}$-q.s. convergence does not imply the convergence in $\mathscr{C}^{\mathscr{P}}$, as we see in Example 10.1.4.

Since the $\mathscr{P}$-q.s. convergence can be decomposed into $\mathbb{P}$-a.s. convergence for every $\mathbb{P} \in \mathscr{P}$, it does not add significant difficulty in the nonlinear case. However, the convergence in $\mathbb{L}^1(\mathscr{F}_T, \mathscr{P})$ (and $\mathscr{C}^{\mathscr{P}}$) requires uniform convergence in $\mathbb{L}^1(\mathscr{F}_T, \mathbb{P})$, uniformly in $\mathbb{P} \in \mathscr{P}$ which could be mutually singular without a dominating measure. This indeed involves more technicality and requires additional conditions.

First, since $\mathscr{E}^{\mathscr{P}}$ is sublinear, the following partial convergence results are obvious.

Proposition 10.1.7 *Let* $\xi_n, \xi \in \mathbb{L}^0(\mathscr{F}_T)$, $n \geq 1$.

(i) *Assume* $\xi_n \uparrow \xi$ *and* $\xi_1 \in \mathbb{L}^1(\mathscr{F}_T, \mathscr{P})$, *then* $\mathscr{E}^{\mathscr{P}}[\xi] = \lim_{n \to \infty} \mathscr{E}^{\mathscr{P}}[\xi_n]$.
(ii) *Assume* $\xi_n \geq \xi$ *and* $\xi \in \mathbb{L}^1(\mathscr{F}_T, \mathscr{P})$, *then* $\mathscr{E}^{\mathscr{P}}[\liminf_{n \to \infty} \xi_n] \leq \liminf_{n \to \infty} \mathscr{E}^{\mathscr{P}}[\xi_n]$.

To obtain the full convergence theorem, we need some regularity on the random variables.

Theorem 10.1.8 *Let* $\mathscr{P} \subset \mathscr{P}_L$ *for some* $L > 0$. *Assume* ξ_n, $n \geq 1$, *are uniformly continuous in* ω *under* $\| \cdot \|_T$, *uniformly on* n, *and* $\xi_n \to \xi$, $\mathscr{P}$-*q.s. Then* $\xi_n \to \xi$ *in* $\mathbb{L}^1(\mathscr{F}_T, \mathscr{P})$, *namely*

$$\lim_{n \to \infty} \mathscr{E}^{\mathscr{P}}[|\xi_n - \xi|] = 0.$$

Proof Let ρ denote the common modulus of continuity function of ξ_n, $n \geq 1$. First, denote $\tilde{\xi} := \limsup_{n \to \infty} \xi_n$. It is clear that $\tilde{\xi}$ is also uniformly continuous with the same modulus of continuity function ρ and $\tilde{\xi} = \xi$, $\mathscr{P}$-q.s. Thus, without loss of generality we may identify ξ and $\tilde{\xi}$, and thus ξ is also uniformly continuous.

Next, fix an arbitrary $\omega^0 \in \Omega$ such that $\xi_n(\omega^0) \to \xi(\omega^0)$. By Problem 10.5.2 (i) we have

$$|\xi_n(\omega)| \leq |\xi_n(\omega^0)| + \rho(\|\omega - \omega^0\|_T) \leq |\xi_n(\omega^0)| + \rho(1)[1 + \|\omega^0\|_T + \|\omega\|_T].$$

$$(10.1.4)$$

Noting that $\{\xi_n(\omega^0)\}_{n \geq 1}$ are bounded, then it follows from (9.2.16) that

$$\sup_n \mathscr{E}^{\mathscr{P}}[|\xi_n|] \leq \sup_n |\xi_n(\omega^0)| + \rho(1)[\|\omega^0\|_T + 1] + \rho(1)\mathscr{E}^{\mathscr{P}_L}[\|X\|_T] < \infty.$$

Now for $\varepsilon > 0$ and $m \geq 1$, let Ω_m^ε be defined by (9.2.2). By the arguments in Lemma 9.2.1, in particular (9.2.3), there exist a finite sequence $\{\omega^i, i \in I\}$ such that $\Omega_m^\varepsilon \subset \cup_{i \in I} O_{3\varepsilon}(\omega^i)$. Denote $I_1 := \{i \in I : \mathscr{E}^{\mathscr{P}}[E_i] > 0\}$, where $E_i := O_{3\varepsilon}(\omega^i) \cap \Omega_m^\varepsilon$. Then, for each $i \in I_1$, there exists $\tilde{\omega}^i \in E_i$ such that $\lim_{n \to \infty} \xi_n(\tilde{\omega}^i) = \xi(\tilde{\omega}^i)$. Now for each $i \in I_1$ and $\omega \in E_i \subset O_{6\varepsilon}(\tilde{\omega}^i)$,

$$|\xi_n(\omega) - \xi(\omega)| \leq |\xi_n(\omega) - \xi_n(\tilde{\omega}^i)| + |\xi_n(\tilde{\omega}^i) - \xi(\tilde{\omega}^i)| + |\xi(\tilde{\omega}^i) - \xi(\omega)|$$

$$\leq 2\rho(6\varepsilon) + |\xi_n(\tilde{\omega}^i) - \xi(\tilde{\omega}^i)|.$$

Since I and hence I_1 is finite, we have

$$\limsup_{n \to \infty} \mathscr{E}^{\mathscr{P}}\Big[|\xi_n - \xi| 1_{\Omega_m^\varepsilon}\Big] \leq \limsup_{n \to \infty} \Big[2\rho(6\varepsilon) + \max_{i \in I_1} |\xi_n(\tilde{\omega}^i) - \xi(\tilde{\omega}^i)|\Big] = 2\rho(6\varepsilon).$$

$$(10.1.5)$$

Moreover, clearly ξ satisfies the same estimate (10.1.4). Then by (9.2.2) we have

$$\mathscr{E}^{\mathscr{P}}\Big[|\xi_n - \xi| 1_{(\Omega_m^\varepsilon)^c}\Big] \leq C\mathscr{E}^{\mathscr{P}}\Big[[1 + \|X\|_T] 1_{(\Omega_m^\varepsilon)^c}\Big]$$

$$\leq C\mathscr{E}^{\mathscr{P}}\Big[[1 + \|X\|_T][\frac{\|X\|_T}{m} + \varepsilon^{-1} \mathrm{OSC}_{\frac{1}{m}}(X)]\Big]$$

$$\leq \frac{C}{m}\mathcal{E}^{\mathscr{P}_L}[1 + \|X\|_T^2] + C\varepsilon^{-1}\left(\mathcal{E}^{\mathscr{P}_L}[1 + \|X\|_T^2]\right)^{\frac{1}{2}}\left(\mathcal{E}^{\mathscr{P}_L}\left[|\mathrm{OSC}_{\frac{1}{m}}(X)|^2\right]\right)^{\frac{1}{2}}$$

$$\leq \frac{C}{m} + \frac{C}{\sqrt{m\varepsilon}}.$$

thanks to Proposition 10.1.3 and (9.2.16). Then, together with (10.1.5), we have

$$\limsup_{n\to\infty}\mathcal{E}^{\mathscr{P}}[|\xi_n - \xi|] \leq 2\rho(6\varepsilon) + \frac{C}{m} + \frac{C}{\sqrt{m\varepsilon}}.$$

First send $m \to \infty$ and then $\varepsilon \to 0$, we obtain the desired convergence immediately. ∎

We remark that the uniform continuity in the above theorem is a very strong requirement, for example indicators and stopping times typically violate this regularity. The next monotone limit theorem weakens this condition.

Definition 10.1.9 *Let* $\xi \in \mathbb{L}^0(\mathscr{F}_T)$.

(i) *We say* ξ *is* $\mathscr{P}$-*uniformly integrable if* $\lim_{n\to\infty}\mathcal{E}^{\mathscr{P}}[|\xi|\mathbf{1}_{\{|\xi|\geq n\}}] = 0$.

(ii) *We say* ξ *is* $\mathscr{P}$-*quasi-surely continuous (abbreviated as* $\mathscr{P}$-*q.s. continuous) if, for any* $\varepsilon > 0$, *there exists a closed set* $\Omega_\varepsilon \subset \Omega$ *such that* $\mathscr{C}^{\mathscr{P}}[\Omega_\varepsilon^c] < \varepsilon$ *and* ξ *is continuous on* Ω_ε.

We note that, when $p > 1$, any element in $\mathbb{L}^p(\mathscr{F}_T, \mathscr{P})$ is $\mathscr{P}$-uniformly integrable.

Theorem 10.1.10 *Let* $\mathscr{P}$ *be weakly compact.*

(i) *Assume* Ω_n *is a sequence of open sets such that* $\Omega_n \uparrow \Omega$. *Then* $\mathscr{C}^{\mathscr{P}}[\Omega_n^c] \downarrow 0$.

(ii) *Assume, for each* $n \geq 1$, $\xi_n \in \mathbb{L}^0(\mathscr{F}_T)$ *is* $\mathscr{P}$-*uniformly integrable and* $\mathscr{P}$-*q.s. continuous. If* $\xi_n \downarrow \xi$, $\mathscr{P}$-*q.s. then* $\mathcal{E}^{\mathscr{P}}[\xi_n] \downarrow \mathcal{E}^{\mathscr{P}}[\xi]$.

To prove the theorem, we first need a lemma.

Lemma 10.1.11 *Let* $\mathbb{P}_n, \mathbb{P} \in \mathscr{P}$ *such that* $\mathbb{P}_n \to \mathbb{P}$ *weakly. Assume* ξ *is* $\mathscr{P}$-*uniformly integrable and* $\mathscr{P}$-*q.s. continuous, then* $\lim_{n\to\infty}\mathbb{E}^{\mathbb{P}_n}[\xi] = \mathbb{E}^{\mathbb{P}}[\xi]$.

Proof Fix $\varepsilon > 0$ and $m \geq 1$. Let Ω_ε be a closed set as in Definition 10.1.9 (ii), and denote $\Omega_m^\varepsilon := \Omega_\varepsilon \cap \{|\xi| \leq m\}$. Since ξ is continuous on Ω_ε, then Ω_m^ε is also closed. Applying the Tietze extension theorem, or see Problem 10.5.3, there exists $\tilde{\xi} \in C_b^0(\Omega)$ such that $\tilde{\xi} = \xi$ on Ω_m^ε and $\sup_{\omega\in\Omega}|\tilde{\xi}(\omega)| = \sup_{\omega\in\Omega_m^\varepsilon}|\xi(\omega)| \leq m$. By the weak convergence, $\lim_{n\to\infty}\mathbb{E}^{\mathbb{P}_n}[\tilde{\xi}] = \mathbb{E}^{\mathbb{P}}[\tilde{\xi}]$. Then

$$\limsup_{n\to\infty}\left|\mathbb{E}^{\mathbb{P}_n}[\xi] - \mathbb{E}^{\mathbb{P}}[\xi]\right| \leq \limsup_{n\to\infty}\left[\mathbb{E}^{\mathbb{P}_n}[|\xi - \tilde{\xi}|] + \mathbb{E}^{\mathbb{P}}[|\xi - \tilde{\xi}|]\right]$$

$$\leq 2\limsup_{n\to\infty}\mathcal{E}^{\mathscr{P}}\left[|\xi - \tilde{\xi}|\mathbf{1}_{(\Omega_m^\varepsilon)^c}\right].$$

Note that

$$|\xi - \tilde{\xi}|\mathbf{1}_{(\Omega_m^\varepsilon)^c} = |\xi - \tilde{\xi}|[\mathbf{1}_{\{|\xi| \geq m\}} + \mathbf{1}_{\Omega_\varepsilon^c \cap \{|\xi| \leq m\}}] \leq 2|\xi|\mathbf{1}_{\{|\xi| \geq m\}} + 2m\mathbf{1}_{\Omega_\varepsilon^c}$$

(10.1.6)

Then

$$\limsup_{n \to \infty} \left| \mathbb{E}^{\mathbb{P}_n}[\xi] - \mathbb{E}^{\mathbb{P}}[\xi] \right| \leq 4\mathscr{E}^{\mathscr{P}}[|\xi|\mathbf{1}_{\{|\xi| \geq m\}}] + 4m\mathscr{C}^{\mathscr{P}}[\Omega_\varepsilon^c].$$

By first sending $\varepsilon \to 0$ and then $m \to \infty$, it follows from Definition 10.1.9 that $\lim_{n \to \infty} \mathbb{E}^{\mathbb{P}_n}[\xi] = \mathbb{E}^{\mathbb{P}}[\xi]$. ∎

Proof of Theorem 10.1.10. We shall prove (ii) first. For each $n \geq 1$, by the definition of $\mathscr{E}^{\mathscr{P}}$, there exists $\mathbb{P}_n \in \mathscr{P}$ such that $\mathbb{E}^{\mathbb{P}_n}[\xi_n] \geq \mathscr{E}^{\mathscr{P}}[\xi_n] - \frac{1}{n}$. Since $\mathscr{P}$ is weakly compact, $\{\mathbb{P}_n, n \geq 1\}$ has a weak limit $\mathbb{P}^* \in \mathscr{P}$, and without loss of generality, we assume $\mathbb{P}_n \to \mathbb{P}^*$ weakly. Then, for $n \geq m$, by Lemma 10.1.11,

$$\mathscr{E}^{\mathscr{P}}[\xi_n] \leq \mathbb{E}^{\mathbb{P}_n}[\xi_n] + \frac{1}{n} \leq \mathbb{E}^{\mathbb{P}_n}[\xi_m] + \frac{1}{n} \to \mathbb{E}^{\mathbb{P}^*}[\xi_m], \quad \text{as } n \to \infty.$$

Thus $\limsup_{n \to \infty} \mathscr{E}^{\mathscr{P}}[\xi_n] \leq \mathbb{E}^{\mathbb{P}^*}[\xi_m]$ for any $m \geq 1$. Send $m \to \infty$ and apply the standard monotone convergence theorem under $\mathbb{P}^*$, we have $\limsup_{n \to \infty} \mathscr{E}^{\mathscr{P}}[\xi_n] \leq \mathbb{E}^{\mathbb{P}^*}[\xi] \leq \mathscr{E}^{\mathscr{P}}[\xi]$. This, together with the fact that $\xi_n \geq \xi$, proves the desired convergence.

We now prove (i). Set $\xi_n(\omega) := \left[1 - n\inf_{\omega' \in \Omega_n^c} \|\omega - \omega'\|_T\right]^+$, $n \geq 1$. One can easily check that

- ξ_n is bounded and thus $\mathscr{P}$-uniformly integrable;
- ξ_n is continuous and thus $\mathscr{P}$-q.s. continuous;
- $\mathbf{1}_{\Omega_n^c}(\omega) \leq \xi_n(\omega)$ and $\xi_n(\omega) \downarrow 0$ for all $\omega \in \Omega$.

Then it follows from (ii) that $\mathscr{C}^{\mathscr{P}}[\Omega_n^c] \leq \mathscr{E}^{\mathscr{P}}[\xi_n] \to 0$. ∎

10.1.2 Quasi-Sure Continuity

In this subsection we provide a characterization of quasi-sure continuity. Denote

$$\mathbb{L}_0^p(\mathscr{F}_T, \mathscr{P}) := \left\{\xi \in \mathbb{L}^p(\mathscr{F}_T, \mathscr{P}) : \exists \xi_n \in UC_b(\Omega) \text{ s.t. } \lim_{n \to \infty} \mathscr{E}^{\mathscr{P}}[|\xi_n - \xi|^p] = 0\right\}.$$

(10.1.7)

We remark that when $\mathscr{P}$ is a singleton, $\mathbb{L}^p(\mathscr{F}_T, \mathscr{P}) = \mathbb{L}_0^p(\mathscr{F}_T, \mathscr{P})$. However, in general the inclusion $\mathbb{L}_0^p(\mathscr{F}_T, \mathscr{P}) \subset \mathbb{L}^p(\mathscr{F}_T, \mathscr{P})$ is typically strict.

Example 10.1.12 Assume $\mathbb{P}^\theta := \mathbb{P}_0 \circ ((1 + \theta)B)^{-1} \in \mathscr{P}$ for θ small enough, and let $E := \{\langle X \rangle_T = T\}$. Then $\mathbf{1}_E \in \mathbb{L}^1(\mathscr{F}_T, \mathscr{P}) \backslash \mathbb{L}_0^1(\mathscr{F}_T, \mathscr{P})$. In particular, by Proposition 10.1.13 below, $\langle X \rangle$ is not $\mathscr{P}$-q.s. continuous.

Proof It is clear that $\mathbf{1}_E \in \mathbb{L}^1(\mathscr{F}_T, \mathscr{P})$. Assume by contradiction that $\mathbf{1}_E \in \mathbb{L}_0^1(\mathscr{F}_T, \mathscr{P})$, then there exist $\xi_n \in UC(\Omega)$ such that $\mathscr{E}^{\mathscr{P}}[|\xi_n - \mathbf{1}_E|] \to 0$. Note that,

$$\mathbb{E}^{\mathbb{P}_0}[\mathbf{1}_E] = 1, \quad \mathbb{E}^{\mathbb{P}^\theta}[\mathbf{1}_E] = 0.$$

On the other hand, for each n, it follows from the continuity of ξ_n that

$$\lim_{\theta \to 0} \mathbb{E}^{\mathbb{P}^\theta}[\xi_n] = \mathbb{E}^{\mathbb{P}_0}[\xi_n].$$

Then

$$\mathscr{E}^{\mathscr{P}}[|\xi_n - \mathbf{1}_E|] \geq \limsup_{\theta \to 0} \mathbb{E}^{\mathbb{P}^\theta}[\xi_n - \mathbf{1}_E] = \mathbb{E}^{\mathbb{P}_0}[\xi_n] \to \mathbb{E}^{\mathbb{P}_0}[\mathbf{1}_E] = 1, \quad \text{as } n \to \infty.$$

Contradiction. ∎

Proposition 10.1.13 Let $\xi \in \mathbb{L}^p(\mathscr{F}, \mathscr{P})$ such that $|\xi|^p$ is $\mathscr{P}$-uniformly integrable, for some $p \geq 1$.

(i) If $\xi \in \mathbb{L}_0^p(\mathscr{F}_T, \mathscr{P})$, then ξ is $\mathscr{P}$-q.s. continuous;

(ii) Assume $\mathscr{P} \subset \mathscr{P}_L$ for some $L > 0$. If ξ is $\mathscr{P}$-q.s. continuous, then $\xi \in \mathbb{L}_0^p(\mathscr{F}_T, \mathscr{P})$.

Proof

(i) Assume $\xi \in \mathbb{L}_0^p(\mathscr{F}_T, \mathscr{P})$, namely $\lim_{n \to \infty} \mathscr{E}^{\mathscr{P}}[|\xi_n - \xi|^p] = 0$ for some $\xi_n \in UC_b(\Omega)$, which implies $\lim_{n \to \infty} \mathscr{E}^{\mathscr{P}}[|\xi_n - \xi|] = 0$. By otherwise choosing a subsequence, we may assume without loss of generality that $\mathscr{E}^{\mathscr{P}}[|\xi_n - \xi|] \leq 2^{-n}$, $n \geq 1$. Denote $E_n := \{|\xi_n - \xi_{n+1}| > \frac{1}{n^2}\}$ and $\Omega_n := \cap_{m \geq n} E_m^c$. Since ξ_n is continuous, then E_n is open and thus Ω_n is closed. Note that on Ω_n,

$$|\xi_m - \xi_k| \leq \sum_{i=k}^{m-1} |\xi_i - \xi_{i+1}| \leq \sum_{i=k}^{m-1} \frac{1}{i^2} \to 0, \quad \text{as } m, k \to \infty.$$

That is, ξ_m converges to ξ (more precisely, a $\mathscr{P}$-q.s. modification of ξ) uniformly on Ω_n, and thus ξ is (uniformly) continuous on Ω_n. On the other hand,

$$\mathscr{C}^{\mathscr{P}}[\Omega_n^c] = \mathscr{C}^{\mathscr{P}}[\cup_{m \geq n} E_m] \leq \sum_{m \geq n} \mathscr{E}^{\mathscr{P}}[\mathbf{1}_{E_m}] \leq \sum_{m \geq n} m^2 \mathscr{E}^{\mathscr{P}}[|\xi_m - \xi_{m+1}|]$$

$$\leq \sum_{m \geq n} m^2 \mathscr{E}^{\mathscr{P}} \Big[|\xi_m - \xi| + |\xi_{m+1} - \xi| \Big]$$

$$\leq \sum_{m \geq n} m^2 [2^{-m} + 2^{-m-1}] \to 0, \quad \text{as } n \to \infty.$$

This implies that ξ is $\mathscr{P}$-q.s. continuous.

(ii) We next assume $\mathscr{P} \subset \mathscr{P}_L$ and ξ is $\mathscr{P}$-q.s. continuous. For any $\varepsilon > 0, m \geq 1$, denote by $\xi_m^\varepsilon \in C_b^0(\Omega)$ the $\tilde{\xi}$ constructed in the proof of Lemma 10.1.11. Then, by (10.1.6),

$$\mathscr{E}^{\mathscr{P}}[|\xi_m^\varepsilon - \xi|^p] = \mathscr{E}^{\mathscr{P}} \Big[|\xi_m^\varepsilon - \xi|^p \mathbf{1}_{(\Omega_m^\varepsilon)^c} \Big] \leq C_p \mathscr{E}^{\mathscr{P}} \Big[|\xi|^p \mathbf{1}_{\{|\xi| \geq m\}} \Big] + C_p m \mathscr{C}^{\mathscr{P}} [\Omega_\varepsilon^c].$$

By first sending $\varepsilon \to 0$ and then $m \to \infty$, wee see that ξ is in the closure of $C_b^0(\Omega)$. Finally, applying Lemma 9.2.14 (ii) we see that ξ_m^ε is in the closure of $UC_b(\Omega)$. ∎

Moreover, one has the following result.

Theorem 10.1.14 *Assume $\mathscr{P}$ is weakly compact and $\xi \in \mathbb{L}_0^1(\mathscr{F}_T, \mathscr{P})$. Then there exists $\mathbb{P}^* \in \mathscr{P}$ such that $\mathscr{E}^{\mathscr{P}}[\xi] = \mathbb{E}^{\mathbb{P}^*}[\xi]$.*

Proof Let $\xi_n \in UC_b(\Omega)$ such that $\lim_{n\to\infty} \mathscr{E}^{\mathscr{P}}[|\xi_n - \xi|] = 0$ and $\mathbb{P}_m \in \mathscr{P}$ such that $\lim_{m\to\infty} \mathbb{E}^{\mathbb{P}_m}[\xi] = \mathscr{E}^{\mathscr{P}}[\xi]$. Since $\mathscr{P}$ is weakly compact, we may assume without loss of generality that $\mathbb{P}_m \to \mathbb{P}^*$ weakly for some $\mathbb{P}^* \in \mathscr{P}$. Now for any $\varepsilon > 0$, there exists n such that $\mathscr{E}^{\mathscr{P}}[|\xi_n - \xi|] < \varepsilon$. Then

$$\mathbb{E}^{\mathbb{P}^*}[\xi] \geq \mathbb{E}^{\mathbb{P}^*}[\xi_n] - \varepsilon = \lim_{m\to\infty} \mathbb{E}^{\mathbb{P}_m}[\xi_n] - \varepsilon = \lim_{m\to\infty} \mathbb{E}^{\mathbb{P}_m}[\xi + \xi_n - \xi] - \varepsilon$$

$$\geq \liminf_{m\to\infty} \mathbb{E}^{\mathbb{P}_m}[\xi] - \sup_m \mathbb{E}^{\mathbb{P}_m}[|\xi_n - \xi|] - \varepsilon \geq \mathscr{E}^{\mathscr{P}}[\xi] - 2\varepsilon.$$

By the arbitrariness of ε we prove the result. ∎

10.1.3 Some Hitting Times

Stopping time plays an important role in stochastic analysis. In this subsection we introduce three different hitting times: for any $\varepsilon > 0$ and $L > 0$,

$$\mathrm{H}_\varepsilon := \mathrm{H}_{L,\varepsilon} := \inf \{ t \geq 0 : |X_t| + L_1 t \geq \varepsilon \} \wedge T \text{ where } L_1 := L + 1;$$

$$\overline{\mathrm{H}}_\varepsilon := \inf \{ t \geq 0 : |X_t| \geq \varepsilon \} \wedge T; \quad \underline{\mathrm{H}}_\varepsilon := \inf \{ t \geq 0 : X_t^* + t \geq \varepsilon \} \wedge T. \tag{10.1.8}$$

It is clear that

$$\underline{\mathrm{H}}_\varepsilon, \mathrm{H}_\varepsilon, \overline{\mathrm{H}}_\varepsilon \downarrow 0 \text{ as } \varepsilon \downarrow 0, \quad \underline{\mathrm{H}}_\varepsilon \leq \mathrm{H}_{L_1 \varepsilon} \leq \overline{\mathrm{H}}_\varepsilon, \quad \text{and} \quad \overline{\mathrm{H}}_{\frac{\varepsilon}{2}} \leq \underline{\mathrm{H}}_\varepsilon. \tag{10.1.9}$$

As we see in the previous subsection, the regularity is important for convergence under nonlinear expectation. The above hitting times enjoy the following regularity.

Theorem 10.1.15

(i) $\underline{H}_\varepsilon$ *is Lipschitz continuous in ω with Lipschitz constant 1. In particular,*

$$\left|\underline{H}_\varepsilon^{t,\omega^1} - \underline{H}_\varepsilon^{t,\omega^2}\right| \le \|\omega^1 - \omega^2\|_t, \quad (t,\omega^1),(t,\omega^2) \in \overline{\Lambda}_{\underline{H}_\varepsilon}. \quad (10.1.10)$$

(ii) H_ε *is Lipschitz continuous in (t,ω) in the following sense: for any $L > 0$,*

$$\mathscr{E}^{\mathscr{P}_L}\left[\left|H_\varepsilon^{t_1,\omega^1} - H_\varepsilon^{t_2,\omega^2}(X_{t_2-t_1,t_2-t_1+\cdot})\right|\right] \le |\omega_{t_1}^1 - \omega_{t_2}^2| + C_L\sqrt{t_2-t_1}, \quad (10.1.11)$$

$$\text{for all } (t_1,\omega^1),(t_2,\omega^2) \in \overline{\Lambda}_{H_\varepsilon} \text{ such that } t_1 \le t_2.$$

Proof

(i) For any $\omega^1,\omega^2 \in \Omega$, assume without loss of generality that $t_1 := \underline{H}_\varepsilon(\omega^1) < \underline{H}_\varepsilon(\omega^2) =: t_2$. By the definition of $\underline{H}_\varepsilon$ in (10.1.8), we see that

$$\|\omega^1\|_{t_1} + t_1 = \varepsilon \ge \|\omega^2\|_{t_2} + t_2,$$

where the strict inequality is possible only when $t_2 = T$. This implies that

$$\left|\underline{H}_\varepsilon(\omega^1) - \underline{H}_\varepsilon(\omega^2)\right| = t_2 - t_1 \le \|\omega^1\|_{t_1} - \|\omega^2\|_{t_2} \le \|\omega^1\|_{t_1} - \|\omega^2\|_{t_1} \le \|\omega^1$$
$$-\omega^2\|_{t_1} \le \|\omega^1 - \omega^2\|_T.$$

Moreover, for any $(t,\omega^1),(t,\omega^2) \in \overline{\Lambda}_{H_\varepsilon}$ and any $\tilde{\omega} \in \Omega$,

$$\left|\underline{H}_\varepsilon^{t,\omega^1}(\tilde{\omega}) - \underline{H}_\varepsilon^{t,\omega^2}(\tilde{\omega})\right| = \left|[\underline{H}_\varepsilon(\omega^1 \otimes_t \tilde{\omega}) - t] - [\underline{H}_\varepsilon(\omega^2 \otimes_t \tilde{\omega}) - t]\right|$$
$$= \left|\underline{H}_\varepsilon(\omega^1 \otimes_t \tilde{\omega}) - \underline{H}_\varepsilon(\omega^2 \otimes_t \tilde{\omega})\right| \le \|\omega^1 \otimes_t \tilde{\omega} - \omega^2 \otimes_t \tilde{\omega}\|_T$$
$$= \|\omega^1 - \omega^2\|_t.$$

(ii) Denote $s := t_2 - t_1$, $\tau_1 := H_\varepsilon^{t_1,\omega^1}$, $\tau_2 := s + H_\varepsilon^{t_2,\omega^2}(X_{s,s+\cdot})$. On $\{\tau_1 \le \tau_2\} \in \mathscr{F}_{\tau_1}$, we have

$$|\omega_{t_1}^1 + X_{\tau_1}| + L_1(t_1 + \tau_1) = H_\varepsilon(\omega^1 \otimes_{t_1} X) = \varepsilon$$
$$\ge H_\varepsilon(\omega^2 \otimes_{t_2} X_{s,s+\cdot}) = |\omega_{t_2}^2 + X_{s,\tau_2}| + L_1(t_1 + \tau_2).$$

This implies that for any $\mathbb{P} \in \mathscr{P}_L$ and again on $\{\tau_1 \le \tau_2\}$,

$$L_1\left[\mathbb{E}_{\tau_1}^{\mathbb{P}}[\tau_2] - \tau_1\right] \le |\omega_{t_1}^1 + X_{\tau_1}| - \mathbb{E}_{\tau_1}^{\mathbb{P}}[|\omega_{t_2}^2 + X_{s,\tau_2}|]$$
$$\le |\omega_{t_1}^1 + X_{\tau_1}| - |\omega_{t_2}^2 + \mathbb{E}_{\tau_1}^{\mathbb{P}}[X_{s,\tau_2}]|$$

$$\leq |\omega_{t_1}^1 - \omega_{t_2}^2| + |\mathbb{E}_{\tau_1}^{\mathbb{P}}[X_{\tau_1,\tau_2}]| + \mathbb{E}_{\tau_1}[|X_s|]$$
$$\leq |\omega_{t_1}^1 - \omega_{t_2}^2| + L\big[\mathbb{E}_{\tau_1}^{\mathbb{P}}[\tau_2] - \tau_1\big] + \mathbb{E}_{\tau_1}[|X_s|].$$

This implies

$$\mathbb{E}_{\tau_1}^{\mathbb{P}}[\tau_2] - \tau_1 \leq |\omega_{t_1}^1 - \omega_{t_2}^2| + \mathbb{E}_{\tau_1}[|X_s|] \text{ on } \{\tau_1 \leq \tau_2\} \in \mathscr{F}_{\tau_1},$$

and thus

$$\mathbb{E}^{\mathbb{P}}\Big[(\tau_2 - \tau_1)\mathbf{1}_{\{\tau_1 \leq \tau_2\}}\Big] \leq |\omega_t^1 - \omega_t^2|\mathbb{P}(\tau_1 \leq \tau_2) + \mathbb{E}^{\mathbb{P}}\big[|X_s|\mathbf{1}_{\{\tau_1 \leq \tau_2\}}\big].$$

Similarly we have

$$\mathbb{E}^{\mathbb{P}}\Big[(\tau_2 - \tau_1)\mathbf{1}_{\{\tau_1 > \tau_2\}}\Big] \leq |\omega_t^1 - \omega_t^2|\mathbb{P}(\tau_1 > \tau_2) + \mathbb{E}^{\mathbb{P}}\big[|X_s|\mathbf{1}_{\{\tau_1 > \tau_2\}}\big].$$

Then

$$\mathbb{E}^{\mathbb{P}}\Big[|\mathbf{H}_\varepsilon^{t_1,\omega^1} - \mathbf{H}_\varepsilon^{t_2,\omega^2}(X_{s,s+\cdot})|\Big] = \mathbb{E}^{\mathbb{P}}[|\tau_2 - \tau_1|] \leq |\omega_t^1 - \omega_t^2|$$
$$+\mathbb{E}^{\mathbb{P}}[|X_s|] \leq |\omega_t^1 - \omega_t^2| + C_L\sqrt{s}.$$

By the arbitrariness of $\mathbb{P} \in \mathscr{P}_L$, we obtain (10.1.11). ∎

The hitting time $\overline{\mathbf{H}}_\varepsilon$, unfortunately, does not have the desired regularity, as we see in the following example. As a consequence, in this book we will not use $\overline{\mathbf{H}}_\varepsilon$.

Example 10.1.16 *Let $d = 1$, $t \in (0,T)$, $\omega_s = \frac{\varepsilon s}{t}$ and $\omega_s^\delta := \frac{(1-\delta)\varepsilon s}{t}$, $0 \leq s \leq t$. Then*

$$\lim_{\delta \to 0} \|\omega^\delta - \omega\|_t = 0, \quad \text{but} \quad \lim_{\delta \to 0} \mathscr{E}^{\mathbb{P}_L}[|\mathbf{H}_\varepsilon^{t,\omega^\delta} - \mathbf{H}_\varepsilon^{t,\omega}|] > 0.$$

Proof First, it is clear that $\|\omega^\delta - \omega\|_t = \varepsilon\delta \to 0$ as $\delta \to 0$. Next, denote $\mathbb{P}^0 \in \mathscr{P}_L$ correspond to $b' = 0, \sigma' = 0$ in Definition 9.2.9. Then, $X_s = 0$, $0 \leq s \leq T$, $\mathbb{P}^0$-a.s. Thus $\mathbf{H}_\varepsilon^{t,\omega} = t$ and $\mathbf{H}_\varepsilon^{t,\omega^\delta} = T$, $\mathbb{P}^0$-a.s. Therefore, $\mathscr{E}^{\mathbb{P}_L}[|\mathbf{H}_\varepsilon^{t,\omega^\delta} - \mathbf{H}_\varepsilon^{t,\omega}|] \geq \mathbb{E}^{\mathbb{P}^0}[|\mathbf{H}_\varepsilon^{t,\omega^\delta} - \mathbf{H}_\varepsilon^{t,\omega}|] = T - t$. ∎

We remark that $\mathbf{H}_\varepsilon$ enjoys the Markov property in the following sense. For any $t \in [0,T)$, $R > 0$, $x \in \mathbb{R}^d$ such that $|x| \leq R$, define

$$\mathbf{H}^{t,x,R} := \inf\{s \geq 0 : |x + X_s| + L_1 s \geq R\} \wedge (T - t). \tag{10.1.12}$$

The following result is obvious.

Proposition 10.1.17

(i) $H_\varepsilon = H^{0,0,\varepsilon}$, *and* $H_\varepsilon^{t,\omega} = H^{t,\omega_t,\varepsilon - L_1 t}$ *for all* $(t,\omega) \in \Lambda_{H_\varepsilon}$. *In particular,* $H_\varepsilon^{t,\omega}$ *depends only on the current value* ω_t, *not on the past values* $\{\omega_s\}_{0 \leq s < t}$.

(ii) *For any* (t, x, R) *and any stopping time* $\tau \in \mathscr{T}_{H_\varepsilon^{t,x,R}}$, *it holds that*

$$H_\varepsilon^{t,x,R}(X.) = H^{\tau, x + X_\tau, R - L_1(\tau - t)}(X_{\tau,\tau+}.).$$

While $\underline{H}_\varepsilon$ has better regularity than H_ε, it does not share this Markov property. Such Markov property will be quite useful in Chapter 11 below, so in this book we will mainly use H_ε, which enjoys both the desired regularity and the Markov property. In the rest of this subsection we present some further properties of H_ε, which will be used later in the book.

We start with some simple but crucial estimates.

Lemma 10.1.18

(i) $H_\varepsilon(\omega) > 0$ *for all* $\omega \in \Omega$ *and the set* Λ_{H_ε} *is open (under* **d**).

(ii) *For any* $\varepsilon > 0$, $0 < \delta < T$, $p \geq 1$, $\mathscr{C}^{\mathscr{P}_L}(H_\varepsilon \leq \delta) \leq C_{p,L}\varepsilon^{-2p}\delta^p$.

(iii) *For any* $L > 0$, $\inf_{\mathbb{P} \in \mathscr{P}_L} \mathbb{E}^{\mathbb{P}}[H_\varepsilon] > 0$.

Proof Denote $\eta_t(\omega) := |\omega_t| + L_1 t$.

(i) Since $|\omega_0| = 0 < \varepsilon$ and ω is continuous, it is clear that $H_\varepsilon(\omega) > 0$ for all $\omega \in \Omega$. Moreover, the mapping $(t, \omega) \mapsto \eta_t(\omega)$ is continuous under **d**, then Λ_{H_ε} is open.

(ii) Since $\delta < T$, we see that $\{H_\varepsilon \leq \delta\} = \{\eta_\delta^* \geq \varepsilon\}$. Then, for any $\mathbb{P} \in \mathscr{P}_L$,

$$\mathbb{P}(H_\varepsilon \leq \delta) = \mathbb{P}(\eta_\delta^* \geq \varepsilon) \leq \varepsilon^{-2p}\mathbb{E}^{\mathbb{P}}[|\eta_\delta^*|^{2p}]$$

$$\leq C\varepsilon^{-2p}\mathbb{E}^{\mathbb{P}}\Big[|X_\delta^*|^{2p} + |L_1\delta|^{2p}\Big] \leq C\varepsilon^{-2p}\mathbb{E}^{\mathbb{P}}\Big[\delta^2 + \delta^{2p}\Big] \leq C\varepsilon^{-2p}\delta^p.$$

(iii) By (ii), there exists $\delta_\varepsilon > 0$ such that $\mathscr{C}^{\mathscr{P}_L}(H_\varepsilon \leq \delta_\varepsilon) \leq \frac{1}{2}$. Then, for any $\mathbb{P} \in \mathscr{P}_L$, $\mathbb{P}(H_\varepsilon > \delta_\varepsilon) \geq \frac{1}{2}$ and thus $\mathbb{E}^{\mathbb{P}}[H_\varepsilon] \geq \delta_\varepsilon \mathbb{P}(H_\varepsilon > \delta_\varepsilon) \geq \frac{1}{2}\delta_\varepsilon$. By the arbitrariness of $\mathbb{P} \in \mathscr{P}_L$, we prove the result.

∎

Next, H_ε is previsible. Indeed, denote

$$H_{m,\varepsilon} := \inf\{t \geq 0 : |X_t| + L_1 t \geq \varepsilon - \frac{1}{m}\} \wedge (T - \frac{1}{m}), \quad m \geq \frac{1}{\varepsilon \wedge T}. \quad (10.1.13)$$

Lemma 10.1.19

(i) $H_{m,\varepsilon} \uparrow H_\varepsilon$ *as* $m \uparrow \infty$.

(ii) $H_{m-1,\varepsilon}(\omega) \leq H_{m,\varepsilon}(\tilde{\omega}) \leq H_{m+1,\varepsilon}(\omega)$ *for all* $\omega, \tilde{\omega} \in \Omega$ *such that* $\|\tilde{\omega} - \omega\|_T \leq \frac{1}{m(m+1)}$.

(iii) There exist open set $\Omega_m \subset \Omega$ and constant $\theta_m > 0$ such that

$$\mathscr{C}^{\mathscr{P}_L}[\Omega_m^c] \le 2^{-m}, \quad H_{m+1,\varepsilon} - H_{m,\varepsilon} \ge \theta_m \text{ on } \Omega_m.$$

Proof

(i) is obvious.

(ii) Fix arbitrary $\omega, \tilde{\omega} \in \Omega$ such that $\|\tilde{\omega} - \omega\|_T \le \frac{1}{m(m+1)}$. Denote $t := H_{m+1,\varepsilon}(\omega)$. If $t \ge T - \frac{1}{m}$, then $H_{m,\varepsilon}(\tilde{\omega}) \le t$. Now assume $t < T - \frac{1}{m} < T - \frac{1}{m+1}$. Then $|\omega_t| + L_1 t = \varepsilon - \frac{1}{m+1}$, and thus $|\tilde{\omega}_t| + L_1 t \ge \varepsilon - \frac{1}{m+1} - \frac{1}{m(m+1)} = \varepsilon - \frac{1}{m}$. This implies that $H_{m,\varepsilon}(\tilde{\omega}) \le t = H_{m+1,\varepsilon}(\omega)$. Similarly we can prove that $H_{m-1,\varepsilon}(\omega) \le H_{m,\varepsilon}(\tilde{\omega})$ whenever $\|\tilde{\omega} - \omega\|_T \le \frac{1}{m(m-1)}$.

(iii) For any $\delta > 0$, by (9.2.16) we have

$$\mathscr{C}^{\mathscr{P}_L}[\text{OSC}_{\delta^3}(X) \ge \delta] \le \delta^{-2} \mathscr{E}^{\mathscr{P}_L}[|\text{OSC}_{\delta^3}(X)|^2] \le C\delta.$$

Let $\delta_m \le \frac{1}{C2^m}$ and set $\Omega_m := \{\text{OSC}_{\delta_m^3}(X) < \delta_m\}$. Then clearly Ω_m is empty and $\mathscr{C}^{\mathscr{P}_L}[\Omega_m^c] \le C\delta_m \le 2^{-m}$. Moreover, assume further that $L_1\delta_m^3 + \delta_m \le \frac{1}{m(m+1)}$, then we claim that

$$\Omega_m \subset \{H_{m+1,\varepsilon} - H_{m,\varepsilon} \ge \delta_m^3\} \tag{10.1.14}$$

and thus the result holds for $\theta_m := \delta_m^3$.

We prove (10.1.14) by contradiction. First, if $H_{m+1,\varepsilon} = T - \frac{1}{m+1}$, then

$$H_{m+1,\varepsilon} - H_{m,\varepsilon} \ge [T - \frac{1}{m+1}] - [T - \frac{1}{m}] = \frac{1}{m(m+1)} > \delta_m^3.$$

So on $\Omega_m \cap \{H_{m+1,\varepsilon} - H_{m,\varepsilon} < \delta_m^3\}$, we must have $H_{m+1,\varepsilon}(\omega) < T - \frac{1}{m+1}$, and thus $|X_{H_{m+1,\varepsilon}}| + L_1 H_{m+1,\varepsilon} = \varepsilon - \frac{1}{m+1}$. Note that $|X_{H_{m,\varepsilon}}| + L_1 H_{m,\varepsilon} \le \varepsilon - \frac{1}{m}$, then on $\Omega_m \cap \{H_{m+1,\varepsilon} - H_{m,\varepsilon} < \delta_m^3\}$,

$$\frac{1}{m(m+1)} \le \left[|X_{H_{m+1,\varepsilon}}| + L_1 H_{m+1,\varepsilon}\right] - \left[|X_{H_{m,\varepsilon}}| + L_1 H_{m,\varepsilon}\right]$$
$$\le |X_{H_{m,\varepsilon}, H_{m+1,\varepsilon}}| + L_1[H_{m+1,\varepsilon} - H_{m,\varepsilon}]$$
$$\le \text{OSC}_{\delta_m^3}(X) + L_1\delta_m^3 < \delta_m + L_1\delta_m^3 \le \frac{1}{m(m+1)}.$$

This is an obvious contradiction. Then $\Omega_m \cap \{H_{m+1,\varepsilon} - H_{m,\varepsilon} < \delta_m^3\} = \emptyset$ and thus (10.1.14) holds. $\blacksquare$

Finally, H_ε can be used to discretize the path ω as follows. Denote $H_0^\varepsilon := 0$, and for $n \geq 0$,

$$H_{n+1}^\varepsilon := \inf \left\{ t \geq H_n^\varepsilon : |X_{H_n^\varepsilon, t}| + L_1(t - H_n^\varepsilon) \geq \varepsilon \right\} \wedge T. \qquad (10.1.15)$$

Proposition 10.1.20

(i) $H_{n+1}^\varepsilon = H^{H_n^\varepsilon, 0, \varepsilon}(X_{H_n^\varepsilon, H_n^\varepsilon +} \cdot)$. *In particular,* $H_1^\varepsilon = H_\varepsilon$.

(ii) H_n^ε *is increasing in* n, *and* $|X_{H_n^\varepsilon, H_{n+1}^\varepsilon}| + L_1[H_{n+1}^\varepsilon - H_n^\varepsilon] = \varepsilon$ *on* $\{H_{n+1}^\varepsilon < T\}$.

(iii) $\bigcap_{n\geq 1}\{H_n^\varepsilon < T\} = \emptyset$, *and* $\mathcal{C}^{\mathcal{P}_L}[H_n^\varepsilon < T] \leq \frac{C_L}{n\varepsilon^2}$.

(iv) $(X^\varepsilon - X)_T^* \leq 2\varepsilon$, *where* X^ε *is the linear interpolation of* $(H_n^\varepsilon, X_{H_n^\varepsilon})_{n\geq 0}$, *namely*

$$X_t^\varepsilon := X_{H_n^\varepsilon} + \frac{t - H_n^\varepsilon}{H_{n+1}^\varepsilon - H_n^\varepsilon} X_{H_n^\varepsilon, H_{n+1}^\varepsilon}, \quad \forall n \geq 0, t \in [H_n^\varepsilon, H_{n+1}^\varepsilon]. \qquad (10.1.16)$$

Proof

(i) and (ii) are obvious.

(iii) First, assume $\omega \in \bigcap_{n\geq 1}\{H_n^\varepsilon < T\}$, and denote $t_n := H_n^\varepsilon(\omega) < T$. Then by (ii) we have $t_n \uparrow t_\infty \leq T$ and $|\omega_{t_n, t_{n+1}}| + L_1[t_{n+1} - t_n] = \varepsilon$ for all $n \geq 0$. By the convergence of t_n, there exists N such that $t_{n+1} - t_n \leq \frac{\varepsilon}{2L_1}$ for all $n \geq N$. Then $|\omega_{t_n, t_{n+1}}| \geq \frac{\varepsilon}{2}$ for all $n \geq N$. This contradicts with the fact that $\lim_{n\to\infty} \omega_{t_n} = \omega_{t_\infty}$. Thus $\bigcap_{n\geq 1}\{H_n^\varepsilon < T\} = \emptyset$.

Next, for any $m \geq 1$, similar to the above arguments we have

$$\{H_m^\varepsilon < T\} = \bigcap_{0 \leq n < m} \{H_n^\varepsilon < T\} = \bigcap_{0 \leq n < m} \left\{ |X_{H_n^\varepsilon, H_{n+1}^\varepsilon}| + L_1[H_{n+1}^\varepsilon - H_n^\varepsilon] = \varepsilon \right\}$$

Then, on $\{H_m^\varepsilon < T\}$ we have

$$m\varepsilon^2 = \sum_{n=0}^{m-1} \left| |X_{H_n^\varepsilon, H_{n+1}^\varepsilon}| + L_1[H_{n+1}^\varepsilon - H_n^\varepsilon] \right|^2 \leq \sum_{n=0}^{m-1} \left[2|X_{H_n^\varepsilon, H_{n+1}^\varepsilon}|^2 + C[H_{n+1}^\varepsilon - H_n^\varepsilon]^2 \right].$$

Now for any $\mathbb{P} \in \mathcal{P}_L$, recalling Definition 9.2.9 one can easily estimate:

$$\mathbb{P}(H_m^\varepsilon < T) \leq \frac{1}{m\varepsilon^2} \mathbb{E}^{\mathbb{P}} \left[\sum_{n=0}^{m-1} \left[2|X_{H_n^\varepsilon, H_{n+1}^\varepsilon}|^2 + C[H_{n+1}^\varepsilon - H_n^\varepsilon]^2 \right] \right] \leq \frac{C}{m\varepsilon^2}.$$

Since $\mathbb{P} \in \mathcal{P}_L$ is arbitrary, we obtain the desired estimate.

(iv) For any $n \geq 0$ and $t \in [H_n^\varepsilon, H_{n+1}^\varepsilon]$, note that $X_{H_n^\varepsilon}^\varepsilon = X_{H_n^\varepsilon}$ and $|X_{H_n^\varepsilon, t}| \leq \varepsilon$. Then

$$|X_t^\varepsilon - X_t| \leq |X_{H_n^\varepsilon, t}^\varepsilon| + |X_{H_n^\varepsilon, t}| = \frac{t - H_n^\varepsilon}{H_{n+1}^\varepsilon - H_n^\varepsilon} |X_{H_n^\varepsilon, H_{n+1}^\varepsilon}| + |X_{H_n^\varepsilon, t}| \leq 2\varepsilon.$$

This implies immediately that $(X^\varepsilon - X)_T^* \leq 2\varepsilon$.   ∎

Remark 10.1.21

(i) $\overline{H}_\varepsilon$ also satisfies the Markov property in the following sense:

$$\overline{H}^{x,R} = \overline{H}^{x+X_\tau,R} \text{ for all } \tau \in \mathscr{T}_{\overline{H}^{x,R}},$$

where $\overline{H}^{x,R} := \inf\left\{t \geq 0 : |x + X_t| \geq R\right\} \wedge T.$ (10.1.17)

(ii) Both $\overline{H}_\varepsilon$ and $\underline{H}_\varepsilon$ satisfy all the properties in Lemmas 10.1.18, 10.1.19, and Proposition 10.1.20, after obvious modifications whenever necessary.

(iii) $\overline{H}_\varepsilon$ also satisfies (10.1.11) when $\mathscr{P} \subset \mathscr{P}_L$ is uniformly nondegenerate. That is, there exists a constant matrix $\underline{\sigma} > 0$ such that, for all $\mathbb{P} \in \mathscr{P}$ with corresponding $\mathbb{P}'$ and b', σ' as in Definition 9.2.9, $\sigma' \geq \underline{\sigma}$, $\mathbb{P}'$-a.s. Then

$$\mathscr{E}^\mathscr{P}\left[\left|\overline{H}_\varepsilon^{t_1,\omega^1} - \overline{H}_\varepsilon^{t_2,\omega^2}(X_{t_2-t_1,t_2-t_1+\cdot})\right|\right] \leq C\left[|\omega_{t_1}^1 - \omega_{t_2}^2| + \sqrt{t_2 - t_1}\right], \quad (10.1.18)$$

for all $(t_1, \omega^1), (t_2, \omega^2) \in \overline{\Lambda}_{\overline{H}_\varepsilon}$ such that $t_1 \leq t_2$. ∎

10.2 Pathwise Conditional Nonlinear Expectation

Let $\mathscr{P}$ satisfy Assumption 9.3.3 with corresponding $\mathscr{P}_t$, and $\xi \in \mathbb{L}^1(\mathscr{F}_T, \mathscr{P})$. Define the pathwise conditional nonlinear expectation of ξ as:

$$\mathscr{E}_{t,\omega}^\mathscr{P}[\xi] := \mathscr{E}^{\mathscr{P}_t}[\xi^{t,\omega}] = \sup_{\mathbb{P} \in \mathscr{P}_t} \mathbb{E}^\mathbb{P}[\xi^{t,\omega}], \quad \forall(t, \omega) \in \overline{\Lambda}. \quad (10.2.1)$$

To obtain the desired measurability of $\mathscr{E}_{t,\omega}^\mathscr{P}[\xi]$, we need regularity of ξ. Our main result of this section is

Theorem 10.2.1 *Let $\mathscr{P}$ satisfy Assumption 9.3.3, and $\xi \in UC(\Omega)$. Denote $Y_t(\omega) := \mathscr{E}_{t,\omega}^\mathscr{P}[\xi]$. Then*

(i) $Y \in UC(\overline{\Lambda})$.

(ii) Y satisfies the pathwise dynamic programming principle: recalling (9.2.25)

$$Y_t(\omega) = \mathscr{E}^{\mathscr{P}_t}[Y_\tau^{t,\omega}], \quad \forall(t, \omega) \in \overline{\Lambda}, \tau \in \mathscr{T}_{T-t}. \quad (10.2.2)$$

That is, the pathwise conditional nonlinear expectation satisfies the tower property:

$$\mathscr{E}_{t,\omega}^\mathscr{P}[\xi] = \mathscr{E}_{t,\omega}^\mathscr{P}\left[\mathscr{E}_{\tau,\cdot}^\mathscr{P}[\xi]\right], \quad \forall(t, \omega) \in \Lambda, \tau \in \mathscr{T} \text{ such that } \tau \geq t. \quad (10.2.3)$$

Proof We proceed in several steps. Let ρ denote the modulus of continuity function of ξ.

Step 1. First, by Problem 10.5.2 (iii) $\xi^{t,\omega} \in \mathbb{L}^1(\mathscr{F}_T^t, \mathscr{P}_t)$, and thus $Y_t(\omega)$ is well defined. We claim that,

$$|Y_t(\omega^1) - Y_t(\omega^2)| \leq \rho(\|\omega^1 - \omega^2\|_t), \quad \text{for all } t \in [0, T], \omega^1, \omega^2 \in \Omega. \quad (10.2.4)$$

Indeed,

$$|Y_t(\omega^1) - Y_t(\omega^2)| \leq \sup_{\mathbb{P}\in\mathscr{P}_t} \mathbb{E}^{\mathbb{P}}[|\xi^{t,\omega^1} - \xi^{t,\omega^2}|] = \sup_{\mathbb{P}\in\mathscr{P}_t} \mathbb{E}^{\mathbb{P}}\left[|\xi(\omega^1 \otimes_t X) - \xi(\omega^2 \otimes_t X)|\right]$$

$$\leq \sup_{\mathbb{P}\in\mathscr{P}_t} \mathbb{E}^{\mathbb{P}}\left[\rho(\|\omega^1 \otimes_t X - \omega^2 \otimes_t X\|_T)\right] = \rho(\|\omega^1 - \omega^2\|_t).$$

Step 2. We next prove the dynamic programming principle for deterministic times: for any $0 \leq t_1 < t_2 \leq T$ and any $\omega \in \Omega$,

$$Y_{t_1}(\omega) = \mathscr{E}^{\mathscr{P}_{t_1}}[Y_{t_2-t_1}^{t_1,\omega}]. \quad (10.2.5)$$

Without loss of generality, we prove it only in the case $t_1 = 0$. That is, by denoting $t := t_2$,

$$Y_0 = \mathscr{E}^{\mathscr{P}}[Y_t]. \quad (10.2.6)$$

First, for any $\mathbb{P} \in \mathscr{P}$, by Assumption 9.3.3 we see that the r.c.p.d. $\mathbb{P}^{t,\omega} \in \mathscr{P}_t$ for $\mathbb{P}$-a.e. ω. Then

$$\mathbb{E}^{\mathbb{P}}[\xi] = \mathbb{E}^{\mathbb{P}}\left[\mathbb{E}^{\mathbb{P}^{t,\omega}}[\xi^{t,\omega}]\right] \leq \mathbb{E}^{\mathbb{P}}\left[\mathscr{E}^{\mathscr{P}_t}[\xi^{t,\omega}]\right] = \mathbb{E}^{\mathbb{P}}[Y_t].$$

Taking supremum over $\mathbb{P} \in \mathscr{P}$ we obtain

$$Y_0 \leq \mathscr{E}^{\mathscr{P}}[Y_t]. \quad (10.2.7)$$

On the other hand, let $\{\omega^i, i \geq 1\} \subset \Omega$ be the dense sequence in Lemma 9.2.1. For any $\varepsilon > 0$, denote $E_0 := \emptyset$ and define repeatedly $E_i := \{\omega \in \Omega : \|\omega - \omega^i\|_t \leq \varepsilon\} \setminus \cup_{j=1}^{i-1} E_j$, $i \geq 1$. Then $\{E_i, i \geq 1\} \subset \mathscr{F}_t$ is a partition of Ω. For each i, by definition of $Y_t(\omega^i)$, there exists $\mathbb{P}_i \in \mathscr{P}_t$ such that

$$Y_t(\omega^i) \leq \mathbb{E}^{\mathbb{P}_i}[\xi^{t,\omega^i}] + \varepsilon.$$

Now for any $\mathbb{P} \in \mathscr{P}$, by Assumption 9.3.3, we may construct the following $\widehat{\mathbb{P}} \in \mathscr{P}$:

$$\widehat{\mathbb{P}} := \mathbb{P} \otimes_t \sum_{i=1}^{\infty} \mathbb{P}_i \mathbf{1}_{E_i}. \quad (10.2.8)$$

Then, by (10.2.4) and the uniform continuity of ξ,

$$Y_t(\omega) = \sum_{i=1}^{\infty} Y_t(\omega)\mathbf{1}_{E_i}(\omega) \le \sum_{i=1}^{\infty} Y_t(\omega^i)\mathbf{1}_{E_i}(\omega) + \rho(\varepsilon)$$

$$\le \sum_{i=1}^{\infty} \mathbb{E}^{\mathbb{P}_i}[\xi^{t,\omega^i}]\mathbf{1}_{E_i}(\omega) + \varepsilon + \rho(\varepsilon) \le \sum_{i=1}^{\infty} \mathbb{E}^{\mathbb{P}_i}[\xi^{t,\omega}]\mathbf{1}_{E_i}(\omega) + \varepsilon + 2\rho(\varepsilon)$$

$$= \sum_{i=1}^{\infty} \mathbb{E}^{\widehat{\mathbb{P}}^{t,\omega}}[\xi^{t,\omega}]\mathbf{1}_{E_i}(\omega) + \varepsilon + 2\rho(\varepsilon) = \mathbb{E}^{\widehat{\mathbb{P}}^{t,\omega}}[\xi^{t,\omega}] + \varepsilon + 2\rho(\varepsilon). \quad (10.2.9)$$

Thus

$$\mathbb{E}^{\mathbb{P}}[Y_t] = \mathbb{E}^{\widehat{\mathbb{P}}}[Y_t] \le \mathbb{E}^{\widehat{\mathbb{P}}}[\xi] + \varepsilon + 2\rho(\varepsilon) \le Y_0 + \varepsilon + 2\rho(\varepsilon).$$

Sending $\varepsilon \to 0$, we obtain $\mathbb{E}^{\mathbb{P}}[Y_t] \le Y_0$. Since $\mathbb{P} \in \mathscr{P}_0 = \mathscr{P}$ is arbitrary, together with (10.2.7), this implies (10.2.6).

Step 3. We now prove the full regularity. Let $0 \le t < \tilde{t} \le T$, and $\omega, \tilde{\omega} \in \Omega$. Combining Steps 1 and 2 we have

$$|Y_t(\omega) - Y_{\tilde{t}}(\tilde{\omega})| = \left| \mathscr{E}^{\mathscr{P}_t}[Y_{\tilde{t}-t}^{t,\omega}] - Y_{\tilde{t}}(\tilde{\omega}) \right| = \left| \mathscr{E}^{\mathscr{P}_t}\left[Y_{\tilde{t}}(\omega \otimes_t X) - Y_{\tilde{t}}(\tilde{\omega}) \right] \right|$$

$$\le \mathscr{E}^{\mathscr{P}_t}\left[\rho(\|\omega \otimes_t X - \tilde{\omega}\|_{\tilde{t}}) \right] = \mathscr{E}^{\mathscr{P}_t}\left[\rho\big(\mathbf{d}((t,\omega),(\tilde{t},\tilde{\omega})) + \|X\|_{\tilde{t}-t}\big) \right]$$

$$\le \mathscr{E}^{\mathscr{P}_L}\left[\rho\big(\mathbf{d}((t,\omega),(\tilde{t},\tilde{\omega})) + \|X\|_{\tilde{t}-t}\big) \right] \le \overline{\rho}\big(\mathbf{d}((t,\omega),(\tilde{t},\tilde{\omega}))\big),$$

$$(10.2.10)$$

where, recalling the definition of $\mathbf{d}$ in (9.2.1),

$$\overline{\rho}(\delta) := \mathscr{E}^{\mathscr{P}_L}\left[\rho(\delta + \|X\|_{\delta^2}) \right]. \quad (10.2.11)$$

Now as a direct application of the dominated convergence Theorem 10.1.8, we see that $\lim_{\delta \to 0} \overline{\rho}(\delta) = 0$. That is, $\overline{\rho}$ is a modulus of continuity function. This completes the proof of (i).

Step 4. In this step we prove (10.2.2) in the case that $\tau > 0$ and takes only finitely many values: $0 < t_1 < \cdots < t_n \le T-t$. Again, without loss of generality we assume $t = 0$. First, by (10.2.5), $Y_{t_{n-1}}(\omega) = \mathscr{E}^{\mathscr{P}_{n-1}}[Y_{t_n-t_{n-1}}^{t_{n-1},\omega}]$. Following the arguments in Step 2 and noting that $\{\tau = t_n\} = \{\tau \le t_{n-1}\}^c \in \mathscr{F}_{t_{n-1}}$, one can easily show that

$$\mathscr{E}^{\mathscr{P}}[Y_\tau] = \mathscr{E}^{\mathscr{P}}\left[Y_\tau \mathbf{1}_{\{\tau \le t_{n-1}\}} + Y_{t_n}\mathbf{1}_{\{\tau = t_n\}} \right] = \mathscr{E}^{\mathscr{P}}\left[Y_\tau \mathbf{1}_{\{\tau \le t_{n-1}\}} + Y_{t_{n-1}}\mathbf{1}_{\{\tau = t_n\}} \right]$$

$$= \mathscr{E}^{\mathscr{P}}[Y_{\tau \wedge t_{n-1}}].$$

Repeat the arguments we obtain

$$\mathcal{E}^{\mathscr{P}}[Y_\tau] = \mathcal{E}^{\mathscr{P}}[Y_{\tau \wedge t_1}] = \mathcal{E}^{\mathscr{P}}[Y_{t_1}] = Y_0.$$

Step 5. Finally we prove (10.2.2) in the general case. Again, without loss of generality we assume $t = 0$ and thus $\tau \in \mathscr{T}$. For each $n \geq 1$, denote $h := h_n := \frac{T}{n}$, $t_i^n := ih$, $i = 0, \cdots, n$, and $\tau_n := \sum_{i=1}^n t_i^n \mathbf{1}_{\{t_{i-1}^n < \tau \leq t_i^n\}}$. Then $\tau_n \in \mathscr{T}$ and takes finitely many values. Thus by Step 4, $Y_0 = \mathcal{E}^{\mathscr{P}}[Y_{\tau_n}]$. Note that $0 \leq \tau_n - \tau \leq h$, then it follows from (10.2.10) that

$$|Y_0 - \mathcal{E}^{\mathscr{P}}[Y_\tau]| = \left|\mathcal{E}^{\mathscr{P}}[Y_{\tau_n}] - \mathcal{E}^{\mathscr{P}}[Y_\tau]\right| \leq \mathcal{E}^{\mathscr{P}}\left[|Y_{\tau_n} - Y_\tau|\right]$$

$$\leq \mathcal{E}^{\mathscr{P}}\left[\bar{\rho}\big(\mathbf{d}((\tau, X), (\tau_n, X))\big)\right] \leq \mathcal{E}^{\mathscr{P}}\left[\bar{\rho}(\sqrt{h} + \mathrm{OSC}_h(X))\right].$$

Note that $\eta_h(\omega) := \bar{\rho}(\sqrt{h} + \mathrm{OSC}_h(\omega))$ is uniformly continuous in ω, uniformly on h. Moreover, as $n \to \infty$ and thus $h \to 0$, we have $\eta_h(\omega) \to 0$ for all ω. Then it follows from Theorem 10.1.8 again that $\mathcal{E}^{\mathscr{P}}[\eta_h] \to 0$. Therefore, $Y_0 = \mathcal{E}^{\mathscr{P}}[Y_\tau]$. ∎

Remark 10.2.2

(i) The dynamic programming principle is also called tower property or time consistency. It appears in many different contexts. The proof here relies on the regularities. This can be weakened by using the so-called measurable selection.

(ii) As we will see in Chapter 11, the dynamic programming principle is crucial for deriving the PDE or PPDE. For the purpose of PDE or PPDE, actually one may weaken the dynamic programming principle to the so-called weak dynamic programming principle, proposed by Bouchard & Touzi [19]. ∎

We remark that the dynamic programming principle (10.2.2) exactly means that the pathwise conditional nonlinear expectation $\mathcal{E}^{\mathscr{P}}_{t,\omega}[\xi]$ is a pathwise $\mathcal{E}^{\mathscr{P}}$-martingale in the following sense.

Definition 10.2.3 *Let $\mathscr{P}$ satisfy Assumption 9.3.3, and $Y \in \mathbb{L}^1(\mathbb{F}, \mathscr{P})$. We say Y is a pathwise $\mathcal{E}^{\mathscr{P}}$-martingale (resp. $\mathcal{E}^{\mathscr{P}}$-submartingale, $\mathcal{E}^{\mathscr{P}}$-supermartingale) if, for any $(t, \omega) \in \Lambda$ and $\tau \in \mathscr{T}$ satisfying $\tau \geq t$,*

$$Y_t(\omega) = (\text{resp.} \geq, \leq)\ \mathcal{E}^{\mathscr{P}_t}[Y_{\tau^{t,\omega}-t}^{t,\omega}].$$

Lemma 10.2.4 *Let $\mathscr{P}$ satisfy Assumption 9.3.3. If Y is a pathwise $\mathcal{E}^{\mathscr{P}}$-supermartingale, then it is a $\mathbb{P}$-supermartingale for all $\mathbb{P} \in \mathscr{P}$.*

The proof is quite simple and is left to readers. However, we shall emphasize that a pathwise $\mathcal{E}^{\mathscr{P}}$-submartingale Y may not be a $\mathbb{P}$-submartingale for all $\mathbb{P} \in \mathscr{P}$. In fact, it is even not clear whether or not Y is a $\mathbb{P}$-semimartingale, see Pham & Zhang [191] for some study along this direction.

10.3 Optimal Stopping Under Nonlinear Expectation

This section extends the Snell envelope theory, Proposition 6.3.2, to the nonlinear expectation setting. The arguments are quite involved, and the result will mainly be used in Subsection 11.1.3 and Section 11.4 below. Readers who are not interested in those sections can skip this section.

Let $\mathscr{P}$ satisfy Assumption 9.3.3 and $\mathrm{H} \in \mathscr{T}$. Recall (9.3.5) that $\mathrm{H}^{t,\omega} - t \in \mathscr{T}$ for $t < \mathrm{H}(\omega)$. For a scalar process $u \in UC_b(\overline{\Lambda}_{\mathrm{H}})$, define the nonlinear Snell envelope of u on $[0, \mathrm{H}]$:

$$Y_t(\omega) := \sup_{\tau \in \mathscr{T}} \mathscr{E}^{\mathscr{P}_t}[u^{t,\omega}_{\tau \wedge (\mathrm{H}^{t,\omega}-t)}] = \sup_{\tau \in \mathscr{T}_{\mathrm{H}^{t,\omega}-t}} \mathscr{E}^{\mathscr{P}_t}[u^{t,\omega}_{\tau}], \quad (t,\omega) \in \overline{\Lambda}_{\mathrm{H}};$$
$$\tau^* := \inf\{t \geq 0 : Y_t = u_t\}. \tag{10.3.1}$$

We shall assume H satisfies:

Assumption 10.3.1

(i) *The mapping* $(t, \omega) \mapsto \mathrm{H}^{t,\omega}$ *is uniformly continuous under* $\mathscr{E}^{\mathscr{P}_L}$ *in the following sense: for some modulus of continuity function* ρ_1 *and for all* $(t_1, \omega^1), (t_2, \omega^2) \in \overline{\Lambda}_{\mathrm{H}}$ *with* $t_1 \leq t_2$,

$$\mathscr{E}^{\mathscr{P}_L}[|\mathrm{H}^{t_1,\omega^1} - \mathrm{H}^{t_2,\omega^2}(X_{t_2-t_1,t_2-t_1+\cdot})|] \leq \rho_1\big(\mathbf{d}((t_1, \omega^1), (t_2, \omega^2))\big) \tag{10.3.2}$$

(ii) *There exist* $\mathrm{H}_m \in \mathscr{T}$, *open set* $\Omega^{\mathrm{H}}_m \subset \Omega$, *and* $\delta_m, \theta_m \downarrow 0$ *such that* $\mathrm{H}_m \uparrow \mathrm{H}$ *as* $m \uparrow \infty$, *and*

$$\mathrm{H}_{m-1}(\omega) \leq \mathrm{H}_m(\tilde{\omega}) \leq \mathrm{H}_{m+1}(\omega) \quad \text{for all } \omega, \tilde{\omega} \in \Omega \text{ satisfying}$$
$$\|\omega - \tilde{\omega}\|_T \leq \delta_m; \mathscr{E}^{\mathscr{P}_L}[(\Omega^{\mathrm{H}}_m)^c] \leq 2^{-m}, \quad \mathrm{H}_{m+1} - \mathrm{H}_m \geq \theta_m \text{ on } \Omega^{\mathrm{H}}_m.$$

$$\tag{10.3.3}$$

The typical examples of such H include the deterministic times t and the $\mathrm{H}_\varepsilon, \underline{\mathrm{H}}_\varepsilon$ defined in (10.1.8), thanks to Theorem 10.1.15, Lemma 10.1.19, and Remark 10.1.21. Our main result of this section is the following so-called *nonlinear Snell envelope* theory, which extends Proposition 6.3.2 to the nonlinear expectation setting.

Theorem 10.3.2 *Let* $\mathscr{P}$ *satisfy Assumption 9.3.3 and be weakly compact,* $\mathrm{H} \in \mathscr{T}$ *satisfies Assumption 10.3.1, and* $u \in UC_b(\overline{\Lambda}_{\mathrm{H}})$ *with bound* C_0 *and modulus of continuity function* ρ_0. *Then*

(i) $Y \in UC_b(\overline{\Lambda}_{\mathrm{H}})$ *and* $Y_{\tau^*} = u_{\tau^*}$.
(ii) Y *satisfies the dynamic programming principle: for any* $(t, \omega) \in \Lambda_{\mathrm{H}}$ *and* $\tau \in \mathscr{T}_{\mathrm{H}^{t,\omega}-t}$,

$$Y_t(\omega) = \sup_{\tilde{\tau} \in \mathscr{T}} \mathscr{E}^{\mathscr{P}_t}\left[u_{\tilde{\tau}}^{t,\omega} \mathbf{1}_{\{\tilde{\tau} < \tau\}} + Y_\tau^{t,\omega} \mathbf{1}_{\{\tilde{\tau} \geq \tau\}}\right]. \tag{10.3.4}$$

(iii) $Y_{H\wedge\cdot}$ *is a pathwise* $\mathscr{E}^{\mathscr{P}}$*-supermartingale, and* $Y_{\tau^*\wedge\cdot}$ *a pathwise* $\mathscr{E}^{\mathscr{P}}$*-martingale, in the sense of Definition 10.2.3.*

(iv) τ^* *is an optimal stopping time for the problem* Y_0*, namely* $Y_0 = \mathscr{E}^{\mathscr{P}}[u_{\tau^*}]$*.*

Remark 10.3.3

(i) By slightly more involved arguments, Ekren, Touzi, & Zhang [74] proved Theorem 10.3.2 under a weaker condition than $u \in UC_b(\overline{\Lambda}_H)$: u is bounded, càdlàg, and

$$u(t, \omega) - u(t', \tilde{\omega}) \leq \rho_0\Big(\mathbf{d}\big((t, \omega), (\tilde{t}, \tilde{\omega})\big)\Big) \text{ whenever } t \leq \tilde{t}. \tag{10.3.5}$$

(ii) In the cases $H = t_0$ or $H = \underline{H}_\varepsilon$, the arguments can be simplified due to their stronger regularity.

(iii) All the results in Theorem 10.3.2 still hold true if we set $H = \overline{H}_\varepsilon$, or even set H as the hitting time of uniformly continuous processes (instead of the canonical process X), except that Y will only be in $C_b^0(\Lambda_H)$. See Bayraktar & Yao [11]. However, in this case the arguments become much more involved, mainly due to the lack of the desired regularity of $\overline{H}_\varepsilon$.   ∎

Throughout this section, we assume without loss of generality that $\rho_0 \leq 2C_0$. It is obvious that

$$|Y| \leq C_0, \quad Y \geq u, \quad \text{and} \quad Y_H = u_H. \tag{10.3.6}$$

We introduce two additional modulus of continuity functions:

$$\rho_2(\delta) := \rho_0\big(\delta + 2\rho_1^{\frac{1}{7}}(\delta)\big) + 2\rho_1^{\frac{1}{7}}(\delta), \quad \rho_3(\delta) := \rho_2(\delta + \sqrt{\delta}) + \delta. \tag{10.3.7}$$

10.3.1   Regularity and Dynamic Programming Principle

In this subsection we follow the arguments in Theorem 10.2.1, starting with the regularity of Y in ω.

Lemma 10.3.4 *For any* $\omega, \tilde{\omega} \in \Omega$ *and* $t \leq H(\omega) \wedge H(\tilde{\omega})$, $|Y_t(\omega) - Y_t(\tilde{\omega})| \leq C\rho_2(\|\omega - \tilde{\omega}\|_t)$.

Proof Denote $\delta := \|\omega - \tilde{\omega}\|_t$. First, for any $\tau \in \mathscr{T}$, we have

$$\left| u^{t,\omega}_{\tau \wedge (\mathrm{H}^{t,\omega}-t)} - u^{t,\tilde{\omega}}_{\tau \wedge (\mathrm{H}^{t,\tilde{\omega}}-t)} \right| = \left| u((t+\tau) \wedge \mathrm{H}^{t,\omega}, \omega \otimes_t X) - u((t+\tau) \wedge \mathrm{H}^{t,\tilde{\omega}}, \tilde{\omega} \otimes_t X)) \right|$$

$$\leq \rho_0\Big(\mathbf{d}\big(((t+\tau) \wedge \mathrm{H}^{t,\omega}, \omega \otimes_t X), ((t+\tau) \wedge \mathrm{H}^{t,\tilde{\omega}}, \tilde{\omega} \otimes_t X))\big)\Big)$$

$$\leq \rho_0\Big(\sqrt{|\mathrm{H}^{t,\omega} - \mathrm{H}^{t,\tilde{\omega}}|} + \delta + \mathrm{OSC}_{|\mathrm{H}^{t,\omega} - \mathrm{H}^{t,\tilde{\omega}}|}(X) \Big).$$

Then, for any constants $\varepsilon, \theta > 0$, and any $\mathbb{P} \in \mathscr{P}_t$,

$$\mathbb{E}^{\mathbb{P}}\left[\left| u^{t,\omega}_{\tau \wedge (\mathrm{H}^{t,\omega}-t)} - u^{t,\tilde{\omega}}_{\tau \wedge (\mathrm{H}^{t,\tilde{\omega}}-t)} \right| \right]$$

$$\leq \mathbb{E}^{\mathbb{P}}\left[\rho_0\big(\sqrt{\theta} + \delta + \mathrm{OSC}_\theta(X)\big) + 2C_0 \mathbf{1}_{\{|\mathrm{H}^{t,\omega} - \mathrm{H}^{t,\tilde{\omega}}| \geq \theta\}} \right]$$

$$\leq \mathbb{E}^{\mathbb{P}}\left[\rho_0\big(\sqrt{\theta} + \delta + \varepsilon\big) + 2C_0 \mathbf{1}_{\{|\mathrm{H}^{t,\omega} - \mathrm{H}^{t,\tilde{\omega}}| \geq \theta\}} + 2C_0 \mathbf{1}_{\{\mathrm{OSC}_\theta(X) \geq \varepsilon\}} \right]$$

$$\leq \rho_0\big(\sqrt{\theta} + \delta + \varepsilon\big) + 2C_0 \mathbb{E}^{\mathbb{P}}\left[\frac{1}{\theta^2} |\mathrm{H}^{t,\omega} - \mathrm{H}^{t,\tilde{\omega}}| + \frac{1}{\varepsilon^2} |\mathrm{OSC}_\theta(X)|^2 \right]$$

$$\leq \rho_0\big(\sqrt{\theta} + \delta + \varepsilon\big) + C\left[\frac{\rho_1(\delta)}{\theta^2} + \frac{\theta}{\varepsilon^2} \right].$$

Setting $\theta := \rho_1^{\frac{3}{7}}(\delta)$ and $\varepsilon := \rho_1^{\frac{1}{7}}(\delta)$, we have

$$\mathbb{E}^{\mathbb{P}}\left[\left| u^{t,\omega}_{\tau \wedge (\mathrm{H}^{t,\omega}-t)} - u^{t,\tilde{\omega}}_{\tau \wedge (\mathrm{H}^{t,\tilde{\omega}}-t)} \right| \right] \leq C\rho_2(\delta). \tag{10.3.8}$$

Since $\mathbb{P}$ and τ are arbitrary, this proves uniform continuity of Y_t in ω. ∎

We next establish the dynamic programming principle at deterministic times.

Lemma 10.3.5 *For any* $(t, \omega) \in \Lambda_{\mathrm{H}}$ *and* $0 < s \leq T - t$, *it holds*

$$Y_t(\omega) = \sup_{\tau \in \mathscr{T}_{\mathrm{H}^{t,\omega}-t}} \mathscr{E}^{\mathscr{P}_t}\left[u^{t,\omega}_\tau \mathbf{1}_{\{\tau < s\}} + Y^{t,\omega}_s \mathbf{1}_{\{\tau \geq s\}} \right]. \tag{10.3.9}$$

Proof We proceed in two steps. For simplicity, we assume $(t, \omega) = (0, 0)$.

Step 1. We first prove "$\leq$." For any $\tau \in \mathscr{T}_{\mathrm{H}}$ and $\mathbb{P} \in \mathscr{P}$:

$$\mathbb{E}^{\mathbb{P}}[u_\tau] = \mathbb{E}^{\mathbb{P}}\left[u_\tau \mathbf{1}_{\{\tau < s\}} + \mathbb{E}^{\mathbb{P}}_s[u_\tau] \mathbf{1}_{\{\tau \geq s\}} \right].$$

By the definition of r.c.p.d., and noting that $\mathbb{P}^{s,\omega} \in \mathscr{P}_s$ and $\tau^{s,\omega} \leq \mathrm{H}^{s,\omega}$, we have

$$\mathbb{E}^{\mathbb{P}}_s[u_\tau](\omega) = \mathbb{E}^{\mathbb{P}^{s,\omega}}[u^{s,\omega}_{\tau^{s,\omega}}] \leq Y_s(\omega), \quad \text{for } \mathbb{P}\text{-a.e. } \omega \in \{\tau \geq s\}.$$

Then,

$$\mathbb{E}^{\mathbb{P}}[u_\tau] \leq \mathbb{E}^{\mathbb{P}}\left[u_\tau \mathbf{1}_{\{\tau < s\}} + Y_s \mathbf{1}_{\{\tau \geq s\}} \right].$$

By taking supremum over τ and $\mathbb{P}$ we have

$$Y_0 = \sup_{\tau \in \mathscr{T}_H} \mathscr{E}^{\mathscr{P}}[u_\tau] \le \sup_{\tau \in \mathscr{T}_H} \mathscr{E}^{\mathscr{P}}\left[u_\tau \mathbf{1}_{\{\tau < s\}} + Y_s \mathbf{1}_{\{\tau \ge s\}}\right].$$

Step 2. We next prove "$\ge$." It suffices to prove: for any $\tau \in \mathscr{T}_H$ and $\mathbb{P} \in \mathscr{P}$,

$$\mathbb{E}^{\mathbb{P}}\left[u_\tau \mathbf{1}_{\{\tau < s\}} + Y_s \mathbf{1}_{\{\tau \ge s\}}\right] \le Y_0. \tag{10.3.10}$$

Let $\varepsilon > 0$, and $\{E_i\}_{i \ge 1}$ be an $\mathscr{F}_s$-measurable partition of the event $\{\tau \ge s\} \in \mathscr{F}_s$ such that $\|\omega - \tilde{\omega}\|_s \le \varepsilon$ for all $\omega, \tilde{\omega} \in E_i$. For each i, fix an $\omega^i \in E_i$, and by the definition of Y we have

$$Y_s(\omega^i) \le \mathbb{E}^{\mathbb{P}^i}\left[u^{s,\omega^i}_{\tau^i \wedge (H^{s,\omega^i} - s)}\right] + \varepsilon \text{ for some } (\tau^i, \mathbb{P}^i) \in \mathscr{T} \times \mathscr{P}_s.$$

By Lemma 10.3.4 and (10.3.8), we have

$$|Y_s(\omega) - Y_s(\omega^i)| \le C\rho_2(\varepsilon),$$

$$\mathbb{E}^{\mathbb{P}^i}\left[|u^{s,\omega}_{\tau^i \wedge (H^{s,\omega} - s)} - u^{s,\omega^i}_{\tau^i \wedge (H^{s,\omega^i} - s)}|\right] \le C\rho_2(\varepsilon), \text{ for all } \omega \in E_i.$$

Thus, for $\omega \in E_i$,

$$Y_s(\omega) \le Y_s(\omega^i) + C\rho_2(\varepsilon) \le \mathbb{E}^{\mathbb{P}^i}\left[u^{s,\omega^i}_{\tau^i \wedge (H^{s,\omega^i} - s)}\right] + \varepsilon + C\rho_2(\varepsilon)$$

$$\le \mathbb{E}^{\mathbb{P}^i}\left[u^{s,\omega}_{\tau^i \wedge (H^{s,\omega} - s)}\right] + \varepsilon + C\rho_2(\varepsilon). \tag{10.3.11}$$

Thanks to Assumption 9.3.3 (ii) and Problem 10.5.9, we may define the following pair $(\hat{\tau}, \widehat{\mathbb{P}}) \in \mathscr{T} \times \mathscr{P}$:

$$\hat{\tau} := \mathbf{1}_{\{\tau < s\}}\tau + \mathbf{1}_{\{\tau \ge s\}} \sum_{i \ge 1} \mathbf{1}_{E_i}[t + \tau^i(X_{s,s+}.)]; \quad \widehat{\mathbb{P}} := \mathbb{P} \otimes_s \left[\sum_{i \ge 1} \mathbf{1}_{E_i}\mathbb{P}^i + \mathbf{1}_{\{\tau < s\}}\mathbb{P}\right].$$

It is obvious that $\{\tau < s\} = \{\hat{\tau} < s\}$. Then, by (10.3.11),

$$\mathbb{E}^{\mathbb{P}}\left[u_\tau \mathbf{1}_{\{\tau < s\}} + Y_s \mathbf{1}_{\{\tau \ge s\}}\right] = \mathbb{E}^{\mathbb{P}}\left[u_\tau \mathbf{1}_{\{\tau < s\}} + \sum_{i \ge 1} Y_s \mathbf{1}_{E_i}\right]$$

$$\le \mathbb{E}^{\mathbb{P}}\left[u_\tau \mathbf{1}_{\{\tau < s\}} + \sum_{i \ge 1} \mathbb{E}^{\mathbb{P}^i}[u^{s,\omega}_{\tau^i \wedge (H^{s,\omega} - s)}]\mathbf{1}_{E_i}\right] + \varepsilon + C\rho_2(\varepsilon)$$

$$= \mathbb{E}^{\mathbb{P}}\left[u_\tau \mathbf{1}_{\{\hat{\tau} < s\}} + \sum_{i \ge 1} \mathbb{E}^{(\widehat{\mathbb{P}})^{t,\omega}}[u^{s,\omega}_{(\hat{\tau})^{s,\omega} \wedge H^{s,\omega} - s}]\mathbf{1}_{E_i}\right] + \varepsilon + C\rho_2(\varepsilon)$$

$$= \mathbb{E}^{\widehat{\mathbb{P}}}[u_{\hat{\tau}}] + \varepsilon + C\rho_2(\varepsilon) \le Y_0 + \varepsilon + C\rho_2(\varepsilon),$$

where the $\mathbb{E}^{\mathbb{P}}$ in the second and third lines are with respect to ω. Send $\varepsilon \to 0$, this proves (10.3.10) and hence completes the proof. ∎

We now derive the regularity of Y in t.

Lemma 10.3.6 *For each* $\omega \in \Omega$ *and* $0 \le t_1 < t_2 \le \mathrm{H}(\omega)$,

$$|Y_{t_1}(\omega) - Y_{t_2}(\omega)| \le C\rho_3\Big(\mathbf{d}\big((t_1,\omega),(t_2,\omega)\big)\Big).$$

Proof Denote $\delta := \mathbf{d}\big((t_1,\omega),(t_2,\omega)\big)$ and $s := t_2 - t_1 \le \delta^2$. If $\delta \ge \frac{1}{8}$, then clearly $|Y_{t_1,t_2}(\omega)| \le 2C_0 \le C\rho_3(\delta)$. So we continue the proof by assuming $\delta \le \frac{1}{8}$. First, by setting $\tau = s \wedge (\mathrm{H}^{t_1,\omega} - t_1)$ in Lemma 10.3.5,

$$Y_{t_1,t_2}(\omega) \le Y_{t_2}(\omega) - \mathcal{E}^{\mathscr{P}_{t_1}}\big[Y^{t_1,\omega}_{s\wedge(\mathrm{H}^{t_1,\omega}-t_1)}\big] \le \mathcal{E}^{\mathscr{P}_L}\big[Y_{t_2}(\omega) - Y^{t_1,\omega}_{s\wedge(\mathrm{H}^{t_1,\omega}-t_1)}\big].$$

On the other hand, by Lemma 10.3.5, the regularity of u, and the inequality $u \le Y$, we have

$$-Y_{t_1,t_2}(\omega) = \sup_{\tau \in \mathscr{T}_{\mathrm{H}^{t_1,\omega}-t_1}} \mathcal{E}^{\mathscr{P}_{t_1}}\big[u^{t_1,\omega}_{\tau}\mathbf{1}_{\{\tau<s\}} + Y^{t_1,\omega}_{s}\mathbf{1}_{\{\tau \ge s\}}\big] - Y_{t_2}(\omega)$$

$$\le \sup_{\tau \in \mathscr{T}_{\mathrm{H}^{t_1,\omega}-t_1}} \mathcal{E}^{\mathscr{P}_L}\Big[\big[u^{t_1,\omega}_{s\wedge(\mathrm{H}^{t_1,\omega}-t_1)}+\rho_0\big(\sqrt{s}+\mathrm{OSC}_s(X)\big)\big]\mathbf{1}_{\{\tau<s\}} + Y^{t_1,\omega}_{s}\mathbf{1}_{\{\tau \ge s\}}\Big] - Y_{t_2}(\omega)$$

$$\le \mathcal{E}^{\mathscr{P}_L}\Big[Y^{t_1,\omega}_{s\wedge(\mathrm{H}^{t_1,\omega}-t_1)} - Y_{t_2}(\omega) + \rho_0\big(\delta + \mathrm{OSC}_{\delta^2}(X)\big)\Big].$$

Then

$$|Y_{t_1,t_2}(\omega)| \le \mathcal{E}^{\mathscr{P}_L}\Big[|Y^{t_1,\omega}_{s\wedge(\mathrm{H}^{t_1,\omega}-t_1)} - Y_{t_2}(\omega)| + \rho_0\big(\delta + \mathrm{OSC}_{\delta^2}(X)\big)\Big].$$

Note that, by (10.2.11) and similar to Problem 10.5.2 (ii),

$$\mathcal{E}^{\mathscr{P}_L}\Big[\rho_0\big(\delta + \mathrm{OSC}_{\delta^2}(X)\big)\Big] \le C\rho_3(\delta);$$

$$\mathcal{E}^{\mathscr{P}_L}\Big[|Y^{t_1,\omega}_{s\wedge(\mathrm{H}^{t_1,\omega}-t_1)}-Y_{t_2}(\omega)|\Big] \le C\mathcal{E}^{\mathscr{P}_L}\Big[\rho_2\big(\mathbf{d}((t_2,\omega),((t_1+s)\wedge \mathrm{H}^{t_1,\omega},\omega \otimes_{t_1} X))\big)\Big]$$

$$\le C\mathcal{E}^{\mathscr{P}_{t_1}}\Big[\rho_2\big(\mathbf{d}((t_2,\omega),(t_1,\omega)) + \|X\|_s\big)\Big] \le C\mathcal{E}^{\mathscr{P}_L}\Big[\rho_2\big(\delta + \|X\|_{\delta^2}\big)\Big] \le C\rho_3(\delta).$$

Then $|Y_{t_1,t_2}(\omega)| \le C\rho_3(\delta)$. ∎

We are now ready to prove the dynamic programming principle for stopping times.

Lemma 10.3.7 *The equality* (10.3.4) *holds, and consequently,* $Y_{\mathrm{H}\wedge\cdot}$ *is a pathwise* $\mathcal{E}^{\mathscr{P}}$-*supermartingale.*

Proof First, for the notations in (10.3.4), follow the arguments in Lemma 10.3.5 Step 1 and note (9.3.8), one can prove straightforwardly that

$$Y_t(\omega) \le \sup_{\tilde{\tau} \in \mathscr{T}} \mathscr{E}^{\mathscr{P}_t} \left[u_{\tilde{\tau}}^{t,\omega} \mathbf{1}_{\{\tilde{\tau} < \tau\}} + Y_\tau^{t,\omega} \mathbf{1}_{\{\tilde{\tau} \ge \tau\}} \right].$$

On the other hand, construct $\tau_n \in \mathscr{T}$ as in Theorem 10.2.1 Step 5 such that $0 \le \tau_n - \tau \le \frac{T}{n}$ and τ_n takes only finitely many values. By Lemma 10.3.5 and following the arguments in Theorem 10.2.1 Step 4, one can easily show that (10.3.4) holds for $\tau_n \wedge (\mathrm{H}^{t,\omega} - t)$. Then for any $\mathbb{P} \in \mathscr{P}_t$ and $\tilde{\tau} \in \mathscr{T}$, denoting $\tilde{\tau}_m := \tilde{\tau} + \frac{1}{m}$ for arbitrary $m \ge 1$, we have

$$\mathbb{E}^{\mathbb{P}} \left[u_{\tilde{\tau}_m}^{t,\omega} \mathbf{1}_{\{\tilde{\tau}_m < \tau_n \wedge (\mathrm{H}^{t,\omega} - t)\}} + Y_{\tau_n \wedge (\mathrm{H}^{t,\omega} - t)}^{t,\omega} \mathbf{1}_{\{\tilde{\tau}_m \ge \tau_n \wedge (\mathrm{H}^{t,\omega} - t)\}} \right] \le Y_t(\omega).$$

Sending $n \to \infty$, by Lemma 10.3.6 and the dominated convergence theorem (under $\mathbb{P}$) we have

$$\mathbb{E}^{\mathbb{P}} \left[u_{\tilde{\tau}_m}^{t,\omega} \mathbf{1}_{\{\tilde{\tau}_m \le \tau\}} + Y_\tau^{t,\omega} \mathbf{1}_{\{\tilde{\tau}_m > \tau\}} \right] \le Y_t(\omega).$$

Here we used the fact $u \le Y$ when $\tau_n = \tau$. Now sending $m \to \infty$, it follows from the (right) continuity of u and the dominated convergence theorem (under $\mathbb{P}$) that

$$\mathbb{E}^{\mathbb{P}} \left[u_{\tilde{\tau}}^{t,\omega} \mathbf{1}_{\{\tilde{\tau} < \tau\}} + Y_\tau^{t,\omega} \mathbf{1}_{\{\tilde{\tau} \ge \tau\}} \right] \le Y_t(\omega).$$

which provides the required result by the arbitrariness of $\mathbb{P}$ and $\tilde{\tau}$. ∎

10.3.2 Local Pathwise $\mathscr{E}^{\mathscr{P}}$-Martingale Property

All the statements of Theorem 10.3.2 obviously hold true if $Y_0 = u_0$ which implies $\tau^* = 0$. Therefore, we focus on the nontrivial case $Y_0 > u_0$. We continue following the proof of the Snell envelope characterization in the standard linear expectation context. Denote

$$\tau_n := \inf\{t \ge 0 : Y_t - u_t \le \frac{1}{n}\} \wedge T, \text{ for } n > (Y_0 - u_0)^{-1}. \quad (10.3.12)$$

Since $Y_0 - u_0 > 0$ and $Y_\mathrm{H} - u_\mathrm{H} = 0$, it is clear that $\tau_n \in \mathscr{T}_\mathrm{H}$ and $Y_{\tau_n} - u_{\tau_n} = \frac{1}{n}$.

Lemma 10.3.8 *The process Y is a pathwise $\mathscr{E}^{\mathscr{P}}$-martingale on $[0, \tau_n]$, in the sense of Definition 10.2.3.*

Proof By the dynamic programming principle of Lemma 10.3.7,

$$Y_0 = \sup_{\tau \in \mathscr{T}} \mathscr{E}^{\mathscr{P}} \left[u_\tau \mathbf{1}_{\{\tau < \tau_n\}} + Y_{\tau_n} \mathbf{1}_{\{\tau \ge \tau_n\}} \right].$$

For any $\varepsilon > 0$, there exist $\tau_\varepsilon \in \mathcal{T}$ and $\mathbb{P}_\varepsilon \in \mathcal{P}$ such that

$$Y_0 \leq \mathbb{E}^{\mathbb{P}_\varepsilon}\Big[u_{\tau_\varepsilon}\mathbf{1}_{\{\tau_\varepsilon < \tau_n\}} + Y_{\tau_n}\mathbf{1}_{\{\tau_\varepsilon \geq \tau_n\}}\Big] + \varepsilon \leq \mathbb{E}^{\mathbb{P}_\varepsilon}\Big[Y_{\tau_\varepsilon \wedge \tau_n} - \frac{1}{n}\mathbf{1}_{\{\tau_\varepsilon < \tau_n\}}\Big] + \varepsilon,$$

(10.3.13)

where we used the fact that $Y_t - u_t > \frac{1}{n}$ for $t < \tau_n$, by the definition of τ_n. On the other hand, it follows from the pathwise $\mathscr{E}^{\mathcal{P}}$-supermartingale property of Y in Lemma 10.3.7 that $\mathbb{E}^{\mathbb{P}_\varepsilon}\Big[Y_{\tau_\varepsilon \wedge \tau_n}\Big] \leq \mathscr{E}^{\mathcal{P}}[Y_{\tau_\varepsilon \wedge \tau_n}] \leq Y_0$. This, together with (10.3.13), implies that $\mathbb{P}_\varepsilon[\tau_\varepsilon < \tau_n] \leq n\varepsilon$. We then get from the first inequality of (10.3.13) that:

$$Y_0 \leq \mathbb{E}^{\mathbb{P}_\varepsilon}\Big[(u_{\tau_\varepsilon} - Y_{\tau_n})\mathbf{1}_{\{\tau_\varepsilon < \tau_n\}} + Y_{\tau_n}\Big] + \varepsilon \leq C\mathbb{P}_\varepsilon[\tau_\varepsilon < \tau_n] + \mathbb{E}^{\mathbb{P}_\varepsilon}[Y_{\tau_n}] + \varepsilon$$

$$\leq \mathscr{E}^{\mathcal{P}}[Y_{\tau_n}] + (Cn + 1)\varepsilon.$$

Since ε is arbitrary, we obtain $Y_0 \leq \mathscr{E}^{\mathcal{P}}[Y_{\tau_n}]$. By the pathwise $\mathscr{E}^{\mathcal{P}}$-supermartingale property of Y established in Lemma 10.3.7, we see $Y_0 = \mathscr{E}^{\mathcal{P}}[Y_{\tau_n}]$. Similarly we may obtain the desired equality at any $(t, \omega) \in \Lambda_{\tau_n}$, and thus Y is a pathwise $\mathscr{E}$–martingale on $[0, \tau_n]$.   ∎

Remark 10.3.9 By Lemmas 10.3.6 and 10.3.8 we have

$$Y_0 - \mathscr{E}^{\mathcal{P}}[Y_{\tau^*}] = \mathscr{E}^{\mathcal{P}}[Y_{\tau_n}] - \mathscr{E}^{\mathcal{P}}[Y_{\tau^*}] \leq C\mathscr{E}^{\mathcal{P}_L}\Big[\rho_3\big(\mathbf{d}((\tau_n, \omega), (\tau^*, \omega))\big)\Big]. \quad (10.3.14)$$

Clearly, $\tau_n \uparrow \tau^*$ and $\rho_3\big(\mathbf{d}((\tau_n, \omega), (\tau^*, \omega))\big) \downarrow 0$, as $n \to \infty$. However, in general it is not clear that the stopping times τ_n, τ^* will be $\mathcal{P}$-q.s. continuous, so we cannot apply the dominated convergence Theorem 10.1.8 or the monotone convergence Theorem 10.1.10 (ii) to conclude $Y_0 \leq \mathscr{E}[Y_{\tau^*}]$. To overcome this difficulty, we need to approximate τ_n by continuous random variables.   ∎

10.3.3   Continuous Approximation of Stopping Times

The following lemma is crucial for us.

Lemma 10.3.10 *Let $\underline{\tau} \leq \tau \leq \overline{\tau}$ be three random variables on Ω, with values in a compact interval $I \subset \mathbb{R}$, such that for some $\Omega_0 \subset \Omega$ and $\delta > 0$:*

$$\underline{\tau}(\omega) \leq \tau(\tilde{\omega}) \leq \overline{\tau}(\omega) \text{ for all } \omega \in \Omega_0 \text{ and } \|\tilde{\omega} - \omega\|_T \leq \delta.$$

Then for any $\varepsilon > 0$, there exists a uniformly continuous function $\hat{\tau} : \Omega \to I$ and an open subset $\Omega_\varepsilon \subset \Omega$ such that

$$\mathscr{C}^{\mathcal{P}_L}[\Omega_\varepsilon^c] \leq \varepsilon \text{ and } \underline{\tau} - \varepsilon \leq \hat{\tau} \leq \overline{\tau} + \varepsilon \text{ in } \Omega_\varepsilon \cap \Omega_0.$$

Proof If I is a single point set, then τ is a constant and the result is trivial. At below we assume the length $|I| > 0$. Let $\{\omega^j\}_{j\geq 1} \subset \Omega$ be a dense sequence as in Lemma 9.2.1. Denote $E_j := \{\omega \in \Omega : \|\omega - \omega^j\|_T < \frac{\delta}{2}\}$ and $\Omega_n := \cup_{j=1}^n E_j$. It is clear that Ω_n is open and $\Omega_n \uparrow \Omega$ as $n \to \infty$. Let $f_n : [0, \infty) \to [0, 1]$ be defined as follows: $f_n(x) = 1$ for $x \in [0, \frac{\delta}{2}]$, $f_n(x) = \frac{1}{n^2|I|}$ for $x \geq \delta$, and f_n is linear in $[\frac{\delta}{2}, \delta]$. Define

$$\tau_n(\omega) := \phi_n(\omega) \sum_{j=1}^n \tau(\omega^j)\varphi_{n,j}(\omega)$$

where $\varphi_{n,j}(\omega) := f_n(\|\omega - \omega^j\|_T)$ and $\phi_n := \left(\sum_{j=1}^n \varphi_{n,j}\right)^{-1}$.

Then clearly τ_n is uniformly continuous and takes values in I. For each $\omega \in \Omega_n \cap \Omega_0$, $\phi_n(\omega) \leq 1$ and the set $J_n(\omega) := \{1 \leq j \leq n : \|\omega - \omega^j\|_T \leq \delta\} \neq \emptyset$. Then, by our assumption,

$$\tau_n(\omega) - \overline{\tau}(\omega) = \phi_n(\omega)\Big[\sum_{j \in J_n(\omega)} [\tau(\omega_j) - \overline{\tau}(\omega)]\varphi_{n,j}(\omega) + \sum_{j \notin J_n(\omega)} [\tau(\omega_j) - \overline{\tau}(\omega)]\varphi_{n,j}(\omega) \Big]$$

$$\leq \phi_n(\omega) \sum_{j \notin J_n(\omega)} |I|\varphi_{n,j}(\omega) \leq \phi_n(\omega) \sum_{j \notin J_n(\omega)} \frac{1}{n^2} \leq \frac{1}{n}.$$

Similarly one can show that $\underline{\tau} - \frac{1}{n} \leq \tau_n$ in $\Omega_n \cap \Omega_0$. Finally, since $\Omega_n \uparrow \Omega$ as $n \to \infty$ and $\mathscr{P}_L$ is weakly compact, it follows from the monotone convergence Theorem 10.1.10 (i) that $\lim_{n\to\infty} \mathscr{C}^{\mathscr{P}_L}[\Omega_n^c] = 0$. ∎

As an application of the previous lemma, we may approximate the H_m in Assumption 10.3.1 (ii) and the τ_n in (10.3.12) with continuous random variables.

Lemma 10.3.11 *Recall the notations in Assumption 10.3.1 (ii) and let $\eta_n > 0$, $n \geq 1$.*

(i) *For any m, by possibly choosing a smaller $\delta_m > 0$, there exist $\hat{\mathsf{H}}_m \in UC_b(\Omega)$ and open set $\Omega_m \subset \Omega$ such that*

$$\mathsf{H}_{m-1} - \eta_m \leq \hat{\mathsf{H}}_m \leq \mathsf{H}_{m+1} + \eta_m \text{ on } \Omega_m; \tag{10.3.15}$$

$$|\hat{\mathsf{H}}_m(\omega) - \hat{\mathsf{H}}_m(\tilde{\omega})| \leq \eta_m, \quad \text{whenever } \|\omega - \tilde{\omega}\|_T \leq \delta_m; \tag{10.3.16}$$

$$\mathscr{C}^{\mathbb{P}_L}[\Omega_{m,\delta_m}^c] \leq 2^{-m}, \quad \text{where } \Omega_{m,\delta} := \{\omega \in \Omega_m : \mathbf{d}(\omega, \Omega_m^c) > \delta\}. \tag{10.3.17}$$

(ii) *Denote $\tau_n^m := \tau_n \wedge \hat{\mathsf{H}}_m$. There exist $\hat{\tau}_n^m \in UC_b(\Omega)$ and open set $\hat{\Omega}_n^m \subset \Omega_{m,\delta_m}$ such that*

$$\mathscr{C}^{\mathscr{P}_L}[(\hat{\Omega}_n^m)^c] \leq 2^{-m} + 2^{-n}, \ \tau_{n-1}^m - 2\eta_m - \eta_n \leq \hat{\tau}_n^m \leq \tau_{n+1}^m + 2\eta_m + \eta_n \text{ on } \hat{\Omega}_n^m.$$

$$\tag{10.3.18}$$

Proof

(i) By Assumption 10.3.1 (ii) and applying Lemma 10.3.10 with $\underline{\tau} = H_{m-1}$, $\tau = H_m$, $\overline{\tau} = H_{m+1}$, and $\Omega_0 = \Omega$, $\varepsilon := \eta_m \wedge 2^{-1-m}$, there exist $\hat{H}_m \in UC_b(\Omega)$ and open subset $\Omega_m \subset \Omega$ such that $\mathscr{C}^{\mathbb{P}L}[\Omega_m^c] \leq 2^{-1-m}$ and (10.3.15) holds. Since $\hat{H}_m$ is uniformly continuous, by otherwise choosing a smaller $\delta_m > 0$, we may assume (10.3.16) holds. Moreover, denote

$$\xi_{m,\delta}(\omega) := \varphi_\delta(\mathbf{d}(\omega, \Omega_m^c)), \quad \text{where } \varphi_\delta(x) := \mathbf{1}_{[0,\delta]}(x) - \frac{x-\delta}{\delta}\mathbf{1}_{(\delta,2\delta)}(x).$$

Then clearly $\xi_{m,\delta} \in UC_b(\Omega)$, $\xi_{m,\delta} = 1$ on $\Omega_{m,\delta}^c$, and $\xi_{m,\delta} \downarrow \mathbf{1}_{\Omega_m^c}$, as $\delta \downarrow 0$. Applying Theorem 10.1.10 (ii), we have

$$\limsup_{\delta \to 0} \mathscr{C}^{\mathbb{P}L}[\Omega_{m,\delta}^c] \leq \lim_{\delta \to 0} \mathscr{E}^{\mathscr{P}_L}[\xi_{m,\delta}] = \mathscr{C}^{\mathbb{P}L}[\Omega_m^c] \leq 2^{-1-m}.$$

Then, by otherwise choosing a smaller $\delta_m > 0$, we obtain (10.3.17).

(ii) Let $\delta_n^m \leq \delta_m$ be small enough such that $(\rho_0 + C\rho_2)(\delta_n^m) \leq \frac{1}{n(n+1)}$ for the constant C in Lemma 10.3.4. We first claim that, for any $m \geq 1$,

$$\tau_{n-1}^m(\omega) - 2\eta_m \leq \tau_n^m(\tilde{\omega}) \leq \tau_{n+1}^m(\omega) + 2\eta_m, \quad \forall \omega \in \Omega_{m,\delta_m} \text{ and } \|\tilde{\omega} - \omega\|_T \leq \delta_n^m.$$

$$(10.3.19)$$

Then (10.3.18) follows from Lemma 10.3.10 immediately.

We prove the right inequality of (10.3.19) in two cases. The left one can be proved similarly.

Case 1. $\tau_{n+1}(\omega) \geq \hat{H}_m(\tilde{\omega}) - 2\eta_m$. By (10.3.16) we see that $\hat{H}_m(\tilde{\omega}) \leq \hat{H}_m(\omega) + \eta_m$. Then

$$\tau_n^m(\tilde{\omega}) \leq \hat{H}_m(\tilde{\omega}) \leq \tau_{n+1}^m(\omega) + 2\eta_m.$$

Case 2. $\tau_{n+1}(\omega) < \hat{H}_m(\tilde{\omega}) - 2\eta_m$. Since $\omega \in \hat{\Omega}_m \subset \Omega_m$, by (10.3.16) and (10.3.15) we have

$$\tau_{n+1}(\omega) < \hat{H}_m(\omega) - \eta_m \leq H_{m+1}(\omega) \leq H(\omega),$$

$$\text{implying } (Y - u)_{\tau_{n+1}}(\omega) = \frac{1}{n+1}. \quad (10.3.20)$$

Moreover, since $\omega \in \hat{\Omega}_m$ and $\|\tilde{\omega} - \omega\|_T \leq \delta_n^m \leq \delta_m$, it follows from (10.3.17) that $\tilde{\omega} \in \Omega_m$. Then by (10.3.15) we have

$$\tau_{n+1}(\omega) < \hat{H}_m(\tilde{\omega}) - 2\eta_m < H_{m+1}(\tilde{\omega}) \leq H(\tilde{\omega}), \text{ namely } (\tau_{n+1}(\omega), \tilde{\omega}) \in \Lambda_H.$$

Applying Lemma 10.3.4, we have

$$|Y_{\tau_{n+1}(\omega)}(\omega) - Y_{\tau_{n+1}(\omega)}(\tilde{\omega})| \le C\rho_2(\|\omega - \tilde{\omega}\|_T) \le C\rho_2(\delta_n^m).$$

This, together with the uniform continuity of u and (10.3.20), implies that

$$(Y - u)_{\tau_{n+1}(\omega)}(\tilde{\omega}) \le (Y - u)_{\tau_{n+1}(\omega)}(\omega) + (\rho_0 + C\rho_2)(\delta_n^m)$$

$$\le \frac{1}{n+1} + \frac{1}{n(n+1)} = \frac{1}{n}.$$

Then $\tau_n(\tilde{\omega}) \le \tau_{n+1}(\omega)$. This, together with (10.3.16), proves the right inequality of (10.3.19).

∎

10.3.4 Proof of Theorem 10.3.2

Denote $\bar{\tau}_n := \tau_n \wedge H_n \in \mathscr{T}$. By Lemma 10.3.8, for each n large, there exists $\mathbb{P}_n \in \mathscr{P}$ such that

$$Y_0 = \mathscr{E}^{\mathscr{P}}[Y_{\bar{\tau}_n}] \le \mathbb{E}^{\mathbb{P}_n}[Y_{\bar{\tau}_n}] + 2^{-n}$$

Since $\mathscr{P}$ is weakly compact, there exists a subsequence $\{n_j\}$ and $\mathbb{P}^* \in \mathscr{P}$ such that $\mathbb{P}_{n_j}$ converges weakly to $\mathbb{P}^*$. Now for any n large and any $n_j \ge n$, note that $\tau_{n_j} \ge \tau_n$. Since Y is a pathwise $\mathscr{E}^{\mathscr{P}}$-supermartingale and thus a $\mathbb{P}_{n_j}$-supermartingale, we have

$$Y_0 - 2^{-n_j} \le \mathbb{E}^{\mathbb{P}_{n_j}}[Y_{\bar{\tau}_{n_j}}] \le \mathbb{E}^{\mathbb{P}_{n_j}}[Y_{\bar{\tau}_n}]. \tag{10.3.21}$$

The idea is to send $j \to \infty$ and use the weak convergence of $\mathbb{P}_{n_j}$. For this purpose, naturally we approximate $\bar{\tau}_n$ with $\hat{\tau}_n^n$ which is continuous. However, note that Y is continuous only in $\overline{\Lambda}_H$ and it is possible that $\hat{\tau}_n^n > H$, so our approximation is a little more involved.

We first note that, on $\overline{\Omega}_n^1 := \hat{\Omega}_n^n \cap \Omega_{n+1} \cap \Omega_{n+1}^H$, by (10.3.18), (10.3.15), and (10.3.3), we have

$$\hat{\tau}_n^n \le \tau_{n+1}^n + 3\eta_n \le \hat{H}_n + 3\eta_n \le H_{n+1} + 4\eta_n \le H_{n+2} - \theta_{n+1} + 4\eta_n.$$

Setting $\eta_n := \frac{\theta_{n+1}}{4}$, we get $\hat{\tau}_n^n \le H_{n+2} \le H$ on $\overline{\Omega}_n^1$. Then it follows from Lemma 10.3.6 that

$$|Y_{\bar{\tau}_n} - Y_{\hat{\tau}_n^n}| \le C\rho_3\Big(\mathbf{d}\big((\bar{\tau}_n, X), (\hat{\tau}_n^n, X)\big)\Big), \quad \text{on } \overline{\Omega}_n^1. \tag{10.3.22}$$

Next, on $\hat{\Omega}_{n+1}^{n+1} \subset \Omega_{n+1}$, by (10.3.15) and (10.3.18) we have

$$\bar{\tau}_n \leq \tau_n \wedge \hat{H}_{n+1} + \eta_{n+1} \leq \hat{\tau}_{n+1}^{n+1} + 4\eta_{n+1} = \hat{\tau}_{n+1}^{n+1} + \theta_{n+2};$$

and on $\hat{\Omega}_{n-1}^{n-1} \subset \Omega_{n-1}$, by (10.3.15) and (10.3.18) again we have

$$\bar{\tau}_n \geq \tau_n \wedge \hat{H}_{n-1} - \eta_{n-1} \geq \hat{\tau}_{n-1}^{n-1} - 4\eta_{n-1} = \hat{\tau}_{n-1}^{n-1} - \theta_n. \qquad (10.3.23)$$

Thus, on $\overline{\Omega}_n := \overline{\Omega}_n^1 \cap \hat{\Omega}_{n+1}^{n+1} \cap \hat{\Omega}_n^n \cap \hat{\Omega}_{n-1}^{n-1}$, where $\hat{\Omega}_n^n$ is included for later purpose, we have

$$|Y_{\bar{\tau}_n} - Y_{\hat{\tau}_n^n}| \leq C\eta_n, \text{ where } \eta_n := \rho_3\Big(\mathbf{d}\big((\hat{\tau}_{n-1}^{n-1} \wedge \hat{\tau}_n^n - \theta_n, X), (\hat{\tau}_{n+1}^{n+1} \vee \hat{\tau}_n^n + \theta_{n+2}, X))\big)\Big).$$
$$(10.3.24)$$

Note that $\overline{\Omega}_n$ is open. For any $\delta > 0$, denote $\xi_\delta(\omega) := \frac{1}{\delta}[\delta \wedge \mathbf{d}(\omega, (\overline{\Omega}_n)^c))]$. Then

$$\xi_\delta \in UC_b(\Omega), \quad 0 \leq \xi_\delta \leq 1, \quad \{\xi_\delta > 0\} \subset \overline{\Omega}_n, \quad \xi_\delta \uparrow \mathbf{1}_{\overline{\Omega}_n} \text{ as } \delta \downarrow 0.$$

Applying Theorem 10.1.10 (ii), we have $\lim_{\delta \to 0} \mathscr{E}^{\mathscr{P}_L}[1 - \xi_\delta] = \mathscr{C}^{\mathbb{P}_L}[(\overline{\Omega}_n)^c]$. Then there exists $\bar{\delta}_n$ small enough such that

$$\mathscr{E}^{\mathscr{P}_L}[1 - \xi_{\bar{\delta}_n}] \leq \mathscr{C}^{\mathbb{P}_L}[(\overline{\Omega}_n)^c] + 2^{-n} \leq C2^{-n}. \qquad (10.3.25)$$

Now combining (10.3.21), (10.3.24), and (10.3.25), we obtain

$$Y_0 - 2^{-n_j} \leq \mathbb{E}^{\mathbb{P}_{n_j}}\big[Y_{\bar{\tau}_n}\big] = \mathbb{E}^{\mathbb{P}_{n_j}}\Big[Y_{\hat{\tau}_n^n}\xi_{\bar{\delta}_n} + [Y_{\bar{\tau}_n} - Y_{\hat{\tau}_n^n}]\xi_{\bar{\delta}_n} + Y_{\bar{\tau}_n}[1 - \xi_{\bar{\delta}_n}]\Big]$$

$$\leq \mathbb{E}^{\mathbb{P}_{n_j}}\Big[[Y_{\hat{\tau}_n^n} + C\eta_n]\xi_{\bar{\delta}_n}\Big] + C2^{-n}.$$

Note that the integrand under above $\mathbb{E}^{\mathbb{P}_{n_j}}$ is continuous. Sending $j \to \infty$, it follows from the weak convergence of $\mathbb{P}_{n_j}$ that

$$Y_0 \leq \mathbb{E}^{\mathbb{P}^*}\Big[[Y_{\hat{\tau}_n^n} + C\eta_n]\xi_{\bar{\delta}_n}\Big] + C2^{-n}. \qquad (10.3.26)$$

Since $\{\xi_{\bar{\delta}_n} > 0\} \subset \overline{\Omega}_n \subset \hat{\Omega}_{n+1}^{n+1} \cap \hat{\Omega}_n^n \cap \hat{\Omega}_{n-1}^{n-1}$, by (10.3.22) and (10.3.23) we have

$$\bar{\tau}_{n-2} - \theta_n \leq \hat{\tau}_{n-1}^{n-1} \leq \bar{\tau}_n + \theta_n, \ \bar{\tau}_{n-1} - \theta_{n+1} \leq \hat{\tau}_n^n \leq \bar{\tau}_{n+1} + \theta_{n+1}, \ \bar{\tau}_n - \theta_{n+2} \leq \hat{\tau}_{n+1}^{n+1}$$
$$\leq \bar{\tau}_{n+2} + \theta_{n+2}.$$

Then (10.3.26) leads to

$$Y_0 \leq \mathbb{E}^{\mathbb{P}^*} \Big[Y_{\overline{\tau}_n} - Y_{\overline{\tau}_n}[1 - \xi_{\underline{\delta}_n}] + [Y_{\hat{\tau}^n} - Y_{\overline{\tau}_n}]\xi_{\underline{\delta}_n} + C\eta_n\xi_{\underline{\delta}_n} \Big] + C2^{-n}$$

$$\leq \mathbb{E}^{\mathbb{P}^*} \Big[Y_{\overline{\tau}_n} + C\eta_n\xi_{\underline{\delta}_n} \Big] + C2^{-n}$$

$$\leq \mathbb{E}^{\mathbb{P}^*} \Big[Y_{\overline{\tau}_n} + C\rho_3\big(\mathbf{d}((\overline{\tau}_{n-2} - 2\theta_n, X), (\overline{\tau}_{n+2} + 2\theta_{n+1}, X))\big) \Big] + C2^{-n}.$$

Sending $n \to \infty$, note that $\theta_n \downarrow 0$, $\overline{\tau}_n \uparrow \tau^*$, then $\mathbf{d}((\overline{\tau}_{n-2} - 2\theta_n, X), (\overline{\tau}_{n+2} + 2\theta_{n+1}, X)) \to 0$. Therefore, it follows from the dominated convergence theorem under $\mathbb{P}^*$ that

$$Y_0 \leq \mathbb{E}^{\mathbb{P}^*}[Y_{\tau^*}] \leq \mathscr{E}^{\mathscr{P}}[Y_{\tau^*}].$$

Similarly $Y_t(\omega) \leq \mathscr{E}^{\mathscr{P}_t}[Y_{\tau^*}^{t,\omega}]$ for $t < \tau^*(\omega)$. By the pathwise $\mathscr{E}^{\mathscr{P}}$-supermartingale property of Y established in Lemma 10.3.7, this implies that Y is a pathwise $\mathscr{E}^{\mathscr{P}}$-martingale on $[0, \tau^*]$. ∎

10.4 Bibliographical Notes

In the semilinear case, roughly speaking all the measures in class $\mathscr{P}$ are equivalent, and the nonlinear expectation is essentially the g-expectation of Peng [176, 179]. In this chapter our main focus is on the nonlinear expectation in fully nonlinear case, where some measures in $\mathscr{P}$ could be mutually singular and there is no dominating measure. The most important example is the G-expectation developed by Peng [180, 181], see also the survey paper by Peng [182]. While in this book we focus on the backward problems, there are systematic studies on forward problems under G-framework in [181].

The main treatment in [181] is to define nonlinear expectation through certain PDE, while in this book we use the representation through the quasi sure stochastic analysis, introduced by Denis & Martini [63]. The quasi-sure representation of G-expectation is due to Denis, Hu, & Peng [61], and that of conditional G-expectation is due to Soner, Touzi, & Zhang [212]. In this book we assume the set $\mathscr{P}_t$ of the regular conditional probability distribution is independent of ω. The general case is much more involved, and we refer to Nutz [162] and Nutz & van Handel [164] for studies in this direction. We shall also mention that the second order BSDE of Chapter 12 can be viewed as backward SDE under nonlinear expectation.

The dynamic programming principle is an important tool in stochastic control literature. Both in Sections 10.2 and 10.3 we use regularities to prove dynamic programming principle. This is in the line of the stochastic backward semigroup in Peng [175]. As mentioned in Remark 10.2.2, dynamic programming principle may hold under weak conditions. We refer to Wagner [230] for a survey paper on measurable selection, and there have been several works in more general cases related to our contexts, see, e.g., Nutz & van Handel [164], Tang & Zhang [225], El Karouri & Tan [82, 83], and Hu & Ji [106]. We also refer to Bouchard & Touzi [19] for the weak dynamic programming principle.

The hitting times in Section 10.1.3 are important. The works Ekren, Touzi, & Zhang [74–76] used $\overline{H}_\varepsilon$, Bayraktar & Yao [11] introduced $\underline{H}_\varepsilon$, and Ekren & Zhang [77] proposed H_ε. The optimal stopping problem of Section 10.3 is based on [74], where the cases $H = T$ and $H = \overline{H}_\varepsilon$ are studied. The H_ε used in Section 10.3 has regularity in between T and $\overline{H}_\varepsilon$, so the technicality involved here is also in between the two cases in [74]. The result is further extended by [11] to the case that H is a hitting time of certain uniformly continuous processes (instead of the canonical process), motivated by their study for the related Dynkin game under nonlinear expectation in Bayraktar & Yao [12]. We would also like to mention two related works: Nutz & Zhang [165] and Bayraktar & Yao [10], which study the optimal stopping problem in the opposite direction: $\inf_{\tau \in \mathcal{T}} \mathcal{E}^{\mathcal{P}}[u_\tau]$.

10.5 Exercises

Problem 10.5.1 Prove Propositions 10.1.2, 10.1.3, 10.1.6, and 10.1.7. ∎

Problem 10.5.2

(i) Assume $(\mathcal{M}.d)$ is a metric space, and $f : \mathcal{M} \to \mathbb{R}$ is uniformly continuous with modulus of continuity function ρ. Show that, modifying ρ if necessary,

$$\rho(x) \le \rho(1)[x+1], \quad \forall x \ge 0. \tag{10.5.1}$$

(ii) Show that the $\overline{\rho}$ defined by (10.2.11) satisfies:

$$\overline{\rho}(\delta) \le \rho(\delta + \sqrt{\delta}) + C\delta. \tag{10.5.2}$$

(iii) Show that $UC(\Omega) \subset \mathbb{L}^p(\mathbb{F}, \mathcal{P}_L)$ for any $p \ge 1$ and $L > 0$.
(iv) For any $\xi \in UC(\Omega)$, there exists $\xi_n \in UC_b(\Omega)$ such that $\lim_{n \to \infty} \mathcal{E}^{\mathcal{P}_L}[|\xi_n - \xi|^p] = 0$ for all $p \ge 1$ and $L > 0$. ∎

Problem 10.5.3 Let $\Omega_0 \subset \Omega$ be closed and $\xi : \Omega_0 \to \mathbb{R}$ be continuous and $|\xi| \le C_0$.

(i) Show that there exists $\xi_1 \in C_b^0(\Omega)$ such that $|\xi - \xi_1| \le \frac{2C_0}{3}$ on Ω_0. (Hint: if

$$A_+ := \{\omega \in \Omega_0 : \xi \ge \frac{C_0}{3}\}, \quad A_- := \{\omega \in \Omega_0 : \xi \le -\frac{C_0}{3}\}$$

are nonempty, we may define

$$\xi_1(\omega) := \frac{C_0}{3} \frac{d(\omega, A_+) - d(\omega, A_-)}{d(\omega, A_+) + d(\omega, A_-)} \quad \text{where} \quad d(\omega, A) := \inf\{\|\omega - \tilde{\omega}\|_T : \tilde{\omega} \in A\}.$$

Modify the definition if A_+ and/or A_- is empty.)

(ii) Apply (i) repeatedly to obtain $\tilde{\xi} \in C_b^0(\Omega)$ such that $\|\tilde{\xi}\|_\infty \leq C_0$ and $\tilde{\xi} = \xi$ on Ω_0. ∎

Problem 10.5.4 Let $\mathbb{P} \in \mathscr{P}_\infty$ and $\mathscr{P} = \{\mathbb{P}\}$ be a singleton. Show that $\mathbb{L}_0^p(\mathscr{F}_T, \mathscr{P}) = \mathbb{L}^p(\mathscr{F}_T, \mathbb{P})$. ∎

Problem 10.5.5

(i) Construct a counterexample such that $\underline{H}_\varepsilon$ is not Markov. To be precise, find $(t, \omega^1), (t, \omega^2) \in \Lambda_{\underline{H}_\varepsilon}$ such that $\omega_t^1 = \omega_t^2$ but $\underline{H}_\varepsilon^{t,\omega^1} \neq \underline{H}_\varepsilon^{t,\omega^2}$.

(ii) Construct a counterexample such that H_ε is not continuous in ω. That is, find $\omega, \omega^n \in \Omega$ such that $\lim_{n \to \infty} \|\omega^n - \omega\|_T = 0$, but $H_\varepsilon(\omega^n)$ does not converge to $H_\varepsilon(\omega)$. ∎

Problem 10.5.6

(i) Construct a counterexample such that $\tau \in \mathscr{T}$ but τ is not $\mathscr{P}_L$-q.s. continuous.

(ii) Construct a counterexample such that $\tau \in \mathscr{T}$ but Λ_τ is not open. ∎

Problem 10.5.7 Prove all the statements in Remark 10.1.21. ∎

Problem 10.5.8

(i) Prove Lemma 10.2.4.

(ii) Construct a counterexample such that Y is a $\mathscr{E}^{\mathscr{P}_L}$-submartingale, but there exists $\mathbb{P} \in \mathscr{P}_L$ such that Y is not a $\mathbb{P}$-submartingale. ∎

Problem 10.5.9 Assume $0 < t < T$, $\tau, \tau_i \in \mathscr{T}$, $i \geq 1$, and $\{E_i\}_{i \geq 1} \subset \mathscr{F}_t$ is a partition of $\{\tau \geq t\}$. Define

$$\hat{\tau} := \tau \mathbf{1}_{\{\tau < t\}} + \sum_{i \geq 1} [t + \tau_i(X_{t,t+\cdot})] \mathbf{1}_{E_i}.$$

Show that $\hat{\tau} \in \mathscr{T}$ and $(\hat{\tau})^{t,\omega} - t = \tau_i$ for all $i \geq 1$ and $\omega \in E_i$. ∎

Chapter 11
Path Dependent PDEs

The Path Dependent PDEs (PPDE, for short) is a powerful and convenient tool for non-Markov problems. The nonlinear Feynman-Kac formula in Section 5.1 requires Markov structure. With the new notion of PPDEs, we may view the general BSDE (4.0.3) as a probabilistic representation of solutions to a semilinear PPDE, and thus extend the nonlinear Feynman-Kac formula to non-Markov case. Another example is the decoupling field for coupled FBSDE with random coefficients, as introduced in Section 8.3, which extends the PDE to PPDEs in quasilinear case. We are most interested in fully nonlinear parabolic PPDEs, typically path dependent Hamilton-Jacobi-Bellman equations and path dependent Bellman-Isaacs equations, due to their important applications in stochastic control and stochastic differential games. In this chapter, the PPDEs are always backward, in the sense that their terminal conditions are given. Indeed, the value functions arising in many applications are naturally backward. Moreover, for PPDEs we will mainly focus on viscosity solutions. As we will see in Example 11.1.3 below, in path dependent case, even a heat equation may not have a classical solution.

PPDEs are defined on the canonical space $\overline{\Lambda} = [0, T] \times \Omega$, with the canonical process X. The major difficulty for the viscosity theory of PPDEs is that the state space $\overline{\Lambda}$ is not locally compact. As we saw in Proposition 5.5.10, the local compactness plays a crucial role in the proof of comparison principle, which leads to the uniqueness. To overcome this major difficulty, we shall introduce a new notion of viscosity solutions. Consequently, in this chapter,

$$u \text{ is always 1-dimensional.}$$

© Springer Science+Business Media LLC 2017
J. Zhang, *Backward Stochastic Differential Equations*, Probability Theory and Stochastic Modelling 86, DOI 10.1007/978-1-4939-7256-2_11

11.1 The Viscosity Theory of Path Dependent Heat Equations

To motivate the new notion of viscosity solution of PPDEs, in this section we focus on the path dependent heat equation with appropriate terminal condition:

$$\mathscr{L}u(t,\omega) := \partial_t u + \frac{1}{2}\text{tr}[\partial^2_{\omega\omega}u] = 0. \tag{11.1.1}$$

Remark 11.1.1 We emphasize again that the PPDE (11.1.1) is backward with appropriate terminal condition, say $u(T,\omega) = \xi(\omega)$. In the Markovian case: $u(t,\omega) = v(t,\omega_t)$, $\xi(\omega) = g(\omega_T)$, (11.1.1) reduces to a standard heat equation with terminal condition:

$$\partial_t v + \frac{1}{2}\text{tr}[\partial^2_{xx}v] = 0, \quad v(T,x) = g(x). \tag{11.1.2}$$

In this case, introduce the time change as in Remark 5.5.3 (i): $\tilde{v}(t,x) := v(T-t,x)$, then $\tilde{v}$ satisfies the following heat equation with initial condition:

$$\partial_t \tilde{v} - \frac{1}{2}\text{tr}[\partial^2_{xx}\tilde{v}] = 0, \quad \tilde{v}(0,x) = g(x). \tag{11.1.3}$$

However, in the path dependent case, such time change technique fails due to the progressive measurability of u. In particular, one cannot transform the PPDE (11.1.1) into a forward equation with initial condition. ∎

11.1.1 Classical Solutions

In light of the functional Itô formula (9.4.1), for $u \in C^{1,2}(\Lambda)$ we have

$$du = \mathscr{L}u dt + \partial_\omega u dX_t, \quad \mathbb{P}_0\text{-a.s.}$$

where, again, $\mathbb{P}_0$ is the Wiener measure. Then clearly,

u satisfies (11.1.1) if and only if u is a $\mathbb{P}_0$-local martingale.

Consequently, given a terminal condition $\xi \in C^0(\Omega)$ with appropriate integrability, the candidate solution to the PPDE (11.1.1) should be the pathwise conditional expectation:

$$u^0(t,\omega) := \mathbb{E}^{\mathbb{P}_0}[\xi^{t,\omega}], \quad (t,\omega) \in \overline{\Lambda}. \tag{11.1.4}$$

Example 11.1.2 *Let $d = 1$ for simplicity, then the u^0 corresponding to the following ξ are classical solutions to PPDE (11.1.1):*

(i) *$\xi = g(X_T)$ for some continuous function g with polynomial growth.*
(ii) *$\xi = \int_0^T X_t dt$;*
(iii) *$\xi = X_T^*$.*

Proof

(i) In this case, we have $u^0(t, \omega) = v(t, \omega_t)$, where $v(t, x) := \mathbb{E}^{\mathbb{P}_0}[g(x + X_{T-t})]$. By standard theory, we know v is a classical solution to the (standard) heat equation: $\partial_t v + \frac{1}{2}\partial_{xx}^2 v = 0$. Then the result follows from Example 9.4.4 (i).

(ii) By straightforward computation we have

$$u^0(t, \omega) = \mathbb{E}^{\mathbb{P}_0}\Big[\int_0^t \omega_s ds + \int_0^{T-t}(\omega_t + X_s)ds\Big] = \int_0^t \omega_s ds + (T - t)\omega_t.$$

Then it follows from the standard Itô formula that

$$du^0 = (T - t)dX_t, \quad \mathscr{P}_\infty\text{-q.s.}$$

Compare this with the functional Itô formula (9.4.1) we see that $u^0 \in C^{1,2}(\overline{\Lambda})$ with

$$\partial_t u^0(t, \omega) = 0, \quad \partial_\omega u^0(t, \omega) = T - t, \quad \partial_{\omega\omega}^2 u^0(t, \omega) = 0.$$

This implies (11.1.1) immediately.

(iii) Note that, denoting by Φ the cdf of the standard normal distribution,

$$u^0(t, \omega) = \mathbb{E}^{\mathbb{P}_0}\Big[\omega_t^* \vee (\omega_t + X_{T-t}^*)\Big] = v(t, \omega_t, \omega_t^*), \quad \text{where}$$

$$v(t, x, y) := \mathbb{E}^{\mathbb{P}_0}\Big[y \vee (x + X_{T-t}^*)\Big] = x + \sqrt{T - t}\psi\Big(\frac{y - x}{\sqrt{T - t}}\Big), \quad x \le y$$

$$\psi(z) := \mathbb{E}^{\mathbb{P}_0}\Big[z \vee X_1^*\Big] = \mathbb{E}^{\mathbb{P}_0}\Big[z \vee |X_1|\Big]$$

$$= z[2\Phi(z) - 1] + \frac{2}{\sqrt{2\pi}}e^{-z^2/2}, \quad z \ge 0.$$

Here we used the fact that, for $\mathbb{P}_0$-Brownian motion X, X_1^* and $|X_1|$ have the same distribution. Clearly v is smooth for $t < T$, and $D_y v(t, x, x) = 0$. Since the support of dX_t^* is in $\{X_t = X_t^*\}$, it follows that $D_y v(t, X_t, X_t^*)dX_t^* = 0$. This implies that

$$du^0(t, \omega) = dv(t, \omega_t, \omega_t^*) = \partial_t v dt + D_x v dX_t + \frac{1}{2}D_{xx}^2 v d\langle X \rangle_t, \quad \mathscr{P}_\infty\text{-q.s.}$$

Then by (9.4.1) see that $u^0 \in C^{1,2}(\overline{\Lambda})$ with

$$\partial_t u^0(t,\omega) = \partial_t v(t,\omega_t,\omega_t^*), \quad \partial_\omega u^0(t,\omega) = D_x v(t,\omega_t,\omega_t^*),$$
$$\partial^2_{\omega\omega} u^0(t,\omega) = D^2_{xx} v(t,\omega_t,\omega_t^*).$$

Now it is straightforward to check that $\partial_t v + \frac{1}{2}\partial^2_{xx} v = 0$ for all $t < T$ and all (x,y) (even when $x \geq y$). Then u^0 is a classical solution to (11.1.1). ∎

We shall emphasize that, unlike in PDE case the heat equation always have a classical solution, the PPDE (11.1.1) may not have a classical solution.

Example 11.1.3 *Let $d = 1$ and $\xi = X_{t_0}$ for some $0 < t_0 < T$. Then $u^0 \notin C^{1,2}(\Lambda)$ and thus PPDE (11.1.1) with terminal condition ξ does not admit a classical solution.*

Proof Clearly $u^0(t,\omega) = \omega_{t\wedge t_0}$, and thus $du^0 = \mathbf{1}_{[0,t_0]}(t)dX_t$. If $u^0 \in C^{1,2}(\Lambda)$, compare this with (9.4.1) we see that $\partial_\omega u^0(t,\omega) = \mathbf{1}_{[0,t_0]}(t)$, which unfortunately is discontinuous at $t = t_0$. So $u^0 \notin C^{1,2}(\Lambda)$. ∎

Notice that the above u^0 is still in $C^0(\overline{\Lambda})$, so it is still a good candidate for viscosity solution, which we study next.

11.1.2 Definition of Viscosity Solutions

We now let $u \in C^0(\overline{\Lambda})$. In light of Section 5.5 for viscosity solutions of PDEs, to define viscosity supersolution of PPDEs one may naively extend the $\mathscr{A}u(t,x)$ in (5.5.4) to the following class of test functions: for any $(t,\omega) \in \Lambda$ and recalling the notations in (10.1.8) and (9.2.25),

$$\left\{\varphi \in C^{1,2}(\Lambda) : \exists \delta > 0 \text{ such that } [\varphi - u^{t,\omega}](0,0) = 0 = \sup_{(\tilde{t},\tilde{\omega})\in\overline{\Lambda}_{H_\delta}} [\varphi - u^{t,\omega}](\tilde{t},\tilde{\omega})\right\}.$$

(11.1.5)

Indeed, under this definition one can easily show that u^0 is a viscosity solution of (11.1.1), following the arguments in Theorem 5.5.8. However, in this path dependent case, $\overline{\Lambda}_{H_\delta}$ is not compact, and then one cannot mimic the arguments in Proposition 5.5.10 to obtain the comparison principle and the uniqueness. We thus need to modify this set of test functions.

We remark that the notion of viscosity solution is completely determined by the choice of the set of test functions. A simple but crucial observation is:

the larger the set of test function is,
the easier the uniqueness will be and the harder the existence will be .

(11.1.6)

Notice that, given the choice of (11.1.5), the main difficulty lies in the uniqueness. Then by the above observation we shall enlarge this set so as to help for the uniqueness. However, if we make this set too large, the existence will fail. The optimal one should be the set of all smooth functions which satisfy the requirements for existence. Recall the proof of Theorem 5.5.8 and Remark 5.5.9 concerning the existence in PDE case, we introduce the following sets of test functions: for any $(t, \omega) \in \Lambda$,

$$\overline{\mathscr{A}}u(t,\omega) := \left\{ \varphi \in C^{1,2}(\Lambda^t) : \exists \delta > 0 \text{ s.t. } [\varphi - u^{t,\omega}]_0 = 0 = \sup_{\tau \in \mathscr{T}_{H_\delta}} \mathbb{E}^{\mathbb{P}_0}\big[(\varphi - u^{t,\omega})_\tau\big] \right\};$$

$$\underline{\mathscr{A}}u(t,\omega) := \left\{ \varphi \in C^{1,2}(\Lambda^t) : \exists \delta > 0 \text{ s.t. } [\varphi - u^{t,\omega}]_0 = 0 = \inf_{\tau \in \mathscr{T}_{H_\delta}} \mathbb{E}^{\mathbb{P}_0}\big[(\varphi - u^{t,\omega})_\tau\big] \right\}.$$

$$(11.1.7)$$

We note that, since u and φ are continuous, by choosing $\delta > 0$ small the expectation involved in $\overline{\mathscr{A}}u(t,\omega)$ and $\underline{\mathscr{A}}u(t,\omega)$ always exist. We shall call H_δ a localization time. We also remark that, for heat equations actually it is fine to replace H_δ with any stopping time $H > 0$. However, for fully nonlinear PPDEs in the next section, it is crucial to use hitting times H_δ, mainly due to the optimal stopping problem Theorem 10.3.2.

Recall Definition 5.5.2 for viscosity solutions of PDEs, we now define

Definition 11.1.4 *Let $u \in C^0(\overline{\Lambda})$.*

(i) *We say u is a viscosity supersolution of PPDE (11.1.1) if, for any $(t, \omega) \in \Lambda$ and any $\varphi \in \overline{\mathscr{A}}u(t, \omega)$, it holds that $\mathscr{L}\varphi(0, 0) \leq 0$.*

(ii) *We say u is a viscosity subsolution of PPDE (11.1.1) if, for any $(t, \omega) \in \Lambda$ and any $\varphi \in \underline{\mathscr{A}}u(t, \omega)$, it holds that $\mathscr{L}\varphi(0, 0) \geq 0$.*

(iii) *We say u is a viscosity solution to (5.5.2) if it is both a viscosity supersolution and a viscosity subsolution.*

11.1.3 Well-Posedness in the Sense of Viscosity Solutions

Note that many results in the previous two chapters, for example the optimal stopping problem in Section 10.3, are prepared for fully nonlinear PPDEs. Since we use the heat equation mainly to illustrate the idea, for the ease of presentation in this subsection we consider only bounded viscosity solutions so that we can apply the results in the previous two chapters directly, by noticing that

$$\mathscr{P} := \{\mathbb{P}_0\} \text{ satisfies Assumption 9.3.3 with } \mathscr{P}_t = \{\mathbb{P}_0\} \text{ and is weakly compact.}$$

$$(11.1.8)$$

For the heat equation case, however, one can easily extend the results, in particular one can relax the boundedness requirement to certain growth conditions.

The well-posedness will be built on the following simple but crucial result, which illustrates the role of the optimal stopping problem in the viscosity theory.

Lemma 11.1.5 *Let $u \in UC_b(\overline{\Lambda})$. If $u(t,\omega) > \mathbb{E}^{P_0}\left[u_{T-t}^{t,\omega}\right]$ for some $(t,\omega) \in \Lambda$, then,*

the constant process $u(t^*,\omega^*) \in \underline{\mathscr{A}}u^{t,\omega}(t^*,\omega^*)$ for some $(t^*,\omega^*) \in \Lambda^t$.

Proof We first note that $u^{t,\omega} \in UC_b(\overline{\Lambda}^t)$, so $\underline{\mathscr{A}}u^{t,\omega}(\tilde{t},\tilde{\omega})$ makes sense. Without loss of generality, we may assume $(t,\omega) = (0,0)$. Consider the optimal stopping problem $V_0 := \sup_{\tau \in \mathscr{T}} \mathbb{E}^{P_0}[u_\tau]$. By (11.1.8) and applying Theorem 10.3.2 with $H = T$, one can find corresponding Y and τ^*. Note that

$$\mathbb{E}^{P_0}[u_{\tau^*}] = \mathbb{E}^{P_0}[Y_{\tau^*}] = Y_0 \ge u_0 > \mathbb{E}^{P_0}[u_T].$$

This implies $\mathbb{P}_0(\tau^* < T) > 0$, and thus there exists $\omega^* \in \Omega$ such that $t^* := \tau^*(\omega^*) < T$. By the definition of Y in (10.3.1), we have

$$u_{t^*}(\omega^*) = Y_{t^*}(\omega^*) = \sup_{\tau \in \mathscr{T}_{T-t^*}} \mathbb{E}^{P_0}[u_\tau^{t^*,\omega^*}] \ge \sup_{\tau \in \mathscr{T}_{H_\delta}} \mathbb{E}^{P_0}[u_\tau^{t^*,\omega^*}],$$

for any $\delta \le \frac{T-t^*}{L_1}$. On the other hand, it is always true that $u_{t^*}(\omega^*) \le \sup_{\tau \in \mathscr{T}_{H_\delta}} \mathbb{E}^{P_0}[u_\tau^{t^*,\omega^*}]$. Then $u_{t^*}(\omega^*) = \sup_{\tau \in \mathscr{T}_{H_\delta}} \mathbb{E}^{P_0}[u_\tau^{t^*,\omega^*}]$, which exactly means the constant process $u_{t^*}(\omega^*) \in \underline{\mathscr{A}}u(t^*,\omega^*)$. ∎

We are now ready to prove the following:

Theorem 11.1.6 *Let $\xi \in UC_b(\Omega)$. Then the u^0 defined by (11.1.4) is the unique viscosity solution of (11.1.1) in $UC_b(\Lambda)$ with terminal condition ξ.*

Proof Existence. First, by (11.1.8) and applying Theorem 10.2.1 we see that $u^0 \in UC_b(\Lambda)$ and satisfies the tower property (10.2.2), which reads in this case $u_t^0(\omega) = \mathbb{E}^{P_0}[(u^0)_\tau^{t,\omega}]$ for any $\tau \in \mathscr{T}_{T-t}$. Without loss of generality, we shall only verify the viscosity supersolution property at $(0,0)$. Assume to the contrary that there exists $\varphi \in \overline{\mathscr{A}}u(0,0)$ with localization time H_δ such that $c := \mathscr{L}\varphi(0,0) > 0$. By the smoothness of φ, $\mathscr{L}\varphi(t,\omega)$ is continuous in (t,ω). Then, for $\delta > 0$ small enough, we have $\mathscr{L}\varphi \ge \frac{c}{2}$ on $\overline{\Lambda}_{H_\delta}$. Now it follows from the definition of $\overline{\mathscr{A}}u(0,0)$ that

$$[\varphi - u^0]_0 \ge \mathbb{E}^{P_0}\left[(\varphi - u^0)_{H_\delta}\right].$$

By the tower property of u^0 and applying the functional Itô formula on φ, we obtain:

$$0 \ge \mathbb{E}^{P_0}\left[\varphi_{H_\delta} - \varphi_0\right] = \mathbb{E}^{P_0}\left[\int_0^{H_\delta} \mathscr{L}\varphi_t dt\right] \ge \frac{c}{2}\mathbb{E}^{P_0}[H_\delta] > 0.$$

Contradiction.

Uniqueness. We shall prove a stronger statement:

$u \in UC_b(\overline{\Lambda})$ is viscosity subsubsolution of PPDE (11.1.1) with terminal condition

$$u_T \leq \xi \implies u(t, \omega) \leq \mathbb{E}^{\mathbb{P}_0}[\xi^{t,\omega}] \text{ for all } (t, \omega) \in \Lambda. \tag{11.1.9}$$

Similar statement holds true for viscosity supersolutions, which clearly implies the comparison principle and hence the uniqueness. Indeed, without loss of generality we shall only verify (11.1.9) for $(t, \omega) = (0, 0)$, namely $u_0 \leq \mathbb{E}^{\mathbb{P}_0}[\xi]$. Assume to the contrary that $c := u_0 - \mathbb{E}^{\mathbb{P}_0}[\xi] > 0$. Denote $\tilde{u}(t, \omega) := u(t, \omega) + \frac{c}{2T}t$. Then clearly $\tilde{u} \in UC_b(\Lambda)$ and

$$\tilde{u}_0 = u_0 = \mathbb{E}^{\mathbb{P}_0}[\xi] + c \geq \mathbb{E}^{\mathbb{P}_0}[u_T] + c = \mathbb{E}^{\mathbb{P}_0}[\tilde{u}_T] + \frac{c}{2} > \mathbb{E}^{\mathbb{P}_0}[\tilde{u}_T].$$

Applying Lemma 11.1.5 there exists $(t^*, \omega^*) \in \Lambda$ such that the constant process $\tilde{u}(t^*, \omega^*) \in \underline{\mathscr{A}}\tilde{u}(t^*, \omega^*)$. This implies that $\varphi \in \underline{\mathscr{A}}u(t^*, \omega^*)$, where $\varphi(t, \omega) := -\frac{c}{2T}t + u(t^*, \omega^*)$ for all $(t, \omega) \in \Lambda^{t^*}$. Then, it follows from the viscosity subsolution property of u that

$$0 \leq \mathscr{L}\varphi(0, 0) = -\frac{c}{2T},$$

which is a desired contradiction. ∎

We remark that in this case the uniform continuity and the boundedness of ξ and u^0 are actually not necessary, because the optimal stopping problem under $\mathbb{P}_0$ can be solved under weaker conditions, as we saw in Proposition 6.3.2.

11.2 Viscosity Solution of General Parabolic PPDEs

We now study the following fully nonlinear parabolic PPDEs (with certain terminal conditions):

$$\mathscr{L}u(t, \omega) := \partial_t u(t, \omega) + G(t, \omega, u, \partial_\omega u, \partial^2_{\omega\omega} u) = 0, \quad (t, \omega) \in \Lambda, \tag{11.2.1}$$

where the generator $G : \overline{\Lambda} \times \mathbb{R} \times \mathbb{R}^{1 \times d} \times \mathbb{S}^d \to \mathbb{R}$ satisfies the following standing assumptions:

Assumption 11.2.1

(i) For fixed (y, z, γ), $G(\cdot, y, z, \gamma) \in C_b^0(\overline{\Lambda})$.
(ii) G is uniformly Lipschitz continuous in (y, z, γ), with a Lipschitz constant L_0.
(iii) G is parabolic, i.e., nondecreasing in $\gamma \in \mathbb{S}^d$.

The typical examples we are interested in are path dependent Hamilton-Jacobi-Bellman equations and Bellman-Isaacs equations, motivated from their applications in stochastic control and stochastic differential equations. We will provide various examples in the next section.

Definition 11.2.2 *Let* $u \in C^{1,2}(\Lambda)$. *We say* u *is a classical solution (resp. supersolution, subsolution) of PPDE* (11.2.1) *if* $\mathscr{L}u(t,\omega) = $ *(resp.* $\leq, \geq$) 0 *for all* $(t,\omega) \in \Lambda$.

Remark 11.2.3 In the Markov case, namely $G(t,\omega,.) = g(t,\omega_t,.)$ and $u(t,\omega) = v(t,\omega_t)$, the PPDE (11.2.1) reduces to the following PDE:

$$\partial_t v(t,x) + g(t,x,v,Dv,D^2v) = 0, \quad (t,x) \in [0,T) \times \mathbb{R}^d. \quad (11.2.2)$$

It is clear that v is a classical solution (resp. supersolution, subsolution) of PPDE (11.2.2) implies that u is a classical solution (resp. supersolution, subsolution) of PPDE (11.2.1). ∎

11.2.1 Definition of Viscosity Solutions

As we see in Example 11.1.3, in general one cannot expect the existence of classical solutions for PPDEs. We thus turn to viscosity solutions. In the fully nonlinear case, we shall replace the $\mathbb{E}^{\mathbb{P}_0}$ in (11.1.7) with nonlinear expectations. To be precise, for any $L > 0$ and $\xi \in \mathbb{L}^1(\mathscr{F}_T, \mathscr{P}_L)$, denote

$$\overline{\mathscr{E}}^L[\xi] := \mathscr{E}^{\mathscr{P}_L}[\xi] := \sup_{\mathbb{P} \in \mathscr{P}_L} \mathbb{E}^{\mathbb{P}}[\xi], \quad \underline{\mathscr{E}}^L[\xi] := -\overline{\mathscr{E}}^L[-\xi] = \inf_{\mathbb{P} \in \mathscr{P}_L} \mathbb{E}^{\mathbb{P}}[\xi], \quad (11.2.3)$$

and we introduce the set of test functions with parameter L: for any $u \in C^0(\overline{\Lambda})$ and $(t,\omega) \in \Lambda$,

$$\overline{\mathscr{A}}^L u(t,\omega) := \Big\{ \varphi \in C^{1,2}(\Lambda^t) : \exists \delta > 0 \text{ s.t. } [\varphi - u^{t,\omega}]_0 = 0 = \sup_{\tau \in \mathscr{T}_{H\delta}} \overline{\mathscr{E}}^L\big[(\varphi - u^{t,\omega})_\tau\big] \Big\};$$

$$\underline{\mathscr{A}}^L u(t,\omega) := \Big\{ \varphi \in C^{1,2}(\Lambda^t) : \exists \delta > 0 \text{ s.t. } [\varphi - u^{t,\omega}]_0 = 0 = \inf_{\tau \in \mathscr{T}_{H\delta}} \underline{\mathscr{E}}^L\big[(\varphi - u^{t,\omega})_\tau\big] \Big\}.$$

$$(11.2.4)$$

Define, for any $(t,\omega) \in \Lambda$ and $\varphi \in C^{1,2}(\Lambda^t)$,

$$\mathscr{L}^{t,\omega}\varphi(\tilde{t},\tilde{\omega}) := \partial_t\varphi(\tilde{t},\tilde{\omega}) + G^{t,\omega}\big(\tilde{t},\tilde{\omega},\varphi,\partial_\omega\varphi,\partial^2_{\omega\omega}\varphi\big), \quad (\tilde{t},\tilde{\omega}) \in \Lambda^t. \quad (11.2.5)$$

We then extend Definition 11.1.4 to fully nonlinear case.

Definition 11.2.4

(i) *For any* $L > 0$, *we say* $u \in C^0(\overline{\Lambda})$ *is a viscosity* L-*subsolution (resp.* L-*supersolution) of PPDE* (11.2.1) *if, for any* $(t,\omega) \in \Lambda$,

$$\mathscr{L}^{t,\omega}\varphi(0,0) \geq \; (resp. \leq) \, 0, \quad for \, all \; \varphi \in \underline{\mathscr{A}}^L u(t,\omega) \; (resp. \varphi \in \overline{\mathscr{A}}^L u(t,\omega)).$$

$$(11.2.6)$$

(ii) We say u is a viscosity subsolution (resp. supersolution) of PPDE (11.2.1) if u is viscosity L-subsolution (resp. L-supersolution) of PPDE (11.2.1) for some L > 0.

(iii) We say u is viscosity solution of PPDE (11.2.1) if it is a viscosity sub- and supersolution.

We may also call u a viscosity semi-solution if it is either a subsolution or a supersolution. In the rest of this subsection, we provide some remarks concerning the definition of viscosity solutions. In most places we will comment on the viscosity subsolution only, but obviously similar properties hold for the viscosity supersolution as well.

Remark 11.2.5

(i) Typically we shall require $u \in UC_b(\overline{\Lambda})$ for viscosity semi-solutions, so that we can apply the optimal stopping Theorem 10.3.2 in appropriate way. This can be weakened in the sense of Remark 10.3.3 (i).

(ii) In first order case or in semilinear case, the optimal stopping theorem becomes much easier, see, e.g., Proposition 6.3.2. Then the requirement for viscosity solutions can also be weakened. ∎

Remark 11.2.6 Similar to the viscosity solutions of PDEs in Section 5.5:

(i) The viscosity property is a local property, namely the viscosity property at (t,ω) involves only the value of $u^{t,\omega}$ in Λ_{H_δ} for some small $\delta > 0$.

(ii) The fact that u is a viscosity solution does not mean that the PPDE must hold with equality at some (t,ω) and φ in some appropriate set. One has to check viscosity subsolution property and viscosity supersolution property separately.

(iii) In general $\underline{\mathscr{A}}^L u(t,\omega)$ could be empty. In this case automatically u satisfies the viscosity subsolution property at (t,ω). ∎

Remark 11.2.7 Consider the Markov setting in Remark 11.2.3.

(i) Note that we have enlarged the set of test functions $\mathscr{A}u(\cdot,\omega)$ in (11.2.4) in order to help for the uniqueness. So u is a viscosity subsolution of PPDE (11.2.1) in the sense of Definition 11.2.4 implies that v is a viscosity subsolution of PDE (11.2.2) in the standard sense in the spirit of Definition 5.5.2. However, the opposite direction is in general not true. We shall point out though, when the PDE is well posed, by uniqueness our definition of viscosity solution of PPDE (11.2.1) is consistent with the viscosity solution of PDE (11.2.2) in the standard sense.

(ii) In principle it should be easier to prove the uniqueness of viscosity solutions under our sense than that under the standard sense. It will be very interesting to see if one can weaken the conditions for the uniqueness of viscosity solutions for PDEs under our definition. ∎

Remark 11.2.8

(i) Consider the path dependent heat equation (11.1.1). For $L \geq 1$, the set $\underline{\mathscr{A}}^L u(t, \omega)$ in (11.2.4) is smaller than the set $\underline{\mathscr{A}} u(t, \omega)$ in (11.1.7). So u is a viscosity subsolution of (11.1.1) in the sense of Definition 11.1.4 implies it is a viscosity subsolution in the sense of Definition 11.2.4, but not vice versa in general. In particular, Theorem 11.1.6 does not imply the uniqueness of viscosity solutions of (11.1.1) in the sense of Definition 11.2.4.

(ii) For $0 < L_1 < L_2$, obviously $\mathscr{P}_{L_1} \subset \mathscr{P}_{L_2}$, $\underline{\mathscr{E}}^{L_2} \leq \underline{\mathscr{E}}^{L_1}$, and $\underline{\mathscr{A}}^{L_2} u(t, \omega) \subset \underline{\mathscr{A}}^{L_1} u(t, \omega)$. Then one can easily check that a viscosity L_1-subsolution must be a viscosity L_2-subsolution. Consequently, u is a viscosity subsolution if and only if

there exists an $L \geq 1$ such that, for all $L' \geq L$, u is a viscosity L'-subsolution.

(iii) While the constant L may vary for different viscosity subsolutions, for fixed viscosity subsolution u, we require the same L for all (t, ω). ∎

11.2.2 Consistency with Classical Solutions

Theorem 11.2.9 *Let Assumption 11.2.1 hold and $u \in C^{1,2}(\Lambda)$. Then u is a classical solution (resp. subsolution, supersolution) of PPDE (11.2.1) if and only if it is a viscosity solution (resp. subsolution, supersolution).*

Proof We prove the supersolution property only, and assume for simplicity that $d = 1$.

"$\Longleftarrow$" Assume u is a viscosity L-supersolution. For any (t, ω), since $u \in C^{1,2}(\Lambda)$, we have $u^{t,\omega} \in C^{1,2}(\Lambda^t)$ and thus $u^{t,\omega} \in \overline{\mathscr{A}}^L u(t, \omega)$ with $\mathrm{H} := T - t$. By definition of viscosity L-supersolution we see that $\mathscr{L} u(t, \omega) = \mathscr{L}^{t,\omega} u^{t,\omega}(t, 0) \leq 0$.

"$\Longrightarrow$" Assume u is a classical supersolution. If u is not a viscosity supersolution, then it is not a viscosity L_0-supersolution. Thus there exist $(t, \omega) \in \Lambda$ and $\varphi \in \overline{\mathscr{A}}^{L_0} u(t, \omega)$ such that $c := \mathscr{L} \varphi(0, 0) > 0$. Without loss of generality, we assume $(t, \omega) = (0, 0)$ and $\mathrm{H} := \mathrm{H}_\delta$ is a localization time. Now recall (9.2.10) and let $\mathbb{P} \in \mathscr{P}_{L_0}$ corresponding to some constants b' and σ' which will be determined later. Then

$$0 \geq \overline{\mathscr{E}}^{L_0}\left[(\varphi - u)_{\mathrm{H}}\right] \geq \mathbb{E}^{\mathbb{P}}\left[(\varphi - u)_{\mathrm{H}}\right].$$

Applying functional Itô formula and noticing that $(\varphi - u)_0 = 0$, we have

$$(\varphi - u)_H = \int_0^H \Big[\partial_t(\varphi - u)_t + \frac{1}{2}\partial_{\omega\omega}^2(\varphi - u)_t|\sigma'|^2 + \partial_\omega(\varphi - u)_t b'\Big]dt$$

$$+ \int_0^H \partial_\omega(\varphi - u)_t dM_t^{\mathbb{P}}$$

$$= \int_0^H (\tilde{\mathscr{L}}\varphi - \tilde{\mathscr{L}}u)_t dt + \int_0^H \partial_\omega(\varphi - u)_t dM_t^{\mathbb{P}},$$

where $M_t^{\mathbb{P}} := X_t - b't$ is a $\mathbb{P}$-martingale and

$$\tilde{\mathscr{L}}\varphi(t,\omega) := \mathscr{L}\varphi(t,\omega) - G(t,\omega,\varphi,\partial_\omega\varphi,\partial_{\omega\omega}^2\varphi) + \frac{1}{2}(\partial_{\omega\omega}^2\varphi)|\sigma'|^2 + \partial_\omega\varphi b'.$$

Taking expected values, this leads to

$$0 \geq \mathbb{E}^{\mathbb{P}}\Big[\int_0^H (\tilde{\mathscr{L}}\varphi - \tilde{\mathscr{L}}u)_t dt\Big].$$

Since $\tilde{\mathscr{L}}\varphi$ and $\tilde{\mathscr{L}}u$ are continuous, for δ small enough we have $|\tilde{\mathscr{L}}\varphi_t - \tilde{\mathscr{L}}\varphi_0| + |\tilde{\mathscr{L}}u_t - \tilde{\mathscr{L}}u_0| \leq \frac{c}{4}$ on $[0, H]$. Then

$$0 \geq \mathbb{E}^{\mathbb{P}}\Big[(\tilde{\mathscr{L}}\varphi_0 - \tilde{\mathscr{L}}u_0 - \frac{c}{2})H\Big]. \tag{11.2.7}$$

Note that $\mathscr{L}u_0 \leq 0$, $\mathscr{L}\varphi_0 = c$, and $\varphi_0 = u_0$. Thus

$$\tilde{\mathscr{L}}\varphi_0 - \tilde{\mathscr{L}}u_0 \geq c + \frac{1}{2}\partial_{\omega\omega}^2(\varphi - u)_0|\sigma'|^2 + \partial_\omega(\varphi - u)_0 b'$$

$$- \Big[G(\cdot, u, \partial_\omega\varphi, \partial_{\omega\omega}^2\varphi) - G(\cdot, u, \partial_\omega u, \partial_{\omega\omega}^2 u)\Big]_0.$$

By Assumption 11.2.1 (iii), there exist constant b' and σ' such that $\mathbb{P} \in \mathscr{P}_{L_0}$ and

$$\Big[G(\cdot, u, \partial_\omega\varphi, \partial_{\omega\omega}^2\varphi) - G(\cdot, u, \partial_\omega u, \partial_{\omega\omega}^2 u)\Big]_0 = \frac{1}{2}\partial_{\omega\omega}^2(\varphi - u)_0|\sigma'|^2 + \partial_\omega(\varphi - u)_0 b'.$$

$$\tag{11.2.8}$$

Then $\tilde{\mathscr{L}}\varphi_0 - \tilde{\mathscr{L}}u_0 \geq c$, and (11.2.7) leads to $0 \geq \mathbb{E}^{\mathbb{P}}[\frac{c}{2}H] > 0$, contradiction. ∎

Remark 11.2.10 One may define viscosity solutions alternatively by replacing the $\overline{\mathscr{E}}^L$ in (11.2.4) with $\mathscr{E}^{\mathscr{P}}$ for certain class of probability measures $\mathscr{P}$. As we see in (11.2.8), the crucial thing is that $\mathscr{P}$ should cover all measures induced from the linearization of G in terms of (z, γ). Given this inclusion, the smaller $\mathscr{P}$ is, the easier for uniqueness to hold.

(i) The standard notion of viscosity solution in PDE literature, as in (5.5.4), amounts to saying that $\mathscr{P}$ contains all measures (not necessarily semimartingale measures), in particular the measures with mass. The notion in (11.2.4) uses only semimartingale measures and thus helps for uniqueness. However, as we see in (11.2.8), in general fully nonlinear case, we need $\mathscr{P} = \mathscr{P}_L$ with L greater than the Lipschitz constant L_0 in Assumption 11.2.1.

(ii) In the first order case, namely G does not depend on γ, one may consider only those $\mathbb{P} \in \mathscr{P}_L$ such that $\sigma' = 0$, where σ' is introduced in Definition 9.2.9.

(iii) In the semilinear case which corresponds to the BSDE (4.0.3), one may consider $\mathscr{P} = \mathscr{P}_L^{drift}$ defined in (9.2.11).   ∎

11.2.3 Equivalent Definition via Semijets

In standard viscosity theory for PDEs, one may define viscosity solution equivalently via semijets, see Remark 5.5.6 and Problem 5.7.9. This is the case for PPDEs as well. For $(c, a, p, q) \in \mathbb{R} \times \mathbb{R} \times \mathbb{R}^{1 \times d} \times \mathbb{S}^d$, define paraboloids:

$$\phi^{c,a,p,q}(\tilde{t}, \tilde{\omega}) := c + a\tilde{t} + p\tilde{\omega}_{\tilde{t}} + \frac{1}{2}q : \tilde{\omega}_{\tilde{t}}(\tilde{\omega}_{\tilde{t}})^{\mathrm{T}}, \quad (\tilde{t}, \tilde{\omega}) \in \overline{\Lambda}. \qquad (11.2.9)$$

We then introduce the corresponding subjets and superjets: for $L > 0$,

$$\overline{\mathscr{J}}^L u(t, \omega) := \{(a, p, q) \in \mathbb{R} \times \mathbb{R}^{1 \times d} \times \mathbb{S}^d : \phi^{u(t,\omega),a,p,q} \in \overline{\mathscr{A}}^L u(t, \omega)\};$$

$$\qquad\qquad\qquad\qquad\qquad\qquad\qquad\qquad\qquad\qquad\qquad\qquad\qquad (11.2.10)$$

$$\underline{\mathscr{J}}^L u(t, \omega) := \{(a, p, q) \in \mathbb{R} \times \mathbb{R}^{1 \times d} \times \mathbb{S}^d : \phi^{u(t,\omega),a,p,q} \in \underline{\mathscr{A}}^L u(t, \omega)\}.$$

Proposition 11.2.11 *Let Assumption 11.2.1 hold and $L > 0$. A process $u \in C^0(\overline{\Lambda})$ is an L-viscosity supersolution (resp. subsolution) of PPDE (11.2.1) if and only if: for any $(t, \omega) \in \Lambda$,*

$$a + G(t, \omega, u_t(\omega), p, q) \leq \ (resp. \ \geq) \ 0,$$

$$\forall (a, p, q) \in \overline{\mathscr{J}}^L u(t, \omega) \ (resp. \ \underline{\mathscr{J}}^L u(t, \omega)). \qquad (11.2.11)$$

Proof Without loss of generality we prove only the supersolution case at $(t, \omega) = (0, 0)$, and for simplicity assume $d = 1$. Note that

$$\text{for } \varphi := \phi^{c,a,p,q} : \quad \partial_t \varphi_0 = a, \ \partial_\omega \varphi_0 = p, \ \partial^2_{\omega\omega} \varphi_0 = q. \qquad (11.2.12)$$

"$\Longrightarrow$" Assume u is an L-viscosity supersolution at $(0, 0)$. For any $(a, p, q) \in \overline{\mathscr{J}}^L u(0, 0)$, we have $\phi^{u_0,a,p,q} \in \overline{\mathscr{A}}^L u(0, 0)$. Then it follows from the viscosity property of u and (11.2.12) that

$$0 \geq \mathscr{L}\phi_0^{u_0,a,p,q} = a + G(0, 0, u_0, p, q).$$

"$\Longleftarrow$" Assume (11.2.11) holds at $(0,0)$ and $\varphi \in \overline{\mathscr{A}}^L u(0,0)$ with localization time H_δ. Denote

$$c := u_0, \quad a := \partial_t \varphi_0, \quad a_\varepsilon := a - \varepsilon(1+2L), \quad p := \partial_\omega \varphi_0, \quad q := \partial^2_{\omega\omega}\varphi_0, \quad \forall \varepsilon > 0.$$

(11.2.13)

Then, for any $\tau \in \mathscr{T}_{\mathrm{H}_\delta}$ and $\mathbb{P} \in \mathscr{P}_L$,

$$\phi^{c,a_\varepsilon,p,q}_\tau - \varphi_\tau = \int_0^\tau [a_\varepsilon - \partial_t \varphi_t] dt + \int_0^\tau [p + qX_t - \partial_\omega \varphi_t] dX_t$$

$$+ \frac{1}{2} \int_0^\tau [q - \partial^2_{\omega\omega} \varphi_t] d\langle X \rangle_t, \quad \mathbb{P}\text{-a.s.}$$

By choosing $\delta > 0$ small, we may assume without loss of generality that

$$|\partial_t \varphi_t - a| \le \varepsilon, \quad |\partial_\omega \varphi_t - p - qX_t| \le \varepsilon, \quad |\partial^2_{\omega\omega}\varphi_t - q| \le \varepsilon, \quad 0 \le t \le \mathrm{H}_\delta.$$

(11.2.14)

Then,

$$\mathbb{E}^\mathbb{P}\Big[\phi^{c,a_\varepsilon,p,q}_\tau - \varphi_\tau\Big]$$

$$= \mathbb{E}^\mathbb{P}\Big[-\varepsilon(1+2L)\tau + \int_0^\tau [a - \partial_t \varphi_t] dt + \int_0^\tau [p + qX_t - \partial_\omega \varphi_t] dX_t$$

$$+ \frac{1}{2} \int_0^\tau [q - \partial^2_{\omega\omega}\varphi_t] d\langle X \rangle_t\Big]$$

$$\le \mathbb{E}^\mathbb{P}\Big[-\varepsilon(1+2L)\tau + \int_0^\tau |a - \partial_t \varphi_t| dt + L\int_0^\tau |p + qX_t - \partial_\omega \varphi_t| dt$$

$$+ L\int_0^\tau |q - \partial^2_{\omega\omega}\varphi_t| dt\Big] \le 0.$$

This implies that

$$\mathscr{E}^L[\phi^{c,a_\varepsilon,p,q}_\tau - u_\tau] \le \mathscr{E}^L[\varphi_\tau - u_\tau] \le 0.$$

Then $\phi^{c,a_\varepsilon,p,q} \in \overline{\mathscr{A}}^L u(0,0)$, and thus $(a_\varepsilon, p, q) \in \overline{\mathscr{J}}^L u(0,0)$. By our assumption we have $a_\varepsilon + G(0,0,c,p,q) \le 0$. Send $\varepsilon \to 0$, we obtain $\mathscr{L}\varphi_0 = a + G(0,0,c,p,q) \le 0$. That is, u is an L-viscosity supersolution at $(0,0)$. ∎

Remark 11.2.12

(i) As we will see quite often later, it is a common trick to introduce the ε in (11.2.13) to cancel the small errors in (11.2.14).

(ii) One may easily modify the value of $\phi^{c,a,p,q}$ outside of $\overline{\Lambda}_{H_\delta}$ for some $\delta > 0$, denoted as $\tilde{\phi}$, so that $\tilde{\phi} \in C^{1,2}(\Lambda^t) \cap UC_b(\overline{\Lambda}^t)$. Since $\tilde{\phi} = \phi^{c,a,p,q}$ on $\overline{\Lambda}_{H_\delta}$, then $\tilde{\phi} \in \overline{\mathscr{A}}^L u(t, \omega)$ whenever $\phi^{c,a,p,q} \in \overline{\mathscr{A}}^L u(t, \omega)$. So by Proposition 11.2.11, when necessary we may assume $\varphi \in C^{1,2}(\Lambda^t) \cap UC_b(\overline{\Lambda}^t)$ in (11.2.4). This uniform continuity is particularly useful for the optimal stopping problem.

∎

11.2.4 A Change Variable Formula

Proposition 11.2.13 *Let Assumption 11.2.1 hold true, $\lambda \in \mathbb{R}$, $u \in \mathscr{U}$, and denote $\tilde{u}_t := e^{\lambda t} u_t$. Then u is a viscosity supersolution of PPDE (11.2.1) if and only if $\tilde{u}$ is a viscosity supersolution of:*

$$\mathscr{L}\tilde{u} := \partial_t \tilde{u} + \tilde{G}(t, \omega, \tilde{u}, \partial_\omega \tilde{u}, \partial^2_{\omega\omega}\tilde{u}) = 0, \qquad (11.2.15)$$

where $\tilde{G}(t, \omega, y, z, \gamma) := -\lambda y + e^{\lambda t} G(t, \omega, e^{-\lambda t}y, e^{-\lambda t}z, e^{-\lambda t}\gamma).$

The proof is quite standard, in particular we shall use the trick in Remark 11.2.12. We leave it to the readers, see Problem 11.7.1.

Remark 11.2.14 Since G is Lipschitz continuous in y with Lipschitz constant, by choosing $\lambda > L$, we see that $\tilde{G}$ is strictly decreasing in y. This will be crucial for proving the comparison principle of viscosity solutions. ∎

11.3   Examples of PPDEs

In this section, we study several special PPDEs which have (semi-)explicit viscosity solutions. These solutions provide probabilistic representations for the PPDEs and thus can be viewed as path dependent nonlinear Feynman-Kac formula. More importantly, as value functions of some stochastic control problems, these examples on one hand illustrate how to check the viscosity properties of processes arising in applied problems, and on the other hand serve as applications of PPDEs. As in the viscosity theory of PDEs, the main tools are the regularity in (t, ω) of the processes and the dynamic programming principle.

11.3.1   First Order PPDEs

Example 11.3.1 *Suppose that $u(t, \omega) = v(\omega_t)$ for all $(t, \omega) \in \Lambda$, where $v : \mathbb{R}^d \to \mathbb{R}$ is bounded and continuous. By (9.4.2) formally we should have $\partial_t u = 0$. We now verify that u is indeed a viscosity solution of the equation $\partial_t u = 0$.*

Proof For $\varphi \in \overline{\mathscr{A}}^L u(t, \omega)$ with localization time H_ε, it follows from our definition that:

$$(\varphi - u^{t,\omega})_0 = 0 \geq \mathbb{E}^{\mathbb{P}^0}\big[(\varphi - u^{t,\omega})_{\delta \wedge \mathrm{H}_\varepsilon}\big] \text{ for all } \delta > 0.$$

where $\mathbb{P}^0 \in \mathscr{P}_L$ corresponds to $b' = 0, \sigma' = 0$ in (9.2.10). Notice that under $\mathbb{P}^0$, the canonical process X is frozen to its initial value. Then $\mathrm{H}_\varepsilon = \frac{\varepsilon}{L_1}$, $\mathbb{P}^0$-a.s. and thus, for $\delta < \frac{\varepsilon}{L_1}$,

$$\varphi(0,0) - v(\omega_t) = (\varphi - u^{t,\omega})_0 \geq \mathbb{E}^{\mathbb{P}^0}\big[(\varphi - u^{t,\omega})_{\delta \wedge \mathrm{H}_\varepsilon}\big] = \varphi(\delta, 0) - v(\omega_t).$$

This implies that $\partial_t \varphi(0,0) \leq 0$. Similarly one can show that $\partial_t \varphi(0,0) \geq 0$ for all $\varphi \in \underline{\mathscr{A}}^L u(t, \omega)$. ∎

Example 11.3.2 *Let $d = 1$ and introduce the following one-sided running maximum process: $\check{\omega}_t := \sup_{0 \leq s \leq t} \omega_s$. Then $u(t, \omega) := 2\check{\omega}_t - \omega_t$ is a viscosity solution of the first order PPDE:*

$$\partial_t u + |\partial_\omega u| - 1 = 0. \tag{11.3.1}$$

By the same arguments as in Example 9.4.4 (iv), u is not smooth, so it is a viscosity solution but not a classical solution.

Proof When $\omega_t < \check{\omega}_t$, we see that $u^{t,\omega}(\tilde{t}, \tilde{\omega}) = 2\check{\omega}_t - \omega_t - \tilde{\omega}_{\tilde{t}}$ for $(\tilde{t}, \tilde{\omega}) \in \Lambda_{\mathrm{H}_\delta}$ where $\delta := \check{\omega}_t - \omega_t$. Clearly $u^{t,\omega} \in C^{1,2}(\Lambda_{\mathrm{H}_\delta})$ with $\partial_t u(t, \omega) = 0$, $\partial_\omega u(t, \omega) = -1$, $\partial^2_{\omega\omega} u(t, \omega) = 0$, and thus satisfies (11.3.1). Applying a local version of Theorem 11.2.9, see Problem 11.7.2, u is a viscosity solution at (t, ω). So it suffices to check the viscosity property when $\check{\omega}_t = \omega_t$. Without loss of generality, we assume $(t, \omega) = (0, 0)$.

(i) We first show that $\underline{\mathscr{A}}^L u(0, 0)$ is empty for $L \geq 1$, and thus u is a viscosity subsolution. Indeed, assume $\varphi \in \underline{\mathscr{A}}^L u(0, 0)$ with localization time H_ε. By choosing $\varepsilon > 0$ small, we may assume $\partial_t \varphi, \partial^2_{\omega\omega} \varphi$ are bounded on $\overline{\Lambda}_{\mathrm{H}_\varepsilon}$. Since $\mathbb{P}_0 \in \mathscr{P}_L$, by definition of $\underline{\mathscr{A}}^L u(0, 0)$ we have, for any $\delta > 0$,

$$0 \leq \mathbb{E}^{\mathbb{P}_0}\big[(\varphi - u)_{\delta \wedge \mathrm{H}_\varepsilon}\big] = \mathbb{E}^{\mathbb{P}_0}\bigg[\int_0^{\delta \wedge \mathrm{H}_\varepsilon} (\partial_t \varphi + \partial^2_{\omega\omega} \varphi)(t, \omega) dt - 2\check{X}_{\delta \wedge \mathrm{H}_\varepsilon}\bigg]$$

$$\leq C\mathbb{E}^{\mathbb{P}_0}[\delta \wedge \mathrm{H}_\varepsilon] - 2\mathbb{E}^{\mathbb{P}_0}[\check{X}_{\delta \wedge \mathrm{H}_\varepsilon}] \leq C\delta - 2\mathbb{E}^{\mathbb{P}_0}[\check{X}_\delta] + 2\mathbb{E}^{\mathbb{P}_0}[\check{X}_\delta \mathbf{1}_{\{\mathrm{H}_\varepsilon \leq \delta\}}],$$

where $\check{X}_t := \sup_{0 \leq s \leq t} X_s$. Denote $c_0 := \mathbb{E}^{\mathbb{P}_0}[\check{X}_1]$. Applying (9.2.16) and Lemma 10.1.18 (ii) with $p = 2$, we have

$$\mathbb{E}^{\mathbb{P}_0}[\check{X}_\delta] = c_0\sqrt{\delta}, \quad \mathbb{E}^{\mathbb{P}_0}[\check{X}_\delta \mathbf{1}_{\{\mathrm{H}_\varepsilon \leq \delta\}}] \leq \big(\mathbb{E}^{\mathbb{P}_0}[|\check{X}_\delta|^2]\big)^{\frac{1}{2}}\big(\mathbb{P}_0(\mathrm{H}_\varepsilon \leq \delta)\big)^{\frac{1}{2}} \leq C\varepsilon^{-2}\delta.$$

Then

$$0 \le C\delta - c_0\sqrt{\delta} + C\varepsilon^{-2}\delta = C[1 + \varepsilon^{-2}]\delta - c_0\sqrt{\delta}.$$

This leads to a contradiction when δ is small enough. Therefore, $\underline{\mathscr{A}}^L u(0,0)$ is empty.

(ii) We next check the viscosity supersolution property. Assume to the contrary that $c := \partial_t\varphi(0,0) + |\partial_\omega\varphi(0,0)| - 1 > 0$ for some $\varphi \in \overline{\mathscr{A}}^L u(0,0)$ with localization time H_ε and some $L \ge 1$. Set constants $b' := \text{sign}(\partial_\omega\varphi(0,0))$ (with the convention $\text{sign}(0) := 1$), $\sigma' := 0$, and $\mathbb{P} \in \mathscr{P}_L$ be determined by (9.2.10). When $b' = 1$, we have $X_t = t, \check{X}_t = t$, $\mathbb{P}$-a.s. When $b' = -1$, we have $X_t = -t$, $\check{X}_t = 0$, $\mathbb{P}$-a.s. In both cases, it holds that $u(t,\omega) = t$, $H_\varepsilon = \varepsilon$, $\mathbb{P}$-a.s. By choosing ε small enough, we may assume

$$|\partial_t\varphi_t - \partial_t\varphi_0| + |\partial_\omega\varphi_t - \partial_\omega\varphi_0| \le \frac{c}{2} \quad \text{for } 0 \le t \le H_\varepsilon.$$

By the definition of $\overline{\mathscr{A}}^L u(0,0)$ we get

$$0 \ge \mathbb{E}^{\mathbb{P}}\left[(\varphi - u)_{H_\varepsilon}\right] = \mathbb{E}^{\mathbb{P}}\left[\int_0^\varepsilon (\partial_t\varphi + b'\partial_\omega\varphi)_t dt - \varepsilon\right]$$

$$\ge \mathbb{E}^{\mathbb{P}}\left[\int_0^\varepsilon \left(\partial_t\varphi_0 + b'\partial_\omega\varphi_0 - \frac{c}{2}\right)dt\right] - \varepsilon$$

$$= \mathbb{E}^{\mathbb{P}}\left[\int_0^\varepsilon \left(\partial_t\varphi_0 + |\partial_\omega\varphi_0| - \frac{c}{2}\right)dt\right] - \varepsilon$$

$$= \int_0^\varepsilon \left(1 + c - \frac{c}{2}\right)dt - \varepsilon = \frac{1}{2}c\varepsilon > 0.$$

This is the required contradiction, and thus u is a viscosity supersolution of (11.3.1). ∎

11.3.2 Semilinear PPDEs

We now consider the following semilinear PPDE which corresponds to BSDE (4.0.3)

$$\mathscr{L}u(t,\omega) := \partial_t u + \frac{1}{2}tr(\partial^2_{\omega\omega}u) + f(t,\omega,u,\partial_\omega u) = 0,$$

$$u(T,\omega) = \xi(\omega), \tag{11.3.2}$$

Assumption 11.3.3

(i) f is uniformly continuous in (t, ω) under $\mathbf{d}$ and ξ is uniformly continuous in ω under $\| \cdot \|_T$, with a common modulus of continuity function ρ_0.

(ii) f is uniformly Lipschitz continuous in (y, z) with Lipschitz constant L_0.

For any $(t, \omega) \in \Lambda$, consider the following BSDE on $[0, T - t]$:

$$Y_s^{t,\omega} = \xi^{t,\omega} + \int_s^{T-t} f^{t,\omega}(r, Y_r^{t,\omega}, Z_r^{t,\omega})dr - \int_s^{T-t} Z_r^{t,\omega} dB_r, \quad \mathbb{P}_0\text{-a.s.} \quad (11.3.3)$$

Here, as before, we denote $B := X$ to emphasize it's a Brownian motion under $\mathbb{P}_0$. Under Assumption 11.3.3, clearly FBSDE (11.3.3) has a unique solution $(Y^{t,\omega}, Z^{t,\omega})$. Moreover, for any fixed (t, ω), by the Blumenthal zero-one law $Y_t^{t,\omega}$ is deterministic.

Proposition 11.3.4 *Under Assumption 11.3.3, $u(t, \omega) := Y_t^{t,\omega}$ is a viscosity solution of PPDE (11.3.2). Moreover, for the solution (Y, Z) to BSDE (4.0.3), it holds that*

$$Y_t(\omega) = u(t, \omega), \ 0 \le t \le T, \quad for \ \mathbb{P}_0\text{-a.e. } \omega \in \Omega. \quad (11.3.4)$$

Proof First, by Theorem 4.2.3 one can see that $u(t, \cdot)$ is uniformly continuous in ω, uniformly on t. Then, following the arguments in Theorems 10.2.1 and 5.1.3, we can show that

$$|u(t_1, \omega^1) - u(t_2, \omega^2)| \le C[1 + \|\omega^1\|_{t_1} + \|\omega^2\|_{t_2}]\sqrt{|t_1 - t_2|}$$
$$+ C\rho\big(\mathbf{d}((t_1, \omega^1), (t_2, \omega^2))\big), \quad (11.3.5)$$

for all $(t_1, \omega^1), (t_2, \omega^2) \in \overline{\Lambda}$ and for some modulus of continuity function ρ, and u satisfies (11.3.4) as well as the dynamic programming principle: for any $(t, \omega) \in \Lambda$ and $\tau \in \mathscr{T}_{T-t}$,

$$Y_s^{t,\omega} = u^{t,\omega}(\tau, B.) + \int_s^\tau f^{t,\omega}(r, Y_r^{t,\omega}, Z_r^{t,\omega})dr - \int_s^\tau Z_r^{t,\omega} dB_r, \quad \mathbb{P}_0\text{-a.s.} \quad (11.3.6)$$

We leave details to the readers, see Problem 11.7.3.

We now show that u is an L-viscosity solution for any $L \ge L_0$. Without loss of generality, we verify only the viscosity supersolution property at $(t, \omega) = (0, 0)$. Assume to the contrary that

$$c := \partial_t \varphi_0 + \frac{1}{2}\text{tr}(\partial_{\omega\omega}^2 \varphi_0) + f(\cdot, u, \partial_\omega \varphi)(0, 0) > 0,$$

for some $\varphi \in \overline{\mathscr{A}}^L u(0, 0)$ with localization time $\mathbf{H} := \mathbf{H}_\varepsilon$. By the regularities of φ, u, and f, we may assume ε is small enough such that

$$\left[\partial_t\varphi + \frac{1}{2}\mathrm{tr}(\partial^2_{\omega\omega}\varphi) + f(\cdot, u, \partial_\omega\varphi)\right](t,\omega) \geq \frac{c}{2} > 0, \quad (t,\omega) \in \overline{\Lambda}_\mathrm{H}.$$

Using the dynamic programming principle (11.3.6) for u and applying the functional Itô formula (9.4.1) on φ, we have:

$$(\varphi - u)_\mathrm{H} = (\varphi - u)_\mathrm{H} - (\varphi - u)_0$$

$$= \int_0^\mathrm{H} [\partial_\omega\varphi_s - Z_s]dB_s + \int_0^\mathrm{H} \left[\partial_t\varphi_s + \frac{1}{2}\mathrm{tr}(\partial^2_{\omega\omega}\varphi_s) + f(s, u_s, Z_s)\right]ds$$

$$\geq \int_0^\mathrm{H} [\partial_\omega\varphi_s - Z_s]dB_s + \int_0^\mathrm{H} \left[\frac{c}{2} + f(s, u_s, Z_s) - f(s, u_s, \partial_\omega\varphi_s)\right]ds$$

$$= \int_0^\mathrm{H} [\partial_\omega\varphi_s - Z_s]dB_s + \int_0^\mathrm{H} \left[\frac{c}{2} - (\partial_\omega\varphi_s - Z_s)\alpha\right]ds$$

$$= \int_0^\mathrm{H} [\partial_\omega\varphi_s - Z_s](dB_s - \alpha_s ds) + \frac{c}{2}\mathrm{H}, \quad \mathbb{P}_0\text{-a.s.}$$

where $|\alpha| \leq L_0 \leq L$. Applying Girsanov Theorem 2.6.4 one sees immediately that there exists $\tilde{\mathbb{P}} \in \mathscr{P}_L$ equivalent to $\mathbb{P}_0$ such that $dB_t - \alpha_t dt$ is a $\tilde{\mathbb{P}}$-Brownian motion. Then the above inequality holds $\tilde{\mathbb{P}}$-a.s., and by the definition of $\overline{\mathscr{A}}^L u$:

$$0 \geq \mathbb{E}^{\tilde{\mathbb{P}}}[(\varphi - u)_\mathrm{H}] \geq \frac{c}{2}\mathbb{E}^{\tilde{\mathbb{P}}}[\mathrm{H}] > 0,$$

which is the required contradiction.   ∎

Remark 11.3.5 Consider the following more general semilinear PPDE:

$$\partial_t u + \frac{1}{2}\sigma\sigma^\top(t,\omega) : \partial^2_{\omega\omega}u + f(t,\omega, u, \partial_\omega u\sigma) = 0, \quad u(T,\omega) = \xi(\omega), \quad (11.3.7)$$

where $\sigma \in UC_b(\overline{\Lambda}, \mathbb{R}^{d\times d})$ is uniformly Lipschitz continuous in ω. Following similar arguments, one can show that $u(t,\omega) := Y_t^{t,\omega}$ is a viscosity solution, where $(X^{t,\omega}, Y^{t,\omega}, Z^{t,\omega})$ is the solution to the following decoupled FBSDE on $[0, T-t]$:

$$\begin{cases} \tilde{X}_s^{t,\omega} = \int_0^s \sigma^{t,\omega}(r, \tilde{X}_\cdot^{t,\omega})dB_r; \\[2mm] Y_s^{t,\omega} = \xi^{t,\omega}(\tilde{X}_\cdot^{t,\omega}) + \int_s^{T-t} f^{t,\omega}(r, \tilde{X}_\cdot^{t,\omega}, Y_r^{t,\omega}, Z_r^{t,\omega})dt - \int_s^{T-t} Z_r^{t,\omega}dB_r; \end{cases} \quad \mathbb{P}_0\text{-a.s.} \quad (11.3.8)$$

This extends the nonlinear Feynman-Kac formula of Theorem 5.5.8 to the path dependent case.   ∎

11.3.3 Path Dependent HJB Equations

Let $\mathbb{K}$ be a Polish space. We now consider the following path dependent HJB equation:

$$\mathscr{L}u(t,\omega) := \partial_t u + G(t,\omega,u,\partial_\omega u, \partial^2_{\omega\omega}u) = 0, \quad u(T,\omega) = \xi(\omega); \ (11.3.9)$$

where $G(t,\omega,y,z,\gamma) := \sup_{k\in\mathbb{K}}\left[\frac{1}{2}\sigma\sigma^\top(t,\omega,k):\gamma + F(t,\omega,y,z\sigma(t,\omega,k),k)\right]$,

where $\sigma \in \mathbb{S}^d$ and F are $\mathbb{F}$-measurable in all variables, and ξ is $\mathscr{F}_T$-measurable. We shall assume

Assumption 11.3.6

(i) σ, $F(t,\omega,0,0,k)$, and ξ are bounded by C_0, and $\sigma > 0$.
(ii) σ is uniformly Lipschitz continuous in ω, and F is uniformly Lipschitz continuous in (y,z), with a Lipschitz constant L_0.
(iii) F and ξ are uniformly continuous in ω with a common modulus of continuity function ρ_0.
(iv) $\sigma(\cdot,k)$, $F(\cdot,y,z,k)$, and $G(\cdot,y,z)$ are continuous in (t,ω) under $\mathbf{d}$ for any (y,z,k).

We first provide a probabilistic representation for the solution of PPDE (11.3.9) in strong formulation. Let $\mathscr{K}$ denote the set of $\mathbb{F}$-measurable $\mathbb{K}$-valued processes on $\overline{\Lambda}$. For any $(t,\omega) \in \overline{\Lambda}$ and $k \in \mathscr{K}$, under Assumption 11.3.6, following the arguments in Chapters 3 and 4 one can easily show that the following decoupled FBSDE on $[0, T - t]$ under $\mathbb{P}_0$ is well posed:

$$\tilde{X}_s = \int_0^s \sigma^{t,\omega}(r,\tilde{X}_\cdot,k_r)dB_r;$$

$$0 \le s \le T - t, \mathbb{P}_0\text{-a.s.} \quad (11.3.10)$$

$$\tilde{Y}_s = \xi^{t,\omega}(\tilde{X}_\cdot) + \int_s^{T-t} F^{t,\omega}(r,\tilde{X}_\cdot,\tilde{Y}_r,\tilde{Z}_r,k_r)dr - \int_s^{T-t}\tilde{Z}_rdB_r;$$

Denote its unique solution as $(\tilde{X}^{t,\omega,k}, \tilde{Y}^{t,\omega,k}, \tilde{Z}^{t,\omega,k})$. We then consider the stochastic control problem in strong formulation:

$$u(t,\omega) := \sup_{k\in\mathscr{K}} \tilde{Y}_0^{t,\omega,k}, \quad (t,\omega) \in \overline{\Lambda}. \quad (11.3.11)$$

The above problem can be rewritten equivalently in weak formulation. Define

$$\mathscr{P}(t,\omega) := \{\mathbb{P}^{t,\omega,k} : k \in \mathscr{K}\} \subset \mathscr{P}_\infty \quad \text{where} \quad (11.3.12)$$

$$\mathbb{P}^{t,\omega,k} := \mathbb{P}_0 \circ (\tilde{X}^{t,\omega,k})^{-1}.$$

Since $\sigma > 0$, as discussed in Remark 9.2.8 (iii) $\tilde{X}^{t,\omega,k}$ and B induce the same $\mathbb{P}_0$-augmented filtration on $[0, T-t]$, and thus there exists $\tilde{k} \in \mathcal{K}$ such that $\tilde{k}(s, \tilde{X}^{t,\omega,k}_\cdot) = k(s, B_\cdot)$, $0 \le s \le T - t$, $\mathbb{P}_0$-a.s. Moreover, for any $\tau \in \mathcal{T}_{T-t}$ and $\eta \in \mathbb{L}^2(\mathcal{F}_\tau, \mathbb{P}^{t,\omega,k})$, let $(\mathscr{Y}^{t,\omega,k}(\tau, \eta), \mathscr{Z}^{t,\omega,k}(\tau, \eta))$ solves the following BSDE on $[0, \tau]$:

$$Y_s = \eta + \int_s^\tau F^{t,\omega}(r, X_\cdot, Y_r, Z_r \sigma^{t,\omega}(r, \tilde{k}_r), \tilde{k}_r) dr \qquad (11.3.13)$$

$$- \int_s^\tau Z_r dX_r, \quad \mathbb{P}^{t,\omega,k}\text{-a.s.}$$

Then it is clear that, denoting $Y^{t,\omega,k} := \mathscr{Y}^{t,\omega,k}(T - t, \xi^{t,\omega}), Z^{t,\omega,k} := \mathscr{Z}^{t,\omega,k}(T - t, \xi^{t,\omega})$,

$$\tilde{Y}^{t,\omega,k}(s, B_\cdot) = Y^{t,\omega,k}(s, \tilde{X}^{t,\omega,k}_\cdot), \quad \tilde{Z}^{t,\omega,k}(s, B_\cdot) = Z^{t,\omega,k}(s, \tilde{X}^{t,\omega,k}_\cdot)\sigma^{t,\omega}(s, B_\cdot, k_s), \quad \mathbb{P}_0\text{-a.s.}$$

$$(11.3.14)$$

and thus

$$u(t, \omega) = \sup_{k \in \mathcal{K}} Y_0^{t,\omega,k}. \qquad (11.3.15)$$

Lemma 11.3.7 *Under Assumption 11.3.6, the process u in (11.3.11) (or equivalently in (11.3.15)) is in $UC_b(\bar{\Lambda})$ and the following dynamic programming principle holds:*

$$u(t, \omega) = \sup_{k \in \mathcal{K}} \mathscr{Y}_0^{t,\omega,k}(\tau, u^{t,\omega}(\tau, \cdot)), \text{ for any } (t, \omega) \in \Lambda, \tau \in \mathcal{T}_{T-t}. (11.3.16)$$

Proof We first remark that the $\mathscr{P}(t, \omega)$ in (11.3.12) depends on ω (unless σ does not depend on ω), and in particular $\mathscr{P} := \mathscr{P}(0, 0)$ does not satisfy Assumption 9.3.3. However, here the value process $u(t, \omega)$ is defined through controls in strong formulation. As pointed out in Remark 9.3.4, in this case the regularity of u in ω is easy. Indeed, by first extending Theorem 3.2.4 to path dependent case and then applying Theorem 4.2.3, one can easily see that $u(t, \cdot)$ is uniformly continuous in ω, uniformly on t. Moreover, one can easily verify the following properties for $\mathscr{P}(t, \omega)$ which are crucial for establishing the dynamic programming principle for u:

- For any $k \in \mathcal{K}$ and $t \in [0, T]$, $\mathscr{Y}_t^{0,0,k}(T, \xi) \le u(t, \cdot)$, and thus $\mathscr{Y}_0^{0,0,k}(T, \xi) \le \mathscr{Y}_0^{0,0,k}(t, u(t, \cdot))$.
- Let $0 < t \le T$, $\{E_i, i \ge 1\} \subset \mathcal{F}_t$ be a partition of Ω, and $k, k^i \in \mathcal{K}$, $i \ge 1$. Then the following concatenation control $\hat{k}$ is also in $\mathcal{K}$:

$$\hat{k}_s := \mathbf{1}_{[0,t)}(s)k_s + \mathbf{1}_{[t,T]}(s)\sum_{i \ge 1} k^i_{s-t}(B_{t,t+\cdot})\mathbf{1}_{E_i}(\tilde{X}^{0,0,k}_\cdot), \quad s \in [0, T].$$

Then, following similar arguments as in Theorem 10.2.1, together with the a priori estimates in Chapters 3 and 4, we may prove the lemma. Since the arguments are lengthy but rather standard, we leave the details to the interest readers, see Problem 11.7.4.   ∎

The next result shows that our notion of viscosity solution is also suitable for this stochastic control problem.

Proposition 11.3.8 *Under Assumption 11.3.6, the process u in (11.3.11) is a viscosity solution of PPDE (11.3.9).*

Proof We first verify the L-viscosity property for some L large enough. Again we shall only prove it at $(t, \omega) = (0, 0)$. We shall assume $d = 1$ and will omit the superscript ^{0,0}. However, since in this case u is defined through a supremum, we need to prove the viscosity subsolution property and supersolution property differently. By Proposition 11.2.13, without loss of generality we assume

$$G, \text{ hence } F, \text{ is increasing in } y. \tag{11.3.17}$$

Viscosity L-subsolution property. Assume to the contrary that,

$$-c := \big[\partial_t\varphi + G(\cdot, u, \partial_\omega\varphi, \partial^2_{\omega\omega}\varphi)\big](0,0) < 0 \text{ for some } \varphi \in \underline{\mathscr{A}}^L u(0,0) \text{ with}$$

$$\text{localization time } H_\varepsilon.$$

By the regularities of φ, u and G in (t, ω) under **d**, we may assume $\varepsilon > 0$ is small enough such that

$$\big[\partial_t\varphi + G(\cdot, u, \partial_\omega\varphi, \partial^2_{\omega\omega}\varphi)\big](t,\omega) \le -\frac{c}{2} < 0, \quad (t,\omega) \in \Lambda_{H_\varepsilon}.$$

By the definition of G, this implies that, for any $(t,\omega) \in \Lambda_{H_\varepsilon}$ and $k \in \mathbb{K}$,

$$\Big[\partial_t\varphi + \frac{1}{2}\sigma^2(t,\omega,k)\partial^2_{\omega\omega}\varphi + F(t,\omega,u,\sigma(\cdot,k)\partial_\omega\varphi,k)\Big](t,\omega) \le -\frac{c}{2} < 0.$$

Now for any $k \in \mathscr{K}$, denote $(Y^k, Z^k) := (\mathscr{Y}^k(H_\varepsilon, u(H_\varepsilon, \cdot)), \mathscr{Z}^k(H_\varepsilon, u(H_\varepsilon, \cdot)))$. One can easily see that $u \ge Y^k$, $\mathbb{P}^k$-a.s. on Λ_{H_ε}. Notice that $d\langle X\rangle_t = \sigma^2(t, X, \tilde{k}_t)dt$, $\mathbb{P}^k$-a.s. For any $\delta > 0$, denoting $H^\delta_\varepsilon := H_\varepsilon \wedge \delta$ and applying functional Itô formula on φ, we see that:

$$(\varphi - Y^k)_0 - (\varphi - u)_{H^\delta_\varepsilon} \ge (\varphi - Y^k)_0 - (\varphi - Y^k)_{H^\delta_\varepsilon}$$

$$= -\int_0^{H^\delta_\varepsilon} \Big[\partial_t\varphi + \frac{1}{2}\sigma^2\partial^2_{\omega\omega}\varphi + F(\cdot, Y^k, Z^k\sigma)\Big](s, X, \tilde{k}_s)ds$$

$$- \int_0^{H^\delta_\varepsilon} (\partial_\omega\varphi - Z^k)(s, X, \tilde{k}_s)dX_s$$

$$\geq \int_0^{H_\varepsilon^\delta} \left[\frac{c}{2} + F(\cdot, u, \sigma\partial_\omega\varphi) - F(\cdot, Y^k, \sigma Z^k)\right](s, X., \tilde{k}_s)ds$$

$$-\int_0^{H_\varepsilon^\delta} (\partial_\omega\varphi - Z^k)(s, X., \tilde{k}_s)dX_s, \quad \mathbb{P}^k\text{-a.s.}$$

Note again that $Y_s^k \leq u(s, B.)$. Then by (11.3.17) we have

$$(u - Y^k)_0 - (\varphi - u)_{H_\varepsilon^\delta}$$

$$\geq \int_0^{H_\varepsilon^\delta} \left[\frac{c}{2} + F(\cdot, u, \sigma\partial_\omega\varphi) - F(\cdot, u, \sigma Z^k)\right](s, X., \tilde{k}_s)ds$$

$$-\int_0^{H_\varepsilon^\delta} (\partial_\omega\varphi - Z^k)(s, X., \tilde{k}_s)dX_s$$

$$= \int_0^{H_\varepsilon^\delta} \left[\frac{c}{2} + \sigma(\partial_\omega\varphi - Z^k)\alpha\right](s, X., \tilde{k}_s)ds - \int_0^{H_\varepsilon^\delta} (\partial_\omega\varphi - \mathscr{Z}^k)(s, X., \tilde{k}_s)dX_s$$

$$= \frac{c}{2}(H_\varepsilon^\delta) - \int_0^{H_\varepsilon^\delta} (\partial_\omega\varphi - Z^k)\sigma(s, X., \tilde{k}_s)[\sigma^{-1}(s, X., \tilde{k}_s)dX_s - \alpha_s ds], \quad \mathbb{P}^k\text{-a.s.}$$

where $|\alpha| \leq L_0 \leq L$ and λ is bounded. Recall (9.2.9) that $\sigma^{-1}(t, X., \tilde{k}_t)dX_t$ is a $\mathbb{P}^k$-Brownian motion. Applying Girsanov Theorem 2.6.4 we may define $\tilde{\mathbb{P}}^k \in \mathscr{P}_L$ equivalent to $\mathbb{P}$ such that $\sigma^{-1}(t, X., \tilde{k}_t)dX_t - \alpha_t dt$ is a $\tilde{\mathbb{P}}^k$-Brownian motion. Then the above inequality holds $\tilde{\mathbb{P}}^k$-a.s., and by the definition of $\mathscr{A}^L u$, we have

$$u_0 - Y_0^k \geq u_0 - Y_0^k - \mathbb{E}^{\tilde{\mathbb{P}}^k}\left[(\varphi - u)_{H_\varepsilon^\delta}\right] \geq \frac{c}{2}\mathbb{E}^{\tilde{\mathbb{P}}^k}[H_\varepsilon^\delta] \geq \frac{c}{2}\delta\left[1 - \tilde{\mathbb{P}}^k[H_\varepsilon \leq \delta]\right].$$

By Lemma 10.1.18 (i), for δ small enough we have

$$u_0 - Y_0^k \geq \frac{c}{2}\delta\left[1 - C\varepsilon^{-2}\delta\right] \geq \frac{c\delta}{4} > 0.$$

This implies that $u_0 - \sup_{k\in\mathscr{K}} Y_0^k \geq \frac{c\delta}{4} > 0$, which is in contradiction with (11.3.16).

Viscosity L-supersolution property. Assume to the contrary that,

$$c := \left[\partial_t\varphi + G(\cdot, u, \partial_\omega\varphi, \partial_{\omega\omega}^2\varphi)\right](0, 0) > 0 \text{ for some } \varphi \in \overline{\mathscr{A}}^L u(0, 0) \text{ with}$$

localization time H_ε.

By the definition of F, there exists $k_0 \in \mathbb{K}$ such that

$$\left[\partial_t\varphi + \frac{1}{2}\sigma^2(\cdot, k_0)\partial_{\omega\omega}^2\varphi + F(\cdot, u, \sigma(\cdot, k_0)\partial_\omega\varphi, k_0)\right](0, 0) \geq \frac{c}{2} > 0$$

Again, by assuming $\varepsilon > 0$ is small enough we may assume

$$\left[\partial_t\varphi+\frac{1}{2}\sigma^2(\cdot,k_0)\partial^2_{\omega\omega}\varphi+F(\cdot,u,\sigma(\cdot,k_0)\partial_\omega\varphi,k_0)\right](t,\omega) \geq \frac{c}{3} > 0, \quad (t,\omega) \in \Lambda_{H_\varepsilon}.$$

Consider the constant process $k := k_0 \in \mathcal{K}$. It is clear that the corresponding $\tilde{k} = k_0$. Follow similar arguments as in the subsolution property, we arrive at the following contradiction:

$$u_0 - \mathcal{Y}_0^k \leq -\frac{c}{3}\mathbb{E}^{\tilde{\mathbb{P}}^k}[\mathrm{H}] < 0. \qquad (11.3.18)$$

The proof is complete now.   ∎

Remark 11.3.9 Consider a special case: $F = 0$, $\sigma = k$, and $\mathbb{K} := \{k \in \mathbb{S}^d : \underline{\sigma} \leq k \leq \overline{\sigma}\}$ for some $0 < \underline{\sigma} < \overline{\sigma}$. Then $u(t,\omega) = \mathcal{E}_{t,\omega}^{\mathscr{P}}[\xi]$ where $\mathscr{P} = \mathscr{P}_{[\underline{\sigma},\overline{\sigma}]}^S$, and the PPDE (11.3.9) becomes the path dependent version of the G-heat equation, also called the Barenblatt equation:

$$\partial_t u + G(\partial^2_{\omega\omega}u) = 0 \quad \text{where} \quad G(\gamma) := \frac{1}{2}\sup_{\underline{\sigma}\leq k\leq\overline{\sigma}} [k^2 : \gamma]. \qquad (11.3.19)$$

In this case, we have another natural representation of the solution: $\tilde{u}(t,\omega) := \mathcal{E}_{t,\omega}^{\tilde{\mathscr{P}}}[\xi]$, where $\tilde{\mathscr{P}} := \mathscr{P}_{[\underline{\sigma},\overline{\sigma}]}^W$ satisfies Assumption 9.3.3. By Theorem 10.2.1, $\tilde{u} \in UC_b(\overline{\Lambda})$ satisfies the dynamic programming principle. Then it follows almost the same arguments as in Theorem 11.3.8 that $\tilde{u}$ is also a viscosity solution to PPDE (11.3.19). Then, by using the uniqueness result which will be established in Subsection 11.4.3 below, we see that $\tilde{u} = u$. Namely the stochastic control problems in strong formulation and in weak formulation induce the same value process. This is consistent with the results in Markov framework, as we discussed in Subsection 9.1.2. However, we should mention again that this equality is due to the uniqueness of the PPDE. We still do not have a direct argument to show that they are equal. In particular, when the viscosity theory fails, for example when ξ is discontinuous, it might be possible that $\tilde{u} \neq u$.   ∎

Remark 11.3.10 Consider the general PPDE (11.3.9) again. We have several different choices for the corresponding control problem. Recall that an open loop control is $\mathbb{F}^B$-measurable.

- Use piecewise constant controls k, and denote the corresponding value as $u^{pc}(t,\omega)$;
- Use general open loop controls and denote the corresponding value as $u^{open}(t,\omega)$. This is exactly the strong formulation and thus u^S is the u in (11.3.11);
- Use close loop controls, namely k is $\mathbb{F}^X$-measurable, and denote the corresponding value as $u^{close}(t,\omega)$. More precisely, for each (t,ω), let $\mathscr{A}(t,\omega)$ denote the set of $(k,\mathbb{P})$ such that k is $\mathbb{F}^X$-measurable and $\mathbb{K}$-valued and $\mathbb{P}$ is a weak solution of the SDE:

$$\tilde{X}_s = \int_0^s \sigma^{t,\omega}(r, \tilde{X}_., k_r(\tilde{X}_.))dB_r, \ \ 0 \le s \le T - t, \ \mathbb{P}_0\text{-a.s.}$$

Then $u^{close}(t, \omega) := \sup_{(k,\mathbb{P}) \in \mathscr{A}(t,\omega)} Y_0^{t,\omega,k,\mathbb{P}}$, where $Y^{t,\omega,k,\mathbb{P}}$ is the solution to the following BSDE:

$$Y_s = \xi^{t,\omega} + \int_s^{T-t} F^{t,\omega}(r, X_., Y_r, Z_r \sigma^{t,\omega}(r, k_r), k_r)dr$$

$$- \int_s^{T-t} Z_r dX_r, \ 0 \le s \le T - t, \mathbb{P}\text{-a.s.}$$

- Use relaxed controls, and denote the corresponding value as $u^{relax}(t, \omega)$. We refer the details to El Karoui & Tan [83].

  In the spirit of Problem 9.6.10, $u^{pc}(t, \omega)$ is also the value for the optimization problem with piecewise constant close loop controls. Notice that the set of relax controls is larger, and we have the following direct inequalities:

$$u^{pc} \le u^{open}, u^{close} \le u^{relax}.$$

Under mild conditions, one can show that $u^{pc}, u^{open}, u^{relax}$ all satisfy the dynamic programming principle and then are viscosity solutions of PPDE (11.3.9). Assume further the uniqueness of viscosity solutions, which we will investigate in the next section, then $u^{pc} = u^{relax}$, and consequently, $u^{close} = u^{open}$ and is also the viscosity solution of PPDE (11.3.9). However, as mentioned in Remark 11.3.9, we do not have a direct argument for the equality $u^{close} = u^{open}$, and under general conditions, it is not clear how to prove directly the dynamic programming principle for u^{close}. As explained in Section 9.1, u^{close} is important in applications. These subtle issues deserve further investigation. The issue will become even more subtle for games, as we already saw in Subsection 9.1.3. ∎

11.3.4   Path Dependent Isaacs Equations

We now extend the path dependent HJB equation (11.3.9) to the following path dependent Isaacs equation with two controls:

$$\mathscr{L}u(t, \omega) : = \partial_t u + G(t, \omega, u, \partial_\omega u, \partial^2_{\omega\omega}u) = 0, \quad u(T, \omega) = \xi(\omega) \tag{11.3.20}$$

where $G(t, \omega, y, z, \gamma) := \sup_{k_1 \in \mathbb{K}_1} \inf_{k_2 \in \mathbb{K}_2}$

$$\left[\frac{1}{2}\sigma^2(t, \omega, k_1, k_2) : \gamma + F(t, \omega, y, z\sigma(t, \omega, k_1, k_2), k_1, k_2) \right],$$

$\mathbb{K}_1$, $\mathbb{K}_2$ are Polish spaces, $\sigma \in \mathbb{S}^d$ and F are $\mathbb{F}$-measurable in all variables, and ξ is $\mathscr{F}_T$-measurable. Denote $\mathbb{K} := \mathbb{K}_1 \times \mathbb{K}_2$ and consider $k = (k_1, k_2) \in \mathbb{K}$. We shall assume

Assumption 11.3.11

(i) σ, $F(t, \omega, 0, 0, k)$, and ξ are bounded by C_0, and $\sigma > 0$.

(ii) σ is uniformly Lipschitz continuous in ω, and F is uniformly Lipschitz continuous in (y, z), with a Lipschitz constant L_0.

(iii) σ, F and ξ are uniformly continuous in (t, ω) with a common modulus of continuity function ρ_0.

We note that for path dependent HJB equation, typically the representations under different formulations induce the same value, as we already saw in Remarks 11.3.9 and 11.3.10. For Isaacs equation, however, the associated games under strong formulation and under weak formulation are fundamentally different, as we explained in Subsection 9.1.3. We may use the game with strategy versus control under strong formulation, in the spirit of Remark 9.1.5, to provide a viscosity solution to PPDE (11.3.20), see Ekren & Zhang [77]. However, since the weak formulation (or precisely, close loop controls) makes more sense from game point of view, as discussed in Subsection 9.1.3, here we shall use the weak formulation.

The problem is actually more subtle. To ensure the regularity of the value process, we first assume

$$\sigma = \sigma(t, k) \text{ does not depend on } \omega, \text{ and denote } \sigma^t(s, k) := \sigma(t + s, k). \quad (11.3.21)$$

Next, for $i = 1, 2$, let $\mathscr{K}^i$ denote the set of $\mathbb{F}$-measurable $\mathbb{K}_i$-valued processes k^i, and denote $\mathscr{K} := \mathscr{K}^1 \times \mathscr{K}^2$ and $k := (k^1, k^2)$. Given $(t, \omega) \in \overline{\Lambda}$ and $k = (k^1, k^2) \in \mathscr{K}$, we shall consider the following SDE on $[0, T - t]$:

$$\tilde{X}_s = \int_0^s \sigma^t(r, \tilde{X}_\cdot, k_r(\tilde{X}_\cdot)) dB_r, \quad 0 \le s \le T - t, \quad \mathbb{P}_0\text{-a.s.} \quad (11.3.22)$$

We emphasize that the k here corresponds to the $\tilde{k}$ in the previous subsection. However, for an arbitrary k, the SDE (11.3.22) may not be well posed. Therefore, we shall impose restrictions on k. Let $\mathscr{K}_0 = (\mathscr{K}_0^1, \mathscr{K}_0^2)$ denote the set of $k = (k^1, k^2) \in \mathscr{K}$ taking the following form:

$$k_s(\omega) = \sum_{i=0}^{n-1} \sum_{j=1}^{n_i} k_{i,j} \mathbf{1}_{[t_i, t_{i+1})}(s) \mathbf{1}_{E_j^i}(\omega) \quad \text{where } 0 = t_0 < \cdots < t_n = T, \quad (11.3.23)$$

$\{E_j^i, 1 \le j \le n_i\} \subset \mathscr{F}_{t_i}$ is a partition of Ω, and $k_{i,j} \in \mathbb{K}$ are constants.

Under Assumption 11.3.11, for any $t \in [0, T]$ and any $k \in \mathscr{K}_0$, by Problem 9.6.10 the SDE (11.3.22) has a unique strong solution, denoted as $X^{t,k}$. We then define $\mathbb{P}^{t,k} := \mathbb{P}_0 \circ (X^{t,k})^{-1}$. We emphasize that $\mathbb{P}^{t,k}$ does not depend on ω, thanks to the assumption (11.3.21). As in the previous subsection, for any $\tau \in \mathscr{T}_{T-t}$ and $\eta \in \mathbb{L}^2(\mathscr{F}_\tau, \mathbb{P}^{t,k})$, let $(\mathscr{Y}^{t,\omega,k}(\tau, \eta), \mathscr{Z}^{t,\omega,k}(\tau, \eta))$ solve the following BSDE on $[0, \tau]$:

$$Y_s = \eta + \int_s^\tau F^{t,\omega}(r, X., Y_r, Z_r\sigma^t(r, X., k_r), k_r)dr - \int_s^\tau Z_r dX_r, \quad \mathbb{P}^{t,k}\text{-a.s.} \quad (11.3.24)$$

We now define:

$$u(t, \omega) := \sup_{k^1 \in \mathscr{K}_0^1} \inf_{k^2 \in \mathscr{K}_0^2} \mathscr{Y}_0^{t,\omega,k^1,k^2}(T - t, \xi^{t,\omega}), \quad (t, \omega) \in \overline{\Lambda}. \quad (11.3.25)$$

We first have

Lemma 11.3.12 *Let Assumption 11.3.11 and (11.3.21) hold. Then $u \in UC_b(\overline{\Lambda})$ and it satisfies the dynamic programming principle: for any $(t, \omega) \in \Lambda$ and $\tau \in \mathscr{T}_{T-t}$,*

$$u(t, \omega) = \sup_{k^1 \in \mathscr{K}_0^1} \inf_{k^2 \in \mathscr{K}_0^2} \mathscr{Y}_0^{t,\omega,k^1,k^2}(\tau, u^{t,\omega}(\tau, \cdot)). \quad (11.3.26)$$

Proof While the main ideas are still similar to those of Theorem 10.2.1, this proof is more involved. We will provide some arguments here. The remaining arguments are somewhat standard and left to interested readers in Problem 11.7.5. We proceed in several steps.

Step 1. We shall show in Problem 11.7.5 that u is uniformly continuous in ω with a modulus of continuity function $\rho_1 \geq \rho_0$. We note that the assumption (11.3.21) is crucial for this step.

Step 2. For any $\varepsilon > 0$ and $t \in (0, T]$, following similar arguments as in Lemma 9.2.1 we shall show in Problem 11.7.5 that there exist a finite partition $\{E_i, 0 \leq i \leq n\} \subset \mathscr{F}_t$ of Ω such that

- $\|\omega - \tilde{\omega}\|_t \leq \varepsilon$ for all $\omega, \tilde{\omega} \in E_i, i = 1, \cdots, n$;
- $\mathscr{C}^{\mathscr{P}_L}[E_0] \leq \varepsilon$.

Step 3. We shall show in Problem 11.7.5 the following partial dynamic programming principle:

$$u(t, \omega) \geq \sup_{k^1 \in \mathscr{K}_0^1} \inf_{k^2 \in \mathscr{K}_0^2} \mathscr{Y}_0^{t,\omega,k^1,k^2}(h, u^{t,\omega}(h, \cdot)), \quad \text{for all } 0 \leq t < t + h \leq T, \omega \in \Omega.$$

$$(11.3.27)$$

Step 4. We shall provide detailed arguments here for the opposite direction of partial dynamic programming principle:

$$u(t, \omega) \leq \sup_{k^1 \in \mathscr{K}_0^1} \inf_{k^2 \in \mathscr{K}_0^2} \mathscr{Y}_0^{t,\omega,k^1,k^2}(h, u^{t,\omega}(h, \cdot)), \quad \text{for all } 0 \leq t < t + h \leq T, \omega \in \Omega.$$

$$(11.3.28)$$

Step 5. Combining Steps 3 and 4, we obtain the dynamic programming principle for deterministic times. This, together with Step 1, implies that $u \in UC_b(\overline{\Lambda})$. The details are left to Problem 11.7.5.

Step 6. Prove the full dynamic programming principle (11.3.26) in Problem 11.7.5.

We now prove (11.3.28), which explains why we require k to take only finitely many values (instead of general piecewise constant processes as in (9.2.14)). Without loss of generality, we shall only prove: denoting $\mathscr{Y}^k := \mathscr{Y}^{0,0,k}$ and for any $0 < t < T$,

$$u_0 \leq \sup_{k^1 \in \mathscr{K}_0^1} \inf_{k^2 \in \mathscr{K}_0^2} \mathscr{Y}_0^{k^1,k^2}(t, u(t, \cdot)). \tag{11.3.29}$$

Clearly, it suffices to prove: for any $k^1 = \sum_{i=0}^{n-1} \sum_{j=1}^{n_i} k_{i,j}^1 \mathbf{1}_{[t_i,t_{i+1})} \mathbf{1}_{E_j^i} \in \mathscr{K}_0^1$ as in (11.3.23),

$$\inf_{k^2 \in \mathscr{K}_0^2} \mathscr{Y}_0^{k^1,k^2}(T, \xi) \leq \inf_{k^2 \in \mathscr{K}_0^2} \mathscr{Y}_0^{k^1,k^2}(t, u(t, \cdot)). \tag{11.3.30}$$

By otherwise adding t into the partition point of k^1, we assume t is one of those t_i. Note that $\xi = u(t_n, \cdot)$. Then (11.3.30) is a direct consequence of the following inequalities:

$$\inf_{k^2 \in \mathscr{K}_0^2} \mathscr{Y}_0^{k^1,k^2}(t_{i+1}, u(t_{i+1}, \cdot)) \leq \inf_{k^2 \in \mathscr{K}_0^2} \mathscr{Y}_0^{k^1,k^2}(t_i, u(t_i, \cdot)), \quad i = 0, \cdots, n. \tag{11.3.31}$$

To see (11.3.31), let us fix arbitrary i, $k^2 \in \mathscr{K}_0^2$, $\varepsilon > 0$, and denote $h := t_{i+1} - t_i$. Note that $(k^1)^{t_i,\omega} \in \mathscr{K}_0^1$. Then by (11.3.27) we have

$$u(t_i, \omega) \geq \inf_{\tilde{k}^2 \in \mathscr{K}_0^2} \mathscr{Y}_0^{t_i,\omega,(k^1)^{t_i,\omega},\tilde{k}^2}(h, u^{t_i,\omega}(h, \cdot)) \tag{11.3.32}$$

Let $\{E_l, 0 \leq l \leq m\} \subset \mathscr{F}_{t_i}$ be as in Step 2. Denote $E_j^l := E_l \cap E_{i,j}$ and fix an arbitrary $\omega^{l,j} \in E_j^l$, for $j = 1, \cdots, n_i$ and $l = 1, \cdots, m$. For each ω_j^l, there exists $k^{2,l,j} \in \mathscr{K}_0^2$ such that

$$u(t_i, \omega_j^l) \geq \mathscr{Y}_0^{t,\omega_j^l,k_{i,j}^1,k^{2,l,j}}(h, u^{t_i,\omega_j^l}(h, \cdot)) - \varepsilon, \tag{11.3.33}$$

where $k_{i,j}^1$ is understood as a constant process. By Theorem 4.2.3, we have

$$|u(t_i, \omega_j^l) - u(t_i, \omega)| \leq \rho_1(\varepsilon),$$

$$\left| \mathscr{Y}_0^{t,\omega_j^l,k_{i,j}^1,k^{2,l,j}}(h, u^{t_i,\omega_j^l}(h, \cdot)) - \mathscr{Y}_0^{t,\omega,k_{i,j}^1,k^{2,l,j}}(h, u^{t_i,\omega}(h, \cdot)) \right| \leq C\rho_1(\varepsilon), \qquad \text{for all } \omega \in E_j^l.$$

Then it follows from (11.3.33) that

$$\mathscr{Y}_0^{t,\omega,k_{i,j}^1,k^{2,l,j}} \left(h, u^{t_i,\omega}(h,\cdot)\right) \le u(t_i,\omega) + \varepsilon + C\rho_1(\varepsilon), \quad \text{for all } \omega \in E_j^l. \quad (11.3.34)$$

For any $k^2 \in \mathscr{K}_0^2$, denote

$$\hat{k}_s^2 := \mathbf{1}_{[0,t_i)}(s)k_s^2 + \mathbf{1}_{[t_i,T]}(s)\left[\sum_{l=1}^m \sum_{j=1}^{n_i} k_{s-i}^{2,l,j}(X_{t,t+\cdot})\mathbf{1}_{E_j^l} + k_s^2\mathbf{1}_{E_0}\right].$$

One can easily check that $\hat{k}^2 \in \mathscr{K}_0^2$, thanks to the fact that each $k^{2,l,j}$ takes only finitely many values. Then (11.3.34) leads to

$$\mathscr{Y}_0^{t,\omega,(k^1)^{t_i,\omega},(\hat{k}^2)^{t,\omega}} \left(h, u^{t_i,\omega}(h,\cdot)\right) \le u(t_i,\omega) + \varepsilon + C\rho_1(\varepsilon), \quad \text{for all } \omega \notin E_0,$$

and thus

$$\mathscr{Y}_0^{t,\omega,(k^1)^{t_i,\omega},(\hat{k}^2)^{t,\omega}} \left(h, u^{t_i,\omega}(h,\cdot)\right) \le u(t_i,\omega) + \varepsilon + C\rho_1(\varepsilon) + C\mathbf{1}_{E_0}(\omega), \text{ for all } \omega \in \Omega.$$

Applying the comparison Theorem 4.4.1 and Theorem 4.2.3 again, we obtain

$$\mathscr{Y}_0^{k^1,\hat{k}^2}\left(t_{i+1}, u(t_{i+1},\cdot)\right) = \mathscr{Y}_0^{k^1,k^2}\left(t_i, \mathscr{Y}_0^{t,\omega,(k^1)^{t_i,\omega},(\hat{k}^2)^{t,\omega}}\left(h, u^{t_i,\omega}(h,\cdot)\right)\right)$$

$$\le \mathscr{Y}_0^{k^1,k^2}\left(t_i, u(t_i,\cdot) + \varepsilon + C\rho_1(\varepsilon) + C\mathbf{1}_{E_0}\right)$$

$$\le \mathscr{Y}_0^{k^1,k^2}\left(t_i, u(t_i,\cdot)\right) + C\left(\mathbb{E}^{\mathbb{P}^{0,k^1,k^2}}\left[(\varepsilon + C\rho_1(\varepsilon) + C\mathbf{1}_{E_0})^2\right]\right)^{\frac{1}{2}}$$

$$\le \mathscr{Y}_0^{k^1,k^2}\left(t_i, u(t_i,\cdot)\right) + C[\varepsilon + \rho_1(\varepsilon)] + C\left(\mathscr{C}^{\mathbb{P}_L}[E_0]\right)^{\frac{1}{2}}$$

$$\le \mathscr{Y}_0^{k^1,k^2}\left(t_i, u(t_i,\cdot)\right) + C[\varepsilon + \rho_1(\varepsilon) + \sqrt{\varepsilon}].$$

By the arbitrariness of k^2 and ε, this implies (11.3.31) immediately. ∎
   Our main result of this subsection is:

Proposition 11.3.13 *Let Assumption 11.3.11 and (11.3.21) hold. Then the process u in (11.3.25) is a viscosity solution of PPDE (11.3.20).*

Proof By Proposition 11.2.13, again we may assume (11.3.17) holds true. We shall only verify the L-viscosity property at $(0,0)$ for some L large enough, and again we assume $d = 1$.
   Viscosity L-supersolution property. Assume by contradiction that there exists $\varphi \in \overline{\mathscr{A}}^L u(0,0)$ with localization time H_ε such that: recalling $k = (k_1, k_2)$,

$$c := \partial_t \varphi_0 + \sup_{k_1 \in \mathbb{K}_1} \inf_{k_2 \in \mathbb{K}_2} \left[\frac{1}{2} \sigma_0^2(k) \partial_{\omega\omega}^2 \varphi_0 + F_0(u_0, \partial_\omega \varphi_0 \sigma_0(k), k) \right] > 0.$$

Then there exists $k_1^* \in \mathbb{K}_1$ such that, for all $k_2 \in \mathbb{K}_2$

$$\partial_t \varphi_0 + \frac{1}{2} \sigma_0^2(k_1^*, k_2) : \partial_{\omega\omega}^2 \varphi_0 + F_0(u_0, \partial_\omega \varphi_0 \sigma_0(k_1^*, k_2), k_1^*, k_2) \geq \frac{c}{2}.$$

By the uniform regularity of σ and F, we may assume without loss of generality that $\varepsilon > 0$ is small enough such that, for any $k_2 \in \mathbb{K}_2$ and $0 \leq t \leq H_\varepsilon$,

$$|(\varphi - u)_t| \leq \varepsilon, \quad \partial_t \varphi_t + \frac{1}{2} \sigma_t^2(k_1^*, k_2) \partial_{\omega\omega}^2 \varphi_t + F_t(\varphi_t, \partial_\omega \varphi_t \sigma_t(k_1^*, k_2), k_1^*, k_2) \geq \frac{c}{3}.$$

$$(11.3.35)$$

Now fix $\delta > 0$ and denote $H_\varepsilon^\delta := H_\varepsilon \wedge \delta$. Let $k_1^* \in \mathcal{K}_0^1$ be the constant process and $k^2 \in \mathcal{K}_0^2$ be arbitrary. Denote $\mathbb{P} := \mathbb{P}^{0,k_1^*,k^2}$ and

$$Y := \mathscr{Y}^{0,0,k_1^*,k^2}\left[H_\varepsilon^\delta, u_{H_\varepsilon^\delta}\right], \quad Z := \mathscr{Z}^{0,0,k_1^*,k^2}\left[H_\varepsilon^\delta, u_{H_\varepsilon^\delta}\right], \quad \Delta Y_t := \varphi(t, X) - Y_t,$$

$$\Delta Z_t := \partial_\omega \varphi(t, X) - Z_t.$$

Then, applying the functional Itô formula we obtain:

$$d\Delta Y_t = \left[\partial_t \varphi_t + \frac{1}{2} \partial_{\omega\omega}^2 \varphi_t \sigma_t^2(k_1^*, k_t^2) + F_t(Y_t, Z_t \sigma_t, k_1^*, k_t^2) \right] dt + \Delta Z_t dX_t$$

$$\geq \left[\frac{c}{3} - F_t(\varphi_t, \partial_\omega \varphi_t \sigma_t(k_1^*, k_2), k_1^*, k_2) + F_t(Y_t, Z_t \sigma_t(k_1^*, k_2), k_1^*, k_t^2) \right] dt + \Delta Z_t dX_t$$

$$= \left[\frac{c}{3} - \alpha_t \Delta Y_t - \Delta Z_t \sigma_t(k_1^*, k_2) \beta_t \right] dt + \Delta Z_t dX_t, \quad 0 \leq t \leq H_\varepsilon^\delta, \ \mathbb{P}\text{-a.s.}$$

where $|\alpha|, |\beta| \leq L_0$. Define

$$\Gamma_t := \exp\left(\int_0^t \alpha_s ds \right), \quad M_t := \exp\left(\int_0^t \beta_s \sigma_s^{-1}(k_1^*, k_s^2) dX_s - \frac{1}{2} \int_0^t \beta_s|^2 ds \right),$$

$$d\overline{\mathbb{P}} := M_{H_\varepsilon^\delta} d\mathbb{P}.$$

Then, for L large, $\overline{\mathbb{P}} \in \mathscr{P}_L$ and

$$\Delta Y_0 \leq \mathbb{E}^{\overline{\mathbb{P}}} \left[\Gamma_{H_\varepsilon^\delta} \Delta Y_{H_\varepsilon^\delta} - \frac{c}{3} \int_0^{H_\varepsilon^\delta} \Gamma_t dt \right]$$

$$= \mathbb{E}^{\overline{\mathbb{P}}} \left[\Delta Y_{H_\varepsilon^\delta} - \frac{c}{3} H_\varepsilon^\delta + [\Gamma_{H_\varepsilon^\delta} - 1] \Delta Y_{H_\varepsilon^\delta} - \frac{c}{3} \int_0^{H_\varepsilon^\delta} [\Gamma_t - 1] dt \right]$$

It is clear that $|\Gamma_t - 1| \le Ct$, and it follows from the first inequality of (11.3.35) that $|\Delta Y_{H_\varepsilon^\delta}| = |(\varphi - u)_{H_\varepsilon^\delta}| \le \varepsilon$. Since $\varphi \in \overline{\mathscr{A}}^L u(0,0)$, we have $\mathbb{E}^{\overline{\mathbb{P}}}[\Delta Y_{H_\varepsilon^\delta}] \le 0$. Then, for $\varepsilon > 0$ small enough,

$$u_0 - Y_0 = \varphi_0 - Y_0 = \Delta Y_0 \le (C\varepsilon - \frac{c}{3})\mathbb{E}^{\overline{\mathbb{P}}}[H_\varepsilon^\delta] \le -\frac{c}{4}\mathbb{E}^{\overline{\mathbb{P}}}[H_\varepsilon^\delta].$$

Moreover, by Lemma 10.1.18 (ii),

$$\mathbb{E}^{\overline{\mathbb{P}}}[H_\varepsilon^\delta] \ge \delta - \delta\overline{\mathbb{P}}(H_\varepsilon \le \delta) \ge \delta - \frac{C\delta^2}{\varepsilon^2} = \frac{\delta}{2}, \quad \text{for } \delta := \frac{\varepsilon^2}{2C}.$$

Then

$$u_0 - \mathscr{Y}_0^{0,0,k_1^*,k^2}(H_\varepsilon^\delta, u_{H_\varepsilon^\delta}) \le -\frac{c\delta}{8}, \quad \text{for all } k^2 \in \mathscr{K}_0^2,$$

and thus

$$u_0 \le \inf_{k^2 \in \mathscr{K}_0^2} \mathscr{Y}_0^{0,0,k_1^*,k^2}(H_\varepsilon^\delta, u_{H_\varepsilon^\delta}) - \frac{c\delta}{8} \le \sup_{k^1 \in \mathscr{K}_0^1} \inf_{k^2 \in \mathscr{K}_0^2} \mathscr{Y}_0^{0,0,k^1,k^2}(H_\varepsilon^\delta, u_{H_\varepsilon^\delta}) - \frac{c\delta}{8}.$$

This contradicts with the dynamic programming principle (11.3.26).

Viscosity L-subsolution property. Assume by contradiction that there exists $\varphi \in \underline{\mathscr{A}}^L u_0$ with localization time H_ε such that

$$-c := \partial_t \varphi_0 + \sup_{k_1 \in \mathbb{K}_1} \inf_{k_2 \in \mathbb{K}_2} \left[\frac{1}{2}\sigma_0^2(k)\partial_{\omega\omega}^2\varphi_0 + F_0(u_0, \partial_\omega\varphi_0\sigma_0(k), k)\right] < 0.$$

Then there exists a mapping (no measurability is involved!) $\psi : \mathbb{K}_1 \to \mathbb{K}_2$ such that, for any $k_1 \in \mathbb{K}_1$,

$$\partial_t \varphi_0 + \frac{1}{2}\sigma_0^2(k_1, \psi(k_1))\partial_{\omega\omega}^2\varphi_0 + F_0(u_0, \partial_\omega\varphi_0\sigma_0(k_1, \psi(k_1)), k_1, \psi(k_1))$$

$$\le -\frac{c}{2}. \tag{11.3.36}$$

For any $k^1 \in \mathscr{K}_0^1$, define $k^2 := \psi(k_t^1)$. By the special structure in (11.3.23) one can easily see that $k^2 \in \mathscr{K}_0^2$. Introduce the same notations as in the previous proof for supersolution property, by replacing (k_1^*, k^2) there with (k^1, k^2), and follow almost the same arguments, we obtain

$$u_0 - \mathscr{Y}_0^{0,0,k^1,\psi(k^1)}(H_\varepsilon^\delta, u_{H_\varepsilon^\delta}) \ge \frac{c\delta}{8}, \quad \forall k^1 \in \mathscr{K}_0^1, \tag{11.3.37}$$

for $\varepsilon > 0$ small enough and $\delta := \frac{\varepsilon^2}{C}$ for some large constant C. This implies that

$$u_0 \geq \sup_{k^1 \in \mathcal{K}_0^1} \mathcal{Y}_0^{0,0,k^1,\psi(k^1)}(\mathrm{H}_\varepsilon^\delta, u_{\mathrm{H}_\varepsilon^\delta}) + \frac{c\delta}{8} \geq \sup_{k^1 \in \mathcal{K}_0^1} \inf_{k^2 \in \mathcal{K}_0^2} \mathcal{Y}_0^{0,0,k^1,k^2}(\mathrm{H}_\varepsilon^\delta, u_{\mathrm{H}_\varepsilon^\delta}) + \frac{c\delta}{8},$$

again contradiction with the dynamic programming principle (11.3.26). ∎

Remark 11.3.14

(i) Following similar arguments, one can show that

$$\overline{u}(t,\omega) := \inf_{k^2 \in \mathcal{K}_0^2} \sup_{k^1 \in \mathcal{K}_0^1} \mathcal{Y}_0^{t,\omega,k^1,k^2}(T, \xi^{t,\omega}), \ (t,\omega) \in \overline{\Lambda} \qquad (11.3.38)$$

is a viscosity solution of the PPDE

$$\partial_t \overline{u} + \overline{G}(t,\omega,\overline{u}, \partial_\omega \overline{u}, \partial_{\omega\omega}^2 \overline{u}) = 0, \quad \overline{u}(T,\omega) = \xi(\omega); \qquad (11.3.39)$$

$$\text{where } \overline{G}(t,\omega,y,z,\gamma) := \inf_{k_2 \in \mathbb{K}_2} \sup_{k_1 \in \mathbb{K}_1} \left[\frac{1}{2}\sigma^2(t,\omega,k_1,k_2) : \gamma\right.$$

$$\left. + F(t,\omega,y,z\sigma(t,\omega,k_1,k_2),k_1,k_2)\right].$$

(ii) The problems (11.3.25) and (11.3.38) are called the zero sum stochastic differential game. When $u = \overline{u}$, we say the game value exists. Assume the following Isaacs condition holds:

$$\overline{G}(t,\omega,y,z,\gamma) = G(t,\omega,y,z,\gamma) \qquad (11.3.40)$$

and the viscosity solution of the PPDE (11.3.20) is unique, then clearly $u = \overline{u}$, namely the game value exists.

∎

11.3.5 Stochastic HJB Equations and Backward Stochastic PDEs

In this subsection we consider an optimization problem with random coefficients. Unlike in (11.3.10) the diffusion σ depends on the path of the state process X, here σ depends on current value X_t and the path of noise $B_{t\wedge}$. In this subsection, we will always use the Wiener measure $\mathbb{P}_0$, and thus we use B (instead of X) to denote the canonical process. Our state space is $\overline{\Lambda} \times \mathbb{R}^{d_1}$. For any $(t,\omega,x) \in \Lambda \times \mathbb{R}^{d_1}$ and $\mathscr{F}_T \times \mathscr{B}(\mathbb{R}^{d_1})$-measurable mapping $\xi : \Omega \times \mathbb{R}^{d_1} \to \mathbb{R}$, denote

$$\xi^{t,\omega,x}(\tilde{\omega},\tilde{x}) := \xi(\omega \otimes_t \tilde{\omega}, x + \tilde{x})$$

which is obviously $\mathscr{F}_T^t \times \mathscr{B}(\mathbb{R}^{d_1})$-measurable. We now define

$$u(t, \omega, x) := \sup_{k \in \mathscr{K}} Y_0^{t,\omega,x,k}, \tag{11.3.41}$$

where

$$X_s^{t,\omega,x,k} = \int_0^s \sigma^{t,\omega,x}(r, X_r^{t,\omega,x,k}, B_., k_r) dB_r;$$

$$Y_t^{t,\omega,x,k} = g^{t,\omega,x}(B_., X_{T-t}^k) + \int_s^{T-t} f^{t,\omega,x}(r, B_., X_r^{t,\omega,x,k}, Y_r^{t,\omega,x,k}, Z_r^{t,\omega,x,k}, k_r) dr \tag{11.3.42}$$

$$- \int_s^{T-t} Z_r^{t,\omega,x,k} dB_r; \qquad 0 \le s \le T - t, \mathbb{P}_0\text{-a.s.}$$

By enlarging the canonical space, we may characterize the above u as a viscosity solution to a PPDE. Indeed, denote $\hat{\Omega} := \Omega \times \tilde{\Omega}$ with elements $\hat{\omega} := (\omega, \tilde{\omega})$, where $\tilde{\Omega} := \{\tilde{\omega} \in C^0([0, T], \mathbb{R}^{d_1}) : \tilde{\omega}_0 = 0\}$. Let $\hat{B} := (B, \tilde{B})$ denote the canonical process, and $\hat{\mathbb{P}}_0$ the Wiener measure. Define

$$\hat{u}(t, \hat{\omega}) := \sup_{k \in \mathscr{K}} \hat{Y}_0^{t,\hat{\omega},k}, \tag{11.3.43}$$

where, denoting $\hat{z} = (z, \tilde{z}) \in \mathbb{R}^{1 \times (d+d_1)}$,

$$\hat{\sigma}(t, \hat{\omega}, k) := \begin{bmatrix} I_d & 0 \\ \sigma(t, \omega, \tilde{\omega}_t, k) & 0 \end{bmatrix}, \hat{f}(t, \hat{\omega}, y, \hat{z}, k) :$$

$$= f(t, \omega, \tilde{\omega}_t, y, z, k), \quad \hat{g}(\hat{\omega}) := g(\omega, \tilde{\omega}_T); \tag{11.3.44}$$

and

$$\hat{X}_s^{t,\hat{\omega},k} = \int_0^s \hat{\sigma}^{t,\hat{\omega}}(r, \hat{X}_.^{t,\hat{\omega},k}, k_r) d\hat{B}_r;$$

$$\hat{\mathbb{P}}_0\text{-a.s. } (11.3.45)$$

$$\hat{Y}_s^{t,\hat{\omega},k} = \hat{g}^{t,\hat{\omega}}(\hat{X}_.) + \int_s^{T-t} \hat{f}(r, \hat{X}_.^{t,\hat{\omega},k}, \hat{Y}_r^{t,\hat{\omega},k}, \hat{Z}_r^{t,\hat{\omega},k}, k_r) dr - \int_s^{T-t} \hat{Z}_r^{t,\hat{\omega},k} d\hat{B}_r;$$

Then it is clear that

$$\hat{u}(t, \hat{\omega}) = u(t, \omega, \tilde{\omega}_t). \tag{11.3.46}$$

Moreover, following the arguments in Section 11.3.3, under natural conditions one can show that $\hat{u}$ is a viscosity solution to the following PPDE:

$$\partial_t \hat{u} + \hat{G}(t, \hat{\omega}, \hat{u}, \partial_{\hat{\omega}} \hat{u}, \partial_{\hat{\omega}\hat{\omega}}^2 \hat{u}) = 0, \quad \hat{u}(T, \hat{\omega}) = \hat{g}(\hat{\omega}); \tag{11.3.47}$$

where, denoting $\hat{\gamma} = \begin{bmatrix} \gamma & \gamma_{21}^\mathsf{T} \\ \gamma_{21} & \tilde{\gamma} \end{bmatrix}$ for $\gamma \in \mathbb{S}^d$, $\tilde{\gamma} \in \mathbb{S}^{d_1}$ and $\gamma_{12} \in \mathbb{R}^{d \times d_1}$,

$$\hat{G}(t, \hat{\omega}, y, \hat{z}, \hat{\gamma}) := \sup_{k \in \mathbb{K}} \left[\frac{1}{2} \hat{\sigma}\hat{\sigma}^\mathsf{T}(t, \hat{\omega}, k) : \hat{\gamma} + \hat{f}(t, \hat{\omega}, y, v\hat{\sigma}(t, \hat{\omega}, k), k) \right]$$

$$= \frac{1}{2}\mathrm{tr}(\gamma) + \sup_{k \in \mathbb{K}} \left[\sigma(t, \omega, \tilde{\omega}_t, k) : \hat{\gamma}_{21} + \frac{1}{2}\sigma\sigma^\mathsf{T}(t, \omega, \tilde{\omega}_t, k) : \tilde{\gamma} \right.$$

$$\left. + f(t, \omega, \tilde{\omega}_t, y, z + \tilde{z}\sigma(t, \omega, \tilde{\omega}_t, k), k) \right]. \tag{11.3.48}$$

Remark 11.3.15 The $\hat{\sigma}$ is neither symmetric nor positive definite, so one cannot apply the results in Section 11.3.3 directly.

(i) The symmetry of σ in Section 11.3.3 is mainly for the convenience in weak formulation. There is no significant difficulty to show that $\hat{u}$ is a viscosity solution of PPDE (11.3.47)–(11.3.48).

(ii) The fact that $\hat{\sigma}$ is not positive definite implies PPDE (11.3.47)–(11.3.48) is degenerate. This is a serious issue in establishing the comparison principle for PPDEs. See Remark (11.4.10) (ii). ∎

Peng [173, 174] characterized the random field u with the following stochastic HJB equation, which is a special type of backward stochastic PDE (BSPDE, for short) with solution pair $(u(t, \omega, x), v(t, \omega, x))$: omitting the notation ω inside the random field,

$$u(t, x) = g(x) + \int_t^T F(s, x, u, \partial_x u, \partial_{xx} u, v, \partial_x v^\mathsf{T}) ds - \int_t^T v(s, x) dB_s, \quad \mathbb{P}_0\text{-a.s.}$$

$$\tag{11.3.49}$$

where

$$F(t, x, y, \tilde{z}, \tilde{\gamma}, p, q) := \sup_{k \in \mathbb{K}} \left[\frac{1}{2}\tilde{\gamma} : \sigma\sigma^\mathsf{T}(t, x, k) + q^\mathsf{T} : \sigma(t, x, k) + f(t, x, y, p + \tilde{z}\sigma, k) \right].$$

$$\tag{11.3.50}$$

We remark that the decoupling field u in Section 8.3.2 can also be characterized as a solution to certain BSPDE, see Ma, Yin, & Zhang [145].

The BSPDE (11.3.49) can be easily transformed into a PPDE in the enlarged canonical space. Indeed, assume for each $x \in \mathbb{R}^{d_1}$, $u(\cdot, x) \in C^{1,2}(\Lambda)$. Then by the functional Itô formula (9.4.1),

$$du(t, \omega, x) = \left[\partial_t u + \frac{1}{2}\mathrm{tr}(\partial_{\omega\omega}^2 u) \right] dt + \partial_\omega u dB_t, \quad \mathbb{P}_0\text{-a.s.}$$

Comparing this with (11.3.49), we see that

$$\partial_t u + \frac{1}{2}\text{tr}(\partial^2_{\omega\omega}u) + F(t,\omega,x,u,\partial_x u,\partial_{xx}u,v,\partial_x v^\top) = 0, \quad v = \partial_\omega u.$$

Plug $v = \partial_\omega u$ into the first equation, we obtain a mixed PPDE:

$$\partial_t u + \frac{1}{2}\text{tr}(\partial^2_{\omega\omega}u) + F(t,\omega,x,u,\partial_x u,\partial_{xx}u,\partial_\omega u,\partial_{x\omega}u) = 0. \qquad (11.3.51)$$

Equivalently, if we enlarge the canonical space and view $x = \tilde\omega_t$, the above mixed PPDE is equivalent to the following PPDE such that (11.3.46) holds:

$$\partial_t \hat u(t,\hat\omega) + \hat G(t,\hat\omega,\hat u,\partial_{\hat\omega}\hat u,\partial^2_{\hat\omega\hat\omega}\hat u) = 0; \qquad (11.3.52)$$

$$\text{where } \hat G(t,\hat\omega,y,\hat z,\hat\gamma) := \tfrac{1}{2}\gamma + F(t,\omega,\tilde\omega_t,y,\tilde z,\tilde\gamma,z,\gamma_{12}) = 0.$$

Here again we denote $\hat z = [z,\tilde z]$ and $\hat\gamma = \begin{bmatrix} \gamma & \gamma^\top_{21} \\ \gamma_{21} & \tilde\gamma \end{bmatrix}$. When F is defined in (11.3.50), the PPDE (11.3.52) reduces back to (11.3.47).

Remark 11.3.16 In the BSPDE literature with semilinear F:

$$F(t,\omega,x,y,\tilde z,\tilde\gamma,p,q) = a(t,\omega,x,y,\tilde z,p):q^\top + f(t,\omega,x,y,\tilde z,\tilde\gamma,p),$$

the BSPDE (11.3.49) is called parabolic if $\frac{1}{2}\partial_{\tilde\gamma}f - a^\top a \ge 0$, and strictly parabolic (also called coercive) if $\frac{1}{2}\partial_{\tilde\gamma}f - a^\top a \ge c_0 I_{d_1}$ for some constant $c_0 > 0$. Note that the parabolicity exactly means $\partial_{\hat\gamma}\hat G$ is nonnegative definite or uniformly positive definite.

Moreover, for the $\hat G$ in (11.3.48) corresponding to stochastic HJB equation, we see that $\hat G$ is increasing in $\hat\gamma$, but not strictly increasing. This means that the PPDE (11.3.47) is parabolic but degenerate. ∎

11.4 Well-Posedness of Fully Nonlinear PPDEs

We first extend the crucial Lemma 11.1.5 to the nonlinear case. Recall Remark 11.2.12 (ii) that we may always require test functions φ to be uniformly continuous.

Lemma 11.4.1 Let $\delta > 0$, $L > 0$, either $H = H_\delta$ or $H = T$, and $u \in UC_b(\overline\Lambda_H)$, $\varphi \in C^{1,2}(\Lambda_H) \cap UC_b(\overline\Lambda_H)$. If $[\varphi - u]_0 > \mathscr{E}^L\big[[\varphi-u]_H\big]$, then there exists $(t^*,\omega^*) \in \Lambda_H$ such that

$$\varphi^{t^*,\omega^*} - [\varphi - u](t^*,\omega^*) \in \overline{\mathscr{A}}^L u(t^*,\omega^*).$$

Proof Denote $\tilde{u} := \varphi - u$. Let Y be the nonlinear Snell envelope of $\tilde{u}$ with $\mathscr{P} = \mathscr{P}_L$ in (10.3.1), and $\tau^* := \inf\{t \geq 0 : Y_t = \tilde{u}_t\}$ the corresponding hitting time. Then all the results in Theorem 10.3.2 hold, and thus

$$\overline{\mathscr{E}}^L[\tilde{u}_H] < \tilde{u}_0 \leq Y_0 = \overline{\mathscr{E}}^L[Y_{\tau^*}] = \overline{\mathscr{E}}^L[\tilde{u}_{\tau^*}].$$

Then there exists $\omega^* \in \Omega$ such that $t^* := \tau^*(\omega^*) < H(\omega^*)$. Note that

$$[\varphi - u](t^*, \omega^*) = \tilde{u}_{\tau^*}(\omega^*) = Y_{\tau^*}(\omega^*) = \sup_{\tau \in \mathscr{T}_{H^{t^*,\omega^*}}} \mathscr{E}^L\big[[\varphi - u]_\tau^{t^*,\omega^*}\big].$$

Clearly there is $\tilde{\delta} > 0$ small enough such that $H_{\tilde{\delta}} \leq H^{t^*,\omega^*}$. Then the above implies immediately that $\varphi^{t^*,\omega^*} - [\varphi - u](t^*, \omega^*) \in \overline{\mathscr{A}}^L u(t^*, \omega^*)$ with localization time $H_{\tilde{\delta}}$. ∎

11.4.1 Stability

For any $(y, z, \gamma) \in \mathbb{R} \times \mathbb{R}^d \times \mathbb{S}^d$ and $\delta > 0$, denote

$$O_\delta(y, z, \gamma) := \qquad\qquad\qquad\qquad\qquad\qquad\qquad\qquad (11.4.1)$$
$$\Big\{(\tilde{t}, \tilde{\omega}, \tilde{y}, \tilde{z}, \tilde{\gamma}) \in \Lambda_{H_\delta} \times \mathbb{R} \times \mathbb{R}^d \times \mathbb{S}^d : |\tilde{y} - y| + |\tilde{z} - z| + |\tilde{\gamma} - \gamma| \leq \delta\Big\}.$$

Theorem 11.4.2 *Let* $L > 0$, *G satisfy Assumption 11.2.1, and* $u \in UC_b(\overline{\Lambda})$. *Assume*

(i) *for any* $\varepsilon > 0$, *there exist* G^ε *and* $u^\varepsilon \in UC_b(\overline{\Lambda})$ *such that* G^ε *satisfies Assumption 11.2.1 and* u^ε *is a viscosity L-subsolution of PPDE (11.3.20) with generator* G^ε;

(ii) *as* $\varepsilon \to 0$, *$(G^\varepsilon, u^\varepsilon)$ converge to (G, u) locally uniformly in the following sense: for any* $(t, \omega, y, z, \gamma) \in \Lambda \times \mathbb{R} \times \mathbb{R}^d \times \mathbb{S}^d$, *there exists* $\delta > 0$ *such that,*

$$\lim_{\varepsilon \to 0} \sup_{(\tilde{t}, \tilde{\omega}, \tilde{y}, \tilde{z}, \tilde{\gamma}) \in O_\delta(y, z, \gamma)} \Big[|(G^\varepsilon - G)^{t,\omega}(\tilde{t}, \tilde{\omega}, \tilde{y}, \tilde{z}, \tilde{\gamma})| + |(u^\varepsilon - u)^{t,\omega}(\tilde{t}, \tilde{\omega})| \Big] = 0.$$

$$(11.4.2)$$

Then u is a viscosity L-subsolution of PPDE (11.3.20) with generator G.

Proof Without loss of generality we shall only prove the viscosity subsolution property at $(0, 0)$. By Remark 11.2.12 (ii), let $\varphi \in \overline{\mathscr{A}}^L u(0, 0)$ with localization time H_{δ_0} such that $\varphi \in UC_b(\overline{\Lambda}_{H_{\delta_0}})$. Moreover, by (11.4.2) we may choose $\delta_0 > 0$ small enough such that

$$\lim_{\varepsilon \to 0} \rho(\varepsilon, \delta_0) = 0, \quad \text{where, denoting } (y_0, z_0, \gamma_0) := (\varphi_0, \partial_\omega \varphi_0, \partial^2_{\omega\omega} \varphi_0),$$

$$\rho(\varepsilon, \delta) := \sup_{(t,\omega,y,z,\gamma) \in O_\delta(y_0,z_0,\gamma_0)} \Big[|G^\varepsilon - G|(t,\omega,y,z,\gamma) + |u^\varepsilon - u|(t,\omega) \Big].$$

$$(11.4.3)$$

For $0 < \delta \le \delta_0$, denote $\varphi_\delta(t,\omega) := \varphi(t,\omega) + \delta t$. By (11.2.4) and Lemma 10.1.18 we have

$$(\varphi_\delta - u)_0 = (\varphi - u)_0 = 0 \le \underline{\mathscr{E}}^L \big[(\varphi - u)_{H_\delta} \big] < \underline{\mathscr{E}}^L \big[(\varphi_\delta - u)_{H_\delta} \big].$$

By (11.4.3), there exists $\varepsilon_\delta > 0$ small enough such that, for any $\varepsilon \le \varepsilon_\delta$,

$$(\varphi_\delta - u^\varepsilon)_0 < \underline{\mathscr{E}}^L \big[(\varphi_\delta - u^\varepsilon)_{H_\delta} \big]. \tag{11.4.4}$$

Applying Lemma 11.4.1, there exists $(t^*_{\varepsilon,\delta}, \omega^*_{\varepsilon,\delta}) \in \Lambda_{H_\delta}$, which depend on (ε, δ), such that

$$\varphi^\varepsilon_\delta := \varphi^{t^*_{\varepsilon,\delta}, \omega^*_{\varepsilon,\delta}}_\delta - [\varphi_\delta - u^\varepsilon](t^*_{\varepsilon,\delta}, \omega^*_{\varepsilon,\delta}) \in \underline{\mathscr{A}}^L u^\varepsilon(t^*_{\varepsilon,\delta}, \omega^*_{\varepsilon,\delta}).$$

Since u^ε is a viscosity L-subsolution of PPDE (11.2.1) with generator G^ε, we have

$$0 \le \Big[\partial_t \varphi^\varepsilon_\delta + (G^\varepsilon)^{t^*_{\varepsilon,\delta}, \omega^*_{\varepsilon,\delta}} (\cdot, \varphi^\varepsilon_\delta, \partial_\omega \varphi^\varepsilon_\delta, \partial^2_{\omega\omega} \varphi^\varepsilon_\delta) \Big](0,0)$$

$$= \Big[\partial_t \varphi + \delta + G^\varepsilon(\cdot, u^\varepsilon, \partial_\omega \varphi, \partial^2_{\omega\omega} \varphi) \Big](t^*_{\varepsilon,\delta}, \omega^*_{\varepsilon,\delta}). \tag{11.4.5}$$

Note that $(t^*_{\varepsilon,\delta}, \omega^*_{\varepsilon,\delta}) \in \overline{\Lambda}_{H_\delta}$, then $|u^\varepsilon - u|(t^*_{\varepsilon,\delta}, \omega^*_{\varepsilon,\delta}) \le \rho(\varepsilon, \delta) \le \rho(\varepsilon, \delta_0)$. By (11.4.2), we may set δ small enough and then $\varepsilon \le \varepsilon_\delta$ small enough so that $(\cdot, u^\varepsilon, \partial_\omega \varphi, \partial^2_{\omega\omega} \varphi)(t^*_{\varepsilon,\delta}, \omega^*_{\varepsilon,\delta}) \in O_{\delta_0}(y_0, z_0, \gamma_0)$. Thus, (11.4.5) leads to

$$0 \le \Big[\partial_t \varphi + \delta + G^\varepsilon(\cdot, u^\varepsilon, \partial_\omega \varphi, \partial^2_{\omega\omega} \varphi) \Big](t^*_{\varepsilon,\delta}, \omega^*_{\varepsilon,\delta})$$

$$\le \Big[\partial_t \varphi + G(\cdot, u^\varepsilon, \partial_\omega \varphi, \partial^2_{\omega\omega} \varphi) \Big](t^*_{\varepsilon,\delta}, \omega^*_{\varepsilon,\delta}) + \delta + \rho(\varepsilon, \delta_0)$$

$$\le \Big[\partial_t \varphi + G(\cdot, u, \partial_\omega \varphi, \partial^2_{\omega\omega} \varphi) \Big](t^*_{\varepsilon,\delta}, \omega^*_{\varepsilon,\delta}) + \delta + C\rho(\varepsilon, \delta_0)$$

$$\le \mathscr{L}\varphi_0 + \sup_{(t,\omega) \in \Lambda_{H_\delta}} \Big| G(t, \omega, y_0, z_0, \gamma_0) - G(0, 0, y_0, z_0, \gamma_0) \Big| + \delta + C\rho(\varepsilon, \delta_0)$$

$$+ C \sup_{(t,\omega) \in \Lambda_{H_\delta}} \Big[|u_t(\omega) - u_0| + |\partial_t \varphi_t(\omega) - \partial_t \varphi_0| + |\partial_\omega \varphi_t(\omega) - \partial_\omega \varphi_0|$$

$$+ |\partial^2_{\omega\omega} \varphi_t(\omega) - \partial^2_{\omega\omega} \varphi_0| \Big],$$

where we used Assumption 11.2.1 for G. Now by first sending $\varepsilon \to 0$ and then $\delta \to 0$ we obtain $\mathscr{L}\varphi_0 \geq 0$. Since $\varphi \in \overline{\mathscr{A}}^L u(0,0)$ is arbitrary, we see that u is a viscosity subsolution of PPDE (11.2.1) with generator G at $(0,0)$ and thus complete the proof. ∎

Remark 11.4.3 We need the same L in the proof of Theorem 11.4.2. If u^ε is only a viscosity subsolution of PPDE (11.2.1) with generator G^ε, but with possibly different L_ε, we are not able to show that u is a viscosity subsolution of PPDE (11.2.1) with generator G. ∎

Remark 11.4.4 Recalling the observation (11.1.6), our choice of test function (11.2.4) is essentially sharp for the comparison principle and uniqueness of viscosity solutions. This is true even for PDEs in Markovian case. However, we shall note that for stability result, both the condition and the conclusion involve the viscosity property, so a priori it is not clear if the enlargement of the set of test functions will help or hurt for the stability. In fact, in PDE case the standard definition using the smaller set of test functions (5.5.4) is more convenient for the stability, especially for the so-called Perron's method (see, e.g., Crandall, Ishii, & Lions [42]). It will be very interesting though challenging to modify our set of test functions (and thus modify the definition of viscosity solutions) which helps for the stability but in the meantime still maintains the comparison principle. ∎

11.4.2 Partial Comparison of Viscosity Solutions

In this subsection, we prove a partial comparison principle, namely a comparison result of a viscosity super- (resp. sub-) solution and a classical sub- (resp. super-) solution. This is an extension of Proposition 5.5.10 to the path dependent case, and is a key step for establishing the full comparison principle in the next subsection. The proof follows similar idea as Lemma 11.4.1 and relies heavily on the optimal stopping problem Theorem 10.3.2.

Proposition 11.4.5 *Let Assumption 11.2.1 hold true, and $u^1, u^2 \in UC_b(\overline{\Lambda})$ be a viscosity subsolution and viscosity supersolution of PPDE (11.2.1), respectively. If $u^1(T,\cdot) \leq u^2(T,\cdot)$ and either u^1 or u^2 is in $C^{1,2}(\Lambda)$, then $u^1 \leq u^2$ on $\overline{\Lambda}$.*

Proof Without loss of generality, we assume u^1 is a viscosity L-subsolution and u^2 a classical supersolution, and we shall only prove $u_0^1 \leq u_0^2$. By Proposition 11.2.13, we may assume that

$$G \text{ is nonincreasing in } y. \tag{11.4.6}$$

Assume to the contrary that $c := \frac{1}{2T}[u_0^1 - u_0^2] > 0$. Denote

$$u_t := (u^1 - u^2)_t^+ + ct.$$

It is clear that $u \in UC_b(\overline{\Lambda})$. Let Y and τ^* be defined as in (10.3.1) with $\mathrm{H} = T$ and $\mathscr{P} = \mathscr{P}_L$. It follows from Theorem 10.3.2 that

$$\overline{\mathscr{E}}^L[u_T] = cT < 2cT = u_0 \leq Y_0 = \overline{\mathscr{E}}^L[Y_{\tau^*}] = \overline{\mathscr{E}}^L[u_{\tau^*}].$$

Then there exists $\omega^* \in \Omega$ such that $t^* := \tau^*(\omega^*) < T$. Note that

$$(u^1 - u^2)^+(t^*, \omega^*) + ct^* = u_{t^*}(\omega^*) = Y_{t^*}(\omega^*) \geq \overline{\mathscr{E}}^L[u_{T-t^*}^{t^*,\omega^*}] \geq cT > 0.$$

Then $(u^1 - u^2)(t^*, \omega^*) > 0$. Since $u^1 - u^2 \in UC_b(\overline{\Lambda})$, there exists $\delta > 0$ such that $(u^1 - u^2)^{t^*,\omega^*} > 0$ on $\Lambda_{\mathrm{H}\delta}$. This implies that

$$(u^1 - u^2)(t^*, \omega^*) + ct^* = u_{t^*}(\omega^*) = Y_{t^*}(\omega^*) = \sup_{\tau \in \mathscr{T}_{\mathrm{H}\delta}} \overline{\mathscr{E}}^L[u_\tau^{t^*,\omega^*}]$$

$$= \sup_{\tau \in \mathscr{T}_{\mathrm{H}\delta}} \overline{\mathscr{E}}^L\big[(u^1 - u^2)_\tau^{t^*,\omega^*} + c(t^* + \tau)\big].$$

Denote $\varphi_t := (u^2)_t^{t^*,\omega^*} + (u^1 - u^2)(t^*, \omega^*) - ct$. Since $u^2 \in C^{1,2}(\Lambda)$, then it follows from the above equality that $\varphi \in \mathscr{A}^L u^1(t^*, \omega^*)$. Now by the viscosity L-subsolution property of u^1 we have

$$0 \leq \big[\partial_t \varphi + G^{t^*,\omega^*}(., u^1, \partial_\omega \varphi, \partial_{\omega\omega}^2 \varphi)\big](0,0)$$

$$= \big[\partial_t u^2 + G(., u^1, \partial_\omega u^2, \partial_{\omega\omega}^2 u^2)\big](t^*, \omega^*) - c$$

$$\leq \big[\partial_t u^2 + G(., u^2, \partial_\omega u^2, \partial_{\omega\omega}^2 u^2)\big](t^*, \omega^*) - c,$$

where the last inequality follows from (11.4.6). Since $c > 0$, this is in contradiction with the supersolution property of u^2. ∎

As a direct consequence of the above partial comparison principle, we have

Proposition 11.4.6 *Let Assumption 11.2.1 hold true. If PPDE (11.2.1) has a classical solution $u \in C^{1,2}(\Lambda) \cap UC_b(\overline{\Lambda})$, then u is the unique viscosity solution of PPDE (11.2.1) in $UC_b(\overline{\Lambda})$ with terminal condition $u(T, \cdot)$.*

11.4.3 Comparison Principle of PPDEs

The full comparison principle is much more involved and requires some additional conditions. In this subsection we establish it in a special case, so as to give readers some idea of our approach. Our main idea is to approximate PPDEs with PDEs, and the path discretization in Proposition 10.1.20 plays an important role.

In light of (10.1.15), for any $0 \leq t_0 < T$ we denote: recalling (10.1.8) that $L_1 :=$ $L + 1$,

$$
\partial D_{t_0,\varepsilon} := \{(t,x) : 0 < t \leq T - t_0, |x| + L_1 t = \varepsilon \text{ or } t = T - t_0\},
$$
$$
D_{t_0,\varepsilon} := \{(t,x) : t < T - t_0, |x| + L_1 t < \varepsilon\}, \quad \overline{D}_{t_0,\varepsilon} := D_{t_0,\varepsilon} \cup \partial D_{t_0,\varepsilon}. \tag{11.4.7}
$$

We emphasize that $\partial D_{t_0,\varepsilon}$ and $\overline{D}_{t_0,\varepsilon}$ do not include $\{(0,x) : |x| \leq \varepsilon\}$ and thus are not closed. We next introduce the path frozen PDE: given $(t_0,\omega) \in \Lambda$,

$$
\partial_t v(t,x) + G^{t_0,\omega}(t, 0, v, \partial_x v, \partial_{xx}^2 v) = 0, \quad (t,x) \in D_{t_0,\varepsilon},
$$
$$
v(t,x) = h(t,x), \quad (t,x) \in \partial D_{t_0,\varepsilon}. \tag{11.4.8}
$$

We now assume

Assumption 11.4.7

(i) *G satisfies Assumption 11.2.1 with uniform Lipschitz constant L_0 and is uniformly continuous in ω with a modulus of continuity function ρ_0;*
(ii) *For any $L \geq L_0$ in (11.4.7), any $\varepsilon > 0$, $(t_0,\omega) \in \Lambda$, and any $h \in C_b^0(\partial D_{t_0,\varepsilon})$, the path frozen PDE (11.4.8) has a classical solution $v \in C^{1,2}(D_{t_0,\varepsilon}) \cap C_b^0(\overline{D}_{t_0,\varepsilon})$.*

Our main result in this subsection is

Theorem 11.4.8 *Let Assumption 11.4.7 hold and $u^1, u^2 \in UC_b(\overline{\Lambda})$ be a viscosity subsolution and viscosity supersolution of PPDE (11.2.1), respectively. If $u^1(T,\cdot) \leq u^2(T,\cdot)$, then $u^1 \leq u^2$ on $\overline{\Lambda}$.*

To prove the theorem, we first introduce a bound generator:

$$
\overline{G}(y,z,\gamma) := \frac{1}{2} \sup_{\sigma \geq 0, |\sigma| \leq \sqrt{2L_0}} [\sigma^2 : \gamma] + L_0[|y| + |z|]. \tag{11.4.9}
$$

Lemma 11.4.9

(i) *For any $\varepsilon > 0$, $0 \leq t_0 < T$, $h \in C_b^0(\partial D_{t_0,\varepsilon})$, and any constant C_0, the PDE*

$$
\partial_t v + \overline{G}(t, v, \partial_x v, \partial_{xx}^2 v) + C_0 = 0, \quad (t,x) \in D_{t_0,\varepsilon};
$$
$$
v(t,x) = h(t,x), \quad (t,x) \in \partial D_{t_0,\varepsilon}, \tag{11.4.10}
$$

admits a unique viscosity solution $v \in C_b^0(\overline{D}_{t_0,\varepsilon})$ with the following representation: recalling (10.1.12),

$$
v(t,x) = \sup_{|\alpha| \leq L_0} \overline{\mathcal{E}}^{L_0}\left[h(\mathrm{H}, X_{\mathrm{H}}) e^{\int_0^{\mathrm{H}} \alpha_r dr} + C_0 \int_0^{\mathrm{H}} e^{\int_0^s \alpha_r dr} ds \right], \tag{11.4.11}
$$

where $\mathrm{H} := \mathrm{H}^{t_0 + t, x, \varepsilon - L_1 t}$.

(ii) *For $i = 1, 2$, let v_i be a classical solution to the path frozen PDE (11.4.8) with boundary condition h_i. Then $v_1 - v_2 \leq v$, where v is the viscosity solution to the bound PDE (11.4.10) with $C_0 := 0, h := h_1 - h_2$.*

Proof We shall only sketch a proof, and leave some details to readers in Problem 11.7.7.

(i) Following the arguments in Proposition 11.3.8 one can easily verify that the v in (11.4.11) is a viscosity solution (in our sense or in standard PDE sense as in Section 5.5). The uniqueness follows from standard PDE literature, and we note that we do not need this uniqueness here.

(ii) Denote $\Delta v := v_1 - v_2$. Note that

$$\partial_t \Delta v = \partial_t v_1 - \partial_t v_2 = -G^{t_0,\omega}(t, 0, v_1, \partial_x v_1, \partial^2_{xx} v_1) + G^{t_0,\omega}(t, 0, v_2, \partial_x v_2, \partial^2_{xx} v_2)$$
$$\geq -\overline{G}(t, \Delta v, \partial_x \Delta v, \partial^2_{xx} \Delta v).$$

That is, Δv is a classical subsolution of PDE (11.4.10) with $C_0 = 0, h = h_1 - h_2$. Then our claim follows from the partial comparison of PDE, in the spirit of Proposition 5.5.10. ∎

We will also need some auxiliary sets: for any $\varepsilon > 0, \delta > 0$ and $n \geq 0$,

$$\Pi^n_\varepsilon := \left\{ \pi_n : = (t_i, x_i)_{1 \leq i \leq n} : (t_1, x_1), (t_{i+1} - t_i, x_{i+1} - x_i) \in \partial D_{t_i,\varepsilon}, 1 \leq i \leq n-1 \right\},$$

$$\Pi^n_{\varepsilon,\delta} := \left\{ \pi_n \in \Pi^n_\varepsilon : t_1, t_{i+1} - t_i \geq \delta, 1 \leq i \leq n-1 \right\},$$

$$\overline{D}^n_\varepsilon := \left\{ (\pi_n; t, x) : \pi_n \in \Pi^n_\varepsilon, (t, x) \in \overline{D}_{t_n,\varepsilon} \right\}. \tag{11.4.12}$$

Here we take the notational convention that $t_0 := 0$. For each $\pi_m \in \Pi^m_\varepsilon$, denote by $\mathbf{x}^{\pi_m} \in \Omega$ the linear interpolation of $(0,0), \pi_m, (T, x_m)$. It is obvious that

the mapping $\pi_m \mapsto \mathbf{x}^{\pi_m}$ is continuous in Π^m_ε and uniformly

continuous on $\Pi^n_{\varepsilon,\delta}, \forall \delta > 0.$ \hfill (11.4.13)

Proof of Theorem 11.4.8 Fix an $L \geq L_0$ such that u^1 is a viscosity L-subsolution and u^2 a viscosity L-supersolution, and assume for notational simplicity that $d = 1$. Note that Assumption 11.4.7 remains true under the transformation in Proposition 11.2.13. Then by Proposition 11.2.13 we may assume (11.4.6) holds. Without loss of generality we will prove only $u^1_0 \leq u^2_0$ in several steps.

Step 1. Fix $\varepsilon > 0, m \geq 1$. Given $\pi_m \in \Pi^m_\varepsilon$ with $t_m < T$, consider the following PDE:

$$\partial_t \overline{v}^m_m + G^{t_m, \mathbf{x}^{\pi_m}}(t, \overline{v}^m_m, \partial_x \overline{v}^m_m, \partial^2_{xx} \overline{v}^m_m) = 0, \ (t, x) \in D_{t_m,\varepsilon};$$
$$\overline{v}^m_m(t, x) = u^1\left(t, \mathbf{x}^{\pi_m,(t,x)}\right) + \rho(2\varepsilon), \ (t, x) \in \partial D_{t_m,\varepsilon}, \tag{11.4.14}$$

where ρ is the modulus of continuity function of u^1. By (11.4.13) and Assumption 11.4.7 (ii), the above PDE has a classical solution $\overline{v}_m^m \in C^{1,2}(D_{t_m,\varepsilon}) \cap C_b^0(\overline{D}_{t_m,\varepsilon})$. Clearly $\overline{v}_m^m$ also depends on π_m, thus we may rewrite it as $\overline{v}_m^m(\pi_m; t, x)$ with domain $\overline{D}_\varepsilon^m$. Moreover, by the uniform continuity of $\mathbf{x}^{\pi_m}$ on $\Pi_{\varepsilon,\delta}^m$ in (11.4.13), following the arguments in Lemma 11.4.9 one can show that

the mapping $\pi_m \mapsto \overline{v}_m^m(\pi_m; 0, 0)$ is uniformly continuous on $\Pi_{\varepsilon,\delta}^m$ for any $\delta > 0$.

$$(11.4.15)$$

In particular, by sending $\delta \to 0$ we see that $\overline{v}_m^m(\pi_m; 0, 0)$ is continuous in Π_ε^m. Now applying the procedure repeatedly, we may define $\overline{v}_i^m : \overline{D}_\varepsilon^i \to \mathbb{R}$ backwardly as the classical solution of the following PDE, $i = m - 1, \cdots, 0$: denoting $t_0 = 0$ when $i = 0$,

$$\partial_t \overline{v}_i^m + G^{t_i, \mathbf{x}^{\pi_i}}(t, \overline{v}_i^m, \partial_x \overline{v}_i^m, \partial_{xx}^2 \overline{v}_i^m) = 0, \ (t, x) \in D_{t_i,\varepsilon};$$
$$\overline{v}_i^m(t, x) = \overline{v}_{i+1}^m(\pi_i, (t, x); (0, 0)), \ (t, x) \in \partial D_{t_i,\varepsilon}.$$
$$(11.4.16)$$

Step 2. Recall the H_n^ε in (10.1.15) and X^ε in (10.1.16). Denote

$$\overline{u}_i^m := \overline{v}_i^m\big((\mathrm{H}_j^\varepsilon, X_{\mathrm{H}_j^\varepsilon})_{1 \le j \le i}; t - \mathrm{H}_i^\varepsilon, X_{\mathrm{H}_i^\varepsilon, t}\big) + \rho_0(2\varepsilon)(T - t), \ t \in [\mathrm{H}_i^\varepsilon, \mathrm{H}_{i+1}^\varepsilon), 0 \le i \le m;$$
$$(11.4.17)$$

and when $i = m$, we extend the above definition to $t = \mathrm{H}_{m+1}^\varepsilon$ as well. By the terminal condition in (11.4.16), we see that $\overline{u}^m$ is continuous in $t \in [0, \mathrm{H}_{m+1}^\varepsilon]$. For any $0 \le i < m$ and $(t, \omega) \in \Lambda$ such that $\mathrm{H}_i^\varepsilon(\omega) \le t < \mathrm{H}_{i+1}^\varepsilon(\omega)$, since $\overline{v}_i^m$ is smooth locally, by Example 9.4.4 (i) we see that $(\overline{u}^m)^{t,\omega} \in C^{1,2}(\Lambda_{(\mathrm{H}_{i+1}^\varepsilon)^{t,\omega} - t})$. Then it follows from Lemma 9.4.9 and (11.4.6) that

$$\partial_t \overline{u}^m(t, \omega) + G(t, \omega, \overline{u}_t^m, \partial_\omega \overline{u}_t^m, \partial_{\omega\omega}^2 \overline{u}_t^m)$$
$$= \Big[\partial_t \overline{v}_i^m + G\big(t, \omega, \overline{v}_i^m + \rho_0(2\varepsilon)(T - t), \partial_x \overline{v}_i^m, \partial_{xx}^2 \overline{v}_i^m\big)\Big]$$
$$\big((\mathrm{H}_j^\varepsilon, X_{\mathrm{H}_j^\varepsilon})_{1 \le j \le i}; t - \mathrm{H}_i^\varepsilon, X_{\mathrm{H}_i^\varepsilon, t}\big) - \rho_0(2\varepsilon)$$
$$\le \Big[\partial_t \overline{v}_i^m + G\big(t, \omega, \overline{v}_i^m, \partial_x \overline{v}_i^m, \partial_{xx}^2 \overline{v}_i^m\big)\Big]\big((\mathrm{H}_j^\varepsilon, X_{\mathrm{H}_j^\varepsilon})_{1 \le j \le i}; t - \mathrm{H}_i^\varepsilon, X_{\mathrm{H}_i^\varepsilon, t}\big) - \rho_0(2\varepsilon)$$
$$= \Big[G\big(t, \omega, \overline{v}_i^m, \partial_x \overline{v}_i^m, \partial_{xx}^2 \overline{v}_i^m\big) - G\big(t, X_{t_i \wedge \cdot}^\varepsilon(\omega), \overline{v}_i^m, \partial_x \overline{v}_i^m, \partial_{xx}^2 \overline{v}_i^m\big)\Big]$$
$$\big((\mathrm{H}_j^\varepsilon, X_{\mathrm{H}_j^\varepsilon})_{1 \le j \le i}; t - \mathrm{H}_i^\varepsilon, X_{\mathrm{H}_i^\varepsilon, t}\big) - \rho_0(2\varepsilon)$$
$$\le \rho_0(\|X_{t_i \wedge \cdot}^\varepsilon(\omega) - \omega\|_t) - \rho_0(2\varepsilon) \le 0.$$

Note that

$$\overline{u}_{\mathrm{H}_{m+1}^\varepsilon}^m = u^1\big(\mathrm{H}_{m+1}^\varepsilon, (X^\varepsilon)_{\mathrm{H}_{m+1}^\varepsilon \wedge \cdot}\big) + \rho(2\varepsilon) \ge u^1(\mathrm{H}_{m+1}^\varepsilon, X).$$

Apply the partial comparison Proposition 11.4.5 repeatedly, but restricted to Λ_{H_e} instead of Λ, we obtain $u_0^1 \leq \overline{u}_0^m$.

Step 3. Similar to (11.4.14) and (11.4.16), we define: for $i = m, \cdots, 0$,

$$\partial_t \underline{v}_i^m + G^{t_i, \mathbf{x}^{\pi_i}}(t, \underline{v}_i^m, \partial_x \underline{v}_i^m, \partial_{xx}^2 \underline{v}_i^m) = 0, \ (t, x) \in D_{t_i, \varepsilon};$$

$$\underline{v}_i^m(t, x) = \underline{v}_{i+1}^m(\pi_i, (t, x); (0, 0)), \ (t, x) \in \partial D_{t_i, \varepsilon}, \tag{11.4.18}$$

where $\underline{v}_m^m((\pi_{m+1}; (0, 0)) := u^2(t_{m+1}, \mathbf{x}^{\pi_{m+1}}) - \rho(2\varepsilon)$ and ρ is also a modulus of continuity function of u^2. Similar to (11.4.17), define further that

$$\underline{u}_t^m := \underline{v}_i^m\big((H_j^\varepsilon, X_{H_j^\varepsilon})_{1 \leq j \leq i}; t - H_i^\varepsilon, X_{H_i^\varepsilon, t}\big) - \rho_0(2\varepsilon)(T - t), \ t \in [H_i^\varepsilon, H_{i+1}^\varepsilon), 0 \leq i \leq m.$$

$$\tag{11.4.19}$$

Following the same arguments in Steps 1 and 2 we have $u_0^2 \geq \underline{u}_0^m$.

Moreover, for $i = m, \cdots, 0$, applying Lemma 11.4.9 (i) we may let v_i^m be the unique viscosity solution of the following PDE:

$$\partial_t v_i^m + \overline{G}(t, v_i^m, \partial_x v_i^m, \partial_{xx}^2 v_i^m) = 0, \ (t, x) \in D_{t_i, \varepsilon};$$

$$v_i^m(t, x) = v_{i+1}^m(\pi_i, (t, x); (0, 0)), \ (t, x) \in \partial D_{t_i, \varepsilon}, \tag{11.4.20}$$

where $v_m^m((\pi_{m+1}; (0, 0)) := [u^1 - u^2](t_{m+1}, \mathbf{x}^{\pi_{m+1}}) + 2\rho(2\varepsilon)$. Similarly define

$$u_t^m := v_i^m\big((H_j^\varepsilon, X_{H_j^\varepsilon})_{1 \leq j \leq i}; t - H_i^\varepsilon, X_{H_i^\varepsilon, t}\big) - \rho_0(2\varepsilon)(T - t), \ t \in [H_i^\varepsilon, H_{i+1}^\varepsilon), 0 \leq i \leq m.$$

$$\tag{11.4.21}$$

Define $\Delta v_i^m := \overline{v}_i^m - \underline{v}_i^m$ and $\Delta u_t^m := \overline{u}_t^m - \underline{u}_t^m$. Applying Lemma 11.4.9 (ii) repeatedly we obtain

$$\Delta v_i^m \leq v_i^m, i = m, \cdots, 0, \ \text{and thus} \ \Delta u_t^m \leq u_t^m + 2\rho_0(2\varepsilon)(T - t), \quad 0 \leq t \leq H_{m+1}^\varepsilon.$$

In particular,

$$u_0^1 - u_0^2 \leq \Delta u_0^m \leq u_0^m + 2T\rho_0(2\varepsilon). \tag{11.4.22}$$

Step 4. Applying Lemma 11.4.9 (i) repeatedly, one can easily prove that

$$u_0^m = \sup_{|\alpha| \leq L_0} \overline{\mathscr{E}}^{L_0}\left[u_{H_{m+1}^\varepsilon}^m e^{\int_0^{H_{m+1}^\varepsilon} \alpha_s ds} \right] \tag{11.4.23}$$

$$= \sup_{|\alpha| \leq L_0} \overline{\mathscr{E}}^{L_0}\left[\big[(u^1 - u^2)(H_{m+1}^\varepsilon, X_{H_{m+1}^\varepsilon \wedge \cdot}^\varepsilon) + 2\rho(2\varepsilon)\big] e^{\int_0^{H_{m+1}^\varepsilon} \alpha_s ds} \right].$$

Since $u^1(T, \cdot) \leq u^2(T, \cdot)$ and they are all bounded, we have

$$u_0^m \leq \sup_{|\alpha| \leq L_0} \overline{\mathscr{E}}^{L_0} \left[\left[C1_{\{H_{m+1}^\varepsilon < T\}} + 2\rho(2\varepsilon) \right] e^{\int_0^{H_{m+1}^\varepsilon} \alpha_s ds} \right]$$

$$\leq C \mathscr{E}^{\mathscr{P}_{L_0}} [H_{m+1}^\varepsilon < T] + C\rho(2\varepsilon) \leq \frac{C}{m\varepsilon^2} + C\rho(2\varepsilon),$$

where the last inequality thanks to Proposition 10.1.20 (iii). Plug this into (11.4.22), we have

$$u_0^1 - u_0^2 \leq \frac{C}{m\varepsilon^2} + C\rho(2\varepsilon) + 2T\rho_0(2\varepsilon).$$

By first sending $m \to \infty$ and then $\varepsilon \to 0$, we obtain $u_0^1 \leq u_0^2$. ∎

Remark 11.4.10

(i) The existence of classical solution in Assumption 11.4.7 typically requires the uniform nondegeneracy of G in γ: for some $c_0 > 0$,

$$G(\cdot, \gamma + A) - G(\cdot, \gamma) \geq c_0 A, \quad \text{for all } \gamma, A \in \mathbb{S}^d \text{ with } A \geq 0. \ (11.4.24)$$

Under the above condition (and some other mild conditions), when reduced to the Markov case the HJB equation (11.3.9) will have a classical solution. We refer to the books of Krylov [131] and Lieberman [136], as well as the works of Wang [232–234], for the regularity results of parabolic PDEs. However, for the Isaacs equation (11.3.20), one typically needs $d \leq 2$ to obtain classical solutions, see, e.g., Pham & Zhang [192]. We also refer to Nadirashvili & Vladut [158] for a counterexample in higher dimensional case.

(ii) Theorem 11.4.8 assumes the existence of classical solution of the path frozen PDE, which is a strong requirement. By additional efforts one may use viscosity theory of path frozen PDEs to establish the comparison principle for PPDEs, which in particular allows us to deal with degenerate PPDEs. See Ren, Touzi, & Zhang [205] and Ekren & Zhang [77] for some study along this direction.

(iii) The uniform continuity of G in ω is violated when the σ in (11.3.9) or (11.3.20) depends on ω, and thus this condition is also not desirable. Ekren & Zhang [77] removes this constraint; however, as a trade off they require certain piecewise Markov structure on the viscosity solutions. ∎

11.5 Monotone Scheme for PPDEs

In this section we turn to numerical methods for PPDEs; particularly the so-called monotone schemes.

11.5.1 Monotone Scheme for PDEs

The convergence of monotone schemes is one of the most important applications of the viscosity solution theory. To illustrate the idea, in this subsection we consider the following PDE:

$$\mathbb{L}u(t,x) := \partial_t u + G(t,x,u,\partial_x u, \partial_{xx}^2 u) = 0, \quad (t,x) \in [0,T) \times \mathbb{R}^d. \quad (11.5.1)$$

We extend Definition 5.5.2 to PDE (11.5.1) in an obvious way. To focus on the main idea we will consider only viscosity semi-solutions in $UC_b([0,T] \times \mathbb{R}^d)$, namely uniformly continuous and bounded solutions, and consequently we will impose slightly stronger conditions.

Assumption 11.5.1

(i) $G(\cdot,y,z,\gamma) \in C_b^0([0,T] \times \mathbb{R}^d)$ for any fixed $(y,z,\gamma) \in \mathbb{R} \times \mathbb{R}^{1\times d} \times \mathbb{S}^d$.
(ii) G is uniformly Lipschitz continuous in (y,z,γ) with a Lipschitz constant L_0.
(iii) G is parabolic, namely G is nondecreasing in $\gamma \in \mathbb{S}^d$.
(iv) The PDE (11.5.1) satisfies the comparison principle in the sense of (5.5.16).
(v) The terminal condition $g : \mathbb{R}^d \to \mathbb{R}$ is bounded and uniformly continuous.

For any $(t,x) \in [0,T) \times \mathbb{R}^d$ and $h \in (0,T-t)$, let $\mathbb{T}_h^{t,x}$ be a discretization operator on $UC_b(\mathbb{R}^d)$, whose conditions will be specified later. For $n \geq 1$, denote

$$h := \frac{T}{n}, \quad t_i := ih, \quad i = 0,1,\cdots,n, \quad (11.5.2)$$

and, given the terminal condition $u(T,\cdot) = g$, define:

$$u^h(t_n,x) := g(x), \quad u^h(t,x) := \mathbb{T}_{t_i-t}^{t,x}\big[u^h(t_i,\cdot)\big], \ t \in [t_{i-1},t_i), \ i=n,\cdots,1. \ (11.5.3)$$

Then we have

Theorem 11.5.2 *Let Assumption 11.5.1 hold. Assume $\mathbb{T}_h^{t,x}$ satisfies the following conditions:*

(i) *Consistency: for any $(t,x) \in [0,T) \times \mathbb{R}^d$ and $\varphi \in (C^{1,2} \cap UC_b)([t,T] \times \mathbb{R}^d)$,*

$$\lim_{(\tilde{x},c)\to(x,0),\ \tilde{t}\downarrow t,\ h\downarrow 0} \frac{\mathbb{T}_h^{\tilde{t},\tilde{x}}\big[c + \varphi(\tilde{t}+h,\cdot)\big] - [c+\varphi](\tilde{t},\tilde{x})}{h} = \mathbb{L}\varphi(t,x). \quad (11.5.4)$$

(ii) *Monotonicity: for any $\varphi, \psi \in UC_b(\mathbb{R}^d)$, there exists a modulus of continuity function ρ_{mon}, which depends only on L_0, d, and the uniform continuity of φ, ψ, but does not depend on the specific φ, ψ, such that*

$$\max_{\tilde{x} \in \mathbb{R}^d}[\varphi - \psi](\tilde{x}) \le 0 \quad implies \quad \mathbb{T}_h^{t,x}[\varphi] \le \mathbb{T}_h^{t,\omega}[\psi] + h\rho_{mon}(h). \quad (11.5.5)$$

(iii) *Stability: u^h is uniformly bounded and uniformly continuous in (t, x), uniformly on h.*

Then PDE (11.5.1) with terminal condition $u(T, x) = g(x)$ has a unique viscosity solution $u \in UC_b([0, T] \times \mathbb{R}^d)$, and u_h converges to u locally uniformly as $h \to 0$.

Proof By the stability assumption (iii), u^h is bounded. Define

$$\underline{u}(t, x) := \liminf_{h \to 0} u^h(t, x), \quad \overline{u}(t, x) := \limsup_{h \to 0} u^h(t, x). \quad (11.5.6)$$

Clearly $\underline{u}(T, x) = g(x) = \overline{u}(T, x)$, $\underline{u} \le \overline{u}$, and $\underline{u}, \overline{u} \in UC_b([0, T] \times \mathbb{R}^d)$. We shall show that $\underline{u}$ (resp. $\overline{u}$) is a viscosity supersolution (resp. subsolution) of PDE (11.5.1). Then by the comparison principle we see that $\overline{u} \le \underline{u}$ and thus $u := \overline{u} = \underline{u}$ is the unique viscosity solution of PDE (11.5.1). The convergence of u^h is obvious now, which, together with the uniform regularity of u^h and u, implies further the locally uniform convergence.

Without loss of generality, we shall only prove that $\underline{u}$ satisfies the viscosity supersolution property at $(0, 0)$. Let $\varphi^0 \in \mathscr{A}\underline{u}(0, 0)$ with corresponding $O_\delta(0, 0)$, as defined in (5.5.4) and (5.5.3). Denote

$$\varphi(t, x) := \varphi^0(t, x) - C_\delta[t^2 + |x|^4]. \quad (11.5.7)$$

Since φ^0 and u^h are uniformly bounded by certain constant C, by choosing C_δ large enough we may assume $\varphi - u^h < -2C$ outside of $O_\delta(0, 0)$ for all h. Let (t_h, x_h) denote a maximum argument of $\varphi - u^h$. Since $[\varphi - u^h](0, 0) \ge -2C$, then $(t_h, x_h) \in O_\delta(0, 0)$.

We now choose a sequence $h_k \downarrow 0$ such that $\lim_{k \to \infty} u^{h_k}(0, 0) = \underline{u}(0, 0)$. Denote $t_k := t_{h_k}, x_k := x_{h_k}$, and $u^k := u^{h_k}$. Note that $\{(t_k, x_k)\}_{k \ge 1} \subset O_\delta(0, 0)$ has a limit point (t_*, x_*). Without loss of generality we assume $(t_k, x_k) \to (t_*, x_*)$. By the uniform continuity of u^k and the definition (5.5.4), we have

$$0 = [\varphi^0 - \underline{u}](0, 0) = \lim_{k \to \infty}[\varphi - u^k](0, 0) \le \liminf_{k \to \infty}[\varphi - u^k](t_k, x_k)$$

$$= \liminf_{k \to \infty}[\varphi - u^k](t_*, x_*)$$

$$\le [\varphi - \underline{u}](t_*, x_*) = [\varphi^0 - \underline{u}](t_*, x_*) - C_\delta[|t_*|^2 + |x_*|^4] \le -C_\delta[|t_*|^2 + |x_*|^4].$$

This implies $(t_*, x_*) = (0, 0)$, and thus

$$\lim_{k \to \infty} (t_k, x_k) = (0, 0). \qquad (11.5.8)$$

Now for each k, assume $(i_k - 1)h_k \leq t_k < i_k h_k$. Denote

$$\tilde{t}_k := i_k h_k, \quad \tilde{h}_k := \tilde{t}_k - t_k, \quad c_k := [u^k - \varphi](t_k, x_k).$$

It is clear that $\lim_{k \to 0} \tilde{h}_k = 0$, and

$$\lim_{k \to \infty} c_k = \lim_{k \to \infty} [u^k - \varphi](0, 0) = [\underline{u} - \varphi^0](0, 0) = 0.$$

By the optimality of (t_k, x_k), we have

$$\varphi(\tilde{t}_k, x) + c_k \leq u^k(\tilde{t}_k, x), \quad \text{for all } x \in \mathbb{R}^d.$$

Then it follows from the monotonicity condition (11.5.5) that

$$\mathbb{T}^{t_k, x_k}_{\tilde{h}_k}\Big[\varphi(\tilde{t}_k, \cdot) + c_k\Big] \leq \mathbb{T}^{t_k, x_k}_{\tilde{h}_k}[u^k(\tilde{t}_k, \cdot)] + \tilde{h}_k \rho_{mon}(\tilde{h}_k)$$

$$= u^k(t_k, x_k) + \tilde{h}_k \rho_{mon}(\tilde{h}_k) = \varphi(t_k, x_k) + c_k + \tilde{h}_k \rho_{mon}(\tilde{h}_k).$$

Now by the consistency condition (11.5.4) we have

$$0 \geq \lim_{k \to \infty} \frac{\mathbb{T}^{t_k, x_k}_{\tilde{h}_k}\Big[\varphi(\tilde{t}_k, \cdot) + c_k\Big] - [\varphi(t_k, x_k) + c_k] - \tilde{h}_k \rho_{mon}(\tilde{h}_k)}{\tilde{h}_k}$$

$$= \mathbb{L}\varphi(0, 0) = \mathbb{L}\varphi^0(0, 0).$$

This implies that $\underline{u}$ is a viscosity supersolution of the PDE (11.5.1) at $(0, 0)$. ∎

Remark 11.5.3

(i) Note that (11.5.3) can be viewed as a discretization of the PDE. Then the monotonicity condition (11.5.5) can be interpreted as the comparison principle for the discrete PDE.

(ii) While Theorem 11.5.2 is very elegant, in general it may not be easy to propose (efficient) monotone schemes for a given PDE. For a reasonable discretization scheme, the consistency condition (11.5.4) is usually quite straightforward. The stability condition typically also holds true, under appropriate technical conditions, although the technical proof could be involved. The monotonicity condition (11.5.5), however, is a serious requirement on the structure of the scheme.

(iii) By more involved arguments, one may also obtain rate of convergence of monotone schemes.

(iv) There are also non-monotone schemes which converge to the true solution. For example, the backward Euler scheme introduced in Subsection 5.3.2 is not monotone, see Remark 5.3.2 (iii). ∎

11.5.2 Monotone Scheme for PPDEs

We now extend Theorem 11.5.2 to PPDE (11.2.1) with terminal condition ξ. We shall assume

Assumption 11.5.4

(i) Assumption 11.2.1 holds and $\xi \in UC_b(\Omega)$.

(ii) Comparison principle for PPDE (11.2.1) holds. That is, if $u^1, u^2 \in UC_b(\overline{\Lambda})$ are bounded viscosity subsolution and viscosity supersolution of PPDE (11.2.1), respectively, and $u_1(T, \cdot) \leq \xi \leq u_2(T, \cdot)$, then $u_1 \leq u_2$ on $\overline{\Lambda}$.

For any $(t, \omega) \in [0, T) \times \Omega$ and $h \in (0, T - t]$, let $\mathbb{T}_h^{t,\omega}$ be an operator on $UC_b(\mathscr{F}_h) := UC(\Omega) \cap \mathbb{L}^0(\mathscr{F}_h)$. For the setting in (11.5.2), define:

$$u^h(t_n, \omega) := \xi(\omega), \quad u^h(t, \omega) := \mathbb{T}_{t_i-t}^{t,\omega}\big[(u^h(t_i, \cdot))^{t,\omega}\big], \quad t \in [t_{i-1}, t_i), \quad i = n, \cdots, 1.$$

$$(11.5.9)$$

Notice that the above notation is slightly different from that of (11.5.3), due to the shift of the space in our path dependent case. Our main result is:

Theorem 11.5.5 Let Assumption 11.5.4 hold. Assume $\mathbb{T}_h^{t,\omega}$ satisfies the following conditions:

(i) Consistency: for any $(t, \omega) \in [0, T) \times \Omega$, $\varphi \in C^{1,2}(\Lambda^t)$, and $(\tilde{t}, \tilde{\omega}) \in \Lambda^t$, $h \in (0, T - t)$, $c \in \mathbb{R}$,

$$\lim_{(\tilde{t}, \tilde{\omega}, h, c) \to (t, 0, 0, 0)} \frac{\mathbb{T}_h^{t+\tilde{t}, \omega \otimes_t \tilde{\omega}}\big[c + \big(\varphi(\tilde{t}+h, \cdot)\big)^{\tilde{t}, \tilde{\omega}}\big] - [c + \varphi](\tilde{t}, \tilde{\omega})}{h} = \mathscr{L}^{t,\omega}\varphi(0,0).$$

$$(11.5.10)$$

(ii) Monotonicity: for some constant $L \geq L_0$ and any $\varphi, \psi \in UC_b(\mathscr{F}_h)$, there exists a modulus of continuity function ρ_{mon}, which depends only on L, d, and the uniform continuity of φ, ψ, but does not depend on the specific φ, ψ, such that

$$\overline{\mathscr{E}}^L[\varphi - \psi] \leq 0 \quad \text{implies} \quad \mathbb{T}_h^{t,\omega}[\varphi] \leq \mathbb{T}_h^{t,\omega}[\psi] + h\rho_{mon}(h), \quad (11.5.11)$$

(iii) Stability: u^h is uniformly bounded and uniformly continuous in (t, ω), uniformly on h.

Then PPDE (11.2.1) *with terminal condition* $u(T, \cdot) = \xi$ *has a unique bounded L-viscosity solution* u, *and* u_h *converges to* u *locally uniformly as* $h \to 0$.

The above consistency and stability conditions are quite straightforward extension of the corresponding conditions in Theorem 11.5.2. Notice that the monotonicity condition (11.5.5) is due to Definition 5.5.2 and (5.5.4) for viscosity solutions of PDEs. To adapt to Definition 11.2.4 and (11.2.4), the monotonicity condition (11.5.11) is quite natural from theoretical point of view. However, as we will mention in the next subsection, this condition is too strong for most implementable numerical schemes and thus we will weaken it in the next subsection.

To prove the theorem, we first need a technical lemma.

Lemma 11.5.6 *Let* $L > 0$, $\varepsilon > 0$, $c > 0$, *and* $u \in UC_b(\overline{\Lambda}_{H_\varepsilon})$ *with modulus of continuity function* ρ. *Assume* $\delta > 0$ *is small enough, and for some* $\tau \in \mathscr{T}_{\tilde{H}_\delta}$ *with* $\tilde{H}_\delta := \delta \wedge H_\varepsilon$,

$$\overline{\mathscr{E}}^L[u_\tau] - \overline{\mathscr{E}}^L[u_{\tilde{H}_\delta}] \geq c. \tag{11.5.12}$$

Then there exist a constant C, *which may depend on* $(L, d, \varepsilon, \delta, c, \rho)$, *and* $\omega^* \in \Omega$ *such that*

$$t_* := \tau(\omega^*) < \tilde{H}_\delta(\omega^*),$$

$$\|\omega^*\|_{t_*} \leq t_*^{\frac{1}{3}}, \quad \mathscr{C}^{\mathscr{P}_L}\big[(\tilde{H}_\delta)^{t*,\omega^*} - t_* \leq h\big] \leq Ch^2 \text{ for all } h > 0. \tag{11.5.13}$$

Proof First, by Problem 11.7.9 (i),

$$\mathscr{C}^{\mathscr{P}_L}\Big[\|X\|_\tau > \tau^{\frac{1}{3}}\Big] \leq \mathscr{C}^{\mathscr{P}_L}\Big[\sup_{0 < t \leq \delta} \frac{|X_t|}{t^{\frac{1}{3}}} > 1\Big] \tag{11.5.14}$$

$$\leq \overline{\mathscr{E}}^L\Big[\sup_{0 < t \leq \delta} \frac{|X_t|^{12}}{t^4}\Big] \leq C_L \delta^2,$$

Then, for $\delta > 0$ small enough, by (11.5.12) we have

$$\frac{c}{2} \leq \overline{\mathscr{E}}^L\Big[u_\tau \mathbf{1}_{\{\|X\|_\tau \leq \tau^{\frac{1}{3}}\}}\Big] - \overline{\mathscr{E}}^L\Big[u_{\tilde{H}_\delta} \mathbf{1}_{\{\|X\|_\tau \leq \tau^{\frac{1}{3}}\}}\Big].$$

By the definition of $\overline{\mathscr{E}}^L$, there exists $\mathbb{P} \in \mathscr{P}_L$ such that

$$\frac{c}{3} \leq \mathbb{E}^{\mathbb{P}}\Big[u_\tau \mathbf{1}_{\{\|X\|_\tau \leq \tau^{\frac{1}{3}}\}}\Big] - \overline{\mathscr{E}}^L\Big[u_{\tilde{H}_\delta} \mathbf{1}_{\{\|X\|_\tau \leq \tau^{\frac{1}{3}}\}}\Big].$$

Let $\mathbb{P}^0 \in \mathscr{P}_L$ be such that $b' = 0, \sigma' = 0$ in (9.2.10), and $\hat{\mathbb{P}} := \mathbb{P} \otimes_\tau \mathbb{P}^0$ in the sense of (9.3.21)–(9.3.22). It is straightforward to check that $\hat{\mathbb{P}} \in \mathscr{P}_L$ and $\hat{\mathbb{P}} = \mathbb{P}$ on $\mathscr{F}_\tau$. Then, by the uniform regularity of u,

$$\frac{c}{3} \le \mathbb{E}^{\hat{\mathbb{P}}}\Big[[u_\tau - u_{\tilde{H}_\delta}] 1_{\{\|X\|_\tau \le \tau^{\frac{1}{3}}\}} \Big] \le C\mathbb{E}^{\hat{\mathbb{P}}}\Big[\rho\big(\mathbf{d}((\tau, X), (\tilde{H}_\delta, X))\big) 1_{\{\|X\|_\tau \le \tau^{\frac{1}{3}}\}} \Big].$$

Note that X is flat on $[\tau, \tilde{H}_\delta]$, $\hat{\mathbb{P}}$-a.s. Recall (10.1.8), we see that $H_\varepsilon = \frac{\varepsilon - |X_\tau|}{L+1}$, $\hat{\mathbb{P}}$-a.s. Then, when $|X_\tau| \le \tau^{\frac{1}{3}} \le \delta^{\frac{1}{3}}$, for δ small enough we have $H_\varepsilon \ge \delta$ and thus $\tilde{H}_\delta = \delta$. This means

$$\frac{c}{3} \le C\mathbb{E}^{\hat{\mathbb{P}}}\Big[\rho\big(\mathbf{d}((\tau, X), (\delta, X_{\cdot \wedge \tau}))\big) 1_{\{\|X\|_\tau \le \tau^{\frac{1}{3}}\}} \Big] = C\mathbb{E}^{\mathbb{P}}\Big[\rho(\delta - \tau) 1_{\{\|X\|_\tau \le \tau^{\frac{1}{3}}\}} \Big].$$

Let $\delta_0 > 0$ be such that $C\rho(\delta_0) \le \frac{c}{3}$. Then there exists $\omega^* \in \Omega$ such that,

$$t_* := \tau(\omega^*) < \delta = \tilde{H}_\delta(\omega^*), \quad \|\omega^*\|_{t_*} \le t_*^{\frac{1}{3}}, \quad \delta - t_* \ge \delta_0.$$

Finally, recall (10.1.8) again, we see that

$$H_\varepsilon^{t_*, \omega^*} - t_* = \inf\Big\{ t \ge 0 : |\omega_{t_*}^* + X_t| + (L+1)(t_* + t) \ge \varepsilon \Big\}$$

$$\ge \inf\Big\{ t \ge 0 : |X_t| + (L+1)t \ge \varepsilon - \delta^{\frac{1}{3}} - (L+1)\delta \Big\} \ge H_{\frac{\varepsilon}{2}},$$

for $\delta > 0$ small enough. Then

$$(\tilde{H}_\delta)^{t_*, \omega^*} - t_* = (\delta \wedge H_\varepsilon)^{t_*, \omega^*} - t_* \ge \delta_0 \wedge H_{\frac{\varepsilon}{2}}.$$

Then the last inequality of (11.5.13) follows directly from Lemma 10.1.18 (i). ∎

Proof of Theorem 11.5.5 We shall follow similar ideas as in the proof of Theorem 11.5.2. However, the arguments are more involved, mainly due to our definition of viscosity solution of PPDEs and the corresponding monotonicity condition (11.5.11). Define

$$\underline{u}(t, \omega) := \liminf_{h \to 0} u^h(t, \omega), \quad \overline{u}(t, \omega) := \limsup_{h \to 0} u^h(t, \omega). \tag{11.5.15}$$

Following the same arguments as in Theorem 11.5.2, it suffices to show that $\underline{u}$ satisfies the viscosity L-supersolution property at $(0, 0)$. Let $\varphi^0 \in \overline{\mathscr{A}}^L \underline{u}(0, 0)$ with localization time H_ε, and define

$$\varphi^\varepsilon(t, \omega) := \varphi^0(t, \omega) - \varepsilon t. \tag{11.5.16}$$

Moreover, let $h_k \downarrow 0$ be a subsequence such that

$$\lim_{k \to \infty} u_0^{h_k} = \underline{u}_0. \quad \text{where} \tag{11.5.17}$$

Throughout this proof, denote

$$u^k := u^{h_k} \quad \text{and} \quad \tilde{u}^k := \varphi^\varepsilon - u^k.$$

By Remark 11.2.12 (ii), we may assume without loss of generality that φ^0 is uniformly continuous, and thus so is φ^ε. By the stability condition, all the processes involved in this proof are uniformly continuous with a common modulus of continuity function denoted as ρ_0.

Step 1. Denote $\tilde{H}_\delta := \delta \wedge H_\varepsilon$. In this step we show that, for $\delta > 0$ small enough,

$$\liminf_{k \to \infty} \left[\tilde{u}_0^k - \overline{\mathscr{E}}^L \big[\tilde{u}_{\tilde{H}_\delta}^k \big] \right] > \frac{1}{2} \varepsilon \delta. \tag{11.5.18}$$

Indeed, by Lemma 10.1.18 (i) we have

$$\mathscr{C}^{\mathscr{P}_L}(\tilde{H}_\delta \neq \delta) = \mathscr{C}^{\mathscr{P}_L}(H_\varepsilon < \delta) \leq C_\varepsilon \delta. \tag{11.5.19}$$

Then, since $\varphi^0 \in \overline{\mathscr{A}}^L \underline{u}(0,0)$, for $\delta > 0$ small enough,

$$[\varphi^\varepsilon - \underline{u}]_0 = [\varphi^0 - \underline{u}]_0 \geq \overline{\mathscr{E}}^L [(\varphi^0 - \underline{u})_{\tilde{H}_\delta}] = \overline{\mathscr{E}}^L \Big[(\varphi^\varepsilon - \underline{u})_{\tilde{H}_\delta} + \varepsilon \tilde{H}_\delta \Big]$$

$$\geq \overline{\mathscr{E}}^L \Big[(\varphi^\varepsilon - \underline{u})_{\tilde{H}_\delta} \Big] + \varepsilon \underline{\mathscr{E}}^L [\tilde{H}_\delta] = \overline{\mathscr{E}}^L \Big[(\varphi^\varepsilon - \underline{u})_{\tilde{H}_\delta} \Big] + \varepsilon \delta - \varepsilon \overline{\mathscr{E}}^L [(\delta - \tilde{H}_\delta)^+]$$

$$\geq \overline{\mathscr{E}}^L \Big[(\varphi^\varepsilon - \underline{u})_{\tilde{H}_\delta} \Big] + \varepsilon \delta - C_\varepsilon \delta^2. \tag{11.5.20}$$

Note that $\sup_{k \geq n} \tilde{u}_\delta^k$ is also uniformly continuous in ω. Then it follows from the dominated convergence Theorem 10.1.8 that

$$\overline{\mathscr{E}}^L \Big[(\varphi^\varepsilon - \underline{u})_\delta \Big] = \overline{\mathscr{E}}^L \Big[\lim_{n \to \infty} \sup_{k \geq n} \tilde{u}_\delta^k \Big] = \lim_{n \to \infty} \overline{\mathscr{E}}^L \Big[\sup_{k \geq n} \tilde{u}_\delta^k \Big] \geq \limsup_{k \to \infty} \overline{\mathscr{E}}^L [\tilde{u}_\delta^k].$$

Denote

$$\rho(\delta) := \frac{1}{\delta} \overline{\mathscr{E}}^L \Big[\rho_0(\|X\|_\delta) \mathbf{1}_{\{\tilde{H}_\delta < \delta\}} \Big] \to 0, \quad \text{as } \delta \to 0, \tag{11.5.21}$$

thanks to Problem 11.7.9 (ii). Then, by (11.5.20) and (11.5.17),

$$\varepsilon \delta - C_\varepsilon \delta^2 \leq [\varphi^\varepsilon - \underline{u}]_0 - \overline{\mathscr{E}}^L \Big[(\varphi^\varepsilon - \underline{u})_\delta \Big] + C \delta \rho(\delta) \leq \lim_{k \to \infty} \tilde{u}_0^k - \limsup_{k \to \infty} \overline{\mathscr{E}}^L [\tilde{u}_\delta^k] + C \delta \rho(\delta)$$

$$\leq \lim_{k \to \infty} \tilde{u}_0^k - \limsup_{k \to \infty} \overline{\mathscr{E}}^L [\tilde{u}_{\tilde{H}_\delta}^k] + C \delta \rho(\delta) = \liminf_{k \to \infty} \Big[\tilde{u}_0^k - \overline{\mathscr{E}}^L [\tilde{u}_{\tilde{H}_\delta}^k] \Big] + C \delta \rho(\delta).$$

and thus

$$\liminf_{k \to \infty} \Big[\tilde{u}_0^k - \overline{\mathscr{E}}^L [\tilde{u}_{\tilde{H}_\delta}^k] \Big] \geq \Big[\varepsilon - C_\varepsilon \delta - \rho(\delta) \Big] \delta,$$

which implies (11.5.18) immediately by choosing $\delta > 0$ small enough.

Step 2. Fix $\delta > 0$ small enough so that (11.5.18) holds. Then

$$\tilde{u}_0^k - \overline{\mathscr{E}}^L\big[\tilde{u}_{\tilde{H}_\delta}^k\big] \geq \frac{1}{2}\varepsilon\delta, \quad \text{for all large } k. \tag{11.5.22}$$

Denote

$$Y_t^{\delta,k}(\omega) := \sup_{\tau \in \mathscr{T}_{\tilde{H}_\delta}^{t,\omega}-t} \overline{\mathscr{E}}^L\big[(\tilde{u}^k)_\tau^{t,\omega}\big], \ (t,\omega) \in \Lambda_{\tilde{H}_\delta}, \quad \tau_k^\delta := \inf\{t \geq 0 : Y_t^{\delta,k} = \tilde{u}_t^k\}.$$

It is straightforward to verify that $\tilde{H}_\delta$ satisfies Assumption 10.3.1. Then by Theorem 10.3.2 we know $\tau_k^\delta \leq \tilde{H}_\delta$ is an optimal stopping time for $Y_0^{\delta,k}$ and thus, for all large k,

$$0 < \frac{1}{2}\varepsilon\delta \leq \tilde{u}_0^k - \overline{\mathscr{E}}^L\big[\tilde{u}_{\tilde{H}_\delta}^k\big] \leq Y_0^k - \overline{\mathscr{E}}^L\big[\tilde{u}_{\tilde{H}_\delta}^k\big] = \overline{\mathscr{E}}^L\big[\tilde{u}_{\tau_k^\delta}^k\big] - \overline{\mathscr{E}}^L\big[\tilde{u}_{\tilde{H}_\delta}^k\big]. \tag{11.5.23}$$

Applying Lemma 11.5.6, for $\delta > 0$ small enough, there exist a constant $C_{\varepsilon,\delta}$ (independent of k) and $\omega^{\delta,k} \in \Omega$ for each k such that

$$\begin{aligned}
&t_k^\delta := \tau_k^\delta(\omega^{\delta,k}) < \tilde{H}_\delta(\omega^{\delta,k}), \quad \|\omega^{\delta,k}\|_{t_k^\delta} \leq (t_k^\delta)^{\frac{1}{3}}, \\
&\mathscr{C}^{\mathscr{P}_L}\Big(\tilde{H}_\delta^k \leq h\Big) \leq C_{\varepsilon,\delta}h^2 \text{ for all } h > 0, \quad \text{where } \tilde{H}_\delta^k := (\tilde{H}_\delta)^{t_k^\delta,\omega^{\delta,k}} - t_k^\delta.
\end{aligned} \tag{11.5.24}$$

Now let $\{t_i^k\}_{0 \leq i \leq n_k}$ be the time partition corresponding to h_k, and assume $t_{i-1}^k \leq t_k^\delta < t_i^k$. Note that

$$\tilde{u}_{t_k^\delta}^k(\omega^{\delta,k}) = Y_{t_k^\delta}^{\delta,k}(\omega^{\delta,k}) \geq \overline{\mathscr{E}}^L\big[(\tilde{u}^k)_\tau^{t_k^\delta,\omega^{\delta,k}}\big], \quad \forall \tau \in \mathscr{T}_{\tilde{H}_\delta^k}.$$

Denote $h_k^\delta := t_i^k - t_k^\delta \leq h_k$ and set $\tau := h_k^\delta$ in the above inequality. Then by (11.5.24) we have

$$\tilde{u}_{t_k^\delta}^k(\omega^{\delta,k}) \geq \overline{\mathscr{E}}^L\Big[(\tilde{u}^k)_{h_k^\delta \wedge \tilde{H}_\delta^k}^{t_k^\delta,\omega^{\delta,k}}\Big] \geq \overline{\mathscr{E}}^L\Big[(\tilde{u}^k)_{h_k^\delta}^{t_k^\delta,\omega^{\delta,k}}\Big] - C_{\varepsilon,\delta}(h_k^\delta)^2.$$

This implies

$$\overline{\mathscr{E}}^L\Big[\big((\varphi^\varepsilon)_{h_k^\delta}^{t_k^\delta,\omega^{\delta,k}} - \tilde{u}_{t_k^\delta}^k(\omega^{\delta,k}) - C_{\varepsilon,\delta}(h_k^\delta)^2\big) - (u^k)_{h_k^\delta}^{t_k^\delta,\omega^{\delta,k}}\Big] \leq 0.$$

Then by the monotonicity condition (11.5.11) we have

$$\begin{aligned}
u^k(t_k^\delta, \omega^{\delta,k}) &= \mathbb{T}_{h_k^\delta}^{t_k^\delta,\omega^{\delta,k}}\big[(u^k)_{h_k^\delta}^{t_k^\delta,\omega^{\delta,k}}\big] \\
&\geq \mathbb{T}_{h_k^\delta}^{t_k^\delta,\omega^{\delta,k}}\Big[(\varphi^\varepsilon)_{h_k^\delta}^{t_k^\delta,\omega^{\delta,k}} - \tilde{u}_{t_k^\delta}^k(\omega^{\delta,k}) - C_{\varepsilon,\delta}(h_k^\delta)^2\Big] - h_k^\delta\rho_{mon}(h_k^\delta).
\end{aligned}$$

For each $\delta > 0$, fix a k_δ large enough such that

$$\lim_{\delta \to 0} k_\delta = \infty \quad \text{and the above } C_{\varepsilon,\delta} h_{k_\delta}^\delta \leq (h_{k_\delta}^\delta)^{\frac{1}{2}}. \tag{11.5.25}$$

Simplify the notations:

$$t_\delta := t_{k_\delta}^\delta, \quad \omega^\delta := \omega^{\delta,k_\delta}, \quad h_\delta := h_{k_\delta}^\delta.$$

Then

$$u^k(t_\delta, \omega^\delta) \geq \mathbb{T}_{h_\delta}^{t_\delta,\omega^\delta} \left[(\varphi^\varepsilon)_{h_\delta}^{t_\delta,\omega^\delta} - \tilde{u}^{k_\delta}(t_\delta, \omega^\delta) - C_{\varepsilon,\delta}(h_\delta)^2 \right] - h_\delta \rho_{mon}(h_\delta). \tag{11.5.26}$$

Step 3. We next use the consistency condition (11.5.10) on $(t, \omega) = (0,0)$. Set

$$c_\delta := -\tilde{u}^{k_\delta}(t_\delta, \omega^\delta) - C_{\varepsilon,\delta}(h_\delta)^2.$$

Send $\delta \to 0$ and recall the first estimate in (11.5.24), we see that

$$h_\delta \leq h_{k_\delta} \to 0, \quad \mathbf{d}((t_\delta, \omega^\delta), (0,0)) \leq h_{k_\delta} + \|\omega^\delta\|_{t_\delta} \leq h_{k_\delta} + t_\delta^{\frac{1}{3}} \to 0,$$

which, together with (11.5.17), (11.5.25), and the uniform continuity of φ^ε and u^k, implies

$$|c_\delta| \leq \left| \tilde{u}^{k_\delta}(t_\delta, \omega^\delta) - \tilde{u}^{k_\delta}(0,0) \right| + |u_0^{k_\delta} - \underline{u}_0| + h_\delta^{\frac{3}{2}} \to 0.$$

Then, by the consistency condition (11.5.10) we obtain from (11.5.26) and (11.5.25) that

$$0 \geq \frac{\mathbb{T}_{h_\delta}^{t_\delta,\omega^\delta} \left[(\varphi^\varepsilon)_{h_\delta}^{t_\delta,\omega^\delta} - \tilde{u}^{k_\delta}(t_\delta, \omega^\delta) - C_{\varepsilon,\delta}(h_\delta)^2 \right] - u^{k_\delta}(t_\delta, \omega^\delta) - h_\delta \rho_{mon}(h_\delta)}{h_\delta}$$

$$= \frac{\mathbb{T}_{h_\delta}^{t_\delta,\omega^\delta} \left[c + (\varphi^\varepsilon)_{h_\delta}^{t_\delta,\omega^\delta} \right] - [c + \varphi^\varepsilon](t_\delta, \omega^\delta)}{h_\delta} - C_{\varepsilon,\delta} h_\delta - \rho_{mon}(h_\delta)$$

$$\to \mathscr{L}\varphi^\varepsilon(0,0) = \mathscr{L}\varphi^0(0,0) - \varepsilon.$$

This implies that $\mathscr{L}\varphi^0(0,0) \leq \varepsilon$. Note that ε can be arbitrarily small, then $\mathscr{L}\varphi^0(0,0) \leq 0$ and thus $\underline{u}$ is a viscosity L-supersolution of PPDE (11.2.1) at $(0,0)$. ∎

11.5.3 Discretization of the Nonlinear Expectation

While the monotonicity condition (11.5.11) looks natural in our framework, the natural extension of the monotone schemes in existing PDE literature typically do not satisfy (11.5.11). For this purpose, we introduce a discrete nonlinear expectation $\overline{\mathcal{E}}_h$ to replace the nonlinear expectation $\overline{\mathcal{E}}^L$.

Definition 11.5.7 *Consider the setting* (11.5.2) *and let K be a subset of a metric space.*

(i) *Let $\Phi_h : K \times [0,1] \to \mathbb{R}$ be Borel measurable and satisfy, for some $L > 0$ and for any $k \in K$,*

$$\left| \int_0^1 \Phi_h(k,x)dx \right| \le Lh, \quad \int_0^1 |\Phi_h(k,x)|^2 dx \le Lh, \quad \int_0^1 |\Phi_h(k,x)|^3 dx \le Lh^{\frac{3}{2}}.$$

$$(11.5.27)$$

(ii) *Let $(\tilde{\Omega}, \tilde{\mathscr{F}}, \tilde{\mathbb{P}})$ be an arbitrary probability space, on which is defined a sequence of independent random variables $\{U_i\}_{1 \le i \le n}$ with uniform distribution on $[0,1]$. Denote the filtration $\tilde{\mathscr{F}}_i := \sigma\{U_1, \cdots, U_i\}$, $i = 1, \cdots, n$, and let $\mathcal{K} := \{k = \{k_i\}_{0 \le i \le n-1} : k_i \in \mathbb{L}^0(\tilde{\mathscr{F}}_i, K), i = 0, \cdots, n-1\}$.*

(iii) *Consider the time partition in* (11.5.2). *For each $k \in \mathcal{K}$, define*

$$\tilde{X}_0^k := 0, \quad \tilde{X}_{t_{i+1}}^k := \tilde{X}_{t_i}^k + \Phi_h(k_i, U_{i+1}), \ i = 0, \cdots, n-1, \quad (11.5.28)$$

and extend $\tilde{X}$ to the whole time interval $[0,T]$ by linear interpolation.

(iv) *Finally we introduce the discrete nonlinear expectation:*

$$\overline{\mathcal{E}}_h[\varphi] := \sup_{k \in \mathcal{K}} \mathbb{E}^{\tilde{\mathbb{P}}}\left[\varphi(\tilde{X}^k_\cdot) \right], \quad \text{for all } \varphi \in UC_b(\Omega). \quad (11.5.29)$$

The discrete nonlinear expectation $\overline{\mathcal{E}}_h$ approximates $\overline{\mathcal{E}}^L$ in the following sense.

Lemma 11.5.8 *Let Φ_h be as in* (11.5.27) *with constant $L > 0$. Then for any $\varphi \in UC_b(\Omega)$,*

$$\lim_{h \to 0} \overline{\mathcal{E}}_h[\varphi] = \overline{\mathcal{E}}^L[\varphi]. \quad (11.5.30)$$

Moreover, there exists a modulus of continuity function ρ, which depends only on T, d, L, and the bound and the modulus of continuity of φ, such that

$$\overline{\mathcal{E}}_h[\varphi] \le \overline{\mathcal{E}}^L[\varphi] + \rho(h). \quad (11.5.31)$$

The proof relies on the invariance principle, see Sakhanenko [207]. We shall skip the proof and refer interested readers to Ren & Tan [202], which in turn relies on Dolinsky [65] and Tan [220].

By using Lemma 11.5.8, we have the following result whose proof is similar to that of Theorem 11.5.5 and thus is omitted.

Theorem 11.5.9 *Theorem 11.5.5 remains true if we replace the monotonicity condition* (11.5.11) *with the following condition: there exist $\overline{\mathscr{E}}_h$ for each $h > 0$, independent of the φ, ψ in Theorem 11.5.5 (ii), such that*

$$\sup_{\alpha \in \mathbb{L}^0(\mathscr{F}_h), 0 \le \alpha \le L} \overline{\mathscr{E}}_h\big[e^{\alpha h}[\varphi - \psi]\big] \le 0 \quad implies \quad \mathbb{T}_h^{t,\omega}[\varphi] \le \mathbb{T}_h^{t,\omega}[\psi] + h\rho_{mon}(h).$$

$$(11.5.32)$$

Remark 11.5.10 By choosing $\alpha = 0$ in the left side of (11.5.32), we see immediately that Theorem 11.5.5 still remains true if we replace (11.5.32) with the following stronger condition:

$$\overline{\mathscr{E}}_h[\varphi - \psi] \le 0 \quad implies \quad \mathbb{T}_h^{t,\omega}[\varphi] \le \mathbb{T}_h^{t,\omega}[\psi] + h\rho_{mon}(h). \qquad (11.5.33)$$

However, some numerical schemes satisfy (11.5.32) but not (11.5.33), see Ren & Tan [202]. ∎

We conclude this subsection by showing that the finite difference scheme does satisfy the alternative monotonicity condition (11.5.33), and hence (11.5.32) as well. For notational simplicity, we assume $d = 1$. Recall PPDE (11.2.1) and Assumption 11.2.1. We now introduce $\mathbb{T}_h^{t,\omega}$ as follows. For any $\varphi \in UC_b(\mathscr{F}_h)$, we abuse the notation and denote

$$\varphi(x) := \varphi\big((\frac{t}{h}x)_{0 \le t \le h}\big), \quad \text{for all } x \in \mathbb{R}. \qquad (11.5.34)$$

We then define: using spatial increments Δx,

$$\mathbb{T}_h^{t,\omega}[\varphi] := \varphi(0) + hG\big(t, \omega, \varphi(0), D_h\varphi(0), D_h^2\varphi(0)\big), \qquad (11.5.35)$$

where $D_h\varphi(0) := \dfrac{\varphi(\Delta x) - \varphi(-\Delta x)}{2\Delta x}, \quad D_h^2\varphi(0) := \dfrac{\varphi(\Delta x) + \varphi(-\Delta x) - 2\varphi(0)}{(\Delta x)^2}.$

We first note that the above $\mathbb{T}_h^{t,\omega}$ typically does not satisfy (11.5.11). See Problem 11.7.10 for a counterexample. We next show that $\mathbb{T}_h^{t,\omega}$ satisfies (11.5.33).

Proposition 11.5.11 *Let Assumption 11.2.1 hold, and assume there exist $c_0 > 0$ such that*

$$G(t, \omega, y, z, \gamma + \Delta\gamma) - G(t, \omega, y, z, \gamma) \ge c_0 \text{tr}(\Delta\gamma) \text{ for all } (t, \omega, y, z, \gamma) \text{ and } \Delta\gamma \ge 0.$$

$$(11.5.36)$$

Then, by choosing $\Delta x = C_1 \sqrt{h}$ *for some* $C_1 > \sqrt{2L_0}$, *one may construct* $\overline{\mathscr{E}}_h$ *so that* (11.5.33) *holds.*

Proof We first note that, since one may modify $\rho_{mon}(h)$ for large h, it is sufficient to verify (11.5.32) for h small enough. For $\varphi, \psi \in UC_b(\mathscr{F}_h)$ and denoting $\zeta := \varphi - \psi$,

$$\mathbb{T}_h^{t,\omega}[\varphi] - \mathbb{T}_h^{t,\omega}[\psi] = \zeta(0) + h\Big[a_1\zeta(0) + a_2 D_h\zeta(0) + a_3 D_h^2\zeta(0)\Big]$$

$$= \Big[1 + k_1 h - \frac{2k_3}{C_1^2}\Big]\zeta(0) + \Big[\frac{k_3}{C_1^2} + \frac{k_2\sqrt{h}}{2C_1}\Big]\zeta(\Delta x)$$

$$+ \Big[\frac{k_3}{C_1^2} - \frac{k_2\sqrt{h}}{2C_1}\Big]\zeta(-\Delta x),$$

where $|k_1|, |k_2| \leq L_0$ and $c_0 \leq k_3 \leq L_0$. Since $C_1 > \sqrt{2L_0}$, when h is small enough, we see that

$$p_1 := 1 + k_1 h - \frac{2k_3}{C_1^2} > 0, \quad p_2 := \frac{k_3}{C_1^2} + \frac{k_2\sqrt{h}}{2C_1} > 0, \quad p_3 := \frac{k_3}{C_1^2} - \frac{k_2\sqrt{h}}{2C_1} > 0,$$

and their sum is $1 + k_1 h$. Let $k = (k_1, k_2, k_3)$ be the parameter set, and define

$$\Phi_h(k, x) := \Delta x \mathbf{1}_{\{x < \frac{p_2}{1+k_1 h}\}} - \Delta x \mathbf{1}_{\{\frac{p_2}{1+k_1 h} < x < \frac{p_2+p_3}{1+k_1 h}\}}. \tag{11.5.37}$$

Then, for standard uniform distribution U,

$$\mathbb{T}_h^{t,\omega}[\varphi] - \mathbb{T}_h^{t,\omega}[\psi] = (1 + k_1 h)\mathbb{E}\Big[\zeta\big(\Phi_h(k, U)\big)\Big].$$

Then clearly (11.5.33) holds with $\rho_{mon} = 0$ (for small h).

It remains to verify (11.5.27). Indeed,

$$\mathbb{E}\big[\Phi_h(k, U)\big] = \Delta x \frac{p_2}{1 + k_1 h} - \Delta x \frac{p_3}{1 + k_1 h} = \Delta x \frac{k_2\sqrt{h}}{C_1} = k_2 h;$$

$$\mathbb{E}\big[|\Phi_h(k, U)|^2\big] = |\Delta x|^2 \frac{p_2 + p_3}{1 + k_1 h} = |\Delta x|^2 \frac{2k_3}{C_1^2(1 + k_1 h)} = \frac{2k_3}{1 + k_1 h} h;$$

$$\mathbb{E}\big[|\Phi_h(k, U)|^3\big] = |\Delta x|^3 \frac{p_2 + p_3}{1 + k_1 h} = |\Delta x|^3 \frac{2k_3}{C_1^2(1 + k_1 h)} = \frac{2k_3 C_1}{1 + k_1 h} h^{\frac{3}{2}}.$$

Now by choosing L large, obviously (11.5.27) holds true. ∎

11.6   Bibliographical Notes

The notion of PPDE is proposed by Peng [182, 183]. In the linear case, its classical
solution is essentially the functional Itô formula of Dupire [71], see also Cont &
Fournier [36, 37] and Cont [35]. Peng & Wang [186] studied classical solutions of
semilinear PPDE (11.3.2).

The notion of viscosity solution for PPDEs in this chapter is introduced by
Ekren, Keller, Touzi, & Zhang [73] for semilinear PPDEs and Ekren, Touzi,
& Zhang [75, 76] for fully nonlinear PPDEs. In particular, [76] established the
comparison principle for fully nonlinear PPDEs by utilizing the classical solution
of certain path frozen PDEs, as described in Section 11.4.3. As pointed out in
Section 11.4.3, in this approach it is more or less necessary to assume the PPDE
is uniformly nondegenerate. There are two recent works dealing with degenerate
fully nonlinear PPDEs, by using the viscosity theory of PDEs: Ren, Touzi, & Zhang
[205] introduced certain regularization technique and established the comparison
principle under some stronger regularities; and Ekren & Zhang [77] established
the comparison principle under mild conditions, in particular without requiring the
uniform continuity of G in ω, but only in the class of viscosity solutions with certain
piecewise Markov structure.

There have been many works following this notion of viscosity solutions. Ekren
[72] studied fully nonlinear PPDEs with obstacle. Keller [122] studied semilinear
PPDEs on càdlàg spaces. Ren [200] studied elliptic PPDEs. Ren, Touzi, & Zhang
[204] extended the punctual differentiability of Caffarelli & Cabre [27] to path
dependent case and proved the comparison principle for general (possibly degenerate) semilinear PPDE. In particular, [204] introduced the equivalent definition via
semi-jets. Ren [201] proved the existence of viscosity solution for semilinear PPDEs
by using Perron's approach. Cosso, Federico, Gozzi, Rosestolato, & Touzi [39]
studied PPDEs in infinite dimensions. We refer to Ren, Touzi, & Zhang [203] for a
survey and Ma, Ren, Touzi, & Zhang [142] for an application on large deviations.
We also remark that, when restricted to Markov case, the viscosity solution here is
closely related to the stochastic viscosity solution of Bayraktar & Sirbu [8, 9].

There have been several different approaches for PPDEs. Lukoyanov [139]
studied viscosity solutions for first order PPDEs by using compactness arguments.
Peng & Song [184] proposed Sobolev solutions for path dependent HJB equations
by using the estimates from the nonlinear expectation theory, which is more related
to the materials in Chapter 12 below. Cosso & Russo [40] proposed the so-called
strong-viscosity solution, which is by definition more or less the limit of classical
solutions, for semilinear PPDEs and established the well-posedness by using BSDE
arguments.

Section 11.3.4 on path dependent Isaacs equation is based on Pham & Zhang
[192]. The stochastic HJB equation in Section 11.3.5 is introduced by Peng
[173, 174], which solved the case with drift control. The general case with diffusion
control was proposed by Peng [178] as an open problem, and its well-posedness was
established recently by Qiu [197] in Sobolev sense, and by Ekren & Zhang [77] in

viscosity sense. There have also been many studies on backward SPDEs, many of which are on degenerate BSPDEs induced from the decoupling field for FBSDEs, see for example, Dokuchaev [64], Du, Tang, & Zhang [68], Hu, Ma, & Yong [108], Ma, Yin, & Zhang [145], Ma & Yong [147], Qiu & Tang [198], and Tang & Wei [224].

For the monotone schemes in Section 11.5, Subsection 11.5.1 for PDEs is based on Barles & Souganidis [5], Subsection 11.5.2 for PPDEs is based on Zhang & Zhuo [246], and the extension in Subsection 11.5.3 is due to Ren & Tan [202]. We note that the work of Henry-Labordere, Tan, & Touzi [105] proposed a completely different approach by using branching processes to solve semilinear PPDEs. Their proof relies on the well-posedness of viscosity solutions for the PPDE. We also note that there has been a large literature on monotone schemes for fully nonlinear parabolic PDEs, see, e.g., Krylov [132], Barles & Jakobsen [4], Fahim, Touzi, & Waxin [88], Guo, Zhang, & Zhuo [99], and Tan [221].

11.7 Exercises

Problem 11.7.1 Prove Proposition 11.2.13. ∎

Problem 11.7.2 Prove the local version of Theorem 11.2.9. That is, let Assumption 11.2.1 hold and $u \in C^{1,2}(\Lambda_{H_\delta})$ for some $\delta > 0$. Then, for any $(t, \omega) \in \Lambda_{H_\delta}$, u is a viscosity subsolution of PPDE (11.2.1) at (t, ω) if and only if $\mathscr{L}u(t, \omega) \geq 0$. ∎

Problem 11.7.3 Under Assumption 11.3.3, prove (11.3.5), (11.3.4), and (11.3.6). ∎

Problem 11.7.4 Let Assumption 11.3.6 hold.

 (i) Prove Lemma 11.3.7;
(ii) Prove (11.3.18) in detail. ∎

Problem 11.7.5 Let Assumption 11.3.11 and (11.3.21) hold.

 (i) Prove the remaining statements of Lemma 11.3.12 rigorously.
(ii) Prove (11.3.37) in detail. ∎

Problem 11.7.6 Let u be the decoupling field for coupled FBSDE introduced in Section 8.3. Derive formally the backward SPDE and the corresponding PPDE u should satisfy. ∎

Problem 11.7.7

 (i) Show that the v defined by (11.4.11) is a viscosity solution of the bound PDE in the sense of Section 5.5;
(ii) Formulate and prove the partial comparison principle needed in Lemma 11.4.9;
(iii) Verify (11.4.13);

(iv) Prove (11.4.15). (Roughly speaking one can prove this by first applying Lemma 11.4.9 (ii) and then Lemma 11.4.9. (i). However, note that the domain of the PDE depends on t_m, so one cannot apply Lemma 11.4.9 directly.) ∎

Problem 11.7.8 Prove (11.4.23) rigorously. ∎

Problem 11.7.9

(i) For any $L > 0$ and $p > 0$, there exists a constant $C_p = C_{L,d,p}$ such that,

$$\overline{\mathcal{E}}^L\left[\sup_{0 < t \le \delta} \frac{|X_t|^{3p}}{t^p} \right] \le C_p \delta^{\frac{p}{2}} \quad \text{for any } \delta > 0.$$

(ii) Prove (11.5.21). ∎

Problem 11.7.10 Consider the finite difference discretization operator in (11.5.35). Assume $G = \frac{1-\varepsilon}{2}\gamma$, $\Delta x = \sqrt{h}$. Then

$$\mathbb{T}_h^{t,\omega}[\varphi] := \varphi(0) + \frac{(1-\varepsilon)h}{2} D_h^2\varphi(0) = \frac{1-\varepsilon}{2}\left[\varphi(\sqrt{h}) + \varphi(-\sqrt{h})\right].$$

(i) Verify that $\mathbb{T}_h^{t,\omega}$ satisfies all the requirements in Proposition 11.5.11 and thus (11.5.32) holds.
(ii) Introduce $\varphi \in UC_b(\mathscr{F}_h)$:

$$\varphi(\omega) := c\sqrt{h} - Lh - \sup_{0 \le t \le h} \left| |\omega_t| - \frac{t}{\sqrt{h}} \right|, \quad \omega \in \Omega.$$

Show that $\overline{\mathcal{E}}^L[\varphi] \le 0$ for $c > 0$ small enough, but $\mathbb{T}_h^{t,\omega}[\varphi] = [1 - \varepsilon][\sqrt{h} - Lh]$. Consequently, φ and $\psi := 0$ violate (11.5.11). (Note that the modulus of continuity of φ does not depend on h.) ∎

Chapter 12
Second Order BSDEs

Recall BSDE (4.0.3) and semilinear PPDE (11.3.2), we have

$$Y = u, \ Z = \partial_\omega u \quad \text{and thus} \quad \|(Y,Z)\|^2 = \mathbb{E}\left[\sup_{0 \le t \le T} |u(t,\omega)|^2 + \int_0^T |\partial_\omega u(t,\omega)|^2 dt \right].$$

For viscosity solutions of PPDEs we emphasize the pathwise properties, while for BSDEs we focus on the norm estimates. So in spirit we may view solutions of BSDEs as Sobolev type of solutions for semilinear PPDEs. The goal of this chapter is to provide Sobolev type of solutions for path dependent HJB equation (11.3.9), which arises naturally in stochastic optimization with diffusion control and financial models with volatility uncertainty. We shall call the corresponding equation second order BSDE, or BSDE under nonlinear expectation. As in BSDE theory, we shall focus on the norm estimates for the solutions. However, due to the involvement of the nonlinear expectation, in this chapter we assume

$$d_2 = 1, \text{ or say } Y \text{ is always 1-dimensional.}$$

Throughout this chapter, we still use the canonical setting in Section 9.2.

12.1 Quasi-Sure Stochastic Analysis

In this section we fix a class $\mathscr{P} \subset \mathscr{P}_\infty$. For the problems we will study later, we typically obtain certain random variables or processes under each $\mathbb{P} \in \mathscr{P}$ first. Thus the following concept is crucial for our analysis.

© Springer Science+Business Media LLC 2017
J. Zhang, *Backward Stochastic Differential Equations*, Probability Theory and Stochastic Modelling 86, DOI 10.1007/978-1-4939-7256-2_12

Definition 12.1.1

(i) *Given a family of random variables $\{\xi^{\mathbb{P}}, \mathbb{P} \in \mathscr{P}\} \subset \mathbb{L}^0(\mathscr{F}_T)$, we say $\xi \in \mathbb{L}^0(\mathscr{F}_T)$ is its $\mathscr{P}$-aggregator if $\xi^{\mathbb{P}} = \xi$, $\mathbb{P}$-a.s. for all $\mathbb{P} \in \mathscr{P}$.*

(ii) *Given a family of processes $\{X^{\mathbb{P}}, \mathbb{P} \in \mathscr{P}\} \subset \mathbb{L}^0(\mathbb{F})$, we say $X \in \mathbb{L}^0(\mathbb{F})$ is its $\mathscr{P}$-aggregator if $X^{\mathbb{P}} = X$, $\mathbb{P}$-a.s. for all $\mathbb{P} \in \mathscr{P}$.*

We remark that a $\mathscr{P}$-aggregator, if it exists, is unique in the $\mathscr{P}$-q.s. sense.

In the following two subsections, we will study the aggregation issue of stochastic integration and conditional nonlinear expectation, respectively. We remark that these can be viewed as the simplest forward and backward problems in the quasi-sure stochastic setting.

12.1.1 Quasi-Sure Stochastic Integration

In this subsection we consider $\mathscr{P} = \mathscr{P}_\infty$. For $\eta \in \bigcap_{\mathbb{P} \in \mathscr{P}_\infty} \mathbb{L}^2_{loc}(\mathbb{F}, \mathbb{P}, \mathbb{R}^{1 \times d})$, denote

$$M^{\mathbb{P}}_t := \int_0^t \eta_s dX_s, \quad \mathbb{P}\text{-a.s.} \tag{12.1.1}$$

It is a natural and important question: can we aggregate $\{M^{\mathbb{P}}, \mathbb{P} \in \mathscr{P}_\infty\}$?

Theorem 12.1.2 *Assume $\eta \in \cap_{\mathbb{P} \in \mathscr{P}_\infty} \mathbb{L}^2_{loc}(\mathbb{F}, \mathbb{P}, \mathbb{R}^{1 \times d})$ is càdlàg , namely right continuous with left limits, $\mathscr{P}_\infty$-q.s. Then the above $\{M^{\mathbb{P}}, \mathbb{P} \in \mathscr{P}_\infty\}$ has a $\mathscr{P}_\infty$-aggregator M, and we shall denote $M_t := \int_0^t \eta_s dX_s$, $\mathscr{P}_\infty$-q.s.*

Proof For any $\varepsilon > 0$, define $\tau_0^\varepsilon := 0$ and, for $n \geq 0$,

$$\tau_{n+1}^\varepsilon := \inf\{t \geq \tau_n^\varepsilon : |\eta_{\tau_n^\varepsilon, t}| \geq \varepsilon\} \wedge T. \tag{12.1.2}$$

Then clearly $\tau_n^\varepsilon \in \mathscr{T}$ for all $n \geq 0$. We claim that $\tau_n^\varepsilon = T$ when n is large enough, $\mathscr{P}_\infty$-q.s. Indeed, $\tau_\infty^\varepsilon := \lim_{n \to \infty} \tau_n^\varepsilon \leq T$. Note that $|\eta_{\tau_{n-1}^\varepsilon, \tau_n^\varepsilon}| \geq \varepsilon$ whenever $\tau_n^\varepsilon < T$. Then, on $\cap_{n \geq 1}\{\tau_n^\varepsilon < \tau_\infty^\varepsilon\} \subset \cap_{n \geq 1}\{\tau_n^\varepsilon < T\}$, we have $|\eta_{\tau_{n-1}^\varepsilon, \tau_n^\varepsilon}| \geq \varepsilon$ for all $n \geq 1$, and hence $\eta_{\tau_\infty^\varepsilon-}$ does not exist. Since η has left limits, $\mathscr{P}_\infty$-q.s., we see that $\mathbb{P}(\cup_{n \geq 1}\{\tau_n^\varepsilon = \tau_\infty^\varepsilon\}) = 0$ for all $\mathbb{P} \in \mathscr{P}_\infty$.

Now define

$$M_t^\varepsilon := \sum_{n=0}^\infty \eta_{\tau_n^\varepsilon} X_{\tau_n^\varepsilon \wedge t, \tau_{n+1}^\varepsilon \wedge t}, \quad M := \limsup_{m \to \infty} M^{2^{-m}}. \tag{12.1.3}$$

Then clearly M^ε is continuous and $\mathbb{F}$-measurable, and consequently M is $\mathbb{F}$-measurable.

We show that M is the desired $\mathscr{P}_\infty$-aggregator of $\{M^{\mathbb{P}}, \mathbb{P} \in \mathscr{P}_\infty\}$. Indeed, for any $\mathbb{P} \in \mathscr{P}_\infty$, there exists $L > 0$ such that $\mathbb{P} \in \mathscr{P}_L$. Recall the notations in Definition 9.2.9 (i). For any $\varepsilon > 0$, by standard arguments we have

$$\mathbb{E}^{\mathbb{P}}\Big[|(M^\varepsilon - M^{\mathbb{P}})^*_T|^2\Big] = \mathbb{E}^{\mathbb{P}'}\Big[|(M^\varepsilon(X') - M^{\mathbb{P}}(X'))^*_T|^2\Big]$$

$$\le C\mathbb{E}^{\mathbb{P}'}\Big[\sum_{n=0}^{\infty} \int_{\tau_n^\varepsilon(X')}^{\tau_{n+1}^\varepsilon(X')} |\eta_t(X') - \eta_{\tau_n^\varepsilon(X')}(X')|^2 [|b'_t|^2 + |\sigma'_t|^2] dt\Big]$$

$$\le C_L\mathbb{E}^{\mathbb{P}'}\Big[\sum_{n=0}^{\infty} \int_{\tau_n^\varepsilon(X')}^{\tau_{n+1}^\varepsilon(X')} |\eta_t(X') - \eta_{\tau_n^\varepsilon(X')}(X')|^2 dt\Big] = C\mathbb{E}^{\mathbb{P}}\Big[\sum_{n=0}^{\infty} \int_{\tau_n^\varepsilon}^{\tau_{n+1}^\varepsilon} |\eta_{\tau_n^\varepsilon, t}|^2 dt\Big] \le C_L\varepsilon^2.$$

Then, for any $m \ge 0$,

$$\mathbb{E}^{\mathbb{P}}\Big[|(M^{2^{-m}} - M^{2^{-m-1}})^*_T|\Big] \le \mathbb{E}^{\mathbb{P}}\Big[|(M^{2^{-m}} - M^{\mathbb{P}})^*_T| + |(M^{2^{-m-1}} - M^{\mathbb{P}})^*_T|\Big] \le C_L 2^{-m},$$

and thus

$$\mathbb{E}^{\mathbb{P}}\Big[\sum_{m \ge 0} |(M^{2^{-m}} - M^{2^{-m-1}})^*_T|\Big] < \infty. \tag{12.1.4}$$

This implies that M is a limit and $M = M^{\mathbb{P}}$, $\mathbb{P}$-a.s. ∎

Remark 12.1.3

(i) As in the proof of Theorem 12.1.2, quite often we use $\limsup$ or $\liminf$ to construct explicitly a candidate aggregator first, and then verify its desired properties under each $\mathbb{P}$.

(ii) The construction of an aggregator may involve estimates under each $\mathbb{P}$, but typically we do not need the estimates to be uniform in $\mathbb{P}$. For example, in (12.1.4) it is quite possible that $\mathscr{E}^{\mathscr{P}_\infty}\Big[\sum_{m \ge 0} |(M^{2^{-m}} - M^{2^{-m-1}})^*_T|\Big] = \infty$. ∎

Remark 12.1.4

(i) The existence of this aggregator is closely related to the pathwise integration, as discussed in Remark 2.2.6, Section 2.8.3, and Problem 2.10.14. In particular, if the pathwise integration exists, it is automatically the candidate aggregator (one may still need to verify the measurability of the pathwise integration).

(ii) If we assume further that Continuum Hypothesis in set theory holds, then one does not need to assume η has càdlàg paths. By Nutz [160], one may construct pathwise integration and hence the aggregator M for all η which is only measurable with appropriate integrability. However, without the Continuum Hypothesis, it is not clear that such aggregator could exist for general measurable η.

(iii) The proof of Theorem 12.1.2, as well as the approach in Section 2.2, uses approximations of the integrand η. An alternative approach for pathwise integration is the rough path theory which uses approximations of the underlying semimartingale X, see, e.g., Friz & Hairer [94].   ∎

The following results are direct consequence of Theorem 12.1.2.

Corollary 12.1.5

(i) *The quadratic variation process*

$$\langle X \rangle_t := X_t X_t^\top - 2 \int_0^t X_s dX_s^\top \text{ can be defined } \mathscr{P}_\infty\text{-q.s.} \tag{12.1.5}$$

(ii) $\langle X \rangle$ *is absolutely continuous in t,* $\mathscr{P}_\infty$*-q.s. More precisely, denote:*

$$\hat{\sigma}_t := \limsup_{n \to \infty} \left(n[\langle X \rangle_t - \langle X \rangle_{(t - \frac{1}{n})}] \mathbf{1}_{\mathbb{S}^d_+} \left(\langle X \rangle_t - \langle X \rangle_{(t - \frac{1}{n})} \right) \right)^{\frac{1}{2}}. \tag{12.1.6}$$

where $\mathbb{S}^d_+ := \{A \in \mathbb{S}^d : A \geq 0\}$ *and the* $\limsup$ *is taken component wise, then*

$$d\langle X \rangle_t = \hat{\sigma}_t^2 dt, \quad \mathscr{P}_\infty\text{-q.s.} \tag{12.1.7}$$

(iii) $\hat{\sigma}$ *is the* $\mathscr{P}_\infty$*-aggregator of the class* $\{\sigma^{\mathbb{P}}, \mathbb{P} \in \mathscr{P}_\infty\}$ *in* (9.2.8).

Remark 12.1.6

(i) Recalling Definition 9.2.7 and Remark 9.2.8, we see that every $\mathbb{P} \in \mathscr{P}^W_{[\underline{\sigma},\overline{\sigma}]}$ is a weak solution to the following path dependent SDE:

$$\tilde{X}_t = \int_0^t \hat{\sigma}(s, \tilde{X}.) dB_s, \quad \mathbb{P}_0\text{-a.s.} \tag{12.1.8}$$

Moreover, for each $\mathbb{P} \in \mathscr{P}^S_{[\underline{\sigma},\overline{\sigma}]}$, SDE (12.1.8) admits a strong solution $\tilde{X}^{\mathbb{P}}$ corresponding to $\mathbb{P}$ (namely $\tilde{X}^{\mathbb{P}}$ is $\mathbb{F}^B$-measurable and $\mathbb{P}_0 \circ (\tilde{X}^{\mathbb{P}})^{-1} = \mathbb{P}$).

(ii) Clearly uniqueness does not hold for SDE (12.1.8). Note that $\langle X \rangle$ and $\hat{\sigma}$ are not $\mathscr{P}^W_{[\underline{\sigma},\overline{\sigma}]}$-q.s. continuous in ω, so this nonuniqueness does not violate the uniqueness results in the literature which typically require certain regularity (and Markov structure), see, e.g., Stroock & Varadhan [218].   ∎

12.1.2 Quasi-Sure Conditional Nonlinear Expectation

In this subsection, we extend the pathwise conditional nonlinear expectation in Section 10.2 by weakening the uniform regularity. Fix $\mathscr{P} \subset \mathscr{P}_L$ for some $L > 0$. For any $\mathbb{P} \in \mathscr{P}$ and $\tau \in \mathscr{T}$, denote

$$\mathscr{P}(\tau, \mathbb{P}) := \{\mathbb{P}' \in \mathscr{P} : \mathbb{P}' = \mathbb{P} \text{ on } \mathscr{F}_\tau\}, \qquad \mathbb{E}_\tau^{\mathbb{P}}[\cdot] := \mathbb{E}^{\mathbb{P}}[\cdot | \mathscr{F}_\tau]. \quad (12.1.9)$$

Let $\xi \in \mathbb{L}^1(\mathscr{F}_T, \mathscr{P})$, $\mathbb{P} \in \mathscr{P}$ and $\tau \in \mathscr{T}$. For any $\tilde{\mathbb{P}} \in \mathscr{P}(\tau, \mathbb{P})$, note that $\mathbb{E}_\tau^{\tilde{\mathbb{P}}}[\xi]$ is defined in $\tilde{\mathbb{P}}$-a.s. sense and is $\mathscr{F}_\tau$-measurable. Since $\mathbb{P} = \tilde{\mathbb{P}}$ on $\mathscr{F}_\tau$, we see that $\mathbb{E}_\tau^{\tilde{\mathbb{P}}}[\xi]$ is defined in $\mathbb{P}$-a.s. sense. Then, recall Definition 1.1.3, we may define the $\mathbb{P}$-essential supremum of $\{\mathbb{E}_\tau^{\tilde{\mathbb{P}}}[\xi] : \tilde{\mathbb{P}} \in \mathscr{P}(\tau, \mathbb{P})\}$:

$$\mathscr{E}_{\mathbb{P}, \tau}^{\mathscr{P}}[\xi] := \operatorname*{ess\,sup}_{\tilde{\mathbb{P}} \in \mathscr{P}(\tau, \mathbb{P})}^{\mathbb{P}} \mathbb{E}_\tau^{\tilde{\mathbb{P}}}[\xi]. \quad (12.1.10)$$

We first adapt Assumption 9.3.3 (ii) to this setting. We remark that here we do not need Assumption 9.3.3 (i).

Assumption 12.1.7 $\mathscr{P} \subset \mathscr{P}_L$ *for some* $L > 0$, *and for any* $\tau \in \mathscr{T}$, $\mathbb{P} \in \mathscr{P}$, $\mathbb{P}_n \in \mathscr{P}(\tau, \mathbb{P})$, *and any partition* $\{E_n, n \geq 1\} \subset \mathscr{F}_\tau$ *of* Ω, *the following* $\hat{\mathbb{P}}$ *is also in* $\mathscr{P}$ *and hence in* $\mathscr{P}(\tau, \mathbb{P})$:

$$\hat{\mathbb{P}} := \sum_{n \geq 1} \mathbb{P}_n \mathbf{1}_{E_n}, \quad namely\ \hat{\mathbb{P}}(A) := \sum_{n \geq 1} \mathbb{P}_n(A \cap E_n), \quad \forall A \in \mathscr{F}_T^*. \quad (12.1.11)$$

Lemma 12.1.8 *The classes* $\mathscr{P}_{[\underline{\sigma}, \overline{\sigma}]}^S$, $\mathscr{P}_{[\underline{\sigma}, \overline{\sigma}]}^W$, *and* $\mathscr{P}_L$ *satisfy Assumption 12.1.7.*

Proof We prove the result only for $\mathscr{P}_{[\underline{\sigma}, \overline{\sigma}]}^S$, which will be used in later sections. The other cases can be proved similarly.

For $\mathbb{P} \in \mathscr{P}_{[\underline{\sigma}, \overline{\sigma}]}^S$, recall Remark 12.1.6 (i) and let $\tilde{X}$ be a strong solution of (12.1.8) corresponding to $\mathbb{P}$. Note that by our convention of notation $\mathbb{F} = \mathbb{F}^X = \mathbb{F}^B$. Denote $\tilde{\tau} := \tau(\tilde{X})$. Since $\tau \in \mathscr{T}$, $\{\tau \leq t\} \in \mathscr{F}_t^X$ for any t, and thus $\{\tilde{\tau} \leq t\} \in \mathscr{F}_t^{\tilde{X}} \subset \mathscr{F}_t^B = \mathscr{F}_t$. That is, $\tilde{\tau}$ is an $\mathbb{F}$-stopping time. By definition of $\mathbb{P}$, the $\mathbb{P}$-distribution of $(X, \tau, \hat{\sigma})$ is equal to the $\mathbb{P}_0$-distribution of $(\tilde{X}, \tilde{\tau}, \hat{\sigma}(\tilde{X}))$, which implies further that the $\mathbb{P}$-distribution of $(X_{\tau \wedge \cdot}, \hat{\sigma} \mathbf{1}_{[0, \tau)})$ is equal to the $\mathbb{P}_0$-distribution of $(\tilde{X}_{\tilde{\tau} \wedge \cdot}, \hat{\sigma}(\tilde{X}) \mathbf{1}_{[0, \tilde{\tau})})$. Similarly, the $\mathbb{P}^n$-distribution of $(X_{\tau \wedge \cdot}, \hat{\sigma} \mathbf{1}_{[0, \tau)})$ is equal to the $\mathbb{P}_0$-distribution of $(\tilde{X}_{\tilde{\tau}^n \wedge \cdot}^n, \hat{\sigma}(\tilde{X}^n) \mathbf{1}_{[0, \tilde{\tau}^n)})$, where $\tilde{X}^n$ is a strong solution of (12.1.8) corresponding to $\mathbb{P}^n$ and $\tilde{\tau}^n := \tau(\tilde{X}^n)$. Since $\mathbb{P}_n = \mathbb{P}$ on $\mathscr{F}_\tau$ and $(X_{\tau \wedge \cdot}, \hat{\sigma} \mathbf{1}_{[0, \tau)})$ is $\mathscr{F}_\tau$-measurable, we see $(\tilde{X}_{\tilde{\tau}^n \wedge \cdot}^n, \hat{\sigma}(\tilde{X}^n) \mathbf{1}_{[0, \tilde{\tau}^n)})$ and $(\tilde{X}_{\tilde{\tau} \wedge \cdot}, \hat{\sigma}(\tilde{X}) \mathbf{1}_{[0, \tilde{\tau})})$ have the same $\mathbb{P}_0$-distribution. By (12.1.8), this clearly implies that $(\tilde{X}_{\tilde{\tau}^n \wedge \cdot}^n, \hat{\sigma}(\tilde{X}^n) \mathbf{1}_{[0, \tilde{\tau}^n)}, B_{\tilde{\tau}^n \wedge \cdot})$ and $(\tilde{X}_{\tilde{\tau} \wedge \cdot}, \hat{\sigma}(\tilde{X}) \mathbf{1}_{[0, \tilde{\tau})}, B_{\tilde{\tau} \wedge \cdot}))$ also have the same $\mathbb{P}_0$-distribution. Then

$$\tilde{\tau}^n = \tilde{\tau}, \quad \text{and} \quad \tilde{X}_t^n = \tilde{X}_t, 0 \leq t \leq \tilde{\tau}, \quad \mathbb{P}_0\text{-a.s.} \quad (12.1.12)$$

Now denote $\hat{X}_t = \tilde{X}_t \mathbf{1}_{[0, \tilde{\tau})}(t) + \sum_{n \geq 1} \tilde{X}_t^n \mathbf{1}_{E_n}(\tilde{X}) \mathbf{1}_{[\tilde{\tau}, T]}(t)$. It is clear that $\hat{X}$ is also a strong solution of (12.1.8) and $\hat{\mathbb{P}} = \mathbb{P}_0 \circ (\hat{X})^{-1}$. Since $\underline{\sigma} \leq \hat{\sigma} \leq \overline{\sigma}$, $\mathbb{P}^n$-a.s., we have $\underline{\sigma} \leq \hat{\sigma}(\tilde{X}^n) \leq \overline{\sigma}$, $\mathbb{P}_0$-a.s. Then $\underline{\sigma} \leq \hat{\sigma}(\hat{X}) \leq \overline{\sigma}$, $\mathbb{P}_0$-a.s. This means that $\hat{\mathbb{P}} \in \mathscr{P}_{[\underline{\sigma}, \overline{\sigma}]}^S$. Finally, it is obvious that $\hat{\mathbb{P}} = \mathbb{P}$ in $\mathscr{F}_\tau$. ∎

Remark 12.1.9 Let $\mathbb{P}, \mathbb{P}_n \in \mathscr{P}^S_{[\underline{\sigma},\overline{\sigma}]}$ also correspond to σ and σ^n as in Definition 9.2.2. The above proof also implies that when $\mathbb{P}_n = \mathbb{P}$ on $\mathscr{F}_\tau$, then

$$X^\sigma_t = X^{\sigma^n}_t, \quad \sigma_t = \sigma^n_t, \quad 0 \le t \le \tau(X^\sigma), \quad \mathbb{P}_0\text{-a.s.} \tag{12.1.13}$$

and $\hat{\mathbb{P}}$ corresponds to $\sigma_t \mathbf{1}_{[0,\tau(X^\sigma))}(t) + \sum_{n\ge1} \sigma^n_t \mathbf{1}_{E_n}(X^\sigma)\mathbf{1}_{[\tau(X^\sigma),T]}(t)$. ∎

One advantage for using the essential supremum in (12.1.10) is that the dynamic programming principle is almost free, as we see in the following lemma.

Lemma 12.1.10 *Let $\mathscr{P}$ satisfy Assumption 12.1.7 and $\xi \in \mathbb{L}^1(\mathscr{F}_T, \mathscr{P})$.*

(i) For any $\mathbb{P} \in \mathscr{P}$, $\tau \in \mathscr{T}$, and $\varepsilon > 0$, there exists $\mathbb{P}_\varepsilon \in \mathscr{P}(\tau, \mathbb{P})$ such that

$$\mathbb{E}^{\mathbb{P}_\varepsilon}_\tau[\xi] \ge \mathscr{E}^\mathscr{P}_{\mathbb{P},\tau}[\xi] - \varepsilon, \quad \mathbb{P}\text{-a.s.} \tag{12.1.14}$$

(ii) For any $\mathbb{P} \in \mathscr{P}$ and any $\tau_1, \tau_2 \in \mathscr{T}$ such that $\tau_1 \le \tau_2$, we have

$$\mathscr{E}^\mathscr{P}_{\mathbb{P},\tau_1}[\xi] = \underset{\tilde{\mathbb{P}}\in\mathscr{P}(\tau_1,\mathbb{P})}{ess\,sup}{}^{\mathbb{P}} \; \mathbb{E}^{\tilde{\mathbb{P}}}_{\tau_1}\Big[\mathscr{E}^\mathscr{P}_{\tilde{\mathbb{P}},\tau_2}[\xi]\Big], \quad \mathbb{P}\text{-a.s.} \tag{12.1.15}$$

Proof (i) Applying Theorem 1.1.4 (i), there exist $\{\mathbb{P}_n, n \ge 1\} \subset \mathscr{P}(\tau, \mathbb{P})$ such that

$$\max_{n\ge1} \mathbb{E}^{\mathbb{P}_n}_\tau[\xi] = \mathscr{E}^\mathscr{P}_{\mathbb{P},\tau}[\xi], \quad \mathbb{P}\text{-a.s.}$$

For $\varepsilon > 0$, denote

$$E_n := \Big\{\mathbb{E}^{\mathbb{P}_n}_\tau[\xi] \ge \mathscr{E}^\mathscr{P}_{\mathbb{P},\tau}[\xi] - \varepsilon\Big\}, \; n \ge 1; \quad \tilde{E}_n := E_n \setminus \cup_{i=1}^{n-1} E_i, \quad \mathbb{P}_\varepsilon := \sum_{n=1}^\infty \mathbb{P}_n \mathbf{1}_{\tilde{E}_n}.$$

It is clear that $\{\tilde{E}_n, n \ge 1\} \subset \mathscr{F}_\tau$ are disjoint and $\mathbb{P}(\cup_{n\ge1}\tilde{E}_n) = \mathbb{P}(\cup_{n\ge1}E_n) = 1$. By Assumption 12.1.7 we see that $\mathbb{P}_\varepsilon \in \mathscr{P}(\tau, \mathbb{P})$. Then

$$\mathbb{E}^{\mathbb{P}_\varepsilon}_\tau[\xi] = \sum_{n\ge1} \mathbb{E}^{\mathbb{P}_\varepsilon}_\tau[\xi]\mathbf{1}_{\tilde{E}_n} = \sum_{n\ge1} \mathbb{E}^{\mathbb{P}_n}_\tau[\xi]\mathbf{1}_{\tilde{E}_n} \ge \sum_{n\ge1}\Big[\mathscr{E}^\mathscr{P}_{\mathbb{P},\tau}[\xi] - \varepsilon\Big]\mathbf{1}_{\tilde{E}_n} = \mathscr{E}^\mathscr{P}_{\mathbb{P},\tau}[\xi] - \varepsilon, \quad \mathbb{P}\text{-a.s.}$$

This verifies (12.1.14).

(ii) First, for any $\mathbb{P} \in \mathscr{P}$, notice that $\mathbb{P} \in \mathscr{P}(\tau_2, \mathbb{P}) \subset \mathscr{P}(\tau_1, \mathbb{P})$, then it follows from the tower property of standard conditional expectation that:

$$\mathbb{E}^{\mathbb{P}}_{\tau_1}[\xi] = \mathbb{E}^{\mathbb{P}}_{\tau_1}\Big[\mathbb{E}^{\mathbb{P}}_{\tau_2}[\xi]\Big] \le \mathbb{E}^{\mathbb{P}}_{\tau_1}\Big[\mathscr{E}^\mathscr{P}_{\mathbb{P},\tau_2}[\xi]\Big] \le \underset{\tilde{\mathbb{P}}\in\mathscr{P}(\tau_1,\mathbb{P})}{ess\,sup}{}^{\mathbb{P}} \; \mathbb{E}^{\tilde{\mathbb{P}}}_{\tau_1}\Big[\mathscr{E}^\mathscr{P}_{\tilde{\mathbb{P}},\tau_2}[\xi]\Big], \quad \mathbb{P}\text{-a.s.}$$

On the other hand, for any $\tilde{\mathbb{P}} \in \mathscr{P}(\tau_1, \mathbb{P})$ and $\varepsilon > 0$, by (i) there exists $\tilde{\mathbb{P}}_\varepsilon \in \mathscr{P}(\tau_2, \tilde{\mathbb{P}}) \subset \mathscr{P}(\tau_1, \mathbb{P})$ such that

$$\mathscr{E}_{\tilde{\mathbb{P}}, \tau_2}^{\mathscr{P}}[\xi] \le \mathbb{E}_{\tau_2}^{\tilde{\mathbb{P}}_\varepsilon}[\xi] + \varepsilon, \quad \tilde{\mathbb{P}}\text{-a.s.}$$

Thus

$$\mathbb{E}_{\tau_1}^{\tilde{\mathbb{P}}}\Big[\mathscr{E}_{\tilde{\mathbb{P}}, \tau_2}^{\mathscr{P}}[\xi]\Big] \le \mathbb{E}_{\tau_1}^{\tilde{\mathbb{P}}}\Big[\mathbb{E}_{\tau_2}^{\tilde{\mathbb{P}}_\varepsilon}[\xi]\Big] + \varepsilon = \mathbb{E}_{\tau_1}^{\tilde{\mathbb{P}}_\varepsilon}[\xi] + \varepsilon \le \mathscr{E}_{\mathbb{P}, \tau_1}^{\mathscr{P}}[\xi] + \varepsilon, \quad \mathbb{P}\text{-a.s.}$$

Sending $\varepsilon \to 0$, we complete the proof. ∎

The natural and important issue is: can we aggregate $\{\mathscr{E}_{\mathbb{P}, \tau}^{\mathscr{P}}[\xi] : \mathbb{P} \in \mathscr{P}\}$? This leads to the following definition.

Definition 12.1.11 *Let $\mathscr{P}$ satisfy Assumption 12.1.7.*

(i) *We say $\xi \in \mathbb{L}^1(\mathscr{F}_T, \mathscr{P})$ has $\mathscr{E}^{\mathscr{P}}$-conditional expectation if, for any $\tau \in \mathscr{T}$, $\{\mathscr{E}_{\mathbb{P}, \tau}^{\mathscr{P}}[\xi] : \mathbb{P} \in \mathscr{P}\}$ has an $\mathscr{P}$-aggregator, denoted as $\mathscr{E}_\tau^{\mathscr{P}}[\xi]$.*

(ii) *We say $Y \in \mathbb{S}^1(\mathbb{F}, \mathscr{P})$ is an $\mathscr{E}^{\mathscr{P}}$-martingale if $Y_\tau = \mathscr{E}_\tau^{\mathscr{P}}[Y_T]$, $\mathscr{P}$-q.s. for any $\tau \in \mathscr{T}$.*

As a consequence of Lemma 12.1.10, the following results are immediate.

Proposition 12.1.12 *Let $\mathscr{P}$ satisfy Assumption 12.1.7.*

(i) *Assume $\xi \in \mathbb{L}^1(\mathscr{F}_T, \mathscr{P})$ has $\mathscr{E}^{\mathscr{P}}$-conditional expectation $\mathscr{E}_\tau^{\mathscr{P}}[\xi]$. Then $\mathscr{E}_\tau^{\mathscr{P}}[\xi]$ is unique in $\mathscr{P}$-q.s. sense and satisfies the tower property: for any $\tau_1, \tau_2 \in \mathscr{T}$ such that $\tau_1 \le \tau_2$, we have $\mathscr{E}_{\tau_2}^{\mathscr{P}}[\xi]$ has $\mathscr{E}^{\mathscr{P}}$-conditional expectation and*

$$\mathscr{E}_{\tau_1}^{\mathscr{P}}[\xi] = \mathscr{E}_{\tau_1}^{\mathscr{P}}\Big[\mathscr{E}_{\tau_2}^{\mathscr{P}}[\xi]\Big]. \tag{12.1.16}$$

(ii) *Let Y be a $\mathscr{E}^{\mathscr{P}}$-martingale, then $Y_{\tau_1} = \mathscr{E}_{\tau_1}^{\mathscr{P}}[Y_{\tau_2}]$ for any $\tau_1, \tau_2 \in \mathscr{T}$ such that $\tau_1 \le \tau_2$. Moreover, Y is a supermartingale under any $\mathbb{P} \in \mathscr{P}$.*

Remark 12.1.13

(i) There is actually another aggregation issue. Assume $\mathscr{E}_\tau^{\mathscr{P}}[\xi]$ exists for all $\tau \in \mathscr{T}$, can we aggregate the family of random variables $\{\mathscr{E}_\tau^{\mathscr{P}}[\xi] : \tau \in \mathscr{T}\}$ into a $\mathscr{E}^{\mathscr{P}}$-martingale Y? In this section, we shall construct Y directly for $\mathscr{P}$-q.s. continuous ξ, then this aggregation becomes trivial.

(ii) For each $\mathbb{P} \in \mathscr{P}$, by (12.1.15) clearly

$$\mathbb{E}_{\tau_1}^{\mathbb{P}}\Big[\mathscr{E}_{\mathbb{P}, \tau_2}^{\mathscr{P}}[\xi]\Big] \le \mathscr{E}_{\mathbb{P}, \tau_1}^{\mathscr{P}}[\xi], \quad \text{for all } \tau_1, \tau_2 \in \mathscr{T} \text{ such that } \tau_1 \le \tau_2.$$

Then one can always aggregate $\{\mathscr{E}_{\mathbb{P}, \tau}^{\mathscr{P}}[\xi] : \tau \in \mathscr{T}\}$ into a $(\mathbb{P}, \overline{\mathbb{F}}^{\mathbb{P}})$-supermartingale $Y^{\mathbb{P}}$ with càdlàg paths in the sense that $\mathscr{E}_{\mathbb{P}, \tau}^{\mathscr{P}}[\xi] = Y_\tau^{\mathbb{P}}$, $\mathbb{P}$-a.s. for all $\tau \in \mathscr{T}$. The arguments are similar to Karatzas & Shreve [118] Appendix D. Since we do not need this result in this chapter, we skip the details.

(iii) Note that the $Y^{\mathbb{P}}$ in (ii) above is $\overline{\mathbb{F}}^{\mathbb{P}}$-measurable, rather than $\mathbb{F}$-measurable. For $\xi \in \mathbb{L}^1(\mathscr{F}_T, \mathscr{P})$, the theory involves some subtle measurability issue, which will be discussed in Section 12.3 below.

∎

When $\xi \in UC(\Omega)$, the notion of $\mathscr{E}^{\mathscr{P}}$-conditional expectation is equivalent to the pathwise conditional nonlinear expectation in Section 10.2.

Proposition 12.1.14 *Let $\mathscr{P}$ satisfy Assumptions 9.3.3 and 12.1.7, and $\xi \in UC(\Omega)$. Denote $Y_t(\omega) := \mathscr{E}^{\mathscr{P}}_{t,\omega}[\xi]$. Then $Y \in \mathbb{S}^1(\mathbb{F}, \mathscr{P})$ is a $\mathscr{E}^{\mathscr{P}}$-martingale. In particular, $Y_\tau = \mathscr{E}^{\mathscr{P}}_\tau[\xi]$, $\mathscr{P}$-q.s. for any $\tau \in \mathscr{T}$.*

Proof First, by Theorem 10.2.1 (i) and Problem 10.5.2 (iii) we see that $Y \in \mathbb{S}^1(\mathbb{F}, \mathscr{P})$. By (9.3.7) and Theorem 9.3.2, it is clear that $Y_\tau \geq \mathbb{E}^{\mathbb{P}}_\tau[\xi]$, $\mathbb{P}$-a.s. for any $\mathbb{P} \in \mathscr{P}$. Then

$$Y_\tau \geq \mathscr{E}^{\mathscr{P}}_{\mathbb{P},\tau}[\xi], \quad \mathbb{P}\text{-a.s.} \tag{12.1.17}$$

To prove the opposite inequality, we first assume $\tau = t$ is deterministic. Notice that the $\widehat{\mathbb{P}}$ defined by (10.2.8) is in $\mathscr{P}(t, \mathbb{P})$, then by sending $\varepsilon \to 0$ in (10.2.9) we obtain

$$Y_t \leq \mathscr{E}^{\mathscr{P}}_{\mathbb{P},t}[\xi], \quad \mathbb{P}\text{-a.s.}$$

Next, assume $\tau \in \mathscr{T}$ takes only finitely many values t_i, $1 \leq i \leq n$. For any $\mathbb{P} \in \mathscr{P}$ and $\tilde{\mathbb{P}} \in \mathscr{P}(\tau, \mathbb{P})$, note that $\mathbb{E}^{\tilde{\mathbb{P}}}_\tau[\xi] = \sum_{i=1}^n \mathbb{E}^{\tilde{\mathbb{P}}}_{t_i}[\xi] \mathbf{1}_{\{\tau=t_i\}}$. This implies immediately that $\mathscr{E}^{\mathscr{P}}_{\mathbb{P},\tau}[\xi] \leq \sum_{i=1}^n \mathscr{E}^{\mathscr{P}}_{\mathbb{P},t_i}[\xi] \mathbf{1}_{\{\tau=t_i\}}$, $\mathbb{P}$-a.s. On the other hand, for any $\varepsilon > 0$ and each i, by Lemma 12.1.10 there exists $\mathbb{P}_i \in \mathscr{P}(\tau, \mathbb{P})$ such that $\mathscr{E}^{\mathscr{P}}_{\mathbb{P},t_i}[\xi] \leq \mathbb{E}^{\mathbb{P}_i}_{t_i}[\xi] + \varepsilon$, $\mathbb{P}$-a.s. Denote $\hat{\mathbb{P}} := \sum_{i=1}^n \mathbb{P}_i \mathbf{1}_{\{\tau=t_i\}} \in \mathscr{P}(\tau, \mathbb{P})$, we see that

$$\sum_{i=1}^n \mathscr{E}^{\mathscr{P}}_{\mathbb{P},t_i}[\xi] \mathbf{1}_{\{\tau=t_i\}} \leq \sum_{i=1}^n \mathbb{E}^{\hat{\mathbb{P}}}_{t_i}[\xi] \mathbf{1}_{\{\tau=t_i\}} + \varepsilon = \mathbb{E}^{\hat{\mathbb{P}}}_\tau[\xi] + \varepsilon \leq \mathscr{E}^{\mathscr{P}}_{\mathbb{P},\tau}[\xi] + \varepsilon.$$

Send $\varepsilon \to 0$, we obtain

$$\mathscr{E}^{\mathscr{P}}_{\mathbb{P},\tau}[\xi] = \sum_{i=1}^n \mathscr{E}^{\mathscr{P}}_{\mathbb{P},t_i}[\xi] \mathbf{1}_{\{\tau=t_i\}} = \sum_{i=1}^n Y_{t_i} \mathbf{1}_{\{\tau=t_i\}} = Y_\tau, \quad \mathbb{P}\text{-a.s.}$$

Finally, for arbitrary $\tau \in \mathscr{T}$, as in Theorem 10.2.1 Step 5, let $\tau_n \in \mathscr{T}$ take finitely many values and $0 \leq \tau_n - \tau \leq \frac{1}{n}$. For any $\varepsilon > 0$ and $n \geq 1$, by Lemma 12.1.10 again there exists $\mathbb{P}^\varepsilon_n \in \mathscr{P}(\tau_n, \mathbb{P}) \subset \mathscr{P}(\tau, \mathbb{P})$ such that

$$Y_{\tau_n} = \mathscr{E}^{\mathscr{P}}_{\mathbb{P},\tau_n}[\xi] \leq \mathbb{E}^{\mathbb{P}^\varepsilon_n}_{\tau_n}[\xi] + \varepsilon.$$

Then

$$Y_\tau - \mathcal{E}^{\mathscr{P}}_{\mathbb{P},\tau}[\xi] \le Y_\tau - \mathbb{E}^{\mathbb{P}^\varepsilon_n}_\tau[Y_{\tau_n}] + \mathbb{E}^{\mathbb{P}^\varepsilon_n}_\tau[\xi] + \varepsilon - \mathcal{E}^{\mathscr{P}}_{\mathbb{P},\tau}[\xi] \le \mathbb{E}^{\mathbb{P}^\varepsilon_n}_\tau[Y_\tau - Y_{\tau_n}] + \varepsilon, \quad \mathbb{P}\text{-a.s.}$$

By the estimate (10.2.10) and recall the $\bar{p}$ in (10.2.11), we have

$$\mathbb{E}^{\mathbb{P}}\big[Y_\tau - \mathcal{E}^{\mathscr{P}}_{\mathbb{P},\tau}[\xi]\big] \le \mathbb{E}^{\mathbb{P}^\varepsilon_n}\Big[\bar{p}\big(\frac{1}{\sqrt{n}} + OSC_{\frac{1}{n}}(X)\big)\Big] + \varepsilon \le \mathcal{E}^{\mathscr{P}_L}\Big[\bar{p}\big(\frac{1}{\sqrt{n}} + OSC_{\frac{1}{n}}(X)\big)\Big] + \varepsilon,$$

Send $\varepsilon \to 0$ and $n \to \infty$, it follows from either Theorem 10.1.8 or Theorem 10.1.10 that $\mathbb{E}^{\mathbb{P}}\big[Y_\tau - \mathcal{E}^{\mathscr{P}}_{\mathbb{P},\tau}[\xi]\big] \le 0$. This, together with (12.1.17), implies that $Y_\tau = \mathcal{E}^{\mathscr{P}}_{\mathbb{P},\tau}[\xi]$, $\mathbb{P}$-a.s. Since $\mathbb{P} \in \mathscr{P}$ is arbitrary, the equality holds $\mathscr{P}$-q.s. ∎

Recall Problem 1.4.13 (iv). For any $\xi \in UC(\Omega)$ we have: for any $\mathbb{P} \in \mathscr{P}$,

$$\sup_{0 \le t \le T} \mathcal{E}^{\mathscr{P}}_t[\xi] = \sup_{0 \le t \le T} \mathcal{E}^{\mathscr{P}}_{\mathbb{P},t}[\xi] = \operatorname*{ess\,sup}_{0 \le t \le T}^{\mathbb{P}} \mathcal{E}^{\mathscr{P}}_{\mathbb{P},t}[\xi], \quad \mathbb{P}\text{-a.s.} \qquad (12.1.18)$$

Note that the right side above is well defined for all $\xi \in \mathbb{L}^1(\mathscr{F}_T, \mathscr{P})$. We now introduce the following norm and space: for any $p \ge 1$,

$$\|\xi\|^p_{\widehat{\mathbb{L}}^p_{\mathscr{P}}} := \sup_{\mathbb{P} \in \mathscr{P}} \mathbb{E}^{\mathbb{P}}\Big[\operatorname{ess\,sup}^{\mathbb{P}}_{0 \le t \le T} \mathcal{E}^{\mathscr{P}}_{\mathbb{P},t}[|\xi|^p]\Big], \quad \xi \in \mathbb{L}^p(\mathscr{F}_T, \mathscr{P});$$

$$\widehat{\mathbb{L}}^p(\mathscr{F}_T, \mathscr{P}) := \Big\{\xi \in \mathbb{L}^p(\mathscr{F}_T, \mathscr{P}) : \|\xi\|^p_{\widehat{\mathbb{L}}^p_{\mathscr{P}}} < \infty\Big\}; \qquad (12.1.19)$$

$$\widehat{\mathbb{L}}^p_0(\mathscr{F}_T, \mathscr{P}) := \Big\{\text{closure of } UC(\Omega) \text{ in } \mathbb{L}^0(\mathscr{F}_T) \text{ under norm } \|\cdot\|_{\widehat{\mathbb{L}}^p_{\mathscr{P}}}\Big\}$$

Recall (10.1.7). The following estimate can be viewed as the counterpart of Doob's maximum inequality in Lemma 2.2.4 for the new norm $\|\cdot\|_{\widehat{\mathbb{L}}^p_{\mathscr{P}}}$.

Proposition 12.1.15 *Let $\mathscr{P}$ satisfy Assumption 12.1.7. For any $1 \le p < q$, there exists a constant $C_{p,q}$ such that*

$$\|\xi\|^p_{\widehat{\mathbb{L}}^p_{\mathscr{P}}} \le C_{p,q}\|\xi\|_{\mathbb{L}^q_{\mathscr{P}}}, \quad \text{for all } \xi \in \widehat{\mathbb{L}}^p(\mathscr{F}_T, \mathscr{P}), \qquad (12.1.20)$$

$$\mathbb{L}^q_0(\mathscr{F}_T, \mathscr{P}) \subset \widehat{\mathbb{L}}^p_0(\mathscr{F}_T, \mathscr{P}) \subset \mathbb{L}^p_0(\mathscr{F}_T, \mathscr{P}). \qquad (12.1.21)$$

Consequently, all $\xi \in \widehat{\mathbb{L}}^1_0(\mathscr{F}_T, \mathscr{P})$ is $\mathscr{P}$-q.s. continuous.

Proof By Problem 10.5.2 (iv), (12.1.20) obviously implies (12.1.21). Then the $\mathscr{P}$-q.s. continuity of $\xi \in \widehat{\mathbb{L}}^1_0(\mathscr{F}_T, \mathscr{P})$ follows from Proposition 10.1.13. So it remains to prove (12.1.20).

To this end, denote $Y_t := \mathcal{E}^{\mathscr{P}}_t[\xi]$ as in Theorem 12.1.16 (ii). Then (12.1.20) is equivalent to

$$\Big(\mathcal{E}^{\mathscr{P}}[|Y^*_T|^p]\Big)^{\frac{1}{p}} \le C_{p,q}\Big(\mathcal{E}^{\mathscr{P}}[|Y_T|^q]\Big)^{\frac{1}{q}}. \qquad (12.1.22)$$

We shall follow the arguments in Lemma 2.2.4. For each $\lambda > 0$, define τ_λ as in (2.2.7):

$$\tau_\lambda := \inf\{t \geq 0 : |Y_t| \geq \lambda\} \wedge T.$$

Recall (2.2.8), for any $\mathbb{P} \in \mathscr{P}$ we have

$$\mathbb{E}^{\mathbb{P}}[|Y_T^*|^p] = \int_0^\infty p\lambda^{p-1}\mathbb{P}(Y_T^* \geq \lambda)d\lambda = \int_0^\infty p\lambda^{p-1}\mathbb{P}(|Y_{\tau_\lambda}| = \lambda)d\lambda.$$

For any $\varepsilon > 0$, by Lemma 12.1.10 there exists $\mathbb{P}_\lambda^\varepsilon \in \mathscr{P}(\tau_\lambda, \mathbb{P})$ such that $\mathbb{E}_{\tau_\lambda}^{\mathbb{P}_\lambda^\varepsilon}[Y_T] \leq Y_{\tau_\lambda} \leq \mathbb{E}_{\tau_\lambda}^{\mathbb{P}_\lambda^\varepsilon}[Y_T] + \varepsilon$, and thus $|Y_{\tau_\lambda}| \leq \mathbb{E}_{\tau_\lambda}^{\mathbb{P}_\lambda^\varepsilon}[|Y_T|] + \varepsilon$. Note that $\{|Y_{\tau_\lambda}| = \lambda\} \in \mathscr{F}_{\tau_\lambda}$ and $\mathbb{P}_\lambda^\varepsilon = \mathbb{P}$ on $\mathscr{F}_{\tau_\lambda}$. Then

$$\mathbb{P}(|Y_{\tau_\lambda}| = \lambda) = \frac{1}{\lambda}\mathbb{E}^{\mathbb{P}}\Big[|Y_{\tau_\lambda}|\mathbf{1}_{\{|Y_{\tau_\lambda}|=\lambda\}}\Big] \leq \frac{1}{\lambda}\mathbb{E}^{\mathbb{P}_\lambda^\varepsilon}\Big[(|Y_T| + \varepsilon)\mathbf{1}_{\{|Y_{\tau_\lambda}|=\lambda\}}\Big]$$

$$\leq \frac{C_{p,q}}{\lambda}\Big(\mathbb{E}^{\mathbb{P}_\lambda^\varepsilon}[|Y_T|^q + \varepsilon^q]\Big)^{\frac{1}{q}}\Big(\mathbb{P}(|Y_{\tau_\lambda}| = \lambda)\Big)^{1-\frac{1}{q}} \leq \frac{C_{p,q}}{\lambda}[\|Y_T\|_{\mathbb{L}_{\mathscr{P}}^q} + \varepsilon]\Big(\mathbb{P}(|Y_{\tau_\lambda}| = \lambda)\Big)^{1-\frac{1}{q}}$$

This implies

$$\mathbb{P}(|Y_{\tau_\lambda}| = \lambda) \leq \frac{C_{p,q}}{\lambda^q}[\|Y_T\|_{\mathbb{L}_{\mathscr{P}}^q}^q + \varepsilon^q],$$

and thus, for any $\lambda_0 > 0$,

$$\mathbb{E}^{\mathbb{P}}[|Y_T^*|^p] = \Big[\int_0^{\lambda_0} + \int_{\lambda_0}^\infty\Big]p\lambda^{p-1}\mathbb{P}(|Y_{\tau_\lambda}| = \lambda)d\lambda$$

$$\leq \int_0^{\lambda_0} p\lambda^{p-1}d\lambda + \int_{\lambda_0}^\infty p\lambda^{p-1}\frac{C_{p,q}}{\lambda^q}[\|Y_T\|_{\mathbb{L}_{\mathscr{P}}^q}^q + \varepsilon^q]d\lambda = \lambda_0^p + C_{p,q}\lambda_0^{p-q}[\|Y_T\|_{\mathbb{L}_{\mathscr{P}}^q}^q + \varepsilon]^q.$$

Setting $\lambda_0 := \|Y_T\|_{\mathbb{L}_{\mathscr{P}}^q} + \varepsilon$, we obtain

$$\mathbb{E}^{\mathbb{P}}[|Y_T^*|^p] \leq C_{p,q}\Big[\|Y_T\|_{\mathbb{L}_{\mathscr{P}}^q} + \varepsilon\Big]^p.$$

Since $\mathbb{P} \in \mathscr{P}$ and $\varepsilon > 0$ are arbitrary, we prove (12.1.22) immediately. $\blacksquare$

We now extend Proposition 12.1.14 to $\xi \in \widehat{\mathbb{L}}_0^1(\mathscr{F}_T, \mathscr{P})$.

Theorem 12.1.16 *Let $\mathscr{P}$ satisfy Assumptions 9.3.3 and 12.1.7, and $\xi \in \widehat{\mathbb{L}}_0^1(\mathscr{F}_T, \mathscr{P})$. Then there exists $\mathscr{E}^{\mathscr{P}}$-martingale $Y \in \mathbb{S}^1(\mathbb{F}, \mathscr{P})$ such that $Y_\tau = \mathscr{E}_\tau^{\mathscr{P}}[\xi]$ for all $\tau \in \mathscr{T}$.*

We note that, by (12.1.18) and the above theorem, we have

$$\|\xi\|_{\mathbb{L}_{\mathscr{P}}^p}^p := \mathscr{E}^{\mathscr{P}}\Big[\sup_{0 \leq t \leq T} \mathscr{E}_t^{\mathscr{P}}[|\xi|^p]\Big], \quad \xi \in \widehat{\mathbb{L}}_0^p(\mathscr{F}_T, \mathscr{P}). \tag{12.1.23}$$

Proof Since $\xi \in \widehat{\mathbb{L}}_0^1(\mathscr{F}_T, \mathscr{P})$, there exists $\xi_n \in UC(\Omega)$ such that $\lim_{n \to \infty} \|\xi_n - \xi\|_{\widehat{\mathbb{L}}_\mathscr{P}^1} = 0$. Denote $Y_t^n := \mathscr{E}_t^\mathscr{P}[\xi_n]$. Then, by (12.1.18),

$$\mathscr{E}^\mathscr{P}[(Y^m - Y^n)_T^*] \le \|\xi_n - \xi_m\|_{\widehat{\mathbb{L}}_\mathscr{P}^1} \to 0, \quad \text{as } m, n \to \infty. \qquad (12.1.24)$$

Thus there exists $Y \in \mathbb{S}^1(\mathbb{F}, \mathscr{P})$ such that

$$\lim_{n \to \infty} \mathscr{E}^\mathscr{P}[(Y^n - Y)_T^*] = 0 \quad \text{and} \quad Y_T = \xi. \qquad (12.1.25)$$

Now it remains to verify that Y is the conditional nonlinear expectation of ξ.

To this end, for any $\mathbb{P} \in \mathscr{P}$ and $\tau \in \mathscr{T}$, we have

$$\mathbb{E}^\mathbb{P}\Big[\Big|Y_\tau - \mathscr{E}_{\mathbb{P},\tau}^\mathscr{P}[\xi]\Big|\Big] = \lim_{n \to \infty} \mathbb{E}^\mathbb{P}\Big[\Big|Y_\tau - Y_\tau^n + \mathscr{E}_{\mathbb{P},\tau}^\mathscr{P}[\xi_n] - \mathscr{E}_{\mathbb{P},\tau}^\mathscr{P}[\xi]\Big|\Big]$$

$$\le \limsup_{n \to \infty} \mathbb{E}^\mathbb{P}\Big[|Y_\tau - Y_\tau^n|\Big] + \limsup_{n \to \infty} \mathbb{E}^\mathbb{P}\Big[\mathscr{E}_{\mathbb{P},\tau}^\mathscr{P}[|\xi_n - \xi|]\Big].$$

By Lemma 12.1.10 (i), for any $\varepsilon > 0$ and $n \ge 1$, there exists $\mathbb{P}_n^\varepsilon \in \mathscr{P}(\tau, \mathbb{P})$ such that

$$\mathscr{E}_{\mathbb{P},\tau}^\mathscr{P}[|\xi_n - \xi|] \le \mathbb{E}_\tau^{\mathbb{P}_n^\varepsilon}[|\xi_n - \xi|] + \varepsilon, \quad \mathbb{P}\text{-a.s.}$$

Then

$$\mathbb{E}^\mathbb{P}\Big[\Big|Y_\tau - \mathscr{E}_{\mathbb{P},\tau}^\mathscr{P}[\xi]\Big|\Big] \le \limsup_{n \to \infty} \mathbb{E}^\mathbb{P}\Big[|Y_\tau - Y_\tau^n|\Big] + \limsup_{n \to \infty} \mathbb{E}^\mathbb{P}\Big[\mathbb{E}_\tau^{\mathbb{P}_n^\varepsilon}[|\xi_n - \xi|] + \varepsilon\Big]$$

$$= \limsup_{n \to \infty} \mathbb{E}^\mathbb{P}\Big[|Y_\tau - Y_\tau^n|\Big] + \limsup_{n \to \infty} \mathbb{E}^{\mathbb{P}_n^\varepsilon}[|Y_T^n - Y_T|] + \varepsilon \le 2\limsup_{n \to \infty} \mathscr{E}^\mathscr{P}[(Y^n - Y)_T^*] + \varepsilon = \varepsilon.$$

Since $\varepsilon > 0$ is arbitrary, we obtain $Y_\tau = \mathscr{E}_{\mathbb{P},\tau}^\mathscr{P}[Y_T]$, $\mathbb{P}$-a.s. for any $\mathbb{P} \in \mathscr{P}$. ∎

12.2 Second Order BSDEs

The second order BSDE (2BSDE in short) is a BSDE under nonlinear expectation. Roughly speaking, 2BSDE provides a Sobolev type solution to path dependent HJB equation (11.3.9). In particular, we shall provide the dynamics and norm estimates for the viscosity solution u defined in (11.3.11), and extend the results to $\mathscr{P}$-q.s. continuous coefficients. From now on we fix a class $\mathscr{P}$ satisfying:

Assumption 12.2.1 *$\mathscr{P}$ satisfies Assumptions 9.3.3 and 12.1.7, and*

$$\mathscr{P} \subset \mathscr{P}_{[\underline{\sigma},\overline{\sigma}]}^S \quad \text{for some } 0 \le \underline{\sigma} \le \overline{\sigma}. \qquad (12.2.1)$$

Consequently, every $\mathbb{P} \in \mathscr{P}$ *satisfies the martingale representation property.*
Let $\hat{\sigma}$ be the universal process defined by (12.1.6). By (9.2.4), clearly

$$\hat{\sigma} > 0, \quad \mathscr{P}\text{-q.s.} \tag{12.2.2}$$

To motivate the definition of 2BSDE, let us take a closer look at the $\mathscr{E}^{\mathscr{P}}$-conditional expectation $Y_t := \mathscr{E}_t^{\mathscr{P}}[\xi]$ for $\xi \in \widehat{\mathbb{L}}^1(\mathscr{F}_T, \mathscr{P})$. For any $\mathbb{P} \in \mathscr{P}$, Y is a $\mathbb{P}$-supermartingale. By the construction of $\mathscr{P}_{[\underline{\sigma},\overline{\sigma}]}^S$, one may apply the Doob-Meyer decomposition Theorem 2.7.1. Then there exist $\mathbb{R}^{1\times d}$-valued $Z^{\mathbb{P}}$ and increasing $K^{\mathbb{P}}$, which may depend on $\mathbb{P}$, such that

$$dY_t = Z_t^{\mathbb{P}} dX_t - dK_t^{\mathbb{P}}, \quad \mathbb{P}\text{-a.s.}$$

Note that, by (12.1.7),

$$d\langle Y, X^{\top}\rangle_t = Z_t^{\mathbb{P}} d\langle X\rangle_t = Z_t^{\mathbb{P}} \hat{\sigma}_t^2 dt, \quad \mathbb{P}\text{-a.s.}$$

In the spirit of (12.1.5), we see that $\langle Y, X\rangle$ can be defined $\mathscr{P}$-q.s. This, together with (12.2.2), implies that $\{Z^{\mathbb{P}}, \mathbb{P} \in \mathscr{P}\}$ can always be aggregated as

$$Z_t := \frac{d\langle Y, X^{\top}\rangle_t}{dt} \cdot \hat{\sigma}_t^{-2} \mathbf{1}_{\{\hat{\sigma}_t > 0\}}, \quad \mathscr{P}\text{-q.s.} \tag{12.2.3}$$

However, in general Z is not càdlàg, and thus we are not able to apply Theorem 12.1.2 to define $Z_t dX_t$ $\mathscr{P}$-q.s. Consequently, the aggregation of $\{K^{\mathbb{P}}, \mathbb{P} \in \mathscr{P}\}$ is in general not clear (unless we assume the Continuum Hypothesis as in Remark 12.1.4 (ii)). Nevertheless, let us assume for simplicity that $\{K^{\mathbb{P}}, \mathbb{P} \in \mathscr{P}\}$ has a $\mathscr{P}$-aggregator K. By (12.2.1), we see that $Z_t dX_t$ is a $\mathbb{P}$-martingale under all $\mathbb{P} \in \mathscr{P}$, then one can easily see that $-K$ is an $\mathscr{E}^{\mathscr{P}}$-martingale. In other words, the $\mathscr{E}^{\mathscr{P}}$-conditional expectation Y satisfies

$$dY_t = Z_t dX_t - dK_t, \quad Y_T = \xi, \quad \text{and} \ -K \text{ is an } \mathscr{E}^{\mathscr{P}}\text{-martingale.} \tag{12.2.4}$$

This can be viewed as a martingale representation under $\mathscr{E}^{\mathscr{P}}$, or say is a linear 2BSDE.

Our general 2BSDE takes the following form:

$$Y_t = \xi + \int_t^T f_s(Y_s, Z_s, \hat{\sigma}_s) ds - \int_t^T Z_s dX_s + K_T - K_t, \quad 0 \le t \le T, \quad \mathscr{P}\text{-q.s.} \tag{12.2.5}$$

where the nonlinear generator $f : \Lambda \times \mathbb{R} \times \mathbb{R}^{1\times d} \times \mathbb{S}^d \to \mathbb{R}$. We remark that the above 2BSDE obviously depends on $\mathscr{P}$ and thus rigorously we shall call it a $\mathscr{P}$-2BSDE. For notational simplicity we denote: recalling (12.2.2),

$$\hat{Z} := Z\hat{\sigma}, \quad \hat{f}_s(\omega, y, z) := f_s(\omega, y, z\hat{\sigma}_s^{-1}(\omega), \hat{\sigma}_s(\omega)), \quad \hat{f}^0 := \hat{f}(0, 0). \tag{12.2.6}$$

Throughout this chapter, we shall assume:

Assumption 12.2.2

(i) $\xi \in \widehat{\mathbb{L}}^2(\mathscr{F}_T, \mathscr{P})$ and $\int_0^T |\widehat{f}_t^0| dt \in \widehat{\mathbb{L}}^2(\mathscr{F}_T, \mathscr{P})$.

(ii) $\widehat{f}$ is uniformly Lipschitz continuous in (y, z).

We remark that since $\widehat{\sigma}$ does not have good regularity in ω, then in general the regularity of f in ω does not imply the regularity of $\widehat{f}$ in ω.

Definition 12.2.3 *Let Assumptions 12.2.1 and 12.2.2 hold. We say (Y, Z) is a solution to 2BSDE (12.2.5) if*

(i) $Y \in \mathbb{S}^2(\mathbb{F}, \mathscr{P})$ *with* $Y_T = \xi$, $\mathscr{P}$-*q.s. and* $\widehat{Z} := Z\widehat{\sigma} \in \mathbb{L}^2(\mathbb{F}, \mathscr{P})$.

(ii) *for each* $\mathbb{P} \in \mathscr{P}$, *the following process* $K^{\mathbb{P}}$ *is increasing,* $\mathbb{P}$-*a.s.*

$$K_t^{\mathbb{P}} := Y_0 - Y_t - \int_0^t f_s(Y_s, Z_s, \widehat{\sigma}_s) ds + \int_0^t Z_s dX_s \quad \mathbb{P}\text{-a.s.} \quad (12.2.7)$$

(iii) *the class* $\{K^{\mathbb{P}}, \mathbb{P} \in \mathscr{P}\}$ *satisfies the following minimum condition:*

$$K_\tau^{\mathbb{P}} = \underset{\widetilde{\mathbb{P}} \in \mathscr{P}(\tau, \mathbb{P})}{\overset{\mathbb{P}}{\mathrm{ess\,inf}}} \mathbb{E}_\tau^{\widetilde{\mathbb{P}}}[K_T^{\widetilde{\mathbb{P}}}], \quad \mathbb{P}\text{-a.s.} \quad \text{for any } \mathbb{P} \in \mathscr{P} \text{ and } \tau \in \mathscr{T}. \quad (12.2.8)$$

Moreover, if $\{K^{\mathbb{P}}, \mathbb{P} \in \mathscr{P}\}$ *has a* $\mathscr{P}$-*aggregator K, then we call* (Y, Z, K) *a solution to (12.2.5).*

We note that, when the $\mathscr{P}$-aggregator K exists, the minimum condition (12.2.8) amounts to saying that $-K$ is a $\mathscr{P}$-martingale, and $\widehat{Z} := Z\widehat{\sigma} \in \mathbb{L}^2(\mathbb{F}, \mathscr{P})$ means that

$$\mathscr{E}^{\mathscr{P}}\left[\int_0^T |Z_t|^2 d\langle X \rangle_t\right] = \mathscr{E}^{\mathscr{P}}\left[\int_0^T |\widehat{Z}_t|^2 dt\right] < \infty. \quad (12.2.9)$$

Moreover, (12.2.7) is equivalent to: for the $\mathbb{P}$-Brownian motion $B_t^{\mathbb{P}}$ defined in (9.2.9),

$$Y_t = \xi + \int_t^T \widehat{f}_s(Y_s, \widehat{Z}_s) ds - \int_t^T \widehat{Z}_s dB_s^{\mathbb{P}} + K_T^{\mathbb{P}} - K_t^{\mathbb{P}}, \quad \mathbb{P}\text{-a.s.} \quad (12.2.10)$$

Remark 12.2.4

(i) When $f = 0$, one can easily see that $Y_t = \mathscr{E}_t^{\mathscr{P}}[\xi]$. Thus conditional nonlinear expectation can be viewed as the solution to a linear 2BSDE. Or the other way around, we may say 2BSDE is a BSDE under nonlinear expectation $\mathscr{E}^{\mathscr{P}}$.

(ii) When $\mathscr{P} = \{\mathbb{P}_0\}$, the minimum condition (12.2.8) implies that $K = 0$. Then the 2BSDE (12.2.5) is reduced back to BSDE (4.0.3).

(iii) In the special case that $\mathscr{P} = \mathscr{P}^S_{[\underline{\sigma},\overline{\sigma}]}$ and f is linear in σ^2: $f_t(y,z,\sigma) = f^1_t(y,z) + \sigma^2 f^2_t(y,z)$, 2BSDE (12.2.5) is equivalent to the so-called G-BSDE proposed by Hu, Ji, Peng, & Song [107]:

$$Y_t = \xi + \int_t^T f^1_s(Y_s, Z_s)ds + \int_t^T f^2_s(Y_s, Z_s)d\langle X \rangle_s$$

$$- \int_t^T Z_s dX_s + K_T - K_t, \quad \mathscr{P}\text{-q.s.} \qquad (12.2.11)$$

∎

Remark 12.2.5

(i) By (12.2.7), once we have the Y-component of the solution, then the Z-component can be obtained for free by (12.2.3). In particular, the uniqueness of Y implies the uniqueness of Z (in $\mathscr{P}$-q.s. sense, of course).

(ii) Note that the aggregation of $\{K^{\mathbb{P}}, \mathbb{P} \in \mathscr{P}\}$ is equivalent to the aggregation of $\{M^{\mathbb{P}}, \mathbb{P} \in \mathscr{P}\}$ where $M^{\mathbb{P}}_t := \int_0^t Z_s dX_s$, $\mathbb{P}$-a.s. As pointed out in Remark 12.1.4 (ii), if we assume further the Continuum Hypothesis, then it follows from Nutz [160] that $\{K^{\mathbb{P}}, \mathbb{P} \in \mathscr{P}\}$ always have an $\mathscr{P}$-aggregator K, as long as the 2BSDE has a solution (Y, Z).

∎

Remark 12.2.6 Let (Y, Z) be a solution to 2BSDE (12.2.5).

(i) If $Y \in C^{1,2}(\Lambda)$, then by (12.2.5) and the functional Itô formula (9.4.1) that

$$Z_t = \partial_\omega Y_t, \quad dK_t = -\left[\partial_t Y_t + \frac{1}{2}\hat{\sigma}^2_t : \partial^2_{\omega\omega} Y_t + f_t(Y_t, \partial_\omega Y_t, \hat{\sigma}_t)\right]dt.$$

$$(12.2.12)$$

In particular, $\{K^{\mathbb{P}}, \mathscr{P} \in \mathscr{P}\}$ has a $\mathscr{P}$-aggregator.

(ii) In the setting of Subsection 11.3.3, and denote $\mathscr{P} := \{\mathbb{P}^{0,0,k} : k \in \mathscr{K}\}$ where $\mathbb{P}^{0,0,k}$ is defined by (11.3.12). One can easily impose appropriate conditions on σ so that $\mathscr{P}$ satisfies Assumption 12.2.1. See Proposition 12.2.13 below for a special case. Assume PPDE (11.3.9) has a classical solution $u \in C^{1,2}(\Lambda)$. By (11.3.11) and the representation formula (12.2.15) below, we see that $Y = u$ and thus (11.3.11) holds. Denote further that $\Gamma := \partial^2_{\omega\omega} u$. Then by (12.2.12) and (11.3.9) we have:

$$dK_t = \left[G(t, \omega, u, \partial_\omega u, \partial^2_{\omega\omega} u) - \left[\frac{1}{2}\hat{\sigma}^2_t : \partial^2_{\omega\omega} u + f_t(u, \partial_\omega u, \hat{\sigma}_t)\right]\right]dt.$$

Plugging this into (12.2.5) we obtain

$$Y_t = \xi + \int_t^T [f_s(Y_s, Z_s, \hat{\sigma}_s) - G_s(Y_s, Z_s, \Gamma_s)]ds + \frac{1}{2}\int_t^T \Gamma_s : d\langle X \rangle_s - \int_t^T Z_s dX_s, \quad \mathscr{P}\text{-q.s.} \quad (12.2.13)$$

Since the above equation involves nonlinearly the component Γ, which corresponds to the second order path derivatives of Y in ω, we call it second order BSDE. Indeed, the original formulation of 2BSDE in Cheridito, Soner, Touzi, & Victoir [32] involves all three solution components (Y, Z, Γ), but in $\mathbb{P}_0$-a.s. sense rather than $\mathscr{P}$-q.s. sense.

(iii) The existence of Γ and its related norm estimate, which can be roughly viewed as the $\mathbb{L}^2$-estimate for $\partial^2_{\omega\omega} u$, is challenging. We refer to Peng, Song, & Zhang [185] for some study along this direction. ∎

Remark 12.2.7 In spirit 2BSDE (12.2.5) shares some properties with RBSDE (6.1.5). For simplicity, let us compare the linear ones (12.2.4) and (6.1.4).

(i) For the linear RBSDE (6.1.4) induced by the American option price (6.1.2), the increasing process K is induced by the time value of the American option. More precisely, since the holder of the American option has the right to choose exercise time τ between $[t, T]$, such a right has a value which is roughly characterized by $K_T - K_t$. When the time passes away, the holder has less choice on the exercise time and thus the time value $K_T - K_t$ decreases, consequently K is increasing. The minimum condition of K is characterized by the Skorohod condition (6.1.3).

(ii) For the linear 2BSDE (12.2.4) induced by the conditional nonlinear expectation $Y_t := \mathscr{E}_t^{\mathscr{P}}[\xi]$, the increasing process K is induced by the uncertainty of the probability measure $\mathbb{P}$, which is viewed as model uncertainty in robust finance, see, e.g., Section 12.4 below. Roughly speaking, assuming $\mathbb{P}$ is the true measure (or say true model) which is unknown to the investors, then $K_T^{\mathbb{P}} - K_t^{\mathbb{P}}$ characterizes the value (or cost) of the uncertainty of all possible $\tilde{\mathbb{P}} \in \mathscr{P}(t, \mathbb{P})$. In this case, the minimum condition of K is characterized by the nonlinear martingale condition (12.2.8). ∎

In the rest of this section, we establish well-posedness of 2BSDEs, including the norm estimates. By (12.2.3) and (12.2.7), our main focus will be the Y-component.

12.2.1 Representation and Uniqueness

For every $\mathbb{P} \in \mathscr{P}$, $\tau \in \mathscr{T}$, and $\eta \in \mathbb{L}^2(\mathscr{F}_\tau, \mathbb{P})$, in light of (12.2.10) we denote by $(\mathscr{Y}^{\mathbb{P}}, \mathscr{Z}^{\mathbb{P}}) := (\mathscr{Y}^{\mathbb{P}}(\tau, \eta), \mathscr{Z}^{\mathbb{P}}(\tau, \eta))$ the solution to the following BSDE:

$$\mathscr{Y}_t^{\mathbb{P}} = \eta + \int_t^\tau \widehat{f}_s(\mathscr{Y}_s^{\mathbb{P}}, \mathscr{Z}_s^{\mathbb{P}}) ds - \int_t^\tau \mathscr{Z}_s^{\mathbb{P}} dB_s^{\mathbb{P}}, \quad 0 \le t \le \tau, \quad \mathbb{P}\text{-a.s.} \quad (12.2.14)$$

We remark that $\overline{\mathbb{F}^{B^{\mathbb{P}}}}^{\mathbb{P}} = \overline{\mathbb{F}}^{\mathbb{P}}$, thanks to (12.2.1) and (9.2.8) (iii). Then the above BSDE is well posed. Our main result of this section is the following representation theorem.

Theorem 12.2.8 *Let Assumptions 12.2.1 and 12.2.2 hold and* (Y, Z) *be a solution to 2BSDE (12.2.5). Then, for any* $\mathbb{P} \in \mathscr{P}$ *and* $\tau \in \mathscr{T}$,

$$Y_\tau = \underset{\tilde{\mathbb{P}} \in \mathscr{P}(\tau, \mathbb{P})}{\text{ess}^{\mathbb{P}}\text{sup}} \, \mathscr{Y}_\tau^{\tilde{\mathbb{P}}}(T, \xi), \quad \mathbb{P}\text{-}a.s. \tag{12.2.15}$$

Consequently, (Y, Z) *is unique in* $\mathscr{P}$*-q.s. sense.*

Proof The uniqueness of Y follows directly from (12.2.15), then we obtain the uniqueness of Z by Remark 12.2.5 (i). We now prove (12.2.15). For notational simplicity, assume $d = 1$ and denote

$$(\mathscr{Y}^{\mathbb{P}}, \mathscr{Z}^{\mathbb{P}}) := (\mathscr{Y}^{\mathbb{P}}(T, \xi), \mathscr{Z}^{\mathbb{P}}(T, \xi)), \quad Y_\tau^{\mathbb{P}} := \underset{\tilde{\mathbb{P}} \in \mathscr{P}(\tau, \mathbb{P})}{\text{ess}^{\mathbb{P}}\text{sup}} \, \mathscr{Y}_\tau^{\tilde{\mathbb{P}}}. \tag{12.2.16}$$

First, for every $\mathbb{P} \in \mathscr{P}$, since $K^{\mathbb{P}}$ is increasing, comparing BSDEs (12.2.10) and (12.2.14) and applying the comparison principle of BSDEs, we have $Y_\tau \geq \mathscr{Y}_\tau^{\mathbb{P}}$, $\mathbb{P}$-a.s. for all $\tau \in \mathscr{T}$. Then

$$Y_\tau \geq Y_\tau^{\mathbb{P}}, \quad \mathbb{P}\text{-}a.s. \tag{12.2.17}$$

To see the opposite inequality, for any $\varepsilon > 0$, applying Lemma 12.1.10 (i) on (12.2.8) there exists $\mathbb{P}_\varepsilon \in \mathscr{P}(\tau, \mathbb{P})$ such that

$$\mathbb{E}_\tau^{\mathbb{P}_\varepsilon}[K_T^{\mathbb{P}_\varepsilon}] \leq K_\tau^{\mathbb{P}} + \varepsilon, \quad \mathbb{P}\text{-}a.s. \tag{12.2.18}$$

Again consider BSDEs (12.2.10) and (12.2.14), but under $\mathbb{P}_\varepsilon$. Denote $\Delta Y := Y - \mathscr{Y}^{\mathbb{P}_\varepsilon}$, $\Delta Z := \widehat{Z} - \mathscr{Z}^{\mathbb{P}_\varepsilon}$. Then

$$\Delta Y_t = \int_t^T [\alpha_s \Delta Y_s + \beta_s \Delta Z_s] ds - \int_t^T \Delta Z_s dB_s^{\mathbb{P}_\varepsilon} + K_{t,T}^{\mathbb{P}_\varepsilon}, \quad \mathbb{P}_\varepsilon\text{-}a.s.$$

where α, β are bounded. Applying Proposition 4.1.2, we have

$$\Delta Y_\tau = \mathbb{E}_\tau^{\mathbb{P}_\varepsilon}\Big[\int_\tau^T \Gamma_t dK_t^{\mathbb{P}_\varepsilon}\Big],$$

where the adjoint process Γ is defined on $[\tau, T]$:

$$\Gamma_t = 1 + \int_\tau^t \alpha_s \Gamma_s ds + \int_\tau^t \beta_s \Gamma_s dB_s^{\mathbb{P}_\varepsilon}, \quad \tau \leq t \leq T, \ \mathbb{P}_\varepsilon\text{-}a.s.$$

This implies

$$\mathbb{E}^{\mathbb{P}}[\Delta Y_\tau] = \mathbb{E}^{\mathbb{P}_\varepsilon}[\Delta Y_\tau] \le \mathbb{E}^{\mathbb{P}_\varepsilon}\Big[\big(\sup_{\tau \le t \le T} \Gamma_t\big) K_{\tau,T}^{\mathbb{P}_\varepsilon}\Big]$$

$$\le \Big(\mathbb{E}^{\mathbb{P}_\varepsilon}\big[\sup_{\tau \le t \le T} |\Gamma_t|^3\big]\Big)^{\frac{1}{3}} \Big(\mathbb{E}^{\mathbb{P}_\varepsilon}[(K_{\tau,T}^{\mathbb{P}_\varepsilon})^2]\Big)^{\frac{1}{3}} \Big(\mathbb{E}^{\mathbb{P}_\varepsilon}[K_{\tau,T}^{\mathbb{P}_\varepsilon}]\Big)^{\frac{1}{3}}. \quad (12.2.19)$$

It is clear that $\mathbb{E}^{\mathbb{P}_\varepsilon}\big[\sup_{\tau \le t \le T} |\Gamma_t|^3\big] \le C$. Moreover, note that

$$K_T^{\mathbb{P}_\varepsilon} := Y_0 - Y_T - \int_0^T \widehat{f}_s(Y_s, \widehat{Z}_s)ds + \int_0^T \widehat{Z}_s dB_s^{\mathbb{P}_\varepsilon} \quad \mathbb{P}_\varepsilon\text{-a.s.}$$

Then

$$\mathbb{E}^{\mathbb{P}_\varepsilon}\big[|K_T^{\mathbb{P}_\varepsilon}|^2\big] \le C\mathbb{E}^{\mathbb{P}_\varepsilon}\Big[|Y_T^*|^2 + \big(\int_0^T |\widehat{f}_t^0|dt\big)^2 + \int_0^T |\widehat{Z}_t|^2 dt\Big] \le C. \quad (12.2.20)$$

Plugging (12.2.18) and (12.2.20) into (12.2.19), we obtain $\mathbb{E}^{\mathbb{P}}[\Delta Y_\tau] \le C\varepsilon^{\frac{1}{3}}$. Then

$$\mathbb{E}^{\mathbb{P}}\Big[Y_\tau - Y_\tau^{\mathbb{P}}\Big] \le \mathbb{E}^{\mathbb{P}}[\Delta Y_\tau] \le C\varepsilon^{\frac{1}{3}}.$$

Now it follows from the arbitrariness of ε that $\mathbb{E}^{\mathbb{P}}\big[Y_\tau - Y_\tau^{\mathbb{P}}\big] = 0$. This, together with (12.2.17), proves (12.2.15). ∎

As a direct consequence of Theorem 12.2.8 and comparison of BSDEs, we have the comparison of 2BSDEs.

Corollary 12.2.9 *Let Assumption 12.2.1 hold. Assume, for $i = 1, 2$, ξ_i, f^i satisfy Assumption 12.2.2, and (Y^i, Z^i) be the solution to 2BSDE (12.2.5) with coefficients (ξ_i, f^i). Assume further that $\xi_1 \le \xi_2$, $f^1 \le f^2$, $\mathscr{P}$-q.s. Then $Y_t^1 \le Y_t^2$, $0 \le t \le T$, $\mathscr{P}$-q.s.*

12.2.2 A Priori Estimates

We establish the a priori estimates based on the representation (12.2.15) and the a priori estimates for BSDE (12.2.14). Recall the norms in (12.1.19).

Theorem 12.2.10 *Let Assumptions 12.2.1 and 12.2.2 hold and (Y, Z) be a solution to 2BSDE (12.2.5). Then*

$$\mathscr{E}^{\mathscr{P}}\Big[|Y_T^*|^2 + \int_0^T Z_t^\top Z_t : d\langle X \rangle_t\Big] + \sup_{\mathbb{P} \in \mathscr{P}} \mathbb{E}^{\mathbb{P}}[|K_T^{\mathbb{P}}|^2] \le CI_0^2, \quad (12.2.21)$$

$$\text{where } I_0^2 := \|\xi\|_{\mathbb{L}_{\mathscr{P}}^2}^2 + \Big\|\int_0^T \widehat{f}_t^0|dt\Big\|_{\widehat{\mathbb{L}}_{\mathscr{P}}^2}^2.$$

Proof Assume $d = 1$ for simplicity. We proceed in two steps.

Step 1. We first estimate Y. For any $\mathbb{P} \in \mathscr{P}$, recalling (12.2.16) and applying Proposition 4.2.1 on BSDE (12.2.14), but on $[t, T]$ with conditional expectation $\mathbb{E}_t^{\mathbb{P}}$, we have

$$|\mathscr{Y}_t^{\mathbb{P}}|^2 \le C\mathbb{E}_t^{\mathbb{P}}\Big[|\xi|^2 + \big(\int_t^T |\widehat{f}_s^0|ds\big)^2\Big], \quad \mathbb{P}\text{-a.s.}$$

This, together with (12.2.15), implies that

$$|Y_t|^2 \le C\mathscr{E}_{\mathbb{P},t}^{\mathscr{P}}\Big[|\xi|^2 + \big(\int_t^T |\widehat{f}_s^0|ds\big)^2\Big], \quad \mathbb{P}\text{-a.s.}$$

and thus it follows from (12.1.19) and (12.1.23) that

$$\mathscr{E}^{\mathscr{P}}\big[|Y_T^*|^2\big] \le CI_0^2. \tag{12.2.22}$$

Step 2. We now estimate Z and K. Applying Itô formula we have

$$d|Y_t|^2 = -2Y_t\widehat{f}_t(Y_t, \widehat{Z}_t)dt + 2Y_t\widehat{Z}_t dB_t^{\mathbb{P}} + |\widehat{Z}_t|^2 dt - 2Y_t dK_t^{\mathbb{P}}.$$

Then

$$\mathbb{E}^{\mathbb{P}}\Big[\int_0^T |\widehat{Z}_t|^2 dt\Big] = \mathbb{E}^{\mathbb{P}}\Big[|\xi|^2 - |Y_0|^2 + 2\int_0^T Y_t\widehat{f}_t(Y_t, \widehat{Z}_t)dt + 2\int_0^T Y_t dK_t^{\mathbb{P}}\Big]$$

$$\le \mathbb{E}^{\mathbb{P}}\Big[C|Y_T^*|^2 + C\big(\int_0^T |\widehat{f}_t^0|dt\big)^2 + \frac{1}{2}\int_0^T |\widehat{Z}_t|^2 dt + 2|Y_T^*|K_T^{\mathbb{P}}\Big]$$

Together with (12.2.22), this implies that, for any $\varepsilon > 0$,

$$\mathbb{E}^{\mathbb{P}}\Big[\int_0^T |\widehat{Z}_t|^2 dt\Big] \le C_\varepsilon I_0^2 + \varepsilon\mathbb{E}^{\mathbb{P}}[|K_T^{\mathbb{P}}|^2]. \tag{12.2.23}$$

On the other hand, by (12.2.20) we see that

$$\mathbb{E}^{\mathbb{P}}[|K_T^{\mathbb{P}}|^2] \le CI_0^2 + C\mathbb{E}^{\mathbb{P}}\Big[\int_0^T |\widehat{Z}_t|^2 dt\Big]. \tag{12.2.24}$$

Combine (12.2.23) and (12.2.24), and by choosing ε small enough we obtain immediately that

$$\mathbb{E}^{\mathbb{P}}\Big[\int_0^T |\widehat{Z}_t|^2 dt + |K_T^{\mathbb{P}}|^2\Big] \le CI_0^2.$$

Since $\mathbb{P}$ is arbitrary, we have

$$\mathscr{E}^{\mathscr{P}}\Big[\int_0^T |Z_t|^2 d\langle X\rangle_t\Big] + \sup_{\mathbb{P}\in\mathscr{P}} \mathbb{E}^{\mathbb{P}}[|K_T^{\mathbb{P}}|^2] \le C I_0^2.$$

This, together with (12.2.22), completes the proof. ∎

The next theorem estimates the difference of two 2BSDEs, which implies further the stability.

Theorem 12.2.11 *Let Assumption 12.2.1 hold. Assume, for $i = 1, 2$, ξ_i, f^i satisfy Assumption 12.2.2, and (Y^i, Z^i) be the solution to 2BSDE (12.2.5) with coefficients (ξ_i, f^i). Denote $\Delta\varphi := \varphi_1 - \varphi_2$, for $\varphi = Y, Z, K^{\mathbb{P}}, \xi, f$. Then*

$$\mathscr{E}^{\mathscr{P}}\Big[|(\Delta Y)_T^*|^2\Big] \le C J_0^2,$$

$$\sup_{\mathbb{P}\in\mathscr{P}} \mathbb{E}^{\mathbb{P}}\Big[\int_0^T (\Delta Z_t)^\top (\Delta Z_t) : d\langle X\rangle_t + |(\Delta K^{\mathbb{P}})_T^*|^2\Big] \le C[I_{1,0} + I_{2,0} + J_0]J_0, \quad (12.2.25)$$

where, for $i = 1, 2$,

$$J_0^2 := \|\Delta\xi\|_{\mathbb{L}_{\mathscr{P}}^2}^2 + \Big\|\int_0^T \sum_{j=1}^2 |\widehat{\Delta f_t}(Y_t^j, \widehat{Z}_t^j)| dt\Big\|_{\mathbb{L}_{\mathscr{P}}^2}^2 ; \quad I_{i,0}^2 := \|\xi_i\|_{\mathbb{L}_{\mathscr{P}}^2}^2 + \Big\|\int_0^T |\widehat{f_t^i}(0,0)| dt\Big\|_{\mathbb{L}_{\mathscr{P}}^2}^2.$$

Proof Again we assume $d = 1$. We proceed in three steps.

Step 1. We first prove the estimate for ΔY, following the proof of Theorem 12.2.8. Indeed, for any $\mathbb{P} \in \mathscr{P}$, denote $\Delta\mathscr{Y}^{\mathbb{P}} := Y^1 - \mathscr{Y}^{2,\mathbb{P}}$, $\Delta\mathscr{Z}^{\mathbb{P}} := \widehat{Z}^1 - \mathscr{Z}^{2,\mathbb{P}}$. Then,

$$\Delta\mathscr{Y}_t^{\mathbb{P}} = \Delta\xi + \int_t^T [\alpha_s \Delta\mathscr{Y}_s^{\mathbb{P}} + \beta_s \Delta\mathscr{Z}_s^{\mathbb{P}} + \widehat{\Delta f_s}(Y_s^1, \widehat{Z}_s^1)] ds - \int_t^T \Delta\mathscr{Z}_s^{\mathbb{P}} dB_s^{\mathbb{P}} + K_{t,T}^{1,\mathbb{P}}, \quad \mathbb{P}\text{-a.s.}$$

where α, β are bounded. Fix t and introduce the adjoint process Γ:

$$\Gamma_s = 1 + \int_t^s \alpha_r \Gamma_r dr + \int_t^s \beta_r \Gamma_r dB_r^{\mathbb{P}}, \quad t \le s \le T, \quad \mathbb{P}\text{-a.s.}$$

Then, similar to the arguments in Theorem 12.2.8 we have

$$\Delta\mathscr{Y}_t^{\mathbb{P}} = \mathbb{E}_t^{\mathbb{P}}\Big[\Gamma_T \Delta\xi + \int_t^T \Gamma_s \widehat{\Delta f_s}(Y_s^1, \widehat{Z}_s^1) ds + \int_t^T \Gamma_s dK_s^{1,\mathbb{P}}\Big]$$

$$\le C\Big(\mathbb{E}_t^{\mathbb{P}}\Big[|\Delta\xi|^2 + \big(\int_t^T |\widehat{\Delta f_s}(Y_s^1, \widehat{Z}_s^1)| ds\big)^2\Big]\Big)^{\frac{1}{2}} + C\big(\mathbb{E}_t^{\mathbb{P}}[|K_{t,T}^{1,\mathbb{P}}|^2]\big)^{\frac{1}{3}} \big(\mathbb{E}_t^{\mathbb{P}}[K_{t,T}^{1,\mathbb{P}}]\big)^{\frac{1}{3}}.$$

Now applying Theorem 12.2.10 on $\mathbb{E}_t^{\mathbb{P}}[|K_T^{1,\mathbb{P}}|^2]$ and recalling the minimum condition (12.2.8), it follows from the representation (12.2.15) that

$$\Delta Y_t = \mathop{\mathrm{ess\,inf}}_{\tilde{\mathbb{P}}\in\mathscr{P}(t,\mathbb{P})}^{\mathbb{P}} \Delta\mathscr{Y}_t^{\tilde{\mathbb{P}}} \le C\Big(\mathbb{E}_{\mathbb{P},t}^{\mathscr{P}}\Big[|\Delta\xi|^2 + \big(\int_t^T |\widehat{\Delta f_s}(Y_s^1, \widehat{Z}_s^1)| ds\big)^2\Big]\Big)^{\frac{1}{2}}, \quad \mathbb{P}\text{-a.s.}$$

Similarly, we have

$$-\Delta Y_t \leq C\Big(\mathbb{E}^{\mathscr{P}}_{\mathbb{P},t}\Big[|\Delta\xi|^2 + \big(\int_t^T |\Delta\widehat{f}_s(Y_s^2,\widehat{Z}_s^2)|ds\big)^2\Big]\Big)^{\frac{1}{2}}, \quad \mathbb{P}\text{-a.s.}$$

Thus

$$|\Delta Y_t|^2 \leq C\mathbb{E}^{\mathscr{P}}_{\mathbb{P},t}\Big[|\Delta\xi|^2 + \big(\int_t^T [|\Delta\widehat{f}_s(Y_s^1,\widehat{Z}_s^1)| + |\Delta\widehat{f}_s(Y_s^2,\widehat{Z}_s^2)|]ds\big)^2\Big], \quad \mathbb{P}\text{-a.s.}$$

This leads to the desired estimate for ΔY immediately.

Step 2. We next prove the estimate for ΔZ, following the arguments in Theorem 12.2.10 Step 2. Note that, for any $\mathbb{P} \in \mathscr{P}$,

$$\Delta Y_t = \Delta\xi + \int_t^T [\widehat{f}_s^1(Y_s^1,\widehat{Z}_s^1) - \widehat{f}_s^2(Y_s^2,\widehat{Z}_s^2)]ds - \int_t^T \Delta\widehat{Z}_s dB_s^{\mathbb{P}} + \Delta K_T^{\mathbb{P}} - \Delta K_t^{\mathbb{P}}, \quad \mathbb{P}\text{-a.s.}$$

Applying Itô formula on $|\Delta Y|^2$ we have

$$\mathbb{E}^{\mathbb{P}}\Big[\int_0^T |\Delta\widehat{Z}_t|^2 dt\Big]$$

$$= \mathbb{E}^{\mathbb{P}}\Big[|\Delta\xi|^2 - |\Delta Y_0|^2 + 2\int_0^T \Delta Y_t[\widehat{f}_t^1(Y_t^1,\widehat{Z}_t^1) - \widehat{f}_t^2(Y_t^2,\widehat{Z}_t^2)]dt + 2\int_0^T \Delta Y_t d(\Delta K_t^{\mathbb{P}})\Big]$$

$$\leq \mathbb{E}^{\mathbb{P}}\Big[C|(\Delta Y)_T^*|^2 + \frac{1}{2}\int_0^T |\Delta\widehat{Z}_t|^2 dt + \big(\int_0^T |\Delta\widehat{f}_t(Y_t^1,\widehat{Z}_t^1)|dt\big)^2 + 2(\Delta Y)_T^*[K_T^{1,\mathbb{P}} + K_T^{2,\mathbb{P}}]\Big].$$

This implies

$$\mathbb{E}^{\mathbb{P}}\Big[\int_0^T |\Delta\widehat{Z}_t|^2 dt\Big] \leq C\mathbb{E}^{\mathbb{P}}\Big[|(\Delta Y)_T^*|^2 + \big(\int_0^T |\Delta\widehat{f}_t(Y_t^1,\widehat{Z}_t^1)|dt\big)^2\Big]$$

$$+ C\Big(\mathbb{E}^{\mathbb{P}}\big[|(\Delta Y)_T^*|^2\big]\Big)^{\frac{1}{2}}\Big(\mathbb{E}^{\mathbb{P}}[|K_T^{1,\mathbb{P}}|^2 + |K_T^{2,\mathbb{P}}|^2]\Big)^{\frac{1}{2}}.$$

Now applying Theorem 12.2.10, it follows from Step 1 that

$$\mathbb{E}^{\mathbb{P}}\Big[\int_0^T |\Delta\widehat{Z}_t|^2 dt\Big] \leq CJ_0^2 + C[I_{1,0} + I_{2,0}]J_0.$$

Then by the arbitrariness of $\mathbb{P} \in \mathscr{P}$ we obtain the desired estimate for ΔZ.

Step 3. Finally, note that

$$\Delta K_t^{\mathbb{P}} = \Delta Y_0 - \int_0^t [\widehat{f}_s^1(Y_s^1,\widehat{Z}_s^1) - \widehat{f}_s^2(Y_s^2,\widehat{Z}_s^2)]ds + \int_0^t \Delta\widehat{Z}_s dB_s^{\mathbb{P}}.$$

Then

$$\lvert(\Delta K^{\mathbb{P}})_T^*\rvert^2 \le C\Big[\lvert\Delta Y_0\rvert^2 + \big(\int_0^T \lvert\Delta\widehat{f}_t(Y_t^1,\widehat{Z}_t^1)\rvert dt\big)^2 + \int_0^T [\lvert\Delta Y_t\rvert^2 + \lvert\Delta\widehat{Z}_t\rvert^2]dt + \sup_{0\le t\le T}\big\lvert\int_0^t \Delta\widehat{Z}_s dB_s^{\mathbb{P}}\big\rvert^2\Big].$$

Applying the Burkholder-Davis-Gundy inequality, we obtain

$$\mathbb{E}^{\mathbb{P}}\Big[\lvert(\Delta K^{\mathbb{P}})_T^*\rvert^2\Big] \le C\mathbb{E}^{\mathbb{P}}\Big[\lvert(\Delta Y)_T^*\rvert^2 + \int_0^T \lvert\Delta\widehat{Z}_t\rvert^2 dt + \big(\int_0^T \lvert\Delta\widehat{f}_t(Y_t^1,\widehat{Z}_t^1)\rvert dt\big)^2\Big].$$

Now by Steps 1 and 2 we obtain the desired estimate for K. ∎

Remark 12.2.12 Note that our estimate for ΔK in Theorem 12.2.11 is in the form $\mathscr{E}^{\mathscr{P}}[\lvert(\Delta K)_T^*\rvert^2]$. We are not able to provide a desired estimate for the stronger norm $\mathscr{E}^{\mathscr{P}}[\lvert\bigvee_0^T(\Delta K)\rvert^2]$, where $\bigvee_0^T(\Delta K)$ denotes the total variation of ΔK. ∎

12.2.3 Existence

By Theorem 12.2.8, clearly the candidate solution is provided by the representation (12.2.15). However, (12.2.15) involves the aggregation of the process Y. To achieve this, we start from the pathwise approach in Chapter 11, under the additional continuity assumption. To ease the presentation, we restrict ourselves to a simple setting.

Proposition 12.2.13 *Assume*

(i) $\mathscr{P} = \mathscr{P}_{[\underline{\sigma},\overline{\sigma}]}^S$; ξ *and* $\widehat{f}^0$ *are bounded; and* $\widehat{f}$ *is uniformly Lipschitz continuous in* (y,z).

(ii) ξ *is uniformly continuous in* ω *and* f *is uniformly continuous in* (t,ω).

Then 2BSDE (12.2.5) admits a unique solution.

Proof Clearly this is a special case of the setting in Subsection 11.3.3 with:

$$\mathbb{K} := \{k \in \mathbb{S}^d : \underline{\sigma} \le k \le \overline{\sigma}, k > 0\}, \quad \sigma(t,\omega,k) = k, \quad F(t,\omega,y,z,k) = f(t,\omega,y,zk^{-1},k).$$

Then the u defined in (11.3.11) is in $UC_b(\overline{\Lambda})$. We shall show that $Y := u$ is a solution to 2BSDE (12.2.5), following the heuristic arguments in the beginning of this section. Again assume $d = 1$.

Indeed, by (11.3.16) we have, for any $0 \le t_1 < t_2 \le T$ and any $\omega \in \Omega$,

$$Y_{t_1}(\omega) = \sup_{\mathbb{P}\in\mathscr{P}_{t_1}} \mathscr{Y}_0^{t_1,\omega,\mathbb{P}}(t_2 - t_1, Y_{t_2-t_1}^{t_1,\omega}),$$

where $(\mathscr{Y},\mathscr{Z})$ satisfies BSDE (11.3.13). Following the arguments in Proposition 12.1.14 and in Theorem 10.2.1 Steps 4 and 5, by the uniform regularity of u

one can show that

$$Y_{\tau_1} = \underset{\tilde{\mathbb{P}} \in \mathscr{P}(\tau_1, \mathbb{P})}{\text{ess sup}}^{\mathbb{P}} \mathscr{Y}_{\tau_1}^{\tilde{\mathbb{P}}}(\tau_2, Y_{\tau_2}), \quad \mathbb{P}\text{-a.s. for all } \mathbb{P} \in \mathscr{P}, \tau_1,$$

$$\tau_2 \in \mathscr{T} \text{ with } \tau_1 \leq \tau_2. \tag{12.2.26}$$

This implies, for any $\mathbb{P} \in \mathscr{P}$, $Y_{\tau_1} \geq \mathscr{Y}_{\tau_1}^{\mathbb{P}}(\tau_2, Y_{\tau_2})$, namely Y is a $\widehat{f}$ supermartingale under $\mathbb{P}$, in the sense of Definition 6.5.1. Then it follows from Theorem 6.5.2 that there exist $\widehat{Z}^{\mathbb{P}} \in \mathbb{L}^2(\mathbb{F}, \mathbb{P})$ and $K^{\mathbb{P}} \in \mathbb{I}^2(\mathbb{F}, \mathbb{P})$ such that

$$Y_t = \xi + \int_t^T \widehat{f}_s(Y_s, \widehat{Z}_s^{\mathbb{P}})ds - \int_t^T \widehat{Z}_s^{\mathbb{P}} dB_s^{\mathbb{P}} + K_{t,T}^{\mathbb{P}}, \quad 0 \leq t \leq T, \mathbb{P}\text{-a.s.}$$

$$\mathbb{E}^{\mathbb{P}}\left[\int_0^T |\widehat{Z}_t^{\mathbb{P}}|^2 dt + |K_T^{\mathbb{P}}|^2 \right] \leq C\mathbb{E}^{\mathbb{P}}\left[|Y_T^*|^2 + \left(\int_0^T |\widehat{f}_t^0| dt \right)^2 \right] \leq C. \tag{12.2.27}$$

Then it suffices to verify the minimum condition (12.2.8).

To see this, we reverse the arguments in Theorem 12.2.15. Fix $\tau \in \mathscr{T}$ and $\mathbb{P} \in \mathscr{P}$. Set $\tau_1 := \tau$ and $\tau_2 := T$ in (12.2.26). For any $\varepsilon > 0$, by Lemma 12.1.10 (i) there exists $\mathbb{P}_\varepsilon \in \mathscr{P}(\tau, \mathbb{P})$ such that

$$\mathscr{Y}_\tau^{\mathbb{P}_\varepsilon}(T, \xi) \geq Y_\tau - \varepsilon, \quad \mathbb{P}\text{-a.s.} \tag{12.2.28}$$

Denote $\Delta Y := Y - \mathscr{Y}^{\mathbb{P}_\varepsilon}(T, \xi)$, $\Delta Z := \widehat{Z}^{\mathbb{P}_\varepsilon} - \mathscr{Z}^{\mathbb{P}_\varepsilon}(T, \xi)$. By (12.2.27) and (12.2.14) we have

$$\Delta Y_t = \int_t^T [\alpha_s \Delta Y_s + \beta_s \Delta Z_s]ds + \int_t^T \Delta Z_s dB_s^{\mathbb{P}_\varepsilon} + K_{t,T}^{\mathbb{P}_\varepsilon}, \quad \tau \leq t \leq T, \mathbb{P}_\varepsilon\text{-a.s.}$$

where α, β are bounded. Denote

$$\Gamma_t = 1 + \int_\tau^t \alpha_s \Gamma_s ds + \int_\tau^t \beta_s \Gamma_s dB_s^{\mathbb{P}_\varepsilon}, \quad \tau \leq t \leq T, \mathbb{P}_\varepsilon\text{-a.s.}$$

Then, by (12.2.28),

$$\varepsilon \geq \Gamma_\tau \Delta Y_\tau = \mathbb{E}_\tau^{\mathbb{P}_\varepsilon}\left[\int_\tau^T \Gamma_t dK_t^{\mathbb{P}_\varepsilon} \right] \geq \mathbb{E}_\tau^{\mathbb{P}_\varepsilon}\left[\underset{\tau \leq t \leq T}{\inf} \Gamma_t K_{\tau,T}^{\mathbb{P}_\varepsilon} \right]$$

Thus, noting that $K_\tau^{\mathbb{P}_\varepsilon} = K_\tau^{\mathbb{P}}$,

$$0 \leq \mathbb{E}_\tau^{\mathbb{P}_\varepsilon}[K_T^{\mathbb{P}_\varepsilon}] - K_\tau^{\mathbb{P}} = \mathbb{E}_\tau^{\mathbb{P}_\varepsilon}[K_{\tau,T}^{\mathbb{P}_\varepsilon}] = \mathbb{E}_\tau^{\mathbb{P}_\varepsilon}\left[[\underset{\tau \leq t \leq T}{\inf} \Gamma_t]^{-\frac{1}{3}} \times [K_{\tau,T}^{\mathbb{P}_\varepsilon}]^{\frac{2}{3}} \times [\underset{\tau \leq t \leq T}{\inf} \Gamma_t]^{\frac{1}{3}} [K_{\tau,T}^{\mathbb{P}_\varepsilon}]^{\frac{1}{3}} \right]$$

$$\leq C\left(\mathbb{E}_\tau^{\mathbb{P}_\varepsilon}[\underset{\tau \leq t \leq T}{\sup} \Gamma_t^{-1}] \right)^{\frac{1}{3}} \left(\mathbb{E}_\tau^{\mathbb{P}_\varepsilon}[[K_{\tau,T}^{\mathbb{P}_\varepsilon}]^2] \right)^{\frac{1}{3}} \left(\mathbb{E}_\tau^{\mathbb{P}_\varepsilon}[\underset{\tau \leq t \leq T}{\inf} \Gamma_t K_{\tau,T}^{\mathbb{P}_\varepsilon}] \right)^{\frac{1}{3}} \leq C\varepsilon^{\frac{1}{3}}.$$

By the arbitrariness of $\varepsilon > 0$ we obtain (12.2.8), and thus completes the proof. ■

We next extend the result to $\mathscr{P}$-q.s. continuous coefficients.

Theorem 12.2.14 *Assume Proposition 12.2.13 (i) holds, and there exist (ξ_n, f^n), $n \geq 1$, such that*

(i) *(ξ_n, f^n) satisfy Proposition 12.2.13 (i) uniformly, uniformly in n;*

(ii) *For each n, ξ_n is uniformly continuous in ω and f^n is uniformly continuous in (t, ω);*

(iii) *$\lim_{n \to \infty} \|\xi_n - \xi\|_{\widehat{\mathbb{L}}^2_{\mathscr{P}}} = 0$ and $\lim_{n \to \infty} \|\int_0^T \sup_{y,z} |\widehat{f}^n_t(y, z) - \widehat{f}_t(y, z)| dt\|_{\widehat{\mathbb{L}}^2_{\mathscr{P}}} = 0$.*

Then 2BSDE (12.2.5) admits a unique solution.

Proof Assume $d = 1$ for simplicity. For each n, by Proposition 12.2.13 2BSDE (12.2.5) with coefficients (ξ_n, f^n) admits a unique solution (Y^n, Z^n) as well as corresponding $\{K^{n,\mathbb{P}}, \mathbb{P} \in \mathscr{P}\}$. By Theorem 12.2.10,

$$\mathscr{E}^{\mathscr{P}}\left[|(Y^n)^*_T|^2 + \int_0^T |Z^n_t \widehat{\sigma}_t|^2 dt\right] + \sup_{\mathbb{P} \in \mathscr{P}} \mathbb{E}^{\mathbb{P}}[|K^{n,\mathbb{P}}_T|^2] \leq C,$$

where C is independent of n. Then, applying Theorem 12.2.11 we have,

$$\mathscr{E}^{\mathscr{P}}\left[|(Y^n - Y^m)^*_T|^2 + \int_0^T |(Z^n_t - Z^m_t)\widehat{\sigma}_t|^2 dt\right] + \sup_{\mathbb{P} \in \mathscr{P}} \mathbb{E}^{\mathbb{P}}[|(K^{n,\mathbb{P}} - K^{m,\mathbb{P}})^*_T|^2]$$

$$\leq C\left[\|\xi_n - \xi_m\|_{\widehat{\mathbb{L}}^2_{\mathscr{P}}} + \|\int_0^T \sup_{y,z} |\widehat{f}^n_t(y, z) - \widehat{f}_t(y, z)| dt\|_{\widehat{\mathbb{L}}^2_{\mathscr{P}}}\right] \to 0, \quad \text{as } m, n \to \infty.$$

Then there exist (Y, Z) and $\{K^{\mathbb{P}}, \mathbb{P} \in \mathscr{P}\}$ such that

$$\mathscr{E}^{\mathscr{P}}\left[|(Y^n - Y)^*_T|^2 + \int_0^T |(Z^n_t - Z_t)\widehat{\sigma}_t|^2 dt\right] + \sup_{\mathbb{P} \in \mathscr{P}} \mathbb{E}^{\mathbb{P}}[|(K^{n,\mathbb{P}} - K^{\mathbb{P}})^*_T|^2] \to 0, \quad \text{as } n \to \infty.$$

Now one can easily verify that (Y, Z) is a solution to 2BSDE (12.2.5) with coefficients (ξ, f). ∎

We conclude this section with an aggregation result of $\{K^{\mathbb{P}}, \mathbb{P} \in \mathscr{P}\}$, without assuming the Continuum Hypothesis required in Nutz [160], see Remark 12.2.5 (ii).

Proposition 12.2.15 *In the setting of Theorem 12.2.14, assume $\underline{\sigma} > 0$, then $\{K^{\mathbb{P}}, \mathbb{P} \in \mathscr{P}\}$ has an $\mathscr{P}$-aggregator K.*

Proof First, by Theorem 12.2.14, and combined with Problem 12.6.4 and extending it to f, there exist time partitions $\pi_n : 0 < t_1 < \cdots < t_n = T$ and smooth functions $g_n(x_1, \cdots, x_n)$ and $f^n_i(x_1, \cdots, x_i; t, x, y, z, \sigma)$, $i = 0, \cdots, n - 1$, such that $\lim_{n \to \infty} |\pi_n| = 0$, g_n and $f^n_i(x_1, \cdots, x_i; t, x, 0, 0, \sigma)$ are uniformly bounded, uniformly in n, and

$$\lim_{n \to \infty}\left[\|\xi_n - \xi\|_{\widehat{\mathbb{L}}^2_{\mathscr{P}}} + \int_0^T \sup_{y,z} |\widehat{f}^n_t(y, z) - \widehat{f}_t(y, z)| dt\|_{\widehat{\mathbb{L}}^2_{\mathscr{P}}}\right] = 0, \quad (12.2.29)$$

where $\xi_n := g_n(X_{t_1}, \cdots, X_{t_n})$, $\widehat{f}_t^n(y, z) := \sum_{i=0}^{n-1} f_i^n(X_{t_1}, \cdots, X_{t_i}; t, X_t, y, z, \widehat{\sigma}_t) \mathbf{1}_{[t_i, t_{i+1})}(t)$.

Denote, for $i = 0, \cdots, n-1$,

$$G_i^n(x_1, \cdots, x_i; t, x, y, z, \gamma) := \sup_{\underline{\sigma} \leq \sigma \leq \overline{\sigma}} \left[\frac{1}{2} \sigma^2 : \gamma - f_i^n(x_1, \cdots, x_i; t, x, y, z, \sigma) \right], \quad (12.2.30)$$

and consider the following PDEs: for $i = n-1, \cdots, 0$,

$$\begin{cases} \partial_t u_i^n(x_1, \cdots, x_i; t, x) + G_i^n(x_1, \cdots, x_i; t, x, u_i^n, \partial_x u_i^n, \partial_{xx}^2 u_i^n) = 0, \quad t_i \leq t \leq t_{i+1}; \\ u_i^n(x_1, \cdots, x_i; t_{i+1}, x) = u_{i+1}^n(x_1, \cdots, x_{n-1}, x; t_{i+1}, x); \end{cases} \quad (12.2.31)$$

where $u_n^n(x_1, \cdots, x_n; T, x) := g_n(x_1, \cdots, x_n)$. Since $\underline{\sigma} > 0$, by PDE literature, see, e.g., Wang [232–234], the mappings $(t, x) \mapsto u_i^n(x_1, \cdots, x_i; t, x)$ are smooth for each i and any $(x_1, \cdots, x_i)$.

Recall Remark 12.2.6 (ii). For each n, the 2BSDE (12.2.5) admits a unique solution:

$$Y_t^n := \sum_{i=0}^{n-1} u_i^n(X_{t_1}, \cdots, X_{t_i}; t, X_t) \mathbf{1}_{[t_i, t_{i+1})}(t) + g_n(X_{t_1}, \cdots, X_{t_n}) \mathbf{1}_{\{T\}}(t),$$

$$Z_t^n := \sum_{i=0}^{n-1} \partial_x u_i^n(X_{t_1}, \cdots, X_{t_i}; t, X_t) \mathbf{1}_{[t_i, t_{i+1})}(t),$$

$$dK_t^n := \sum_{i=0}^{n-1} \left[G_i^n(\cdot, u_i^n, \partial_x u_i^n, \partial_{xx}^2 u_i^n) - \left[\frac{1}{2} \widehat{\sigma}_t^2 : \partial_{xx}^2 u_i^n + \widehat{f}_i^n(\cdot, u_i^n, \partial_x u_i^n) \right] \right](X_{t_1}, \cdots, X_{t_i}; t, X_t) \mathbf{1}_{[t_i, t_{i+1})} dt.$$

We emphasize that the above K^n is aggregated. Now by (12.2.29) and Theorem 12.2.10 we see that

$$\sup_n \mathscr{E}^{\mathscr{P}} \left[|(Y^n)_T^*|^2 + \int_0^T (Z_t^n)^\top Z_t^n : d\langle X \rangle_t + |K_T^n|^2 \right] < \infty.$$

Then it follows from Theorem 12.2.11 and (12.2.29) that, as $n \to \infty$,

$$\mathscr{E}^{\mathscr{P}} \left[|(Y^n - Y)_T^*|^2 + \int_0^T (Z_t^n - Z_t)^\top (Z_t^n - Z_t) : d\langle X \rangle_t \right] + \sup_{\mathbb{P} \in \mathscr{P}} \mathbb{E}^{\mathbb{P}} \left[|(K^n - K^{\mathbb{P}})_T^*|^2 \right] \to 0.$$

This clearly implies that $\{K^{\mathbb{P}}, \mathbb{P} \in \mathscr{P}\}$ has an $\mathscr{P}$-aggregator $K := \limsup_{n \to \infty} K^n$. ∎

Remark 12.2.16 By some more involved approximation, one may remove the assumption $\underline{\sigma} > 0$ and still obtain the $\mathscr{P}$-aggregator K, see Soner, Touzi, & Zhang [212]. ∎

12.3 Extension to the Case with Measurable Coefficients

In this section, we shall extend the results in the previous sections to measurable coefficients without quasi-sure continuity. This involves the very subtle measurability issue. We will only present some partial results, so that the readers can have a taste of the results, and we refer to the original papers for the proofs.

Given a class $\mathscr{P}$ of probability measures on $(\Omega, \mathscr{F}_T)$, we first introduce the following $\mathscr{P}$-universal filtration

$$\mathbb{F}^{\mathscr{P}} := \{\mathscr{F}_t^{\mathscr{P}}\}_{0 \leq t \leq T} := \bigcap_{\mathbb{P} \in \mathscr{P}} \overline{\mathbb{F}^+}^{\mathbb{P}}, \tag{12.3.1}$$

and we extend the $\mathscr{P}$-aggregators in Definition 12.1.1 to $\mathscr{F}_T^{\mathscr{P}}$-measurable random variables and $\mathbb{F}^{\mathscr{P}}$-measurable processes, respectively, in the obvious sense.

Remark 12.3.1

(i) Thanks to Proposition 1.2.1, in Parts I and II of this book we were able to establish the theory for measurable processes (without regularity in ω) under the natural filtration $\mathbb{F}$. However, the version $\tilde{X}$ in Proposition 1.2.1 typically depends on $\mathbb{P}$. In this chapter, we consider all measures $\mathbb{P} \in \mathscr{P}$ simultaneously, and unfortunately we do not have a counterpart result of Proposition 1.2.1 for $\mathscr{P}$. For example, given a process $X \in \mathbb{L}^0(\mathscr{F}_T^{\mathscr{P}})$, in general we do not have $\tilde{X} \in \mathbb{L}^0(\mathscr{F}_T)$ such that $\tilde{X} = X$, $\mathscr{P}$-q.s.

(ii) Clearly we can easily extend each $\mathbb{P} \in \mathscr{P}$ uniquely to $\mathscr{F}_T^{\mathscr{P}}$, and we shall still denote it as $\mathbb{P}$. Moreover, under each fixed $\mathbb{P}$, since $\mathbb{F}^{\mathscr{P}} \subset \overline{\mathbb{F}^+}^{\mathbb{P}}$, all the results in Parts I and II of this book can be extended to $\mathscr{F}_T^{\mathscr{P}}$-measurable random variables and/or $\mathbb{F}^{\mathscr{P}}$-measurable processes in obvious sense (with appropriate modification whenever necessary).

(iii) $\mathscr{F}_0^{\mathscr{P}}$ may not be degenerate, namely $X \in \mathbb{L}^0(\mathscr{F}_0^{\mathscr{P}})$ does not imply X is a constant $\mathscr{P}$-q.s. Indeed, under certain conditions, e.g., $\mathscr{P} \subset \mathscr{P}_{[\underline{\sigma}, \overline{\sigma}]}^S$, for each $\mathbb{P} \in \mathscr{P}$, $X = \mathbb{E}^{\mathbb{P}}[X]$, $\mathbb{P}$-a.s. However, in general $\mathbb{E}^{\mathbb{P}}[X]$ may depend on $\mathbb{P}$ and thus X is not a constant $\mathscr{P}$-q.s.

(iv) We say an event $A \subset \Omega$ is a $\mathscr{P}$-polar set if $\mathbb{P}(A) = 1$ for all $\mathbb{P} \in \mathscr{P}$. Note that $\mathscr{P}$-polar sets are by definition in $\mathscr{F}_T^{\mathscr{P}}$, but not necessarily in $\mathscr{F}_T$. ∎

We next extend Theorem 12.1.2 for quasi-sure stochastic integration. The following result is due to Karandikar [119], see also Bichteler [17] and Follmer [91].

Theorem 12.3.2 *Let $\mathscr{P}$ be the set of all semimartingale measures on $(\Omega, \mathscr{F}_T)$. Assume $\eta \in \cap_{\mathbb{P} \in \mathscr{P}} \mathbb{L}_{loc}^2(\mathbb{F}^{\mathscr{P}}, \mathbb{P}, \mathbb{R}^{1 \times d})$ is càdlàg , $\mathscr{P}$-q.s. Then the $\{M^{\mathbb{P}}, \mathbb{P} \in \mathscr{P}\}$ defined by (12.1.1) has an $\mathbb{F}^{\mathscr{P}}$-measurable $\mathscr{P}$-aggregator M, and we shall still denote it $M_t := \int_0^t \eta_s dX_s$, $\mathscr{P}$-q.s.*

Remark 12.3.3

(i) Theorem 12.3.2 can be extended further to the case that X is a càdlàg semi-martingale, namely we consider $\mathbb{D}([0, T])$ as the canonical space and $\mathscr{P}$ is the class of all semimartingale measures on $\mathbb{D}([0, T])$.

(ii) In the setting of Theorem 12.3.2, even if η is $\mathbb{F}$-measurable, it is not clear if we have an $\mathbb{F}$-measurable $\mathscr{P}$-aggregator.

(iii) If we restrict to $\mathscr{P} = \mathscr{P}_\infty$, then the proof of Theorem 12.3.2 is almost identical to that of Theorem 12.1.2.   ∎

We now turn to the backward problems. The following result is due to Nutz & van Handel [164] for conditional nonlinear expectation, which extends Theorems 10.2.1 and 12.1.16.

Theorem 12.3.4 *Let $\mathscr{P} = \mathscr{P}_L$ and $\xi \in \mathbb{L}^1(\mathscr{F}_T, \mathscr{P})$. Then for any $\tau \in \mathscr{T}$, $\{\mathscr{E}_{\mathbb{P},\tau}^{\mathscr{P}}[\xi] : \mathbb{P} \in \mathscr{P}\}$ has an $\mathscr{F}_\tau^{\mathscr{P}}$-measurable $\mathscr{P}$-aggregator $\mathscr{E}_\tau^{\mathscr{P}}[\xi]$.*

Remark 12.3.5

(i) The assumption $\mathscr{P} = \mathscr{P}_L$ is just for simplicity. It can be weakened significantly to certain compatibility conditions in the spirit of Assumptions 9.3.3 and 12.1.7.

(ii) The aggregation is verified only for $\mathbb{F}$-stopping times τ, not for $\mathbb{F}^{\mathscr{P}}$-stopping times. Recall Proposition 1.2.5, we remark that the measurability of the stopping times is quite subtle in quasi-sure framework, similar to Remark 12.3.1 (i).

(iii) Even though we require ξ to be $\mathscr{F}_T$-measurable, in general $\mathscr{E}_\tau^{\mathscr{P}}[\xi]$ is only $\mathscr{F}_\tau^{\mathscr{P}}$-measurable, not $\mathscr{F}_\tau$-measurable. See Nutz & van Handel [164] for a counterexample.

(iv) For $\xi \in \mathbb{L}^1(\mathscr{F}_T^{\mathscr{P}}, \mathscr{P})$ and $t \in [0, T]$, $\{\mathscr{E}_{\mathbb{P},t}^{\mathscr{P}}[\xi] : \mathbb{P} \in \mathscr{P}\}$ may not have an $\mathscr{F}_t^{\mathscr{P}}$-measurable $\mathscr{P}$-aggregator. Again see Nutz & van Handel [164] for a counterexample.   ∎

Our final result of this section is due to Possamai, Tan, & Zhou [195] for 2BSDEs, which extends Theorem 12.2.14.

Theorem 12.3.6 *Assume*

(i) $\mathscr{P} = \mathscr{P}_{[\underline{\sigma},\overline{\sigma}]}^S$, and ξ and $\widehat{f}^0$ are bounded;

(ii) $\widehat{f}$ is uniformly Lipschitz continuous in (y, z).

(iii) ξ is $\mathscr{F}_T$-measurable and f is $\mathbb{F}$-measurable.

Then 2BSDE (12.2.5) admits a unique $\mathbb{F}^{\mathscr{P}}$-measurable solution (Y, Z).

Remark 12.3.7

(i) The conditions in Theorem 12.3.6 (i) is for simplicity. They can be weakened significantly. In particular, $\mathscr{P} \subset \mathscr{P}_\infty$ can be a general set satisfying certain compatibility conditions in the spirit of Assumptions 9.3.3 and 12.1.7.

(ii) It is crucial to allow for measurable ξ and f (without requiring their quasi-sure regularity) in some applications, see, e.g., Cvitanic, Possamai, & Touzi [48, 49] for an application on Principal-Agent problems.

(iii) There is a large literature on the dynamic programming principle involving measurable coefficients. See some references in Section 9.5. ∎

Remark 12.3.8 While it is always desirable to extend the results requiring only measurable coefficients in semilinear theory to the fully nonlinear theory, we shall notice that in the semilinear theory only one probability $\mathbb{P}$ is involved, and all $\mathscr{F}_T^X$-measurable random variables are in fact $\mathscr{P}$-q.s. continuous for $\mathscr{P} = \{\mathbb{P}\}$. That is, in semilinear case, measurability is equivalent to quasi-sure (actually almost-sure) continuity, and thus all the results in Parts I and II are actually under quasi-sure regularity. In the fully nonlinear case, as we already saw, there is a gap between Borel measurability and quasi-sure continuity. ∎

12.4 An Application in an Uncertain Volatility Model

Recall the notations in Subsection 4.5.1. For simplicity, we assume the interest rate $r = 0$. For a self-financing portfolio (V^π, π), we have

$$dV_t^\pi = \pi_t dS_t, \quad \mathbb{P}\text{-a.s.}$$

Given a terminal payoff $\xi = g(S_.)$, when the market is complete, one can find a hedging portfolio π such that $V_T^\pi = \xi$, $\mathbb{P}$-a.s. and the corresponding V_0^π is the unique arbitrage free price of ξ at time 0. When the market is incomplete, there are various ways to determine an arbitrage free price, and a typical one is the super hedging price which is the largest arbitrage free price:

$$V_0^{\mathbb{P}}(\xi) := \inf\left\{ y : \text{there exists } \pi \text{ such that } y + \int_0^T \pi_t dS_t \geq g(S_.), \ \mathbb{P}\text{-a.s.}\right\}. \quad (12.4.1)$$

We note that at above the probability $\mathbb{P}$ is given. That is, we know the distribution of S, or say we know the model of S. In the situation with model uncertainty, we may need to consider a family of $\mathbb{P} \in \mathscr{P}$. We remark that even if the market is complete under each $\mathbb{P} \in \mathscr{P}$, in general $V_0^{\mathbb{P}}(\xi)$ may vary for different $\mathbb{P}$, and thus we are not able to hedge ξ $\mathscr{P}$-q.s. Then the natural counterpart of (12.4.1) is the $\mathscr{P}$-superhedging price:

$$V_0(\xi) := \inf\{ y : \text{there exists } \pi \text{ such that}$$

$$y + \int_0^T \pi_t dS_t \geq g(S_.), \ \mathbb{P}\text{-a.s. for all } \mathbb{P} \in \mathscr{P}\}. \quad (12.4.2)$$

The goal of this section is to characterize the above $V_0(\xi)$ via second order BSDEs.

To ease the presentation, we simplify the problem slightly. Consider the canonical setting $(\Omega, B, \mathbb{F}^{\mathscr{P}})$ where $\mathscr{P} := \mathscr{P}^S_{[\underline{\sigma}, \overline{\sigma}]}$ for some $0 \le \underline{\sigma} < \overline{\sigma}$. Given $\xi \in \widehat{\mathbb{L}}^2(\mathscr{F}^{\mathscr{P}}_T, \mathscr{P})$, define

$$V_0 := \inf \{ y : \text{there exists } Z \text{ such that } \widehat{\sigma}Z \in \mathbb{L}^2(\mathbb{F}^{\mathscr{P}}, \mathscr{P}) \text{ and}$$
$$y + \int_0^T Z_t dX_t \ge \xi, \ \mathbb{P}\text{-a.s. for all } \mathbb{P} \in \mathscr{P} \}. \tag{12.4.3}$$

By Lemma 9.2.5 (ii) the market is complete under each $\mathbb{P} \in \mathscr{P}$. Moreover, the constraint $\mathscr{P} = \mathscr{P}^S_{[\underline{\sigma}, \overline{\sigma}]}$ means that we consider only risk neutral measures (in terms of X). However, since equivalent measures induce the same super hedging price, so in the spirit of Girsanov theorem this is not a real constraint.

Remark 12.4.1 In (12.4.3) we interpret X as the stock price S itself. In the literature typically we consider models like $dS_t = S_t \sigma_t [dB_t + \mu_t dt]$, $\mathbb{P}_0$-a.s. While we may assume $\mu = 0$ by using risk neutral measure only, it is not reasonable to assume $S_t \sigma_t$ will be bounded. To overcome this, we may interpret the canonical process as $X := \ln \frac{S_t}{S_0}$. Note that

$$\pi_t dS_t = \pi_t S_0 de^{X_t} = \pi_t S_t [dX_t + \frac{1}{2} d\langle X \rangle_t].$$

Then, by denoting $Z := \pi S$, we may reformulate the superhedging problem (12.4.2) as:

$$V_0 := \inf \{ y : \exists Z \text{ s.t. } \widehat{\sigma}Z \in \mathbb{L}^2(\mathbb{F}^{\mathscr{P}}, \mathscr{P}) \text{ and}$$
$$y + \int_0^T Z_t dX_t + \int_0^T \frac{1}{2} \widehat{\sigma}_t^2 Z_t dt \ge \xi, \ \mathbb{P}\text{-a.s. for all } \mathbb{P} \in \mathscr{P} \}. \tag{12.4.4}$$

This still falls into our framework of 2BSDE. Nevertheless, for simplicity at below we consider the simplified problem (12.4.3), and one may easily extend the result to (12.4.4), see Problem 12.6.5.   ∎

Let (Y, Z) be the solution to the 2BSDE:

$$Y_t = \xi - \int_t^T Z_s dX_s + K_T - K_t, \quad \mathscr{P}\text{-q.s.} \tag{12.4.5}$$

Note that $Y_0 \in \mathbb{L}^0(\mathscr{F}^{\mathscr{P}}_0)$ may not be a constant. However, for each $\mathbb{P} \in \mathscr{P}$, by Lemma 9.2.5 (i) and the Blumenthal 0-1 law Theorem 2.1.9, we see that Y_0 is a constant, $\mathbb{P}$-a.s. Therefore, recalling (12.2.14) with $\widehat{f} = 0$ and applying Theorem 12.2.8,

$$Y_0 = Y_0^{\mathbb{P}}, \ \mathbb{P}\text{-a.s.} \quad \text{where} \quad Y_0^{\mathbb{P}} := \mathbb{E}^{\mathbb{P}}[Y_0] = \sup_{\widetilde{\mathbb{P}} \in \mathscr{P}(0, \mathbb{P})} \mathbb{E}^{\widetilde{\mathbb{P}}}[\xi], \tag{12.4.6}$$

Our main result of this section is

Theorem 12.4.2 *Let $\mathscr{P} = \mathscr{P}^S_{[\underline{\sigma},\overline{\sigma}]}$, $\xi \in \widehat{\mathbb{L}}^2(\mathscr{F}^{\mathscr{P}}_T, \mathscr{P})$, and V_0 and (Y,Z) be defined by (12.4.3) and (12.4.5), respectively. Then,*

$$V_0 = \sup_{\mathbb{P}\in\mathscr{P}} Y_0^{\mathbb{P}} = \sup_{\mathbb{P}\in\mathscr{P}} \mathbb{E}^{\mathbb{P}}[\xi], \quad and \quad Z \text{ is a superhedging strategy.} \quad (12.4.7)$$

Proof First, for any $\varepsilon > 0$, by the definition of V_0 there exists a desired Z^ε such that $V_0 + \varepsilon + \int_0^T Z_t^\varepsilon dX_t \geq \xi$, $\mathbb{P}$-a.s. for all $\mathbb{P} \in \mathscr{P}$. Then $V_0 + \varepsilon \geq \mathbb{E}^{\mathbb{P}}[\xi]$ for all $\mathbb{P} \in \mathscr{P}$. By the arbitrariness of ε and $\mathbb{P}$, it follows from (12.4.6) that

$$V_0 \geq \sup_{\mathbb{P}\in\mathscr{P}} \mathbb{E}^{\mathbb{P}}[\xi] = \sup_{\mathbb{P}\in\mathscr{P}} Y_0^{\mathbb{P}} =: \tilde{V}_0.$$

On the other hand, for any $\mathbb{P} \in \mathscr{P}$, we have

$$\tilde{V}_0 + \int_0^T Z_t dX_t \geq Y_0 + \int_0^T Z_t dX_t = Y_T + K_T^{\mathbb{P}} = \xi + K_T^{\mathbb{P}} \geq \xi, \quad \mathbb{P}\text{-a.s.}$$

This implies that $\tilde{V}_0 \geq V_0$. Then $V_0 = \tilde{V}_0$, and Z is a superhedging strategy. ∎

Remark 12.4.3

(i) In Theorem 12.4.2, it is crucial that we have an aggregated version of Z, because the investor needs to determine the superhedging strategy without relying on the probability measure which is unknown due to the model uncertainty. However, we do not need the aggregated version of $K^{\mathbb{P}}$, namely the amount of over hedging $K_T^{\mathbb{P}}$ may depend on the model.

(ii) By Theorem 12.4.3 we see that the superhedging price under uncertain volatility model is equal to the value of the optimization problem with diffusion control. However, we shall emphasize that for the former problem the next focus is the common superhedging strategy Z, while for the latter problem the next focus is the optimal control $\mathbb{P}^*$. ∎

12.5 Bibliographical Notes

This chapter is mainly based on Soner, Touzi, & Zhang [212–215] as well as the survey paper Touzi [228]. The notion of 2BSDE was initiated by Cheridito, Soner, Touzi, & Victoir [32], which uses strong formulation under Wiener measure. The quasi-sure stochastic analysis was introduced by Denis & Martini [63], and was applied to the G-Brownian motion framework in Denis, Hu, & Peng [61]. The estimate Proposition 12.1.15 was obtained independently by Soner, Touzi, &

Zhang [212] and Song [216]. In a very special case, Xu & Zhang [238] obtained the martingale representation for the so-called symmetric G-martingale. Based on Song [217], Peng, Song, & Zhang [185] studied further regularity of the solution to 2BSDEs. We remark that Kharroubi & Pham [124] proposed a different approach in strong formulation to deal with fully nonlinear PDEs. Another highly relevant work is Drapeau, Heyne, & Kupper [67].

There have been many generalizations along the directions of quasi-sure stochastic analysis and 2BSDEs. For example, Cohen [34] extended [213] to general space; Nutz [161] studied 2BSDEs where the underlying probabilities in $\mathscr{P}$ are semimartingales with drift part as well; Possamai & Zhou [193] and Lin [137] studied 2BSDEs with quadratic growth; and Matoussi, Possamai, & Zhou [154] studied 2BSDEs with reflections.

There have also been many applications of the theory, see, e.g., Nutz & Soner [163], Denis & Kervarec [62], Matoussi, Possamai, & Zhou [155], Epstein & Jin [85, 86], Cvitanic, Possamai, & and Touzi [48, 49], and Sung [219]. Moreover, Tan [220] and Possamai & Tan [194] proposed some approximations for 2BSDEs.

Finally, for pathwise stochastic integration we refer to the references mentioned in Section 2.9.

12.6 Exercises

Problem 12.6.1 Let $\{X_i, i \in I\} \subset \mathbb{L}^0(\mathscr{F}_T)$ be a possibly uncountable set of random variables and $\mathbb{P}$ be an arbitrary probability measure on $\mathscr{F}_T$. Assume X_i is uniformly continuous in ω under $\|\cdot\|_T$, uniformly in $i \in I$. Show that $\sup_{i \in I} X_i = \text{ess sup}_{i \in I}^{\mathbb{P}} X_i$, $\mathbb{P}$-a.s.   ∎

Problem 12.6.2 Prove Remark 12.1.6 rigorously.   ∎

Problem 12.6.3 Under the setting of Theorem 10.3.2, show that

$$Y_\tau = \underset{\tilde{\tau} \in \mathscr{T}: \tilde{\tau} \geq \tau}{\text{ess}} \overset{\mathbb{P}}{\sup} \; \mathscr{E}_{\mathbb{P},\tau}^{\mathscr{P}}[X_{\tilde{\tau}}], \quad \text{for all } \mathbb{P} \in \mathscr{P} \text{ and } \tau \in \mathscr{T}.$$

∎

Problem 12.6.4 Let $\mathscr{P} \subset \mathscr{P}_L$ for some $L > 0$ and $\xi \in UC(\Omega)$. Then there exist $\xi_n = g_n(B_{t_1}, \cdots, B_{t_n})$ for some partition $\pi_n : 0 < t_1 < \cdots < t_n = T$ and some bounded smooth function g_n such that $\lim_{n \to \infty} |\pi_n| = 0$ and $\lim_{n \to \infty} \|\xi_n - \xi\|_{\mathbb{L}^2_{\mathscr{P}}} = 0$.   ∎

Problem 12.6.5 Extend Theorem 12.4.2 to the problem (12.4.4).   ∎

References

1. Antonelli, F.: Backward-forward stochastic differential equations. Ann. Appl. Probab. **3**(3), 777–793 (1993)
2. Antonelli, F., Ma, J.: Weak solutions of forward-backward SDE's. Stoch. Anal. Appl. **21**(3), 493–514 (2003)
3. Bally, V., Pages, G., Printems, J.: A quantization tree method for pricing and hedging multidimensional American options. Math. Financ. **15**(1), 119–168 (2005)
4. Barles, G., Jakobsen, E.R.: Error bounds for monotone approximation schemes for parabolic Hamilton-Jacobi-Bellman equations. Math. Comp. **76**(260), 1861–1893 (2007) (electronic)
5. Barles, G., Souganidis, P.E.: Convergence of approximation schemes for fully nonlinear second order equations. Asymptot. Anal. **4**(3), 271–283 (1991)
6. Barlow, M.T.: One-dimensional stochastic differential equations with no strong solution. J. Lond. Math. Soc. (2) **26**(2), 335–347 (1982)
7. Barrieu, P., El Karoui, N.: Monotone stability of quadratic semimartingales with applications to unbounded general quadratic BSDEs. Ann. Probab. **41**(3B), 1831–1863 (2013)
8. Bayraktar, E., Sirbu, M.: Stochastic Perron's method and verification without smoothness using viscosity comparison: the linear case. Proc. Am. Math. Soc. **140**(10), 3645–3654 (2012)
9. Bayraktar, E., Sirbu, M.: Stochastic Perron's method for Hamilton-Jacobi-Bellman equations. SIAM J. Control Optim. **51**(6), 4274–4294 (2013)
10. Bayraktar, E., Yao, S.: On the robust optimal stopping problem. SIAM J. Control Optim. **52**(5), 3135–3175 (2014)
11. Bayraktar, E., Yao, S.: Optimal stopping with random maturity under nonlinear expectations. Stoch. Process. Appl. (accepted). arXiv:1505.07533
12. Bayraktar, E., Yao, S.: On the robust Dynkin game. Ann. Appl. Probab. **27**(3), 1702–1755 (2017)
13. Bender, C., Denk, R.: A forward scheme for backward SDEs. Stoch. Process. Appl. **117**(12), 1793–1812 (2007)
14. Bender, C., Zhang, J.: Time discretization and Markovian iteration for coupled FBSDEs. Ann. Appl. Probab. **18**(1), 143–177 (2008)
15. Billingsley, P.: Convergence of Probability Measures. Wiley Series in Probability and Statistics: Probability and Statistics, 2nd edn., x + 277 pp. A Wiley-Interscience Publication/Wiley Inc., New York (1999)
16. Bismut, J.-M.: Conjugate convex functions in optimal stochastic control. J. Math. Anal. Appl. **44**, 384–404 (1973)

© Springer Science+Business Media LLC 2017

J. Zhang, *Backward Stochastic Differential Equations*, Probability Theory and
Stochastic Modelling 86, DOI 10.1007/978-1-4939-7256-2

17. Bichteler, K.: Stochastic integration and L^p-theory of semimartingales. Ann. Probab. **9**(1), 49–89 (1981)
18. Bouchard, B., Touzi, N.: Discrete-time approximation and Monte-Carlo simulation of backward stochastic differential equations. Stoch. Process. Appl. **111**(2), 175–206 (2004)
19. Bouchard, B., Touzi, N.: Weak dynamic programming principle for viscosity solutions. SIAM J. Control Optim. **49**(3), 948–962 (2011)
20. Bouchard, B., Warin, X.: Monte-Carlo valuation of American options: facts and new algorithms to improve existing methods. In: Numerical Methods in Finance. Springer Proceedings in Mathematics, vol. 12, pp. 215–255. Springer, Heidelberg (2012)
21. Briand, P., Confortola, F.: BSDEs with stochastic Lipschitz condition and quadratic PDEs in Hilbert spaces. Stoch. Process. Appl. **118**(5), 818–838 (2008)
22. Briand, P., Hu, Y.: BSDE with quadratic growth and unbounded terminal value. Probab. Theory Relat. Fields **136**(4), 604–618 (2006)
23. Briand, P., Hu, Y.: Quadratic BSDEs with convex generators and unbounded terminal conditions. Probab. Theory Relat. Fields **141**(3–4), 543–567 (2008)
24. Buckdahn, R., Engelbert, H.-J., Rascanu, A.: On weak solutions of backward stochastic differential equations. Teor. Veroyatn. Primen. **49**(1), 70–108 (2004). Reprinted in Theory Probab. Appl. **49**(1), 16–50 (2005)
25. Buckdahn, R., Li, J.: Stochastic differential games and viscosity solutions of Hamilton-Jacobi-Bellman-Isaacs equations. SIAM J. Control Optim. **47**(1), 444–475 (2008)
26. Buckdahn, R., Ma, J., Zhang, J.: Pathwise Taylor expansions for random fields on multiple dimensional paths. Stoch. Process. Appl. **125**(7), 2820–2855 (2015)
27. Caffarelli, L.A., Cabre, X.: Fully Nonlinear Elliptic Equations. American Mathematical Society Colloquium Publications, vol. 43, vi + 104 pp. American Mathematical Society, Providence, RI (1995)
28. Cardaliaguet, P., Rainer, C.: Stochastic differential games with asymmetric information. Appl. Math. Optim. **59**(1), 1–36 (2009)
29. Carmona, R.: Lectures on BSDEs, Stochastic Control, and Stochastic Differential Games with Financial Applications. SIAM Series on Financial Mathematics. SIAM, Philadelphia (2016)
30. Chen, Z., Epstein, L.: Ambiguity, risk and asset returns in continuous time. Econometrica **70**(4), 1403–1443 (2002)
31. Cheridito, P., Nam, K.: Multidimensional quadratic and subquadratic BSDEs with special structure. Stochastics **87**(5), 871–884 (2015)
32. Cheridito, P., Soner, H.M., Touzi, N., Victoir, N.: Second-order backward stochastic differential equations and fully nonlinear parabolic PDEs. Commun. Pure Appl. Math. **60**(7), 1081–1110 (2007)
33. Cherny, A.S., Engelbert, H.-J.: Singular Stochastic Differential Equations. Lecture Notes in Mathematics, vol. 1858, viii + 128 pp. Springer, Berlin (2005)
34. Cohen, S.N.: Quasi-sure analysis, aggregation and dual representations of sublinear expectations in general spaces. Electron. J. Probab. **17**(62), 15 pp. (2012)
35. Cont, R.: Functional Itô calculus and functional Kolmogorov equations. In: Stochastic Integration by Parts and Functional Itô Calculus. Advanced Courses in Mathematics—CRM Barcelona, pp. 115–207. Birkhäuser/Springer, Cham (2016)
36. Cont, R., Fournie, D.-A.: Change of variable formulas for non-anticipative functionals on path space. J. Funct. Anal. **259**(4), 1043–1072 (2010)
37. Cont, R., Fournie, D.-A.: Functional Itô calculus and stochastic integral representation of martingales. Ann. Probab. **41**(1), 109–133 (2013)
38. Coquet, F., Hu, Y., Memin, J., Peng, S.: Filtration-consistent nonlinear expectations and related g-expectations. Probab. Theory Relat. Fields **123**(1), 1–27 (2002)
39. Cosso, A., Federico, S., Gozzi, F., Rosestolato, M., Touzi, N.: Path-dependent equations and viscosity solutions in infinite dimension. Ann. Probab. (accepted). arXiv:1502.05648
40. Cosso, A., Russo, F.: Strong-viscosity solutions: semilinear parabolic PDEs and path-dependent PDEs. Preprint arXiv:1505.02927

41. Costantini, C., Kurtz, T.G.: Viscosity methods giving uniqueness for martingale problems. Electron. J. Probab. **20**(67), 27 pp. (2015)
42. Crandall, M.G., Ishii, H., Lions, P.-L.: User's guide to viscosity solutions of second order partial differential equations. Bull. Am. Math. Soc. (N.S.) **27**(1), 1–67 (1992)
43. Crepey, S.: Financial modeling. A backward stochastic differential equations perspective. In: Springer Finance Textbooks. Springer Finance, xx + 459 pp. Springer, Heidelberg (2013)
44. Crisan, D., Manolarakis, K.: Solving backward stochastic differential equations using the cubature method. Application to nonlinear pricing. In: Progress in Analysis and Its Applications, pp. 389–397. World Science Publisher, Hackensack, NJ (2010)
45. Cvitanic, J., Karatzas, I.: Backward stochastic differential equations with reflection and Dynkin games. Ann. Probab. **24**(4), 2024–2056 (1996)
46. Cvitanic, J., Karatzas, I., Soner, H.M.: Backward stochastic differential equations with constraints on the gains-process. Ann. Probab. **26**(4), 1522–1551 (1998)
47. Cvitanic, J., Ma, J.: Hedging options for a large investor and forward-backward SDE's. Ann. Appl. Probab. **6**(2), 370–398 (1996)
48. Cvitanic, J., Possamai, D., Touzi, N.: Moral hazard in dynamic risk management. Manage. Sci. (accepted). arXiv:1406.5852
49. Cvitanic, J., Possamai, D., Touzi, N.: Dynamic programming approach to principal-agent problems. Finance Stochastics (accepted). arXiv:1510.07111
50. Cvitanic, J., Wan, X., Zhang, J.: Optimal compensation with hidden action and lump-sum payment in a continuous-time model. Appl. Math. Optim. **59**(1), 99–146 (2009)
51. Cvitanic, J., Zhang, J.: The steepest descent method for forward-backward SDEs. Electron. J. Probab. **10**, 1468–1495 (2005) (electronic)
52. Cvitanic, J., Zhang, J.: Contract Theory in Continuous-Time Models. Springer Finance, xii + 255 pp. Springer, Heidelberg (2013)
53. Delbaen, F., Peng, S., Rosazza Gianin, E.: Representation of the penalty term of dynamic concave utilities. Finance Stochast. **14**(3), 449–472 (2010)
54. Darling, R.W.R., Pardoux, E.: Backwards SDE with random terminal time and applications to semilinear elliptic PDE. Ann. Probab. **25**(3), 1135–1159 (1997)
55. Delarue, F.: On the existence and uniqueness of solutions to FBSDEs in a non-degenerate case. Stoch. Process. Appl. **99**(2), 209–286 (2002)
56. Delarue, F., Guatteri, G.: Weak existence and uniqueness for forward-backward SDEs. Stoch. Process. Appl. **116**(12), 1712–1742 (2006)
57. Delarue, F., Menozzi, S.: A forward-backward stochastic algorithm for quasi-linear PDEs. Ann. Appl. Probab. **16**(1), 140–184 (2006)
58. Delbaen, F., Hu, Y., Richou, A.: On the uniqueness of solutions to quadratic BSDEs with convex generators and unbounded terminal conditions. Ann. Inst. Henri Poincare Probab. Stat. **47**(2), 559–574 (2011)
59. Delbaen, F., Hu, Y., Richou, A.: On the uniqueness of solutions to quadratic BSDEs with convex generators and unbounded terminal conditions: the critical case. Discrete Contin. Dyn. Syst. **35**(11), 5273–5283 (2015)
60. Delong, L.: Backward Stochastic Differential Equations with Jumps and Their Actuarial and Financial Applications. BSDEs with Jumps. European Actuarial Academy (EAA) Series, x + 288 pp. Springer, London (2013)
61. Denis, L., Hu, M., Peng, S.: Function spaces and capacity related to a sublinear expectation: application to G-Brownian motion paths. Potential Anal. **34**(2), 139–161 (2011)
62. Denis, L., Kervarec, M.: Optimal investment under model uncertainty in nondominated models. SIAM J. Control Optim. **51**(3), 1803–1822 (2013)
63. Denis, L., Martini, C.: A theoretical framework for the pricing of contingent claims in the presence of model uncertainty. Ann. Appl. Probab. **16**(2), 827–852 (2006)
64. Dokuchaev, N.: Backward parabolic Ito equations and the second fundamental inequality. Random Oper. Stoch. Equ. **20**(1), 69–102 (2012)
65. Dolinsky, Y.: Numerical schemes for G-expectations. Electron. J. Probab. **17**(98), 15 pp. (2012)

66. Douglas Jr., J., Ma, J., Protter, P.: Numerical methods for forward-backward stochastic differential equations. Ann. Appl. Probab. **6**(3), 940–968 (1996)
67. Drapeau, S., Heyne, G., Kupper, M.: Minimal supersolutions of BSDEs under volatility uncertainty. Stochastic Process. Appl. **125**(8), 2895–2909 (2015)
68. Du, K., Tang, S., Zhang, Q.: $W^{m,p}$-solution ($p \geq 2$) of linear degenerate backward stochastic partial differential equations in the whole space. J. Differ. Equ. **254**(7), 2877–2904 (2013)
69. Duffie, D., Epstein, L.: Stochastic differential utility. Econometrica **60**(2), 353–394 (1992)
70. Duffie, D., Epstein, L.: Asset pricing with stochastic differential utilities. Rev. Financ. Stud. **5**(3), 411–436 (1992)
71. Dupire, B.: Functional Itô calculus. Preprint papers.ssrn.com (2009)
72. Ekren, I.: Viscosity solutions of obstacle problems for fully nonlinear path-dependent PDEs. Stoch. Process. Appl. (accepted). arXiv:1306.3631
73. Ekren, I., Keller, C., Touzi, N., Zhang, J.: On viscosity solutions of path dependent PDEs. Ann. Probab. **42**(1), 204–236 (2014)
74. Ekren, I., Touzi, N., Zhang, J.: Optimal stopping under nonlinear expectation. Stoch. Process. Appl. **124**(10), 3277–3311 (2014)
75. Ekren, I., Touzi, N., Zhang, J.: Viscosity solutions of fully nonlinear parabolic path dependent PDEs: part I. Ann. Probab. **44**(2), 1212–1253 (2016)
76. Ekren, I., Touzi, N., Zhang, J.: Viscosity solutions of fully nonlinear parabolic path dependent PDEs: part II. Ann. Probab. **44**(4), 2507–2553 (2016)
77. Ekren, I., Zhang, J.: Pseudo Markovian viscosity solutions of fully nonlinear degenerate PPDEs. Probab. Uncertainty Quant. Risk **1**, 6 (2016). doi:10.1186/s41546-016-0010-3
78. El Karoui, N., Kapoudjian, C., Pardoux, E., Peng, S., Quenez, M.C.: Reflected solutions of backward SDE's, and related obstacle problems for PDE's. Ann. Probab. **25**(2), 702–737 (1997)
79. El Karoui, N., Huang, S.-J.: A general result of existence and uniqueness of backward stochastic differential equations. In: Backward Stochastic Differential Equations (Paris, 1995–1996). Pitman Research Notes in Mathematics Series, vol. 364, pp. 27–36. Longman, Harlow (1997)
80. El Karoui, N., Mazliak, L.: Backward Stochastic Differential Equations. Pitman Research Notes in Mathematics Series, vol. 364. Longman, Harlow (1997)
81. El Karoui, N., Peng, S., Quenez, M.C.: Backward stochastic differential equations in finance. Math. Financ. **7**(1), 1–71 (1997)
82. El Karoui, N., Tan, X.: Capacities, measurable selection and dynamic programming. Part I: abstract framework. Preprint arXiv:1310.3363
83. El Karoui, N., Tan, X.: Capacities, measurable selection and dynamic programming. Part II: application in stochastic control problems. Preprint arXiv:1310.3364
84. Elliott, R.J., Kalton, N.J.: The Existence of Value in Differential Games. Memoirs of the American Mathematical Society, vol. 126, iv + 67 pp. American Mathematical Society, Providence, RI (1972)
85. Epstein, L.G., Ji, S.: Ambiguous volatility and asset pricing in continuous time. Rev. Financ. Stud. **26**, 1740–1786 (2013)
86. Epstein, L.G., Ji, S.: Ambiguous volatility, possibility and utility in continuous time. J. Math. Econ. **50**, 269–282 (2014)
87. Fabes, E.B., Kenig, C.E.: Examples of singular parabolic measures and singular transition probability densities. Duke Math. J. **48**(4), 845–856 (1981)
88. Fahim, A., Touzi, N., Warin, X.: A probabilistic numerical method for fully nonlinear parabolic PDEs. Ann. Appl. Probab. **21**(4), 1322–1364 (2011)
89. Fleming, W.H., Soner, H.M.: Controlled Markov Processes and Viscosity Solutions. Stochastic Modelling and Applied Probability, vol. 25, 2nd edn., xviii + 429 pp. Springer, New York (2006)
90. Fleming, W.H., Souganidis, P.E.: On the existence of value functions of two-player, zero-sum stochastic differential games. Indiana Univ. Math. J. **38**(2), 293–314 (1989)

91. Follmer, H.: Calcul d'Itô sans probabilites (French). In: Seminar on Probability, XV (Univ. Strasbourg, Strasbourg, 1979/1980) (French). Lecture Notes in Mathematics, vol. 850, pp. 143–150. Springer, Berlin (1981)

92. Frei, C.: Splitting multidimensional BSDEs and finding local equilibria. Stoch. Process. Appl. **124**(8), 2654–2671 (2014)

93. Frei, C., dos Reis, G.: A financial market with interacting investors: does an equilibrium exist? Math. Financ. Econ. **4**(3), 161–182 (2011)

94. Friz, P.K., Hairer, M.: A Course on Rough Paths. With an Introduction to Regularity Structures. Universitext, xiv + 251 pp. Springer, Cham (2014)

95. Fuhrman, M., Tessitore, G.: Nonlinear Kolmogorov equations in infinite dimensional spaces: the backward stochastic differential equations approach and applications to optimal control. Ann. Probab. **30**(3), 1397–1465 (2002)

96. Glasserman, P., Yu, B.: Number of paths versus number of basis functions in American option pricing. Ann. Appl. Probab. **14**(4), 2090–2119 (2004)

97. Gobet, E., Lemor, J.-P., Warin, X.: A regression-based Monte Carlo method to solve backward stochastic differential equations. Ann. Appl. Probab. **15**(3), 2172–2202 (2005)

98. Gubinelli, M.: Controlling rough paths. J. Funct. Anal. **216**(1), 86–140 (2004)

99. Guo, W., Zhang, J., Zhuo, J.: A monotone scheme for high-dimensional fully nonlinear PDEs. Ann. Appl. Probab. **25**(3), 1540–1580 (2015)

100. Hamadene, S.: Reflected BSDE's with discontinuous barrier and application. Stoch. Stoch. Rep. **74**(3–4), 571–596 (2002)

101. Hamadene, S., Hassani, M.: BSDEs with two reflecting barriers: the general result. Probab. Theory Relat. Fields **132**(2), 237–264 (2005)

102. Hamadene, S., Jeanblanc, M.: On the starting and stopping problem: application in reversible investments. Math. Oper. Res. **32**(1), 182–192 (2007)

103. Hamadene, S., Lepeltier, J.-P.: Zero-sum stochastic differential games and backward equations. Syst. Control Lett. **24**(4), 259–263 (1995)

104. Hamadene, S., Zhang, J.: Switching problem and related system of reflected backward SDEs. Stoch. Process. Appl. **120**(4), 403–426 (2010)

105. Henry-Labordere, P., Tan, X., Touzi, N.: A numerical algorithm for a class of BSDEs via the branching process. Stoch. Process. Appl. **124**(2), 1112–1140 (2014)

106. Hu, M., Ji, S.: Dynamic programming principle for stochastic recursive optimal control problem driven by a G-Brownian motion. Stoch. Process. Appl. **127**(1), 107–134 (2017)

107. Hu, M., Ji, S., Peng, S., Song, Y.: Backward stochastic differential equations driven by G-Brownian motion. Stoch. Process. Appl. **124**(1), 759–784 (2014)

108. Hu, Y., Ma, J., Yong, J.: On semi-linear degenerate backward stochastic partial differential equations. Probab. Theory Relat. Fields **123**(3), 381–411 (2002)

109. Hu, Y., Peng, S.: Solution of forward-backward stochastic differential equations. Probab. Theory Relat. Fields **103**(2), 273–283 (1995)

110. Hu, Y., Peng, S.: On the comparison theorem for multidimensional BSDEs. C. R. Math. Acad. Sci. Paris **343**(2), 135–140 (2006)

111. Hu, Y., Tang, S.: Multi-dimensional BSDE with oblique reflection and optimal switching. Probab. Theory Relat. Fields **147**(1–2), 89–121 (2010)

112. Hu, Y., Tang, S.: Multi-dimensional backward stochastic differential equations of diagonally quadratic generators. Stoch. Process. Appl. **126**(4), 1066–1086 (2016)

113. Jacod, J., Shiryaev, A.N.: Limit theorems for stochastic processes. In: Grundlehren der Mathematischen Wissenschaften (Fundamental Principles of Mathematical Sciences), vol. 288, 2nd edn., xx + 661 pp. Springer, Berlin (2003)

114. Jamneshan, A., Kupper, M., Luo, P.: Multidimensional quadratic BSDEs with separated generators. Preprint arXiv:1501.00461

115. Jazaerli, S., Saporito, Y.F.: Functional Itô calculus, path-dependence and the computation of Greeks. Stoch. Process. Appl. (2017, preprint). arXiv:1311.3881

116. Jeanblanc, M., Yu, Z.: Optimal investment problems with uncertain time horizon (2010)

117. Karatzas, I., Shreve, S.E.: Brownian Motion and Stochastic Calculus. Graduate Texts in Mathematics, vol. 113, 2nd edn., xxiv + 470 pp. Springer, New York (1991)

118. Karatzas, I., Shreve, S.E.: Methods of Mathematical Finance. Applications of Mathematics (New York), vol. 39, xvi + 407 pp. Springer, New York (1998)

119. Karandikar, R.L.: On pathwise stochastic integration. Stoch. Process. Appl. **57**(1), 11–18 (1995)

120. C. Kardaras, H. Xing, G. Zitkovic, Incomplete stochastic equilibria with exponential utilities close to Pareto optimality. Preprint arXiv:1505.07224

121. Kazamaki, N.: Continuous Exponential Martingales and BMO. Lecture Notes in Mathematics, vol. 1579, viii + 91 pp. Springer, Berlin (1994)

122. Keller, C.: Viscosity solutions of path-dependent integro-differential equations. Stoch. Process. Appl. **126**(9), 2665–2718 (2016)

123. Keller, C., Zhang, J.: Pathwise Itô calculus for rough paths and rough PDEs with path dependent coefficients. Stoch. Process. Appl. **126**(3), 735–766 (2016)

124. Kharroubi, I., Pham, H.: Feynman-Kac representation for Hamilton-Jacobi-Bellman IPDE. Ann. Probab. **43**(4), 1823–1865 (2015)

125. Kloeden, P.E., Platen, E.: Numerical Solution of Stochastic Differential Equations. Applications of Mathematics (New York), vol. 23, xxxvi + 632 pp. Springer, Berlin (1992)

126. Kobylanski, M.: Backward stochastic differential equations and partial differential equations with quadratic growth. Ann. Probab. **28**(2), 558–602 (2000)

127. Kramkov, D., Pulido, S.: A system of quadratic BSDEs arising in a price impact model. Ann. Appl. Probab. **26**(2), 794–817 (2016)

128. Krylov, N.V.: Itô's stochastic integral equations. Teor. Verojatnost. i Primenen **14**, 340–348 (1969) (Russian)

129. Krylov, N.V.: Controlled Diffusion Processes. Stochastic Modelling and Applied Probability, vol. 14, xii + 308 pp. Springer, Berlin (2009). Translated from the 1977 Russian original by A.B. Aries. Reprint of the 1980 edition

130. Krylov, N.V.: Introduction to the theory of diffusion processes. Translations of Mathematical Monographs, vol. 142, xii + 271 pp. American Mathematical Society, Providence, RI (1995). Translated from the Russian manuscript by Valim Khidekel and Gennady Pasechnik

131. Krylov, N.V.: Nonlinear elliptic and parabolic equations of the second order. Mathematics and its Applications (Soviet Series), vol. 7, xiv + 462 pp. D. Reidel Publishing Co., Dordrecht (1987). Translated from the Russian by P.L. Buzytsky [P.L. Buzytskii]

132. Krylov, N.V.: On the rate of convergence of finite-difference approximations for Bellman's equations. Algebra i Analiz **9**(3), 245–256 (1997); reprinted in St. Petersburg Math. J. **9**(3), 639–650 (1998)

133. Ladyzenskaja, O.A., Solonnikov, V.A., Uralceva, N.N.: Linear and quasilinear equations of parabolic type. Translations of Mathematical Monographs, vol. 23, xi + 648 pp. American Mathematical Society, Providence, RI (1968) (Russian). Translated from the Russian by S. Smith

134. Leao, D., Ohashi, A., Simas, A.B.: Weak Functional Itô calculus and applications. Preprint arXiv:1408.1423

135. Lepeltier, J.P., San Martin, J.: Backward stochastic differential equations with continuous coefficient. Stat. Probab. Lett. **32**(4), 425–430 (1997)

136. Lieberman, G.M.: Second Order Parabolic Differential Equations, xii + 439 pp. World Scientific Publishing Co., Inc., River Edge, NJ (1996)

137. Lin, Y.: A new existence result for second-order BSDEs with quadratic growth and their applications. Stochastics **88**(1), 128–146 (2016)

138. Longstaff, F.A., Schwartz, E.S.: Valuing American options by simulation: a simple least-squares approach. Rev. Financ. Stud. **14**, 113–147 (2001)

139. Lukoyanov, N.Y.: On viscosity solution of functional Hamilton-Jacobi type equations for hereditary systems. Proc. Steklov Inst. Math. **259**(Suppl. 2), 190–200 (2007)

140. Lyons, T.J.: Differential equations driven by rough signals. Rev. Mat. Iberoamericana **14**(2), 215–310 (1998)

141. Ma, J., Protter, P., Yong, J.M.: Solving forward-backward stochastic differential equations explicitly – a four step scheme. Probab. Theory Relat. Fields **98**(3), 339–359 (1994)

142. Ma, J., Ren, Z., Touzi, N., Zhang, J.: Large deviations for non-Markovian diffusions and a path-dependent Eikonal equation. Ann. Inst. Henri Poincare Probab. Stat. **52**(3), 1196–1216 (2016)

143. Ma, J., Shen, J., Zhao, Y.: On numerical approximations of forward-backward stochastic differential equations. SIAM J. Numer. Anal. **46**(5), 2636–2661 (2008)

144. Ma, J., Wu, Z., Zhang, D., Zhang, J.: On well-posedness of forward-backward SDEs – a unified approach. Ann. Appl. Probab. **25**(4), 2168–2214 (2015)

145. Ma, J., Yin, H., Zhang, J.: On non-Markovian forward-backward SDEs and backward stochastic PDEs. Stoch. Process. Appl. **122**(12), 3980–4004 (2012)

146. Ma, J., Yong, J.: Solvability of forward-backward SDEs and the nodal set of Hamilton-Jacobi-Bellman equations. Chin. Ann. Math. Ser. B **16**(3), 279–298 (1995). A Chinese summary appears in Chin. Ann. Math. Ser. A **16**(4), 532 (1995)

147. Ma, J., Yong, J.: Adapted solution of a degenerate backward SPDE, with applications. Stoch. Process. Appl. **70**(1), 59–84 (1997)

148. Ma, J., Yong, J.: Forward-Backward Stochastic Differential Equations and Their Applications. Lecture Notes in Mathematics, vol. 1702, xiv + 270 pp. Springer, Berlin (1999)

149. Ma, J., Zhang, J.: Representation theorems for backward stochastic differential equations. Ann. Appl. Probab. **12**(4), 1390–1418 (2002)

150. Ma, J., Zhang, J.: Path regularity for solutions of backward SDEs. Probab. Theory Relat. Fields **122**(2), 163–190 (2002)

151. Ma, J., Zhang, J.: Representations and regularities for solutions to BSDEs with reflections. Stoch. Process. Appl. **115**(4), 539–569 (2005)

152. Ma, J., Zhang, J.: On weak solutions of forward-backward SDEs. Probab. Theory Relat. Fields **151**(3–4), 475–507 (2011)

153. Ma, J., Zhang, J., Zheng, Z.: Weak solutions for forward-backward SDEs – a martingale problem approach. Ann. Probab. **36**(6), 2092–2125 (2008)

154. Matoussi, A., Possamai, D., Zhou, C.: Second order reflected backward stochastic differential equations. Ann. Appl. Probab. **23**(6), 2420–2457 (2013)

155. Matoussi, A., Possamai, D., Zhou, C.: Robust utility maximization in nondominated models with 2BSDE: the uncertain volatility model. Math. Financ. **25**(2), 258–287 (2015)

156. Meyer, P.-A., Zheng, W.A.: Tightness criteria for laws of semimartingales. Ann. Inst. H. Poincare Probab. Stat. **20**(4), 353–372 (1984)

157. Milstein, G.N., Tretyakov, M.V.: Discretization of forward-backward stochastic differential equations and related quasi-linear parabolic equations. IMA J. Numer. Anal. **27**(1), 24–44 (2007)

158. Nadirashvili, N., Vladut, S.: Nonclassical solutions of fully nonlinear elliptic equations. Geom. Funct. Anal. **17**(4), 1283–1296 (2007)

159. Nualart, D.: The Malliavin Calculus and Related Topics. Probability and its Applications (New York), 2nd edn., xiv + 382 pp. Springer, Berlin (2006)

160. Nutz, M.: Pathwise construction of stochastic integrals. Electron. Commun. Probab. **17**(24), 7 pp. (2012)

161. Nutz, M.: A quasi-sure approach to the control of non-Markovian stochastic differential equations. Electron. J. Probab. **17**(23), 23 pp. (2012)

162. Nutz, M.: Random G-expectations. Ann. Appl. Probab. **23**(5), 1755–1777 (2013)

163. Nutz, M., Soner, H.M.: Superhedging and dynamic risk measures under volatility uncertainty. SIAM J. Control Optim. **50**(4), 2065–2089 (2012)

164. Nutz, M., van Handel, R.: Constructing sublinear expectations on path space. Stoch. Process. Appl. **123**(8), 3100–3121 (2013)

165. Nutz, M., Zhang, J.: Optimal stopping under adverse nonlinear expectation and related games. Ann. Appl. Probab. **25**(5), 2503–2534 (2015)

166. Oberhauser, H.: The functional Itô formula under the family of continuous semimartingale measures. Stoch. Dyn. **16**(4), 1650010, 26 pp. (2016)

167. Pardoux, E., Peng, S.G.: Adapted solution of a backward stochastic differential equation. Syst. Control Lett. **14**(1), 55–61 (1990)
168. Pardoux, E., Peng, S.: Backward stochastic differential equations and quasilinear parabolic partial differential equations. In: Stochastic Partial Differential Equations and Their Applications (Charlotte, NC, 1991). Lecture Notes in Control and Information Sciences, vol. 176, pp. 200–217. Springer, Berlin (1992)
169. Pardoux, E., Peng, S.G.: Backward doubly stochastic differential equations and systems of quasilinear SPDEs. Probab. Theory Relat. Fields **98**(2), 209–227 (1994)
170. Pardoux, E., Rascanu, A.: Stochastic Differential Equations, Backward SDEs, Partial Differential Equations. Stochastic Modelling and Applied Probability, vol. 69, xviii + 667 pp. Springer, Cham (2014)
171. Pardoux, E., Tang, S.: Forward-backward stochastic differential equations and quasilinear parabolic PDEs. Probab. Theory Relat. Fields **114**(2), 123–150 (1999)
172. Peng, S.G.: Probabilistic interpretation for systems of quasilinear parabolic partial differential equations. Stoch. Stoch. Rep. **37**(1–2), 61–74 (1991)
173. Peng, S.G.: A generalized Hamilton-Jacobi-Bellman equation. In: Control Theory of Distributed Parameter Systems and Applications (Shanghai, 1990). Lecture Notes in Control and Information Sciences, vol. 159, pp. 126–134. Springer, Berlin (1991)
174. Peng, S.G.: Stochastic Hamilton-Jacobi-Bellman equations. SIAM J. Control Optim. **30**(2), 284–304 (1992)
175. Peng, S.: BSDE and stochastic optimizations. In: Topics in Stochastic Analysis. Lecture Notes of Xiangfan Summer School (1995). J. Yan, S. Peng, S. Fang and L.M. Wu, Ch. 2 (Chinese vers.). Science Publication, Beijing (1997)
176. Peng, S.: Backward SDE and related g-expectation. In: Backward Stochastic Differential Equations (Paris, 1995–1996). Pitman Research Notes in Mathematics Series, vol. 364, pp. 141–159. Longman, Harlow (1997)
177. Peng, S.: Monotonic limit theorem of BSDE and nonlinear decomposition theorem of Doob-Meyer's type. Probab. Theory Relat. Fields **113**(4), 473–499 (1999)
178. Peng, S.: Open problems on backward stochastic differential equations. In: Control of Distributed Parameter and Stochastic Systems (Hangzhou, 1998), pp. 265–273. Kluwer Academic Publishers, Boston, MA (1999)
179. Peng, S.: Nonlinear expectations, nonlinear evaluations and risk measures. In: Stochastic Methods in Finance. Lecture Notes in Mathematics, vol. 1856, pp. 165–253. Springer, Berlin (2004)
180. Peng, S.: G-expectation, G-Brownian motion and related stochastic calculus of Itô type. In: Stochastic Analysis and Applications. Abel Symposia, vol. 2, pp. 541–567. Springer, Berlin (2007)
181. Peng, S.: G-Brownian motion and dynamic risk measure under volatility uncertainty. Preprint arXiv:0711.2834
182. Peng, S.: Backward stochastic differential equation, nonlinear expectation and their applications. In: Proceedings of the International Congress of Mathematicians, vol. I, pp. 393–432. Hindustan Book Agency, New Delhi (2010)
183. Peng, S.: Note on viscosity solution of path-dependent PDE and G-martingales. Preprint arXiv:1106.1144
184. Peng, S., Song, Y.: G-expectation weighted Sobolev spaces, backward SDE and path dependent PDE. J. Math. Soc. Jpn. **67**(4), 1725–1757 (2015)
185. Peng, S., Song, Y., Zhang, J.: A complete representation theorem for G-martingales. Stochastics **86**(4), 609–631 (2014)
186. Peng, S., Wang, F.: BSDE, path-dependent PDE and nonlinear Feynman-Kac formula. Sci. China Math. **59**(1), 19–36 (2016)
187. Peng, S., Wu, Z.: Fully coupled forward-backward stochastic differential equations and applications to optimal control. SIAM J. Control Optim. **37**(3), 825–843 (1999)
188. Peng, S., Xu, M.: The smallest g-supermartingale and reflected BSDE with single and double L^2 obstacles. Ann. Inst. H. Poincare Probab. Stat. **41**(3), 605–630 (2005)

189. Peng, S., Xu, M.: Reflected BSDE with a constraint and its applications in an incomplete market. Bernoulli **16**(3), 614–640 (2010)
190. Pham, H.: Continuous-Time Stochastic Control and Optimization with Financial Applications. Stochastic Modelling and Applied Probability, vol. 61, xviii + 232 pp. Springer, Berlin (2009)
191. Pham, T., Zhang, J.: Some norm estimates for semimartingales. Electron. J. Probab. **18**(109), 25 pp. (2013)
192. Pham, T., Zhang, J.: Two person zero-sum game in weak formulation and path dependent Bellman-Isaacs equation. SIAM J. Control Optim. **52**(4), 2090–2121 (2014)
193. Possamai, D., Zhou, C.: Second order backward stochastic differential equations with quadratic growth. Stoch. Process. Appl. **123**(10), 3770–3799 (2013)
194. Possamai, D., Tan, X.: Weak approximation of second-order BSDEs. Ann. Appl. Probab. **25**(5), 2535–2562 (2015)
195. Possamai, D., Tan, X., Zhou, C.: Stochastic control for a class of nonlinear kernels and applications. Ann. Probab. (accepted). arXiv:1510.08439
196. Protter, P.E.: Stochastic Integration and Differential Equations. Version 2.1. Stochastic Modelling and Applied Probability, vol. 21, 2nd edn. (corrected third printing), xiv + 419 pp. Springer, Berlin (2005)
197. Qiu, J.: Weak solution for a class of fully nonlinear stochastic Hamilton-Jacobi-Bellman equations. Stoch. Process. Appl. **127**(6), 1926–1959 (2017)
198. Qiu, J., Tang, S.: Maximum principle for quasi-linear backward stochastic partial differential equations. J. Funct. Anal. **262**(5), 2436–2480 (2012)
199. dos Reis, G.: Some Advances on Quadratic BSDE: Theory - Numerics - Applications. LAP Lambert Academic Publishing, Saarbrücken (2011)
200. Ren, Z.: Viscosity solutions of fully nonlinear elliptic path dependent PDEs. Ann. Appl. Probab. **26**(6), 3381–3414 (2016)
201. Ren, Z.: Perron's method for viscosity solutions of semilinear path dependent PDEs. Stochastics (accepted). arXiv:1503.02169
202. Ren, Z., Tan, X.: On the convergence of monotone schemes for path-dependent PDE. Stoch. Process. Appl. **127**(6), 1738–1762 (2017)
203. Ren, Z., Touzi, N., Zhang, J.: An overview of viscosity solutions of path-dependent PDEs. In: Stochastic Analysis and Applications 2014. Springer Proceedings in Mathematics & Statistics, vol. 100, pp. 397–453. Springer, Cham (2014)
204. Ren, Z., Touzi, N., Zhang, J.: Comparison of viscosity solutions of semilinear path-dependent partial differential equations. Preprint arXiv:1410.7281
205. Ren, Z., Touzi, N., Zhang, J.: Comparison of viscosity solutions of fully nonlinear degenerate parabolic path-dependent PDEs. SIAM J. Math. Anal. (accepted). arXiv:1511.05910
206. Revuz, D., Yor, M.: Continuous Martingales and Brownian Motion. Grundlehren der Mathematischen Wissenschaften (Fundamental Principles of Mathematical Sciences), vol. 293, 3rd edn., xiv + 602 pp. Springer, Berlin (1999)
207. Sakhanenko, A.I.: A new way to obtain estimates in the invariance principle. In: High Dimensional Probability, II (Seattle, WA, 1999). Progress in Probability, vol. 47, pp. 223–245. Birkhäuser, Boston (2000)
208. Saporito, Y.F.: Functional Meyer-Tanaka formula. Stoch. Dyn. (accepted). arXiv:1408.4193
209. Shreve, S.E.: Stochastic Calculus for Finance. I. The Binomial Asset Pricing Model. Springer Finance, xvi + 187 pp. Springer, New York (2004)
210. Shreve, S.E.: Stochastic Calculus for Finance. II. Continuous-Time Models. Springer Finance, xx + 550 pp. Springer, New York (2004)
211. Sirbu, M.: Stochastic Perron's method and elementary strategies for zero-sum differential games. SIAM J. Control Optim. **52**(3), 1693–1711 (2014)
212. Soner, H.M., Touzi, N., Zhang, J.: Martingale representation theorem for the G-expectation. Stoch. Process. Appl. **121**(2), 265–287 (2011)
213. Soner, H.M., Touzi, N., Zhang, J.: Quasi-sure stochastic analysis through aggregation. Electron. J. Probab. **16**(67), 1844–1879 (2011)

214. Soner, H.M., Touzi, N., Zhang, J.: Well-posedness of second order backward SDEs. Probab. Theory Relat. Fields **153**(1–2), 149–190 (2012)
215. Soner, H.M., Touzi, N., Zhang, J.: Dual formulation of second order target problems. Ann. Appl. Probab. **23**(1), 308–347 (2013)
216. Song, Y.: Some properties on G-evaluation and its applications to G-martingale decomposition. Sci. China Math. **54**(2), 287–300 (2011)
217. Song, Y.: Uniqueness of the representation for G-martingales with finite variation. Electron. J. Probab. **17**(24), 15 pp. (2012)
218. Stroock, D.W., Varadhan, S.R.S.: Multidimensional Diffusion Processes. Reprint of the 1997 edition. Classics in Mathematics, xii + 338 pp. Springer, Berlin (2006)
219. Sung, J.: Optimal contracting under mean-volatility ambiguity uncertainties: an alternative perspective on managerial compensation (2015). Preprint papers.ssrn.com
220. Tan, X.: Discrete-time probabilistic approximation of path-dependent stochastic control problems. Ann. Appl. Probab. **24**(5), 1803–1834 (2014)
221. Tan, X.: A splitting method for fully nonlinear degenerate parabolic PDEs. Electron. J. Probab. **18**(15), 24 pp. (2013)
222. Tang, S.: General linear quadratic optimal stochastic control problems with random coefficients: linear stochastic Hamilton systems and backward stochastic Riccati equations. SIAM J. Control Optim. **42**(1), 53–75 (2003) (electronic)
223. Tang, S.J., Li, X.J.: Necessary conditions for optimal control of stochastic systems with random jumps. SIAM J. Control Optim. **32**(5), 1447–1475 (1994)
224. Tang, S., Wei, W.: On the Cauchy problem for backward stochastic partial differential equations in Hölder spaces. Ann. Probab. **44**(1), 360–398 (2016)
225. Tang, S., Zhang, F.: Path-dependent optimal stochastic control and viscosity solution of associated Bellman equations. Discrete Contin. Dyn. Syst. **35**(11), 5521–5553 (2015)
226. Tevzadze, R.: Solvability of backward stochastic differential equations with quadratic growth. Stoch. Process. Appl. **118**(3), 503–515 (2008)
227. Touzi, N.: Optimal Stochastic Control, Stochastic Target Problems, and Backward SDE. With Chapter 13 by Anges Tourin. Fields Institute Monographs, vol. 29, x + 214 pp. Springer, New York; Fields Institute for Research in Mathematical Sciences, Toronto, ON (2013)
228. Touzi, N.: Second order backward SDEs, fully nonlinear PDEs, and applications in finance. Proceedings of the International Congress of Mathematicians, vol. IV, pp. 3132–3150. Hindustan Book Agency, New Delhi (2010)
229. Tsirelson, B.: An example of a stochastic differential equation having no strong solution. Theory Probab. Appl. **20**(2), 416–418 (1975)
230. Wagner, D.H.: Survey of measurable selection theorems. SIAM J. Control Optim. **15**(5), 859–903 (1977)
231. Wang, H., Zhang, J.: Forward backward SDEs in weak formulation, working paper (2017)
232. Wang, L.: On the regularity theory of fully nonlinear parabolic equations. I. Commun. Pure Appl. Math. **45**(1), 27–76 (1992)
233. Wang, L.: On the regularity theory of fully nonlinear parabolic equations. II. Commun. Pure Appl. Math. **45**(2), 141–178 (1992)
234. Wang, L.: On the regularity theory of fully nonlinear parabolic equations. III. Commun. Pure Appl. Math. **45**(3), 255–262 (1992)
235. Willinger, W., Taqqu, M.S.: Pathwise stochastic integration and applications to the theory of continuous trading. Stoch. Process. Appl. **32**(2), 253–280 (1989)
236. Wong, E., Zakai, M.: On the convergence of ordinary integrals to stochastic integrals. Ann. Math. Stat. **36**, 1560–1564 (1965)
237. Wong, E., Zakai, M.: Riemann-Stieltjes approximations of stochastic integrals. Z. Wahrscheinlichkeitstheorie Verw. Gebiete **12**, 87–97 (1969)
238. Xu, J., Zhang, B.: Martingale characterization of G-Brownian motion. Stoch. Process. Appl. **119**(1), 232–248 (2009)
239. Yong, J.: Finding adapted solutions of forward-backward stochastic differential equations: method of continuation. Probab. Theory Relat. Fields **107**(4), 537–572 (1997)

240. Yong, J.: Linear forward-backward stochastic differential equations. Appl. Math. Optim. **39**(1), 93–119 (1999)
241. Yong, J.: Linear forward-backward stochastic differential equations with random coefficients. Probab. Theory Relat. Fields **135**(1), 53–83 (2006)
242. Yong, J., Zhou, X.Y.: Stochastic Controls. Hamiltonian Systems and HJB Equations. Applications of Mathematics (New York), vol. 43, xxii + 438 pp. Springer, New York (1999)
243. Zhang, J.: Some fine properties of backward stochastic differential equations, with applications. PhD thesis, Purdue University (2001)
244. Zhang, J.: A numerical scheme for BSDEs. Ann. Appl. Probab. **14**(1), 459–488 (2004)
245. Zhang, J.: The well-posedness of FBSDEs. Discrete Contin. Dyn. Syst. Ser. B **6**(4), 927–940 (2006) (electronic)
246. Zhang, J., Zhuo, J.: Monotone schemes for fully nonlinear parabolic path dependent PDEs. J. Financ. Eng. **1**, 1450005 (23 pp.) (2014). doi:10.1142/S2345768614500056
247. Zhao, W., Zhang, G., Ju, L.: A stable multistep scheme for solving backward stochastic differential equations. SIAM J. Numer. Anal. **48**(4), 1369–1394 (2010)
248. Zheng, W.A.: Tightness results for laws of diffusion processes application to stochastic mechanics. Ann. Inst. H. Poincare Probab. Stat. **21**(2), 103–124 (1985)

Frequently Used Notation

I. Notation in Deterministic Setting

- $:=$ means "is defined to be"
- T: fixed time horizon
- C: generic constant in various estimates
- $\mathbb{R}^d$: d-dimensional column vectors
- $\mathbb{R}^{n\times m}$: $n \times m$ matrices
- $\mathbb{S}^d$: $d \times d$ symmetric matrices
- I_d: $d \times d$ identity matrix
- For $a, b \in \mathbb{R}$: $a \vee b := \max(a, b)$, $a \wedge b := \min(a, b)$
- For $a, b \in \mathbb{R}^d$: $a \cdot b := \sum_{i=1}^{d} a_i b_i$ and $|a| := \sqrt{a \cdot a}$
- For $A \in \mathbb{R}^{n\times m}$: $A^\top :=$ the transpose of A
- For $A, B \in \mathbb{R}^{n\times m}$: $A : B :=$ trace of $AB^\top$ and $|A| := \sqrt{A : A^\top}$
- For $A \in \mathbb{S}^d$: $\mathrm{tr}(A) :=$ trace of A
- For $A, B \in \mathbb{S}^d$: $A \geq B$ (resp. $A > B$) means $A - B$ is nonnegative (resp. positive) definite
- For a function (or path) $\theta : [0, T] \to \mathbb{R}^n$:

  - $\bigvee_a^b(\theta) :=$ total variation of θ on $[a, b] \subset [0, T]$
  - $\theta_t^* := \sup_{0 \leq s \leq t} |\theta_s|$ (depending on the contexts, the superscript $*$ may refer to the optimal object, not the running maximum)
  - $\theta_{s,t} := \theta_t - \theta_s$
  - $OSC_\delta(\theta) := \sup_{0 \leq s < t \leq T, t-s \leq \delta} |\theta_{s,t}|$

- 0: depending on the contexts, 0 could mean zero vector or matrix, the constant random variable 0, or the constant process 0.
- Δ: difference of two objects, for example, $\Delta x := x_1 - x_2$

© Springer Science+Business Media LLC 2017
J. Zhang, *Backward Stochastic Differential Equations*, Probability Theory and Stochastic Modelling 86, DOI 10.1007/978-1-4939-7256-2

II. Notation in Stochastic Setting

- Ω: sample space

  - in Part III, fixed as the canonical space $\{\omega \in C([0, T], \mathbb{R}^d) : \omega_0 = 0\}$

- $\omega, \tilde{\omega}$: element of Ω
- $\mathscr{F}$: σ-algebra

  - $\mathscr{F}^X$: σ-algebra generated by a random variable (or random vector) X
  - $\sigma(\mathscr{A})$: the σ-algebra generated by $\mathscr{A}$, where $\mathscr{A}$ is a set of subsets of Ω
  - $\mathscr{B}(\mathbb{R}^d)$: Borel sets
  - $\mathscr{F}_1 \vee \mathscr{F}_2 := \sigma(\mathscr{F}_1 \cup \mathscr{F}_2)$
  - $\overline{\mathscr{F}}^{\mathbb{P}}$: the augmented σ-algebra of $\mathscr{F}$ by all $\mathbb{P}$-null sets
  - $\mathscr{F}_\tau$: the σ-algebra up to a stopping time τ, given a filtration $\mathbb{F} = \{\mathscr{F}_t\}_{0 \le t \le T}$

- $\mathbb{F} = \{\mathscr{F}_t\}_{0 \le t \le T}$: filtration

  - $\mathbb{F}^X$: natural filtration generated by process X
  - $\mathbb{F}^+ = \{\mathscr{F}_t^+\}_{0 \le t \le T}$: right limit filtration
  - $\overline{\mathbb{F}}^{\mathbb{P}}$: the augmented filtration of $\mathbb{F}$ by all $\mathbb{P}$-null sets
  - $\mathbb{F} := \mathbb{F}^B$ when the martingale representation is needed in Parts I and II
  - $\mathbb{F} := \mathbb{F}^X$ in part III, where X is the canonical process
  - $\mathbb{F}^{\mathscr{P}} := \cap_{\mathbb{P} \in \mathscr{P}} \overline{\mathbb{F}}^{\mathbb{P}}$: the universal filtration, used in Sections 12.3 and 12.4

- $\mathbb{P}$: probability measure

  - $\mathbb{P}_0$: the Wiener measure in Part III, namely the canonical process X is a $\mathbb{P}_0$-Brownian motion.
  - $\mathbb{E}$ or $\mathbb{E}^{\mathbb{P}}$: expectation under $\mathbb{P}$
  - $\mathrm{ess\,sup}^{\mathbb{P}}$: essential supremum under $\mathbb{P}$
  - $d\mathbb{P}^\theta := M_T^\theta d\mathbb{P}$: the new probability measure induced from Girsanov theorem

- B: d-dimensional Brownian motion

  - $B_t^\theta := B_t - \int_0^t \theta_s ds$: $\mathbb{P}^\theta$-Brownian motion induced from Girsanov theorem
  - $B_t^{\mathbb{P}} := \int_0^t (\sigma_s^{\mathbb{P}})^{-1} dX_s$: $\mathbb{P}$-Brownian motion for $\mathbb{P} \in \mathscr{P}_{[\underline{\sigma}, \overline{\sigma}]}^W$ with $\sigma^{\mathbb{P}} > 0$

- X: state process

  - in Parts I and II, X is the solution to SDE, $\mathbb{R}^{d_1}$-valued
  - in Part III, X is the canonical process, $\mathbb{R}^d$-valued

- Y: value process

  - in Parts I and II, Y is the solution to BSDE, $\mathbb{R}^{d_2}$-valued
  - in Part III, Y is the value process (or solution to 2BSDE), scalar valued

- Z: solution to BSDE or 2BSDE, $\mathbb{R}^{d_2 \times d}$-valued (in particular, Z is a row vector when $d_2 = 1$)

- $\mathcal{X}, \mathcal{Y}, \mathcal{Z}$: solutions to the same equations in (possibly random) subintervals of $[0, T]$, corresponding to X, Y, Z, respectively
- $\mathcal{T}$ or $\mathcal{T}(\mathbb{F})$: set of $\mathbb{F}$-stopping times

  - τ: generic notation for stopping time
  - $\mathcal{T}^t$: subset of $\tau \in \mathcal{T}$ such that $\tau \geq t$, used in Chapter 6
  - $\mathcal{T}_H := \{\tau \in \mathcal{T} : \tau \leq H\}$, given $H \in \mathcal{T}$

- Hitting times $H_\varepsilon, \overline{H}_\varepsilon, \underline{H}_\varepsilon$, see (10.1.8)

  - $L_1 := L + 1$: constant used in the definition of $H_\varepsilon := H_{L,e}$ in (10.1.8)
  - $H^{t,x,R}$: Markov structure of H_ε, see (10.1.12)
  - H_n^ε: sequence of hitting times for path discretization, see (10.1.15)

- $\mathscr{P}$: class of probability measures $\mathbb{P}$

  - $\mathscr{P}^S_{[\underline{\sigma},\overline{\sigma}]}$: martingales measures induced from strong formulation, see Definition 9.2.2
  - $\mathscr{P}^W_{[\underline{\sigma},\overline{\sigma}]}$: martingales measures in weak formulation, see Definition 9.2.7
  - $\mathscr{P}_L$: semimartingales measures whose characteristics bounded by L, see Definition 9.2.9
  - $\mathscr{P}_\infty := \cup_{L>0}\mathscr{P}_L$

- $\mathscr{E}^{\mathscr{P}} := \sup_{\mathbb{P}\in\mathscr{P}} \mathbb{E}^{\mathbb{P}}$: nonlinear expectation

  - $\mathscr{C}^{\mathscr{P}} := \sup_{\mathbb{P}\in\mathscr{P}} \mathbb{P}$: capacity
  - $\overline{\mathscr{E}}^L := \sup_{\mathbb{P}\in\mathscr{P}_L} \mathbb{E}^{\mathbb{P}}$, $\underline{\mathscr{E}}^L := \inf_{\mathbb{P}\in\mathscr{P}_L} \mathbb{E}^{\mathbb{P}}$

- Spaces of random variables ξ for given $\mathbb{P}$ or $\mathscr{P}$: $\mathbb{R}^n$ omitted when $n = 1$

  - $\mathbb{L}^0(\mathscr{F}, \mathbb{R}^n)$: $\mathbb{R}^n$-valued $\mathscr{F}$-measurable random variable
  - $\mathbb{L}^p(\mathscr{F}, \mathbb{P}, \mathbb{R}^n) \subset \mathbb{L}^0(\mathscr{F}, \mathbb{R}^n)$: $\mathbb{E}^{\mathbb{P}}[|\xi|^p] < \infty$
  - $\mathbb{L}^\infty(\mathscr{F}, \mathbb{P}, \mathbb{R}^n) \subset \mathbb{L}^0(\mathscr{F}, \mathbb{R}^n)$: $\mathbb{P}(|\xi| > C) = 0$ for some $C > 0$
  - $\mathbb{L}^p(\mathscr{F}, \mathscr{P}, \mathbb{R}^n) \subset \mathbb{L}^0(\mathscr{F}, \mathbb{R}^n)$: $\mathscr{E}^{\mathscr{P}}[|\xi|^p] < \infty$
  - $\mathbb{L}^\infty(\mathscr{F}, \mathscr{P}, \mathbb{R}^n) \subset \mathbb{L}^0(\mathscr{F}, \mathbb{R}^n)$: $\mathscr{C}^{\mathscr{P}}(|\xi| > C) = 0$ for some $C > 0$

- Spaces of processes for given $\mathbb{P}$ or $\mathscr{P}$: $\mathbb{R}^n$ is omitted when $n = 1$

  - $\mathbb{L}^0(\mathbb{F}, \mathbb{R}^n)$: $\mathbb{R}^n$-valued $\mathbb{F}$-measurable process
  - $\mathbb{L}^{p,q}(\mathbb{F}, \mathbb{P}, \mathbb{R}^n) := \{Z \in \mathbb{L}^0(\mathbb{F}, \mathbb{R}^n) : \left(\int_0^T |Z_t|^p dt\right)^{\frac{1}{p}} \in \mathbb{L}^q(\mathscr{F}_T, \mathbb{P})\}$
  - $\mathbb{L}^p(\mathbb{F}, \mathbb{P}, \mathbb{R}^n) := \mathbb{L}^{p,p}(\mathbb{F}, \mathbb{P}, \mathbb{R}^n)$
  - $\mathbb{S}^p(\mathbb{F}, \mathbb{P}, \mathbb{R}^n) := \{Y \in \mathbb{L}^0(\mathbb{F}, \mathbb{R}^n) : Y \text{ continuous (in } t), \mathbb{P}\text{-a.s. and } Y_T^* \in \mathbb{L}^p(\mathscr{F}_T, \mathbb{P})\}$
  - $\mathbb{L}^{p,q}(\mathbb{F}, \mathscr{P}, \mathbb{R}^n) := \{Z \in \mathbb{L}^0(\mathbb{F}, \mathbb{R}^n) : \left(\int_0^T |Z_t|^p dt\right)^{\frac{1}{p}} \in \mathbb{L}^q(\mathscr{F}_T, \mathscr{P})\}$
  - $\mathbb{L}^p(\mathbb{F}, \mathscr{P}, \mathbb{R}^n) := \mathbb{L}^{p,p}(\mathbb{F}, \mathscr{P}, \mathbb{R}^n)$
  - $\mathbb{S}^p(\mathbb{F}, \mathscr{P}, \mathbb{R}^n) := \{Y \in \mathbb{L}^0(\mathbb{F}, \mathbb{R}^n) : Y \text{ continuous (in } t), \mathscr{P}\text{-q.s. and } Y_T^* \in \mathbb{L}^p(\mathscr{F}_T, \mathscr{P})\}$
  - $\mathbb{I}^p(\mathbb{F}, \mathbb{P}) := \{K \in \mathbb{S}^p(\mathbb{F}, \mathbb{P}, \mathbb{R}): K_0 = 0 \text{ and } K \text{ increasing (in } t), \mathbb{P}\text{-a.s.}\}$
  - $\mathbb{L}^p_{loc}(\mathbb{F}, \mathbb{P}, \mathbb{R}^n) := \{Z \in \mathbb{L}^0(\mathbb{F}, \mathbb{R}^n) : \int_0^T |Z_t|^p dt < \infty, \mathbb{P}\text{-a.s.}\}$

- $\mathbb{L}_0^2(\mathbb{F},\mathbb{P},\mathbb{R}^n) \subset \mathbb{L}^2(\mathbb{F},\mathbb{P},\mathbb{R}^n)$: elementary (or say piecewise constant) processes
- $\mathbb{L}_0^p(\mathscr{F},\mathscr{P},\mathbb{R}^n)$: closure of $UC_b(\Omega)$ under the norm of $\mathbb{L}^p(\mathscr{F},\mathscr{P},\mathbb{R}^n)$
- $\widehat{\mathbb{L}}^p(\mathscr{F}_T,\mathscr{P})$: closure of $UC(\Omega)$ under the norm specified in (12.1.19)

• Conditional distribution

- $\mathbb{E}_t$ or $\mathbb{E}_\tau$: conditional expectation
- $\mathbb{P}^{\tau,\omega}$: r.c.p.d., see Definition 9.3.1
- $\mathscr{P}(t,\omega) := \{\mathbb{P}^{t,\omega} : \mathbb{P} \in \mathscr{P}\}$: set of r.c.p.d.
- $\mathscr{P}_t = \mathscr{P}(t,\omega)$: independent of ω, see Assumption 9.3.3
- $\mathscr{P}(\tau,\mathbb{P}) := \{\tilde{\mathbb{P}} \in \mathscr{P} : \tilde{\mathbb{P}} = \mathbb{P}$ on $\mathscr{F}_\tau\}$, used in Chapter 12
- $\mathscr{E}_{t,\omega}^{\mathscr{P}}[\xi] := \mathscr{E}^{\mathscr{P}_t}[\xi^{t,\omega}]$: pathwise conditional nonlinear expectation
- $\mathscr{E}_{\mathbb{P},\tau}^{\mathscr{P}} := $ ess sup$_{\tilde{\mathbb{P}} \in \mathscr{P}(\tau,\mathbb{P})}^{\mathbb{P}}$ $\mathbb{E}_\tau^{\tilde{\mathbb{P}}}$: quasi-sure conditional nonlinear expectation

• Localized or shifted spaces

- $\Lambda := [0,T) \times \Omega, \overline{\Lambda} := [0,T] \times \Omega$: the original state space
- $\Lambda_{\mathrm{H}} := \{(t,\omega) \in \Lambda : t < \mathrm{H}(\omega)\}, \overline{\Lambda}_{\mathrm{H}} := \{(t,\omega) \in \Lambda : t \le \mathrm{H}(\omega)\}$: localized state space for given $\mathrm{H} \in \mathscr{T}$
- $\Lambda^t := [0,T-t) \times \Omega, \overline{\Lambda}^t := [0,T-t] \times \Omega$: shifted state space for given $t \in [0,T]$
- $\mathbb{F}^t$ and $\mathscr{F}_s^t$: shifted filtration in (5.0.1) or (9.3.4) (they mean slightly differently, use (5.0.1) in Parts I and II, and use (9.3.4) in Part III)

• Concatenation and shifting operators:

- $\omega \otimes \tilde{\omega}$: concatenation of paths, see (9.3.1)
- $\mathbb{P}_1 \otimes \mathbb{P}_2$: concatenation of measures, see (9.3.21)–(9.3.22), see also (12.1.11)
- $\xi^{t,\omega}(\tilde{\omega}) := \xi(\omega \otimes \tilde{\omega})$: shifted random variable
- $\eta_s^{t,\omega}(\tilde{\omega}) := \eta_{t+s}(\omega \otimes \tilde{\omega})$: shifted process
- $(\eta_{\tau \wedge \cdot})_t := \eta_{\tau \wedge t}$: stopped process by given $\tau \in \mathscr{T}$

• Regularity of processes

- $\|\omega\|_t := \sup_{0 \le s \le t} |\omega_s|$: semi-norm on Ω
- $\mathbf{d}$: pseudo-metric on $\overline{\Lambda}$, see (9.2.1)
- $C^0(\Omega)$: continuous random variables under $\|\cdot\|_T$
- $C^0(\overline{\Lambda})$: continuous processes under $\mathbf{d}$
- C_b^0 is the subset of bounded elements, UC is the subset of uniformly continuous elements, and $UC_b := UC \cap C_b^0$
- $C^{1,2}(\Lambda)$: smooth processes, see Definition 9.4.1
- $u : \overline{\Lambda} \to \mathbb{R}$ is Markovian means: there exists $v : [0,T] \times \mathbb{R}^d \to \mathbb{R}$ such that $u(t,\omega) = v(t,\omega_t)$

• Differentiation of smooth function $u : [0,T] \times \mathbb{R}^d \to \mathbb{R}$

- $\partial_x u$: first order derivatives, $\mathbb{R}^{1 \times d}$-valued (row vector!)
- $\partial_{xx} u := \partial_x((\partial_x u)^\top)$: second order derivatives (Hessian), $\mathbb{S}^d$-valued

- $\partial_t u$: time derivative, scalar valued
- $\mathbb{L}u$: the differential operator for PDE

• Differentiation of smooth process $u : \Lambda \to \mathbb{R}$

- $\partial_\omega u$: first order path derivatives, $\mathbb{R}^{1 \times d}$-valued (row vector!)
- $\partial_{\omega\omega} u$: second order path derivatives, $\mathbb{S}^d$-valued (equal to $\partial_\omega ((\partial_\omega u)^\top)$ when smooth enough)
- $\partial_t u$: time derivative, scalar valued
- $\mathbb{L}u$: the differential operator for PPDE
- $\mathscr{L}^{t,\omega} u$: shifted differential operator, see (11.2.5)

• Test functions for viscosity solutions

- We use φ to denote smooth test functions
- $\overline{\mathscr{A}}u(t,x), \underline{\mathscr{A}}u(t,x)$: test functions for viscosity semi-solutions of PDEs, see (5.5.4)
- $\overline{\mathscr{A}}u(t,\omega), \underline{\mathscr{A}}u(t,\omega)$: test functions for viscosity solutions of path dependent heat equation, see (11.1.7)
- $\overline{\mathscr{A}}^L u(t,\omega), \underline{\mathscr{A}}^L u(t,\omega)$: test functions for viscosity solutions of general PPDE, see (11.2.4)
- $\phi^{c,a,p,q}$: paraboloid test function, see (11.2.9)
- $\overline{\mathscr{J}}^L u(t,\omega), \underline{\mathscr{J}}^L u(t,\omega)$: semi-jets for PPDE, see (11.2.10)

III. Additional Miscellaneous Notation

• $M_t^\theta := \exp(\int_0^T \theta_t dB_t - \frac{1}{2} \int_0^T |\theta_t|^2 dt)$: the exponential process used in Girsanov theorem
• $\langle X \rangle$: quadratic variation process of $\mathbb{R}^d$-valued semimartingale X, $\mathbb{S}^d$-valued
• $\sigma^{\mathbb{P}}$: diffusion coefficient of martingale measure $\mathbb{P}$, see (9.2.8)
• $\hat{\sigma}$: aggregation of $\{\sigma^{\mathbb{P}}, \mathbb{P} \in \mathscr{P}^W_{[\underline{\sigma},\overline{\sigma}]}\}$, see Corollary 12.1.5.
• $(\Omega', \mathbb{F}', \mathbb{P}', b', \sigma')$: the setting in enlarged space used to define $\mathscr{P}_L$, see Definition 9.2.9
• $\widehat{Z}, \widehat{f}$: see (12.2.6), used in Chapter 12
• L: depending on the contexts

- Lipschitz constant
- the lower barrier process for RBSDE
- the bound of the characteristics of the semimartingale measures in $\mathscr{P}_L$ in Part III

• ρ: modulus of continuity functions

- ρ_0: specified in some assumptions
- ρ_1: specified in Assumption 10.3.1 (i)
- ρ_2, ρ_3: defined by (10.3.7)
- $\overline{\rho}$: defined by (10.2.11)

- Domains for path frozen PDEs in Section 11.4.3

  - $D_{t,\varepsilon}, \partial D_{t,\varepsilon}, \overline{D}_{t,\varepsilon}$: see (11.4.7)
  - $\Pi_{\varepsilon}^n, \Pi_{\varepsilon,\delta}^n, \overline{D}_{\varepsilon}^n$: see (11.4.12)

- $\pi : 0 = t_0 < \cdots < t_n = T$: time partition
- k: control process (occasionally may also be used as integer index)

IV. Some Abbreviations

- u.i.: uniformly integrable
- a.s.: almost surely
- q.s.: quasi surely
- r.c.p.d.: regular conditional probability distribution
- PDE: partial differential equations
- SDE: stochastic differential equations
- BSDE: backward SDE
- FBSDE: forward-backward SDE
- PPDE: path dependent PDE
- 2BSDE: second order backward SDE
- HJB: Hamilton-Jacobi-Bellman

Index

A

Aggregation, 336
American option, 133
Arbitrage free market, 53
Arbitrage free price (fair price), 53
Arzela-Ascoli theorem, 221

B

Backward stochastic differential, 79
    G-BSDE, 348
    linear, 80
    Markov, 104
    quadratic, 161
    reflected, 133
    second order, 335
Backward stochastic PDE, 309
Black-Scholes model, 52
Blumenthal 0-1 law, 25
Brownian motion, 21
Burkholder-Davis-Gundy inequality, 39

C

Canonical
    process, 213
    space, 213
Capacity, 245
Central limit theorem, 5
Classical solution (sub-solution, super-solution)
    of path dependent heat equation, 278

of PDE, 121
of PPDE, 283
Comparison principle
    for BSDE, 87
    for 2BSDE, 350
    for FBSDE, 187
    for quadratic BSDE, 174
    for RBSDE, 137
    for SDE, 70
    for viscosity solution of PDEs, 121
    for viscosity solution of PPDE, 314
Complete market, 53
Concatenation
    of measures, 230
    of paths, 225
Conditional expectation, 3
    pathwise conditional nonlinear expectation, 258
    quasi-sure conditional nonlinear expectation, 339
Conjugates, 13
Consistency of viscosity solution, 286
Contraction mapping, 181

D

Decoupling field, 186
Dominated convergence theorem (DCT):, 14
    under nonlinear expectation, 247
Dominating measure, 216
Doob-Dynkin's lemma, 3
Doob-Meyer decomposition, 50
    semilinear, 158

© Springer Science+Business Media LLC 2017
J. Zhang, *Backward Stochastic Differential Equations*, Probability Theory and
Stochastic Modelling 86, DOI 10.1007/978-1-4939-7256-2

Printed in the United States
By Bookmasters